T0382645

GLACIALLY-TRIGGERED FAULTING

Glacially triggered faulting describes movement of pre-existing faults caused by a combination of tectonic and glacially induced isostatic stresses. The most impressive fault scarps are found in Northern Europe, assumed to have been reactivated at the end of deglaciation. However, this view has been challenged as new faults have been discovered globally with advanced techniques such as LiDAR, and fault activity dating has shown several phases of reactivation thousands of years after deglaciation ended. This book summarizes the current state-of-the-art research in glacially triggered faulting, discussing the theoretical aspects that explain the presence of glacially induced structures and reviews the geological, geophysical, geodetic and geomorphological investigation methods. Written by a team of international experts, it provides the first global overview of confirmed and proposed glacially induced faults and provides an outline for modelling these stresses and features. It is a go-to reference for geoscientists and engineers interested in ice sheet–solid earth interaction.

HOLGER STEFFEN is a geophysicist by training who joined the Geodetic Infrastructure Department at Lantmäteriet, the Swedish Mapping, Cadastral and Land Registration Authority, in 2012 after working for several years as a postdoctoral researcher at universities in Germany, Canada and Sweden. His work deals with glacial isostatic adjustment modelling and corresponding acquisition and/or analysis of geodetic, geophysical and geologic observations. He chairs the Working Group of Geodynamics and Earth Observation of the Nordic Geodetic Commission.

ODLEIV OLESEN is a senior researcher at the Geological Survey of Norway with 40 years of professional experience, including management roles, extensive research and mapping activities. He has taught as Adjunct Professor in Applied Geophysics at the Norwegian University of Science and Technology.

RAIMO SUTINEN is a geo-consultant with 40 years' experience in geoscience. He was previously a senior researcher at the Geological Survey of Finland, where he managed the project 'Postglacial Faults'. He has more than 60 impact and 100 proceedings papers in remote sensing, soil physics and biogeochemistry. His recent research has focused on faults and earthquake-induced landforms.

GLACIALLY-TRIGGERED FAULTING

Edited by

HOLGER STEFFEN
Lantmäteriet

ODLEIV OLESEN
Geological Survey of Norway

RAIMO SUTINEN
Geological Survey of Finland

CAMBRIDGE
UNIVERSITY PRESS

University Printing House, Cambridge CB2 8BS, United Kingdom

One Liberty Plaza, 20th Floor, New York, NY 10006, USA

477 Williamstown Road, Port Melbourne, VIC 3207, Australia

314–321, 3rd Floor, Plot 3, Splendor Forum, Jasola District Centre, New Delhi – 110025, India

103 Penang Road, #05–06/07, Visioncrest Commercial, Singapore 238467

Cambridge University Press is part of the University of Cambridge.

It furthers the University's mission by disseminating knowledge in the pursuit of education, learning, and research at the highest international levels of excellence.

www.cambridge.org
Information on this title: www.cambridge.org/9781108490023
DOI: 10.1017/9781108779906

First published 2022

Printed in the United Kingdom by TJ Books Limited, Padstow Cornwall

A catalogue record for this publication is available from the British Library.

ISBN 978-1-108-49002-3 Hardback

Contents

Colour plates appear between 170 and 171.

Figures

Tables

Contributors

John Adams
Canadian Hazards Information Service, Natural Resources Canada, Ottawa, Ontario, Canada

Omid Ahmadi
AFRY, Solna, Sweden

Maria Ask
Department of Civil, Environmental and Natural Resources Engineering, Division of Geosciences and Environmental Engineering, Luleå University of Technology, Luleå, Sweden

Tobias Bauer
Department of Civil, Environmental and Natural Resources Engineering, Division of Geosciences and Environmental Engineering, Luleå University of Technology, Luleå, Sweden

Ruth Beckel
Department of Earth Sciences, Uppsala University, Uppsala, Sweden

Albertas Bitinas
Nature Research Centre, Vilnius, Lithuania

Christian Brandes
Institute of Geology, Leibniz University Hannover, Hannover, Germany

Gregory R. Brooks
Geological Survey of Canada, Natural Resources Canada, Ottawa, Ontario, Canada

Åke Fagereng
School of Earth and Ocean Sciences, Cardiff University, Cardiff, United Kingdom

Jeffrey T. Freymueller
Department of Earth and Environmental Sciences, Michigan State University, East Lansing, Michigan, USA

Nicolai Gestermann
Federal Institute for Geosciences and Natural Resources (BGR), Geozentrum Hannover, Hannover, Germany

Steven J. Gibbons
NORSAR, Kjeller, Norway
Norwegian Geotechnical Institute, Oslo, Norway

Sofie Gradmann
Geological Survey of Norway (NGU), Geophysics Section, Trondheim, Norway

Søren Gregersen
Geological Survey of Denmark and Greenland, København K, Denmark

Susanne Grigull
Swedish Nuclear Fuel and Waste Management Company (SKB) Solna, Sweden, Sweden
Geological Survey of Sweden, Uppsala, Sweden

Fredrik Høgaas
Geological Survey of Norway (NGU), Quaternary Geology Section, Trondheim, Norway

Eija Hyvönen
Geological Survey of Finland, Rovaniemi, Finland

Tor Arne Johansen
Department of Earth Science, University of Bergen, Bergen, Norway

Christopher Juhlin
Department of Earth Sciences, Uppsala University, Uppsala, Sweden

Marie Keiding
Department of Geophysics, Geological Survey of Denmark and Greenland, København K, Denmark

Annakaisa Korja
Department of Geosciences and Geography, Institute of Seismology, University of Helsinki, Helsinki, Finland

Ilmo Kukkonen
Department of Geosciences and Geography, University of Helsinki, Helsinki, Finland

Tormod Kværna
NORSAR, Kjeller, Norway

Jurga Lazauskienė
Department of Geology and Mineralogy, Institute of Geosciences, Vilnius University, Vilnius, Lithuania

Thomas Lege
Federal Institute for Geosciences and Natural Resources (BGR), Geozentrum Hannover, Hannover, Germany

Conrad Lindholm
SeismoConsult, Vittenbergvn, Fjellhamar, Norway

A. J. (Tom) van Loon
College of Earth Science and Engineering, Shandong University of Science and Technology, Qingdao, China

Henning Lorenz
Department of Earth Sciences, Uppsala University, Uppsala, Sweden

Björn Lund
Swedish National Seismic Network, Department of Earth Sciences, Uppsala University, Uppsala, Sweden

Alireza Malehmir
Department of Earth Sciences, Uppsala University, Uppsala, Sweden

Mira Markovaara-Koivisto
Geological Survey of Finland, Espoo, Finland

Jussi Mattila
Rock Mechanics Consulting Finland Oy, Vantaa, Finland

Maarit Middleton
Geological Survey of Finland, Rovaniemi, Finland

Henrik Mikko
Geological Survey of Sweden, Uppsala, Sweden

Katharina Müller
Institute of Geology, Leibniz University Hannover, Hannover, Germany

Raymond Munier
Terra Mobile Consultants AB, Stockholm, Sweden

Svetlana B. Nikolaeva
Geological Institute, Kola Science Center, Russian Academy of Sciences, Apatity, Murmansk region, Russia

Andrey A. Nikonov
Schmidt Institute of Physics of the Earth, Russian Academy of Sciences, Moscow, Russia

Kati Oinonen
Department of Geosciences and Geography, Institute of Seismology, University of Helsinki, Helsinki, Finland

Antti E. K. Ojala
Geological Survey of Finland, Espoo, Finland
Department of Geography and Geology, University of Turku, Finland

Odleiv Olesen
Geological Survey of Norway (NGU), Geophysics Section, Trondheim, Norway

Lars Olsen
Geological Survey of Norway (NGU), Quaternary Geology Section, Trondheim, Norway

Jukka-Pekka Palmu
Geological Survey of Finland, Espoo, Finland

Małgorzata (Gosia) Pisarska-Jamroży
Institute of Geology, Adam Mickiewicz University, Poland

Thomas Plenefisch
Federal Institute for Geosciences and Natural Resources (BGR), Geozentrum Hannover, Hannover, Germany

Chris Rollins
School of Earth and Environment, University of Leeds, Leeds, United Kingdom

Jan-Erik Rosberg
Engineering Geology LTH, Lund University, Lund, Sweden

Natalia A. Ruppert
Alaska Earthquake Information Center, Geophysical Institute, University of Alaska, Fairbanks, Alaska, USA

Timo Ruskeeniemi
Geological Survey of Finland, Espoo, Finland

Jonny Rutqvist
Energy Geosciences Division, Lawrence Berkeley National Laboratory, Berkeley, California, USA

Bent Ole Ruud
Department of Earth Science, University of Bergen, Bergen, Norway

Peter B. E. Sandersen
Geological Survey of Denmark and Greenland, Department of Groundwater and Quaternary Geology Mapping, Århus C, Denmark

Jeanne Sauber
Geodesy and Geophysics Laboratory, NASA Goddard Space Flight Center, Greenbelt, Maryland, USA

Sergey V. Shvarev
Institute of Geography, Russian Academy of Sciences, Moscow, Russia
Schmidt Institute of Physics of the Earth, Russian Academy of Sciences, Moscow, Russia

Colby A. Smith
Geological Survey of Sweden, Uppsala, Sweden

Thomas Spies
Federal Institute for Geosciences and Natural Resources (BGR), Geozentrum Hannover, Hannover, Germany

Holger Steffen
Lantmäteriet, Geodetic Infrastructure, Gävle, Sweden

Rebekka Steffen
Lantmäteriet, Geodetic Infrastructure, Gävle, Sweden

Raimo Sutinen
Geological Survey of Finland, Rovaniemi, Finland

David C. Tanner
Leibniz Institute for Applied Geophysics (LIAG), Hannover, Germany

Marja Uski
Department of Geosciences and Geography, Institute of Seismology, University of Helsinki, Helsinki, Finland

Peter H. Voss
Geological Survey of Denmark and Greenland, København K, Denmark

Jutta Winsemann
Institute of Geology, Leibniz University Hannover, Hannover, Germany

Piotr Paweł Woźniak
Department of Geomorphology and Quaternary Geology, University of Gdańsk, Gdańsk, Poland

Patrick Wu
Department of Geoscience, University of Calgary, Calgary, Canada

Preface

On cloudless summer days in Northern Europe one may spot a small plane crossing the sky like a farmer tilling a field. This plane may be collecting data for a digital elevation model using laser scanning.

During analysis of novel digital elevation data in the early 2010s, a peculiar feature in the form of an almost north–south-oriented lineation was found near the town of Bollnäs in Sweden, approximately 250 km north of Stockholm. After additional investigation on the ground, a glacially induced fault was identified. This was surprising, since the location is more than 400 km south of the area with the prominent scarps of all the glacially induced faults known at that time.

As more faults were identified later in Sweden and Finland with this type of new data, a series of Postglacial Fault Symposia (2015 in Uppsala, Sweden; 2016 in Turku, Finland; 2018 in Kautokeino, Norway) were organized to provide a fruitful forum for the exchange of the newest research results.

In addition, the Postglacial Fault Drilling Project was initiated in 2010 through a workshop in Skokloster, Sweden, funded by the International Continental Scientific Drilling Program. It was recognized that a glacially induced fault represents a special kind of intraplate fault zone. The full drilling proposal, 'Drilling Active Faults in Northern Europe' (DAFNE), was approved in October 2019.

In 2016, in view of the many new results and the planned drilling, the participants of the Postglacial Fault Symposium decided to summarize the findings of the last 50 years up to the present day in a book to be entitled *Glacially-Triggered Faulting*. Two editors were proposed: Robert Lagerbäck and Odleiv Olesen.

Sadly, Robert Lagerbäck had to decline his editorship and passed away in August 2018. Robert Lagerbäck was employed at the Geological Survey of Sweden (SGU) and was the key researcher on the so-called postglacial faults in northern Fennoscandia for more than three decades. In the 1980s, Robert was one of the key persons to map these faults within the Nordkalott Project. With his death, we lost a renowned geologist, whose expertise was often missed during this book's preparation. We hope he would have liked the final product.

We would like to thank all authors and reviewers of the book and the chapters, and we thank the contributors to the *International Database of Glacially Induced Faults*, by Munier et al. (2020), which is available for download at the *PANGAEA* website, doi.org/10.1594/PANGAEA.922705.

This book could not have been edited without them. In addition, we are very grateful for the excellent assistance by Susan Francis and Sarah Lambert from Cambridge University Press during the preparation of this book.

We hope the reader will find this book the ideal reference in the field of glacially triggered faulting. It should also serve as the best comprehensive start for a new generation of scientists working on glacially induced faults.

And finally, during the writing of this book, the authors welcomed at least five new arrivals. Perhaps they will form the next generation of researchers investigating glacially induced faults. There is still much to do. Data collection and research will continue; there will be many more cloudless days in Northern Europe and elsewhere. The next time you see a small plane crossing the sky like a farmer tilling a field, imagine that it may be collecting the data that will help unravel another secret of glacially triggered faulting.

Robert Lagerbäck leading his 2012 fault excursion at the Lansjärv site, northern Sweden (Photo: Raimo Sutinen).

Part I

Introduction

1

Glacially Triggered Faulting

A Historical Overview and Recent Developments

HOLGER STEFFEN, ODLEIV OLESEN AND RAIMO SUTINEN

ABSTRACT

Glacially triggered faulting, also called glacially induced faulting or postglacial faulting, describes fault movement caused by a combination of tectonic and glacially induced isostatic stresses. This type of faulting is mainly recognized in intraplate regions but is also proposed for some plate boundary areas. Stresses induced by the advance and retreat of an ice sheet are thought to be released during or after ice melting and to reactivate pre-existing faults. Past reactivations were probably accompanied by great-magnitude seismic events triggering hundreds of landslides and seismically induced soft-sediment deformation structures in the region surrounding the faults. Reliable field evidence for reactivated faults in and around many formerly glaciated areas has considerably increased the number of confirmed and probable glacially induced faults in recent years.

We provide a historical overview of dedicated geoscientific investigations from the early reports of this type of faulting until recent findings. Beforehand, we discuss the definition of glacially triggered faulting, suggest a revision of the classification criteria and update the grading criteria for glacially induced fault claims.

1.1 Introduction

Climatic variations have led to repeated glaciations on the Earth. Especially, the Pleistocene glaciations have left many visible traces on the Earth's surface in the form of moraines, striated bedrock, erratics, etc. The Earth's response to the load redistributions of water, ice and sediments on the surface is termed *glacial isostatic adjustment* (GIA), (Wu & Peltier, 1982). Due to the nature of GIA several present-day observations can be related to the last glaciation, which peaked between 26 ka BP and 18 ka BP (Clark et al., 2009). The most prominent process is the ongoing land uplift of formerly glaciated areas such as Fennoscandia, North America and Patagonia.

A generally less appreciated GIA effect is crustal stress release that occurs during and after deglaciation and that can reactivate pre-existing faults and weakness zones through earthquakes. The shaking from these earthquakes can cause landslides and soft-sediment deformation structures (SSDS) (Fenton, 1999; Munier & Fenton, 2004; Olesen et al., 2004; Lund, 2015). Meanwhile, a wealth of such observations is available from around the world, which will be discussed in Chapters 11–21 of this book. Research focused until very recently on Northern Europe, i.e. Lapland, and eastern North America, but stress release due to GIA is now suggested in and around other formerly and presently glaciated areas as well (Brandes et al., 2015). Some studies additionally discuss fault reactivation during the advance of an ice sheet (Munier & Fenton, 2004; Brandes et al., 2011; Pisarska-Jamroży et al., 2018).

Several terms have been used to describe GIA-related stress release in the literature (see e.g. Fenton, 1999; Lund & Näslund, 2009). Perhaps the most common term is *postglacial faulting*, and the reactivated faults are consequently called *postglacial faults* (PGFs). This term might have similarities with, but is not connected to, the term *postglacial rebound* (PGR), a term generally used until the late 1970s to describe the land uplift after the last glaciation. However, the word *postglacial* in PGF does constrain events to the time period after the glaciation.

Peltier and Andrews (1976) introduced the term GIA, which encompasses PGR but also effects prior to deglaciation (i.e. during glaciation) and any consequent processes such as geoidal, rotational and sea-level changes as well as any corresponding effects due to stress changes. Hence, postglacial faulting is also encompassed by the nowadays widely accepted term 'GIA'. Fenton (1999) considered the term 'postglacial faulting' unsatisfactory because it implies a temporal constraint and omits the fault genesis. He suggested that the terms *glacio-isostatic faulting* or *glacial rebound faulting* to be more suitable for faulting due to GIA. Nonetheless, postglacial faulting was still used in the literature but understood in a much broader temporal sense, e.g. also applicable to faulting occurring during glacial advance. Lund and Näslund (2009) introduced the term *glacially induced faulting* and correspondingly *glacially induced fault* (GIF) for the reactivated fault. Especially the latter term, GIF, has been increasingly used in the last decade to describe the faults although the 'classic term' PGF has been retained by many researchers, especially when referring to the prominent faults in Northern Europe (Figure 1.1). Another term that was discussed among the community is *glacially triggered faulting* (GTF) and thus *glacially triggered fault*. The term arose because 'induced' was interpreted by some researchers as meaning either new fault generation (rather than fault reactivation) or faulting associated with human activity, such as anthropogenic earthquakes. For others, GTF is simply the same as glacially induced faulting.

In this book the reader will find that all terms except glacially triggered fault have been used, depending on the taste of the authors. The interchangeable terms glacially triggered faulting, glacially induced faulting and postglacial faulting refer to the mechanism, whereas glacially induced fault and postglacial fault refer to the reactivated fault. Nonetheless, we encourage use of glacially triggered faulting or GTF when referring to the mechanism and glacially induced fault or GIF for the reactivated fault.

Glacially triggered faulting should not be confused with glaciotectonics, which is mostly the near-surface deformation of sediments and sometimes bedrock as a direct consequence of ice movement and which sometimes shows similarities with potential GTF features such as faults and SSDS. We refer the reader to Chapter 4, by Müller et al., who discuss the differences of glaciotectonics and GTF in detail.

Next, we discuss classification criteria for a GIF. This is followed by a brief history of major findings on GTF and the latest developments.

Figure 1.1 Oblique aerial photograph (SE view, taken from Olesen et al., 2004) of the fault scarp developed along the Máze Fault System constituting the central part of the Stuoragurra Fault Complex in Norway. The fault segment is located approximately 10 km to the NNE of the Masi settlement. Groundwater is leaking from the foot of the escarpment (lower right). (A black and white version of this figure will appear in some formats. For the colour version, please refer to the plate section.)

1.2 Classification Criteria for a Glacially Induced Fault

Given the heterogeneous structure of Earth's lithosphere, it is necessary to define criteria for correctly identifying a GIF and distinguishing it from the vast number of other faults around the globe. Such criteria were introduced by Mohr (1986) and have been modified and expanded by Fenton (1991, 1994). The six original criteria of Fenton (1991), briefly summarized, were as follows: (1) a fault must be continuous with a prominent disruption of pre-existing geological units; (2) its scarp face should not be affected by ice or meltwater; (3) it is not generated due to differential erosion; (4) it must displace late Quaternary/Holocene sediments or morphological features (e.g. shorelines); (5) it is not generated due to differential compaction; and (6) it should be trenched to ensure fault activity and determine erosional influence. Features 1 and 4 categorize the geological structure; features 2, 3 and 5 exclude other processes; and the last is a rather technical note on investigation methods. The latter was removed by Fenton (1994), who established seven criteria (e.g. Fenton, 1999; Munier & Fenton, 2004) and which we repeat verbatim, except for the addition of 'F' to the numbering, for clarity in the discussion to come:

F1. Faults should have demonstrable movement since the disappearance of the last ice sheet within the area of concern.

F2. The fault should offset glacial and late-glacial deposits, glacial surfaces or other glacial geomorphic features. Preferably, it should be demonstrated that the fault displaces immediately postglacial stratigraphy and/or geomorphic features, though it need not cut younger features.

F3. Fault scarp faces and rupture planes expressed in bedrock should show no signs of glacial modification, such as striations or ice-plucking. Limited glacial modification, however, may be present on scarps that are late-glacial or inter-glacial in age.

F4. Surface ruptures must be continuous over a distance of at least 1 km, with consistent slip and a displacement/length ratio (D/L) of less than 0.001.

F5. Scarps in superficial material must be shown to be the result of faulting and not due to the effects of differential compaction, collapse due to ice melt, or deposition over pre-existing scarps.

F6. Care must be taken with bedrock scarps controlled by banding, bedding, or schistosity to show that they are not the result of differential erosion, ice-plucking, or meltwater erosion.

F7. In areas of moderate to high relief, the possibility of scarps being the result of having been created by deep-seated slumping driven by gravitational instability must be disproved.

Muir Wood (1993), in work contemporary to Fenton's, provides five classification criteria in the form of a checklist. They can be briefly summarized, following Smith et al. (2014), that:

M1. a displaced sediment layer must have been formerly continuous;
M2. this sediment offset must be directly related to a fault;
M3. the ratio of displacement to length should be less than 1:1,000;
M4. the displacement should be consistent along the fault; and
M5. the movement should have occurred synchronously along the fault.

These can be considered as a more specific refinement of F1, F2 and F4 criteria listed earlier. The M1–M5 checklist has been applied in several dedicated studies, see e.g. Olesen et al. (2004), Smith et al. (2014) and Brooks and Adams (2020). Olesen et al. (2004) merged the M1–M5 checklist with a revised form of the F1–F7 criteria.

As research progresses, new findings warrant a discussion of the criteria, most notably criteria M3 (which is equivalent to F4) and M5. Therefore, we introduce revised classification criteria for GIFs. These are modified from the criteria listed earlier and for easier application expressed as a checklist like that of Olesen et al. (2004). We comment on each criterion and thereafter discuss previous criteria that should no longer be considered definitive.

The herein revised classification criteria are as follows:

1. **Disruption of a formerly continuous geological feature**: There is either an offset of an originally continuous surface or of sediment layer(s) which can be seen on the surface in an outcrop or in seismic reflection profiles, and/or there is an internal disturbance of a sediment, e.g. in the form of soft-sediment deformation structures (SSDS).

   Comment: This criterion revises, combines and extends F2 and M1. Previously, the disruption of a sediment unit was generally thought to be by a fault or fault scarp (F1, F2 and M1). However, SSDS also can be generated due to glacially triggered earthquakes along GIFs (e.g. Munier & Fenton, 2004; Müller et al., see Chapter 4), and thus should be added. The age of the sediment (see F1 and F2, e.g. 'since the last ice sheet', 'glacial' or 'postglacial') is of lesser importance because previous glaciations could have led to GIFs, and under certain conditions faults can be reactivated during ice advance (see Steffen et al., Chapter 2). Smith et al. (see Chapter 12), also hypothesize that some recognized GIFs were reactivated by glaciations that occurred prior to the most recent one.

2. **Relation to a fault that shows demonstrable offset**: A fault with noticeable offset can be connected to the disrupted feature (fault, fault scarp, SSDS, etc.). This fault is the GIF.

   Comment: We merge F1 and M2 and rephrase.

3. **Consistent displacement**: There is a reasonably consistent amount of slip along the length of the GIF.

   Comment: This is M4 and parts of F4. This criterion can be easily applied to GIFs with surface exposures. For faults with only indirect evidence of reactivation, e.g. with SSDS, this must be verified with appropriate methods (trenching, geophysical techniques); see e.g. Beckel et al., in Chapter 7, and Gestermann and Plenefisch, in Chapter 6.

4. **Relation to a formerly glaciated area**: The disturbed feature is found within or near to a formerly glaciated area.

   Comment: As clearly indicated by its name a GIF can be found within a formerly glaciated area. However, the GIA process also affects the region surrounding the ice sheet, most notably the peripheral bulge area. This area extends a few hundred kilometres around the ice sheet and is affected by glacially induced stress changes (Wu et al., Chapter 22). GIFs can thus occur in such a peripheral region, e.g. the Osning Thrust in Germany (Figure 1.2) as suggested by Brandes et al. (2015). Therefore, we suggest adding this criterion to highlight that GIFs are not limited to the formerly glaciated area. Earlier criteria assumed that GIFs were only to be found within the formerly glaciated area. In other words, the 'area of concern' of criterion F1 is extended by understanding the physics of GIA.

5. **Convincing exclusion of trigger mechanisms other than GIA**: As other processes are also able to reactivate faults or generate features that can mimic GIFs, those processes must be meticulously excluded in order to clearly confirm a GIF as such. Hence, investigation must convincingly demonstrate that there are:
   - no signs of gravity sliding as the driving mechanism for fault activity in areas of sufficient relief;
   - no signs of glacial modifications of fault scarps (especially those in metamorphic rocks controlled by schistosity, banding or bedding) implying glacial erosion (differential erosion or ice-plucking) was the cause;
   - no signs of glaciotectonics;
   - no signs of collapse due to melting of buried ice, differential compaction or deposition over a pre-existing erosional scarp resulting in an apparent offset in overburden;
   - in case of SSDS, no signs of other processes, e.g. mass movements, landslides, groundwater-level fluctuations, hydrostatic pressure changes related to lake drainage, water-wave or tsunami passage.

   Comment: These are reformulated criteria F3, F5, F6 and F7 (among others), mainly following the criteria list in Olesen et al. (2004). They are combined into a single criterion to provide a checklist. The last point of this criterion concerning SSDS has

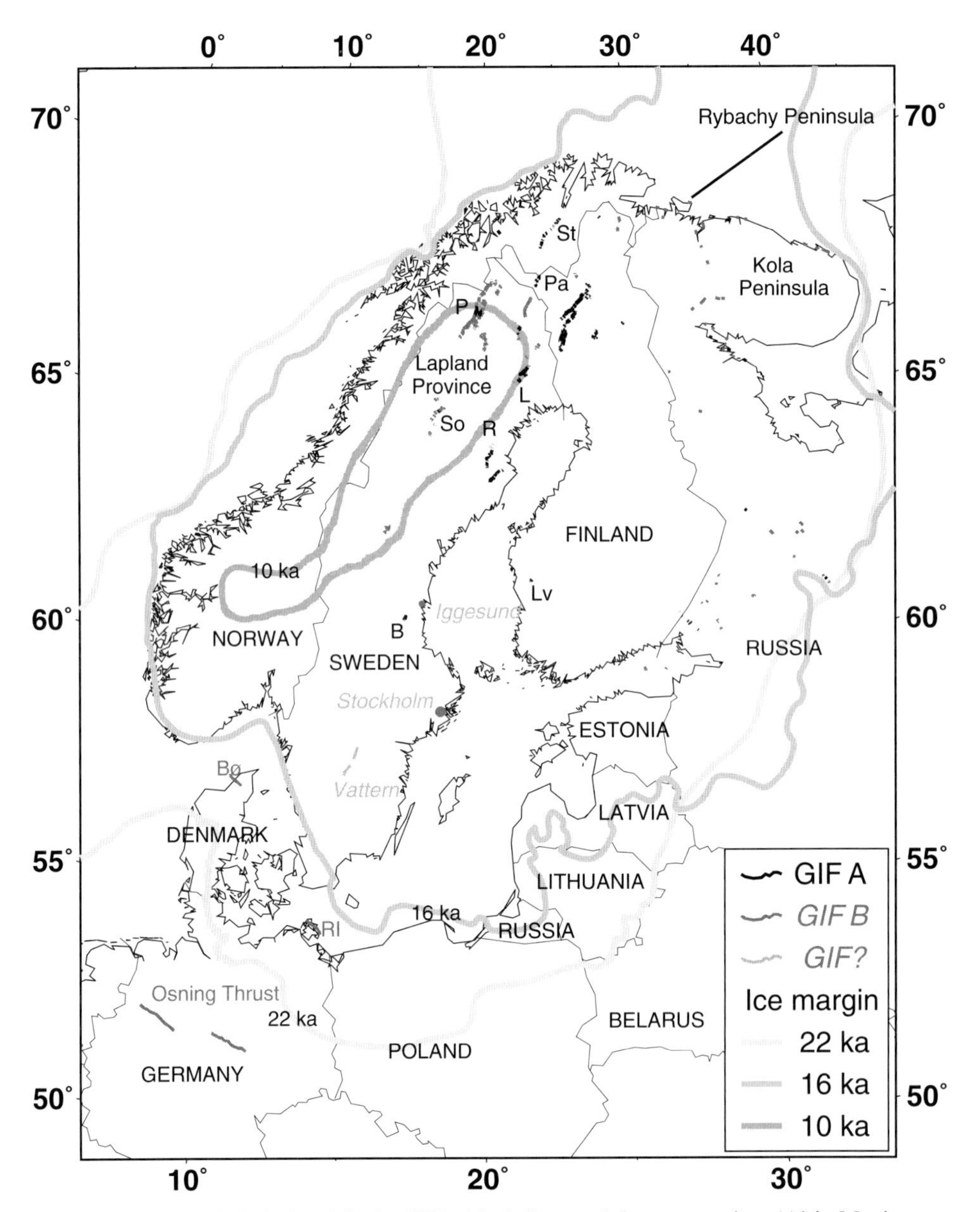

Figure 1.2 Glacially induced faults (GIFs, black lines and dots, uncertainty 'A' in Munier et al., 2020), probable GIFs (dark grey lines, uncertainty 'B' in Munier et al., 2020), suggested GIFs (light grey lines, uncertainty 'C' in Munier et al., 2020) and selected locations of suggested palaeoseismicity (light grey dots) in Northern and Central Europe. Ice limits from DATED-1 (Hughes et al., 2016). B – Bollnäs, Bø – Børglum, L – Lansjärv, Lv – Lauhavuori, P – Pärvie, Pa – Palojärvi, R – Röjnoret, RI – Rügen Island, So – Sorsele, St – Stuoragurra.

been added because these structures are a common feature in the recent literature linked to glacially triggered earthquakes.

These five criteria must all be fulfilled for a fault to qualify as an (almost) certain GIF. There is no longer an age constraint, so GIFs are not limited to just the most recent glaciation and the times shortly before and after local deglaciation. Appropriate investigation methods must be ensured, of course, so that especially criterion 5 is convincingly fulfilled.

We suggest that two previously used criteria, M3 (= F4) and M5, be removed from the list of required criteria, but they might be useful as additional considerations:

1. **Displacement ratio**: The ratio of (usually vertical) displacement to overall length of the fault normally should be less than 1:1,000. For most GIFs this ratio is between 1:1,000 and 1:10,000.

   Comment: This is criterion M3, which can be used for surficial faults like the prominent GIFs in northern Fennoscandia. However, Muir Wood (1993) and Fenton (1994) already noted that this is not a necessary requirement, e.g. because mechanical behavior of some fault materials hampers the development of prominent fault scarps. The Lansjärv Fault in Sweden (Figure 1.2) has a ratio higher than 1:1,000 (Smith et al., 2014) and is thus an exception. This criterion would also limit the term GIF to structures that can be clearly identified on the surface. However, erosional processes and human activity, among other events, in and near formerly glaciated areas, especially those areas around the edge of the former ice sheets where the ice retreated 10,000 years earlier, may have buried, removed or leveled (parts of) surface traces of GIFs (see Sandersen & Sutinen, Chapter 3). Brooks and Adams (2020), for example, argue that the surficial fault traces of the 1989 Ungava earthquake in Canada (average surface offset of 0.8 m; Adams et al., 1991) are likely no longer visible today. Consequently, this criterion appears too strict.
2. **Synchronous displacement**: Reactivation of the fault affected the entire fault.

   Comment: This is M5. It was originally listed to allow an estimation of earthquake magnitude (Raymond Munier, personal communication, 2020). The criterion is also rather strict and was omitted by Olesen et al. (2004). Recent dating results by Olesen et al. (see Chapter 11) show that each of the three systems of the Stuoragurra Fault Complex was reactivated at different times.

Muir Wood (1993) introduced a grading scale for qualifying an observation as neotectonics in view of all information and any uncertainties. This scale is also applied to the GIF database of this book (Munier et al., 2020), but the wording of the scale is slightly altered: the five grades are kept, but neotectonics is replaced with GIF:

A. Almost certainly a GIF
B. Probably a GIF
C. Possibly a GIF
D. Probably not a GIF
E. Very unlikely a GIF

This scale has been used recently by Brooks and Adams (2020) to classify Eastern Canadian GIF claims. The 'classic' postglacial faults in Northern Europe are usually classified as A. Many claims, especially at the edge of and outside the former ice sheet, are currently of grade B or C, mainly because they are not yet fully investigated and thus do not fulfill all classification criteria. Some earlier claims in the literature are classified as D or E when newer investigations could not support the initial claim; see e.g. Olesen et al. (2004). It is therefore fully possible that many currently B- or C-graded claims will receive a D or E after additional investigations in the future or will no longer be listed as GIF. For tracking of claims, we suggest removed claims be documented (ones which can be considered 'F. Not a GIF') together with a rationale for their removal. In turn, a small fraction of D- and E-graded faults may be elevated to A or B after more investigation.

1.3 Brief Historical Overview until the Early 2000s

Munier and Fenton (2004) provided the only previous history of global GTF research. Their review included peer-reviewed publications and limited-distribution reports as well as conference abstracts and personal communications. Some regional overviews were compiled for the Northern European faults (see corresponding Chapters 11–14 for references) and for a few other areas, for example, by Firth and Stewart (2000) for the British Isles and by Fenton (1994) for Eastern Canada and the Eastern United States, here updating work by Oliver et al. (1970) and Adams (1981). The work by Fenton (1994), for example, lists 173 publications. Clearly, our review cannot address all previous studies. In the following, we limit ourselves to the most important contributions, although even those might be different for other researchers. More information can be found though in this book's chapters. Section 1.4 summarizes the developments of the last decade that gave rise to this book.

In the history of GTF research one must distinguish between (i) studies that discuss GTF and use one of the terms (mainly postglacial faulting) mentioned above, (ii) studies that discuss GTF but do not name it as such and (iii) studies that use the term postglacial faulting or similar to describe another process. Fenton (1994) points to Mather (1843) as an early describer of a postglacial faulting feature, which is not named as such. The term was used by Matthew (1894) for the first time. Then, in the late nineteenth/early twentieth century, GTF was widely recognized in investigations of postglacial geomorphic features in Eastern Canada and the Northeastern United States (Munier & Fenton, 2004). However, most of those features have very small offsets. Also, in Europe the term 'postglacial faulting' was used for small features (e.g. Brøgger, 1884; Reusch, 1888; Munthe, 1905), interestingly, far away from the Lapland Province with its prominent, but then-undiscovered, GIFs.

1.3.1 History in Lapland

Tanner (1930) noticed a 10-metre offset in raised shorelines along northern Rybachy (Fisher) Peninsula north of the Kola Peninsula (Figure 1.2) that could be related to fault reactivation in postglacial times. Interestingly, he noted (p. 316) that postglacial faulting in Fennoscandia was not being considered by geologists, and thus no active searches for postglacial faults were conducted. Tanner followed up with a discussion of literature containing hints of GTF all over Fennoscandia. He concluded that although little was known at the time (1930), one should be open to the fact that GIF features could exist anywhere in Fennoscandia where land uplift was occurring.

GIFs were not recognized in Fennoscandia until Kujansuu (1964) discussed remarkable structures in north-western Finland. Since the mid-1970s investigations in Fennoscandia were intensified, especially related to nuclear waste disposal plans, and provided an almost constant stream of new GIF discoveries. Lundqvist and Lagerbäck (1976) informed about the Pärvie Fault west of Kiruna, Sweden, and Olesen (1988) recognized the Stuoragurra Fault in Finnmark, Northern Norway. Lagerbäck and Sundh (2008) listed about 15 GIFs or fault systems that were identified by the mid-2000s. Many of these GIFs were initially found and mapped with the help of aerial photographs. Some of them, the Lansjärv, Sorsele and Röjnoret faults, were investigated in greater detail with additional trenching (Lund et al., 2017). Arvidsson (1996) estimated that the moment magnitude of earthquakes generating the faults could have reached $M_w = 8.2$. Seismic networks in the Nordic countries were installed and extended through the 1980s and '90s and slowly provided a clearer picture of current seismicity around the GIFs (see Gregersen et al., Chapter 10). The dating of the rupture activity was firstly relative to local deglaciation or to land uplift if the highest shoreline was higher than the elevation of the fault, as direct dating, e.g. with cosmogenic (scarps), luminescence (scarp sediments) or radiocarbon (included organic material) methods had not yet been performed (Lagerbäck, 1992; Lagerbäck & Sundh, 2008). For more information on investigations in Finland, Norway and Sweden prior to the early 2000s, the reader is referred to Kuivamäki et al. (1998), Olesen et al. (2013) and Lagerbäck and Sundh (2008), respectively. Karpov (1960) mentioned traces of postglacial tectonic faults in the Khibiny Mountains on the Kola Peninsula, but more detailed investigations in NW Russia did not start until the late 1980s (see Nikolaeva et al., Chapter 14).

1.3.2 History Outside of Lapland

Away from Lapland no GIF had been described, although an overview of studies by Mörner (2005), for example, lists more than 20 different locations all over Sweden that suggest palaeoseismic activity in relation to ice retreat and land uplift – but without identified surface ruptures. Lund et al. (2017) reviewed two of these (Iggesund and Stockholm), which are based on disturbed sediments, and they could not find clear evidence that would further support these findings. Among the 60 claims in mainland Norway reviewed and graded by Olesen et al. (2004, 2013), no potential GIF was discovered in Southern Norway.

In parallel with and as a consequence of Lapland research, suggestions for GIFs were presented for Scotland (see e.g. Fenton, 1991, for an overview), Northern Ireland (Knight, 1999) and Ireland (Mohr, 1986). The claims were subject to discussion: Stewart et al.

(2001) question whether there is any significant postglacial movement on faults in western Scotland, while Munier and Fenton (2004) point to limitations in the Irish studies. A detailed analysis and re-evaluation of the many claims for the British Isles is still pending.

In other parts of formerly glaciated Europe or in its near vicinity, GIFs were not found, but palaeoseismicity or fault reactivation due to GIA was suggested (e.g. Van Vliet-Lanoë et al., 1997). Gregersen et al. (1996) speculated about local reactivation of the northern boundary fault in the Sorgenfrei–Tornquist Zone at the time of deglaciation. Potential palaeoseismic traces were found in Rügen Island (Ludwig 1954/55) and in the Baltic countries (see Bitinas et al., Chapter 18), but these assertions were not confirmed by more detailed field investigations at the time.

Research regarding GTF was sporadically conducted in North America and much material was published (Fenton, 1994). As in Europe, investigations peaked in the 1970s and ‘80s (see summary in Adams, 1996). A strong earthquake (M = 6.3) on 25 December 1989 in Ungava Peninsula produced the first known historical surface rupture in formerly glaciated eastern North America (Adams et al., 1991, 1992). It is sometimes speculated that GIA was the cause, but it is very unlikely the event would have fallen within a period of glacially influenced seismicity (Brooks & Adams, 2020). On another note, Stein et al. (1979) attributed earthquakes in Baffin Island and adjacent Baffin Bay to basement faults reactivated by GIA.

Claims of GTF in North America outside its north-eastern quadrant are rare and often refer to regions with other possible trigger mechanisms, like plate margin deformation and/or volcanism in Alaska (Sauber et al., see Chapter 20), the Sierra Nevada (Greene, 1996) and western Wyoming (Byrd et al., 1994; Hinz et al., 1997) or the failed Reelfoot Rift system with the New Madrid Seismic Zone (Grollimund & Zoback, 2001). These make it difficult to distinguish the GIA contribution to potential fault reactivation and earthquakes.

Some GIFs have been proposed in other parts of the world where ice sheets or larger ice masses were or are present, like Greenland, Iceland and Antarctica, but no GIF has been confirmed to date. An overview of claims can be found in Steffen and Steffen in Chapter 21.

Despite the considerable number of confirmed GIFs and an increasing number of claims, the underlying mechanism remained unknown until Johnston (1987) introduced the concept of earthquake suppression due to earthquakes and increased activity after deglaciation using the Mohr-Coulomb failure theory (as reviewed by Steffen et al. in Chapter 2). This concept also accounted for the low seismicity of Greenland and Antarctica. Wu and Hasegawa (1996a,b) explained the mechanism using finite element modelling of the GIA process (see Wu et al., Chapter 22), and their predicted timing of glacially triggered earthquakes is in reasonable agreement with GIF observations.

In the early 2000s investigations of GIFs continued at varying pace in countries with GIFs or claims of GIFs. In the British Isles, investigations ceased, while nuclear waste authorities in Canada, Sweden and Finland supported on-going investigations. Scientific interest has again increased in the last decade due to several circumstances, which we will illuminate next.

1.4 Recent Developments

Several countries like Sweden, Finland and Norway nowadays generate new elevation models with the help of light detection and ranging (LiDAR) (see Palmu et al., Chapter 5). Inspection of the new data in the early 2010s led to the discovery of the Bollnäs Fault in central Sweden (Smith et al., 2014). LiDAR facilitated more detailed investigations and increased the number of GIFs in Sweden as well as the length and complexity of those already mapped (Mikko et al., 2015; Lund et al., 2017; Smith et al., see Chapter 12). Similarly, in Finland LiDAR reconnaissance has revealed new faults, e.g. the Palojärvi Fault in the north (Sutinen et al., 2014) and Lauhavuori in Southern Finland (Palmu et al., 2015). Additionally, detailed trenching across the scarps has provided evidence of non-stationary seismicity and occurrence of multiple slip events even before the Late Weichselian maximum (see Sutinen et al., Chapter 13).

A dedicated program in Finland enabled dating of multiple independent landslides (see Sutinen et al., Chapter 13). Ojala et al. (2018) reported, based on landslide dating, that some Finnish faults are younger than previously thought and that they had ruptured several times. Similar results were reported from Sweden (Smith et al., 2018; and in this book, see Chapter 8). Recently, dating results from the Stuoragurra Fault Complex showed much younger ages than expected (at least 6,000 years after the end of deglaciation) and that the segments of the fault ruptured at different times (Olsen et al., 2020; see also Olesen et al., Chapter 11).

Several new studies argue for GIFs south of the Lapland Province. Jacobsson et al. (2014) suggested a normal-faulting GIF at the bottom of Lake Vättern in southern Sweden, though there are no traces of this fault on land (Lund et al., 2017). Hence, this finding must be questioned, and further investigations are still needed. Brandes et al. (2018) collected multidisciplinary material that points to postglacial activity of the Børglum Fault in northern Denmark and thus strongly advocated for it being a GIF. Their study is consistent with several others that suggest GTF in Denmark (see Sandersen et al., Chapter 15).

Glacially triggered faulting at the edge of the former ice sheet has been suggested by several groups by analyzing SSDS in Germany (Brandes & Tanner, 2012; Hoffmann & Reicherter, 2012; Pisarska-Jamroży et al., 2018), Poland (van Loon & Pisarska-Jamroży, 2014) and Latvia (van Loon et al., 2016). Pisarska-Jamroży et al. (2018) further claimed to have found evidence for GTF on Rügen Island during the last glacial advance. Based on geological and numerical analysis of the Osning Thrust in Germany, Brandes et al. (2012) suggested this to be the first GIF discovered outside of the glaciated area. In subsequent studies, it was suggested that even historical (Brandes et al., 2015) and recent deep crustal earthquakes in Germany (Brandes et al., 2019) are due to GIA. Brandes et al. (2015) further noted that GTF in Germany due to the Weichselian ice sheet might affect even the northern parts of Central Germany. A summary of these new findings and additional evidence for GIFs at the edge and outside the former ice sheet can be found in Müller et al., in Chapter 16, Pisarska-Jamroży et al., in Chapter 17, and Bitinas et al., in Chapter 18.

In Canada, no certain GIF has yet been determined (see Adams & Brooks, in Chapter 19). However, sub-bottom profiling surveys in eastern Canadian lakes reveal a large number of mass transport deposits pointing to early postglacial seismic shaking due to GIA (Brooks & Adams, 2020). In view of these findings and because of availability of new data imagery sources (such as LiDAR), Brooks and Adams (2020) predict further discoveries of GIF candidates in Canada can be expected in the future.

Based on an increased amount of seismic data in Antarctica, Lough et al. (2018) discovered more earthquakes beneath the ice than previously thought. The authors concluded that Antarctica's seismicity compares to that of the Canadian Shield. They further hypothesized that tectonic intraplate stresses could accumulate long enough over time, and once the suppressing GIA stress has been overcome, seismicity increases and then Antarctica behaves like a typical intraplate area without ice.

In parallel with new field observations, modelling of GTF has progressed. Hetzel and Hampel (2005) introduced a finite element model with a fault structure that showed the behaviour deduced by Johnston (1987), but in the finite element model constant slip is predicted before and after glaciation. Steffen et al. (2014) developed a so-called GIA-fault model, which extends the model from Wu and Hasegawa (1996a,b) with a fault surface and which yields fault slip in a finite amount of time. Such new model development benefitted from progress made in stress modelling (see Gradmann & Steffen, Chapter 23).

Finally, after more than a decade of planning, drilling into the Pärvie Fault has been tentatively scheduled within the International Continental Scientific Drilling Program (ICDP) (Kukkonen et al., 2011; Ask et al., see Chapter 9). The professional community is very curious about any results this drilling will produce; for example, whether it can confirm that the Pärvie Fault is apparently not moving on the surface as found by Mantovani and Scherneck (2013).

1.5 Conclusions

More than two dozen fault scarps a kilometre-plus in length have been identified in northern Fennoscandia since the 1960s and '70s, when extensive investigations began. Their identification is nowadays made easier with the help of LiDAR images. These scarps are categorized as GIFs. Comparable structures were also described in the British Isles and Eastern Canada. In other formerly glaciated areas in Europe – e.g. the southern parts of Sweden, Norway and Finland, the southern Baltic Sea, Denmark, Northern Germany and Poland, and the Baltic countries – GIFs were seldom recognized, but the number of studies with reliable field evidence has considerably increased in recent years. But their estimated fault displacements are not as large as those in northern Fennoscandia. There are also other areas in the world where GTF may play a role, such as northern Alaska, some Arctic islands, the Alps (Jäckli, 1965) and Patagonia (Bentley et al., 2005).

As well as the many new (potential) GIF discoveries and features that point to glacially triggered seismicity, fault reactivation has been identified by dating events on several Fennoscandian GIFs. The results show that at least some of the scarps were generated during more than a single event, and for some it is hypothesized that part of the displacement happened before the last glaciation (see Chapter 12, by Smith et al.).

In view of the many new findings a necessary revision of the well-known classification criteria of Fenton (1994) and Muir Wood (1993) is suggested. The revised criteria are provided in checklist form. Since research is always progressing, we forecast that future findings may lead to further revision(s) of these criteria. Lastly, all these new findings have strong implications for the role of glacially induced faulting on intraplate seismicity. This will be discussed in Chapter 24 by Olesen et al.

Acknowledgements

We thank Raymond Munier, John Adams, Christian Brandes and Conrad Lindholm for their constructive reviews, and Antti Ojala, Lars Olsen and Rebekka Steffen for discussion and remarks on earlier versions of this chapter. Figure 1.2 was drawn with GMT5 (Wessel et al., 2013).

References

Adams, J. (1981). *Postglacial Faulting: A Literature Survey of Occurrences in Eastern Canada and Comparable Glaciated Areas*. Atomic Energy of Canada Limited Technical Report, **TR-142**.

Adams, J. (1996). Paleoseismology in Canada: a dozen years of progress. *Journal of Geophysical Research*, **101**, 6193–6207, doi.org/10.1029/95JB01817.

Adams, J., Wetmiller, R. J., Hasegawa, H. S. and Drysdale, J. (1991). The first surface faulting from a historical intraplate earthquake in North America. *Nature*, **352**, 617–619, doi.org/10.1038/352617a0.

Adams, J., Percival, J. A., Wetmiller, R. J., Drysdale, J. and Robertson, P. B. (1992). Geological controls on the 1989 Ungava surface rupture: a preliminary interpretation. *Papers of the Geological Survey of Canada*, **92-1C**, 147–155.

Arvidsson, R. (1996). Fennoscandian earthquakes: whole crustal rupturing related to postglacial rebound. *Science*, **274**, 744–746, doi.org/10.1126/science.274.5288.744.

Bentley, M. J. and McCulloch, R. D. (2005). Impact of neotectonics on the record of glacier and sea level fluctuations, Strait of Magellan, southern Chile. *Geografiska Annaler: Series A, Physical Geography*, **87**(2), 393–402, doi.org/10.1111/j.0435-3676.2005.00265.x.

Brandes, C. and Tanner, D. (2012). Three-dimensional geometry and fabric of shear deformation-bands in unconsolidated Pleistocene sediments. *Tectonophysics*, **518–521**, 84–92, doi.org/10.1016/j.tecto.2011.11.012.

Brandes, C., Polom, U. and Winsemann, J. (2011). Reactivation of basement faults: interplay of ice-sheet advance, glacial lake formation and sediment loading. *Basin Research*, **23**, 53–64, doi.org/10.1111/j.1365-2117.2010.00468.x.

Brandes, C., Steffen, H., Steffen, R. and Wu, P. (2015). Intraplate seismicity in northern Central Europe is induced by the last glaciation. *Geology*, **43**, 611–614, doi.org/10.1130/G36710.1.

Brandes, C., Steffen, H., Sandersen, P. B. E., Wu, P. and Winsemann, J. (2018). Glacially induced faulting along the NW segment of the Sorgenfrei–Tornquist Zone, northern Denmark: implications for neotectonics and Lateglacial fault-bound basin formation. *Quaternary Science Reviews*, **189**, 149–168, doi.org/10.1016/j.quascirev.2018.03.036.

Brandes, C., Plenefisch, T., Tanner, D. T., Gestermann, N. and Steffen, H. (2019). Evaluation of deep crustal earthquakes in northern Germany – possible tectonic causes. *Terra Nova*, **31**(2), 83–93, doi.org/10.1111/ter.12372.

Brandes, C., Winsemann, J., Roskosch, J. et al. (2012). Activity along the Osning Thrust in Central Europe during the Lateglacial: ice-sheet and lithosphere interactions. *Quaternary Science Reviews*, **38**, 49–62, doi.org/10.1016/j.quascirev.2012.01.021.

Brøgger, W. C. (1884). Spaltenverwerfungen in der Gegend Langesund-Skien [Crevasse faults in the Langesund-Skien area]. *Nyt Magazin for Naturvidenskaberne*, **28**, 253–419.

Brooks, G. R. and Adams, J. (2020). A review of evidence of glacially-induced faulting and seismic shaking in southeastern Canada. *Quaternary Science Reviews*, **228**, 106070, doi.org/10.1016/j.quascirev.2019.106070.

Byrd, J. O. D., Smith, R. B. and Geissman, J. W. (1994). The Teton fault, Wyoming: topographic signature, neotectonics, and mechanisms of deformation. *Journal of Geophysical Research*, **99**(B10), 20095–20122,doi.org/10.1029/94JB00281.

Clark, P. U., Dyke, A. S., Shakun, J. D., et al. (2009). The last glacial maximum. *Science*, **325**(5941), 710–714, doi.org/10.1126/science.1172873.

Fenton, C. (1991). *Neotectonics and Palaeoseismicity in North West Scotland*. PhD thesis, University of Glasgow, Glasgow.

Fenton, C. (1994). *Postglacial Faulting in Eastern Canada*. Geological Survey of Canada Open File, 2774.

Fenton, C. (1999). Glacio-isostatic (postglacial) faulting: criteria for recognition. In K. L. Hanson, K. I. Kelson, M. A. Angell and W. R. Lettis, eds., *Identifying Faults and Determining Their Origins*. U.S. Nuclear Regulatory Commission, pp. A-51–A-99.

Firth, C. R. and Stewart, I. S. (2000). Postglacial tectonics of the Scottish glacio-isostatic uplift centre. *Quaternary Science Reviews*, **19**, 1469–1493, doi.org/10.1016/S0277-3791(00)00074-3.

Greene, D. C. (1996). Quaternary reactivation of the Lost Lakes fault, a brittle fault zone containing pseudotachylite in the Tuolumne intrusive suite, Sierra Nevada, California. *Geological Society of America Cordilleran Section, Spring Meeting, Abstracts with Program*, **28**, p. 70.

Gregersen, S., Leth, J., Lind, G. and Lykke-Andersen, H. (1996). Earthquake activity and its relationship with geologically recent motion in Denmark. *Tectonophysics*, **254**, 265–273, doi.org/10.1016/0040-1951(95)00193-X.

Grollimund, B. and Zoback, M. D. (2001). Did deglaciation trigger intraplate seismicity in the New Madrid seismic zone? *Geology*, **29**, 175–178, doi.org/10.1130/0091-7613 (2001)029%3C0175:DDTISI%3E2.0.CO;2.

Hetzel, R. and Hampel, A. (2005). Slip rate variations on normal faults during glacial-interglacial changes in surface loads. *Nature*, **435**, 81–84, doi.org/10.1038/ nature03562.

Hinz, N. H., Carson, R. J., Gardner, T. W. and McKenna, K. (1997). Late Quaternary deglaciation, flooding, and tectonism (?), upper Clarks Fork Valley, Park County, Wyoming. *Geological Society of America Annual Meeting, Abstracts with Program*, **29**, p. 15.

Hoffmann, G. and Reicherter, K. (2012). Soft-sediment deformation of late Pleistocene sediments along the southwestern coast of the Baltic Sea (NE Germany). *International Journal of Earth Sciences*, **101**, 351–363, doi.org/10.1007/s00531-010-0633-z.

Hughes, A. L. C., Gyllencreutz, R., Lohne, Ø. S., Mangerud, J. and Svendsen, J. I. (2016). The last Eurasian ice sheets – a chronological database and time-slice reconstruction, DATED-1. *Boreas*, **45**, 1–45, doi.org/10.1111/bor.12142.

Jäckli, H. C. A. (1965). Pleistocene glaciation of the Swiss Alps and signs of postglacial differential uplift. *Geological Society of America, Special Paper*, **84**, 153–157, doi .org/10.1130/SPE84-p153.

Jakobsson, M., Björck, S., O'Regan, M. et al. (2014). Major earthquake at the Pleistocene–Holocene transition in Lake Vättern, southern Sweden. *Geology*, **42**, 379–382, doi .org/10.1130/G35499.1.

Johnston, A. C. (1987). Suppression of earthquakes by large continental ice sheets. *Nature*, **330**, 467–469, doi.org/10.1038/330467a0.

Karpov, N. N. (1960). Traces of postglacial tectonic faults in the Khibiny Mountains. *Moscow University Bulletin*, **5**(4), 61.

Knight, J. (1999). Geological evidence for neotectonic activity during deglaciation of the southern Sperrin Mountains, Northern Ireland. *Journal of Quaternary Science*, **14**, 45–57, doi.org/10.1002/(SICI)1099-1417(199902)14:1<45::AID-JQS389>3.0 .CO;2-3.

Kuivamäki, A., Vuorela, P. and Paananen, M. (1998). *Indications of Postglacial and Recent Bedrock Movements in Finland and Russian Karelia*. Geological Survey of Finland Nuclear Waste Disposal Research Report YST-99, Espoo, Finland, 92 pp.

Kujansuu, R. (1964). Nuorista siirroksista Lapissa [English summary: Recent faults in Lapland]. *Geologi*, **16**, 30–36 (in Finnish).

Kukkonen, I. T., Olesen, O., Ask, M. V. S. and the PFDP Working Group (2010). Postglacial faults in Fennoscandia: targets for scientific drilling. *GFF*, **132**(1), 71–81, doi.org/10.1080/11035891003692934.

Lagerbäck, R. (1992). Dating of Late Quaternary faulting in northern Sweden. *Journal of the Geological Society*, **149**, 285–291, doi.org/10.1144/gsjgs.149.2.0285.

Lagerbäck, R. and Sundh, M. (2008). Early Holocene faulting and paleoseismicity in northern Sweden. *Geological Survey of Sweden Research Paper C 836*, 80 pp.

Lough, A. C., Wiens, D. A. and Nyblade, A. (2018). Reactivation of ancient Antarctic rift zones by intraplate seismicity. *Nature Geoscience*, **11**(7), 515–519, doi.org/10.1038/s41561-018-0140-6.

Ludwig, A. O. (1954/1955). Eistektonik und echte Tektonik in Ost-Rügen (Jasmund) [Ice tectonics and real tectonics in East Rügen Island (Jasmund)]. *Wissenschaftliche Zeitschrift der E.-M.-A.-Universität Greifswald*, **4**, 251–288.

Lund, B. (2015). Palaeoseismology of glaciated terrain. In M. Beer et al., eds., *Encyclopedia of Earthquake Engineering*. Springer-Verlag, Berlin/Heidelberg, 1765–1779.

Lund, B. and Näslund, J.-O. (2009). Glacial isostatic adjustment – implications for glacially induced faulting and nuclear waste repositories. In C. B. Connor, N. A. Chapman and L. J. Connor, eds., *Volcanic and Tectonic Hazard Assessment for Nuclear Facilities*. Cambridge University Press, Cambridge, pp. 142–155.

Lund, B., Roberts, R. and Smith, C. A. (2017). Review of paleo-, historical and current seismicity in Sweden and surrounding areas with implications for the seismic analysis underlying SKI report 92:3. *Swedish Radiation Safety Authority Report*, **2017**:35.

Lundqvist, J. and Lagerbäck, R. (1976). The Pärve Fault: a late-glacial fault in the Precambrian of Swedish Lapland. *Geologiska Föreningens i Stockholm Förhandlingar*, **98**, 45–51, doi.org/10.1080/11035897609454337.

Mantovani, M. and Scherneck, H.-G. (2013). DInSAR investigation in the Pärvie end-glacial fault region, Lapland, Sweden. *International Journal of Remote Sensing*, **34**(23), 8491–8502, doi.org/10.1080/01431161.2013.843871.

Mather, W. W. (1843). *Geology of New-york. Part 1, Comprising Geology of the First Geological District*. Carroll & Cook, Albany, New York.

Matthew, G. F. (1894). Movements of the Earth's crust at St. John, N. B., in post-glacial times. *Bulletin of the Natural History Society of New Brunswick*, **12**, 34–42.

Mikko, H., Smith, C. A., Lund, B., Ask, M. V. S. and Munier, R. (2015). LiDAR-derived inventory of post-glacial fault scarps in Sweden. *GFF*, **137**, 334–338, doi.org/10.1080/11035897.2015.1036360.

Mohr, P. (1986). Possible Late Pleistocene faulting in Iar (west) Connacht, Ireland. *Geological Magazine*, **123**, 545–552, doi.org/10.1017/S0016756800035135.

Mörner, N.-A. (2005). An interpretation and catalogue of paleoseismicity in Sweden. *Tectonophysics*, **408**, 265–307, doi.org/10.1016/j.tecto.2005.05.039.

Muir Wood, R. (1993). *A Review of the Seismotectonics of Sweden*. SKB Technical Report TR-93-13, Swedish Nuclear Fuel and Waste Management Co., Stockholm.

Munier, R. and Fenton, C. (2004). Review of postglacial faulting. In R. Munier and H. Hökmark, eds., *Respect Distances*. SKB Technical Report TR-04-17, Swedish Nuclear Fuel and Waste Management Co., Stockholm, pp. 157–218.

Munier, R., Adams, J., Brandes, C. et al. (2020). *International Database of Glacially-Induced Faults, PANGAEA*, doi.org/10.1594/PANGAEA.922705.

Munthe, H. (1905). Om en sen- eller postglacial förkastning vid Allebergsände i Västergötland och om en postglacial rubbning i silurlagren SV om Visby [About a Lateglacial or postglacial fault at Allebergsände in Västergötland and about a post-glacial disturbance in the Silurian deposits SW of Visby]. *Geologiska Föreningens i Stockholm Förhandlingar*, **27**(6), 346.

Ojala, A. E. K., Markovaara-Koivisto, M., Middleton, M. et al. (2018). Dating of paleo-landslides in western Finnish Lapland. *Earth Surface Processes and Landforms*, **43**, 2449–2462, doi.org/10.1002/esp.4408.

Olesen, O. (1988). The Stuoragurra fault, evidence of neotectonics in the Precambrian of Finnmark, northern Norway. *Norsk Geologisk Tidskrift*, **68**, 107–118.

Olesen, O., Blikra, L. H., Braathen, A. et al. (2004). Neotectonic deformation in Norway and its implications: a review. *Norwegian Journal of Geology*, **84**, 3–34.

Olesen, O., Bungum, H., Lindholm, C. et al. (2013). Neotectonics, seismicity and contemporary stress field in Norway – Mechanisms and implications. In L. Olsen, O. Fredin and O. Olesen, eds., *Quaternary Geology of Norway. Geological Survey of Norway Special Publication Vol. 13*, pp. 145–174.

Oliver, J., Johnston, T. and Dorman, J. (1970). Postglacial faulting and seismicity in New York and Quebec. *Canadian Journal of Earth Sciences*, **7**, 579–590, doi.org/10.1139/e70-059.

Olsen, L., Olesen, O. and Høgaas, F. (2020). Dating of the Stuoragurra Fault at Finnmarksvidda, northern Norway. *Abstracts and Proceedings of the Geological Society of Norway 1 – The 34th Nordic Geological Winter Meeting, Oslo*, pp. 157–158.

Palmu, J.-P., Ojala, A. E. K., Ruskeeniemi, T., Sutinen, R. and Mattila, J. (2015). LiDAR DEM detection and classification of postglacial faults and seismically-induced landforms in Finland: a paleoseismic database, *GFF*, **137**(4), 344–352, doi.org/10.1080/11035897.2015.1068370.

Peltier, W. R. and Andrews, J. T. (1976). Glacial-isostatic adjustment – I. The forward problem. *Geophysical Journal of the Royal Astronomical Society*, **46**, 605–646, doi.org/10.1111/j.1365-246X.1976.tb01251.x.

Reusch, H. (1888). *Bømmeløen og Karmøen med omgivelser* [*Bømmeløen and Karmøen with Surroundings*]. Norges geologiske undersøkelse, Norway.

Pisarska-Jamroży, M., Belzyt, S., Börner, A. et al. (2018). Evidence from seismites for glacio-isostatically induced crustal faulting in front of an advancing land-ice mass (Rügen Island, SW Baltic Sea). *Tectonophysics*, **745**, 338–348, doi.org/10.1016/j.tecto.2018.08.004.

Smith, C. A., Sundh, M. and Mikko, H. (2014). Surficial geology indicates early Holocene faulting and seismicity, central Sweden. *International Journal of Earth Sciences*, **103**(6), 1711–1724, doi.org/10.1007/s00531-014-1025-6.

Smith, C. A., Grigull, S. and Mikko, H. (2018). Geomorphic evidence of multiple surface ruptures of the Merasjärvi "postglacial fault", northern Sweden. *GFF*, **140**(4), 318–322, doi.org/10.1080/11035897.2018.1492963.

Steffen, R., Wu, P., Steffen, H. and Eaton, D. W. (2014). On the implementation of faults in finite-element glacial isostatic adjustment models. *Computers & Geosciences*, **62**, 150–159, doi.org/10.1016/j.cageo.2013.06.012.

Stein, S., Sleep, N. H., Geller, R. J., Wang, S.-C. and Kroeger, G. C. (1979). Earthquakes along the passive margin of eastern Canada. *Geophysical Research Letters*, **6**, 537–540, doi.org/10.1029/GL006i007p00537.

Stewart, I. S., Firth, C. R., Rust, D. J., Collins, P. E. F. and Firth, J. A. (2001). Postglacial fault movement and palaeoseismicity in western Scotland: a reappraisal of the Kinloch Hourn fault, Kintail. *Journal of Seismology*, **5**, 307–328, doi.org/10.1023/A:1011467307511.

Sutinen, R., Hyvönen, E., Middleton, M. and Ruskeeniemi, T. (2014). Airborne LiDAR detection of postglacial faults and Pulju moraine in Palojärvi, Finnish Lapland. *Global and Planetary Change* **115**, 24–32, doi.org/10.1016/j.gloplacha.2014.01.007.

Tanner, V. (1930). Om nivåförändringarna och grunddragen av den geografiska utvecklingen efter istiden i Ishavsfinland samt om homotaxin av Fennoskandias kvartära marina avlagringar. *Studier över kvartärsystemet i Fennoskandias nordliga delar – IV*. [On level changes and basic features of the geographical development after the ice age in the Polar Sea of Finland and on the homotaxis of Fennoscandia's Quaternary marine deposits. Studies of the Quaternary System in Northern Fennoscandia – IV]. *Bulletin de la Commission Géologique de Finlande*, **88**, 594 pp.

van Loon, A. J. and Pisarska-Jamroży, M. (2014). Sedimentological evidence of Pleistocene earthquakes in NW Poland induced by glacio-isostatic rebound. *Sedimentary Geology*, **300**, 1–10, doi.org/10.1016/j.sedgeo.2013.11.006.

van Loon, A. J., Pisarska-Jamroży, M., Nartišs, M., Krievāns, M. and Soms, J. (2016). Seismites resulting from high-frequency, high-magnitude earthquakes in Latvia caused by Late Glacial glacio-isostatic uplift. *Journal of Palaeogeography*, **5**, 363–380, doi.org/10.1016/j.jop.2016.05.002.

Van Vliet-Lanoë, B., Bonnet, S., Hallegouët, B. and Laurent, M. (1997). Neotectonic and seismic activity in the Armorican and Cornubian Massifs: regional stress field with glacio-isostatic influence? *Journal of Geodynamics*, **24**(1–4), 219–239, doi.org/10.1016/S0264-3707(96)00035-X.

Wessel, P., Smith, W. H. F., Scharroo, R., Luis, J. F. and Wobbe, F. (2013). Generic Mapping Tools: improved version released. *EOS Transactions American Geophysical Union*, **94**, 409–410, doi.org/10.1002/2013EO450001.

Wu, P. and Hasegawa, H. S. (1996a). Induced stresses and fault potential in eastern Canada due to a disc load: a preliminary analysis. *Geophysical Journal International*, **125**, 415–430, doi.org/10.1111/j.1365-246X.1996.tb00008.x.

Wu, P. and Hasegawa, H. S. (1996b). Induced stresses and fault potential in eastern Canada due to a realistic load: a preliminary analysis. *Geophysical Journal International*, **127**, 215–229, doi.org/10.1111/j.1365-246X.1996.tb01546.x.

Wu, P. and Peltier, W. R. (1982). Viscous gravitational relaxation. *Geophysical Journal of the Royal Astronomical Society*, **70**, 435–486, doi.org/10.1111/j.1365-246X.1982.tb04976.x.

2

Geomechanics of Glacially Triggered Faulting

REBEKKA STEFFEN, PATRICK WU AND BJÖRN LUND

ABSTRACT

Glacially induced faults are found in regions surrounding and beneath formerly glaciated areas. The reactivation of such faults is linked to the increase and decrease of ice mass. But, whether or not faults are reactivated by glacially induced stresses depends to a large degree on the crustal stress field, fault properties and fluid pressures. The background (tectonic and lithostatic) stress field has a major effect on the potential for reactivation, as the varying stresses induced by the ice sheet affect the state of stress around the fault, bringing the fault to more stable or more unstable conditions. Here, we describe the effect of glacially induced stresses on fault reactivation, under three potential background stress regimes of normal, strike-slip and thrust/reverse faulting. The Mohr diagram is used to illustrate how glacially induced stresses affect the location and the size of the Mohr circle – in particular, whether the distance between the circle and the linear failure envelope (the Coulomb failure stress) increases or decreases. In some cases, the Mohr circle is even able to touch the linear failure envelope, leading to unstable conditions. We will show that stress changes beneath the ice sheet during and after unloading favour a reactivation of faults within a thrust-faulting stress regime, while the stress conditions within a normal-faulting stress regime remain stable. Stability also exists in a strike-slip-faulting stress regime if glacially induced horizontal stresses are similar in their magnitudes, a valid scenario for nearly circular ice sheets. An increase in eccentricity could lead to different conditions, and instability of the area within the ice margin might occur during and after deglaciation. Glacially induced stresses in the peripheral bulge lead to increased stability in a thrust-faulting stress regime and unstable conditions exist in normal as well as strike-slip-faulting stress regimes. We review these different cases by applying an analysis of the stress state at different time points in the glacial cycle. In addition, we present an overview of fault properties that affect the reactivation of glacially induced faults, such as pore-fluid pressure and the coefficient of friction.

2.1 Introduction

Stresses in the Earth occur for various reasons, e.g. due to overlying rock mass (e.g. Jaeger et al., 2007), tectonic forces (e.g. Zoback et al., 1989), sedimentation and erosion (e.g. Zoback, 1992) and hydrological effects (e.g. Nur & Booker, 1972); also, in some areas the load of large continental ice sheets of the late Pleistocene (e.g. Walcott, 1970) and today's smaller glaciers (e.g. Sauber et al., see Chapter 20) result in glacially induced stresses. The stresses required to create a new fault in the Earth's crust are typically around 100 MPa (Johnson & DeGraff, 1988), while stress changes of only a few kPa (King et al., 1994) are enough to reactivate an old fault, which is critically stressed. Thus, understanding the geomechanics of glacially triggered faulting involves not only knowing about stresses induced by ice masses but also the style and magnitude of the background stress field and certain fault properties, such as the frictional behaviour. Stress and fault stability are commonly assessed using a Mohr diagram, which consists of the Mohr circle and the Mohr-Coulomb failure envelope.

The Mohr diagram (Figure 2.1A; Coulomb, 1776; Mohr, 1914), plots the normal stress in relation to the shear stress acting on a fault. The normal stress (σ_n) acts perpendicular to the fault plane, pushing the fault surfaces together and thus opposing fault movement, whereas the shear stress (τ) acts parallel to the fault plane and tends to promote fault movement (e.g. Twiss & Moores, 2007). The linear failure envelope (also called the Mohr-Coulomb failure envelope) depends only on the coefficient of friction (μ) and cohesion (C) (Heyman, 1972) and is therefore independent of the fault dimensions. The orientation of the fault plane in the Mohr diagram is defined by the angle between a point on the Mohr circle to the midpoint of the Mohr circle and the horizontal normal stress axis (angle 2θ; Figure 2.1A), and each point on the Mohr circle is related to a different fault orientation. The angle θ is also the angle between the normal of the fault plane and the maximum principal stress direction (Figure 2.1B). The Mohr circle itself is calculated from the principal stresses σ_1 (maximum principal stress), σ_2 (intermediate principal stress) and σ_3 (minimum principal stress). In two-dimensional cases, only the maximum and minimum principal stresses are used, and the Mohr circle is independent of the intermediate principal stress (Figure 2.1A).

Principal stresses depend on the tectonic setting and are subject to changes during external processes (variations in the pore-fluid pressure, glacial loading and unloading, etc.; e.g. Zoback & Zoback, 2015). Three main stress regimes can be distinguished (Anderson, 1951): thrust/reverse-faulting (σ_1 and σ_2 in the horizontal, σ_3 in the vertical, Figure 2.1C); normal-faulting (σ_2 and σ_3 in the horizontal, σ_1 in the vertical, Figure 2.1C); and strike-slip-faulting (σ_1 and σ_3 in the horizontal, σ_2 in the vertical, Figure 2.1C) stress regimes. The principal stress components in a simple background stress regime (present before additional stresses are applied, for example, those due to sediment loading, topographic and gravitational loads, pore-pressure changes or glacial loading/unloading) consist of tectonic stresses and lithostatic stress, the latter acting in all directions and being equal to the overburden pressure.

The Mohr diagram can be used to identify faults that are close to failure. Coulomb failure stress (*CFS*, Figure 2.1A; e.g. Harris, 1998) is commonly applied, which is the vertical distance between the failure envelope and any point on or within the Mohr circle ($CFS = \tau - \mu\sigma_n$, assuming no cohesion). It is common to use ΔCFS, the change in *CFS*, for a specific fault between different points in time, to estimate stress changes rather than using the entire stress field. ΔCFS is often used for active regions, like California (United

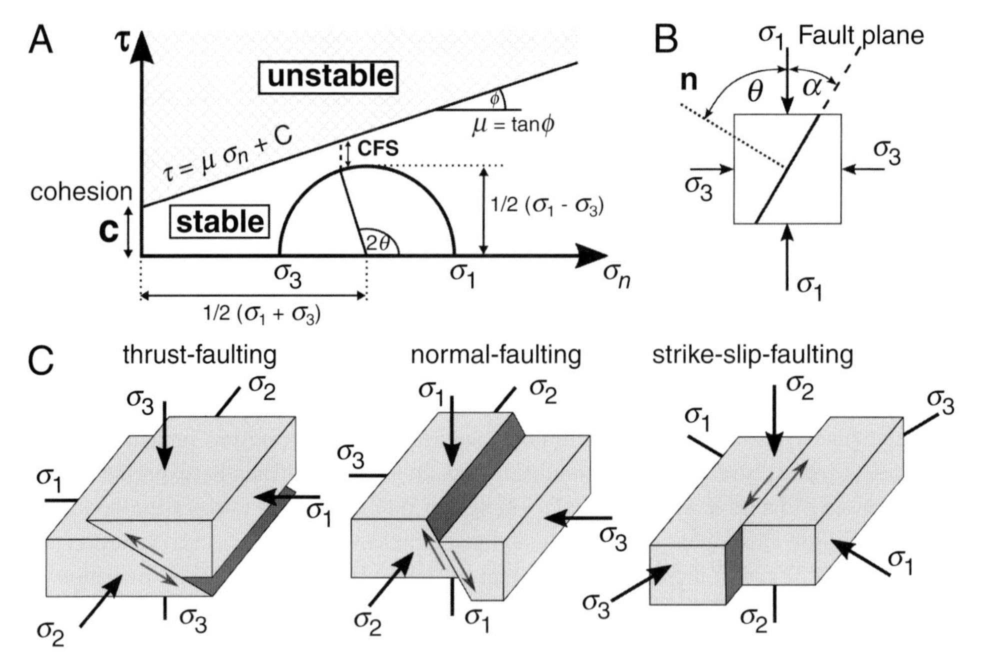

Figure 2.1 Conceptual figure presenting the stresses acting on a fault. A) Mohr diagram showing the line of failure and Mohr circle in two dimensions, with σ_n as the normal stress along the x-axis and τ as shear stress along the y-axis. The vertical distance between the Mohr circle and line of failure is marked *CFS*. The coefficient of friction μ is related to the angle of friction, ϕ, which depends on the angle θ via $|2\theta| = 90^\circ + \phi$ for optimally orientated fault planes. Points in the area below the line of failure are assumed to be stable, while points in the upper part (shaded) are unstable. B) Orientation of the fault plane with respect to the maximum (σ_1) and minimum (σ_3) principal stresses. The vector $\boldsymbol{n}$ is the normal vector of the fault plane. C) The three different stress regimes (thrust-faulting, normal-faulting and strike-slip-faulting) and their respective orientations of principal stresses as well as the orientation of the optimally orientated fault plane and slip direction (shown by the dark grey arrows).

States) or Japan, to identify which faults are prone to reactivation after an earthquake (e.g. Stein, 1999; Ishibe et al., 2011). Positive ΔCFS values are related to a decrease in stability and hence an increased probability of earthquakes, while negative ΔCFS values indicate an increase in stability and thus earthquakes tend to be suppressed (e.g. Harris, 1998; Stein, 1999). Thus, the time variation of ΔCFS informs as to when instability increases or decreases and is therefore essential for evaluating the seismic potential of a region (King et al., 1994). Changes in Coulomb failure stress can be calculated for any fault orientation, as only the change of normal and shear stress between different time points is used. This allows the estimation of ΔCFS for both optimally and non-optimally orientated faults.

An optimally orientated fault is orientated such that it is closer to failure than any other fault orientation θ in the prevailing stress field (e.g. Sibson, 1985). As the angle θ is related to fault orientation (Figure 2.1B), the angle of an optimally orientated fault to the maximum principal stress σ_1 is such that the line in Figure 2.1A from the midpoint of the circle to the point on the circle is perpendicular to the failure envelope. The directions of the principal stresses and the coefficient of friction thus determine the parameters of an optimally orientated fault. For example, if the coefficient of friction is 0.6 (Byerlee, 1978), then the optimal orientation of the normal of the fault plane has an angle of $\sim 60°$ to the direction of the maximum principal stress σ_1 (e.g. Sibson, 1985). Furthermore, the intermediate principal stress σ_2 is parallel to the optimally orientated fault plane (e.g. Twiss & Moores, 2007). This results in fault constellations of $\sim 30°$ dip and $0°/180°$ strike in a thrust-faulting stress regime ($\sigma_1 \| x \|$ east–west, $\sigma_2 \| y \|$ north–south, $\sigma_3 \| z \|$ depth, Figure 2.1C), $\sim 60°$ dip and $0°/180°$ strike in a normal-faulting stress regime ($\sigma_1 \| z \|$ depth, $\sigma_2 \| y \|$ north–south, $\sigma_3 \| x \|$ east–west; Figure 2.1C) and $90°$ dip and $\sim 60°/\sim 120°/\sim 240°/\sim 300°$ strike in a strike-slip-faulting stress regime ($\sigma_1 \| x \|$ east–west, $\sigma_2 \| z \|$ depth, $\sigma_3 \| y \|$ north-south, Figure 2.1C). A lower or higher coefficient of friction would result in a decrease or increase of the angle between the maximum principal stress direction and the normal of the fault plane, respectively ($\mu = 0.6 \rightarrow \sim 60°, \mu = 0.4 \rightarrow \sim 56°,\ \mu = 0.8 \rightarrow \sim 64°$). A value of 0.6, however, is a valid approximation used for non-active faults and rock masses (Scholz, 2019). Stress release along faults and fractures when the Mohr circle touches the linear failure envelope is possible only if the reactivated structures have an optimal orientation. For deviations from the optimal orientation (so-called non-optimally orientated faults/fractures) higher stresses are required, and the Mohr circle for a specific area thus crosses the linear failure envelope before the fault/fracture moves. Non-optimally orientated faults can have any orientation in a specific stress setting. Those faults are more stable and therefore more difficult to move (e.g. Sibson, 1985). The possibility for movement depends on fault orientations (dip and strike values) and relative stress magnitudes.

If optimally orientated faults are assumed to be critically stressed initially, *CFS* is equal to zero before additional stresses are added. This assumption has been shown to be valid in many regions (e.g. Zoback & Townend, 2001; Zoback & Zoback, 2015), such that reactivation of faults only requires a few kPa (e.g. King et al., 1994). In addition, the cohesion C on these faults is assumed to be low enough to be neglected (e.g. Zoback & Townend, 2001). Any additional stresses due to pore-pressure changes, glacial loading and unloading or previous earthquakes are superimposed on the background stress field, and these cause *CFS* to change over time (e.g. Lund et al., 2009; Steffen et al., 2014a), becoming positive (unstable conditions) or negative (under stable conditions). If the background stress field can be assumed to have a *CFS* equal to zero (frictional equilibrium), changes to the stress field in subsequent time steps result in a decrease or increase in stability, which is shown by $\Delta CFS > 0$ or $\Delta CFS < 0$, respectively. The change in *CFS* depends on the properties of the fault, the tectonic stress regime and an additional stress variation, for example, due to a glacial load.

The geomechanics of glacially induced stresses will be discussed in the next section, which is followed by a description of fault stability for all three tectonic stress regimes during a glacial cycle. While this chapter focuses on the generalized theory of fault reactivation during a glacial cycle, its type and timing requires the numerical estimation of glacially induced stresses. These estimations are presented in Chapter 22 by Wu et al.; estimations of background stresses are discussed in Chapter 23 by Gradmann and Steffen.

Resolving the glacially induced stress field in space and time, as calculated, depends on the available Earth and ice models, and a large variety in stress field results is usually obtained (e.g. Steffen et al., 2019). We follow the geologic sign convention in which compressional stresses have a positive sign and tensional stresses have a negative sign (e.g. Twiss & Moores, 2007).

2.2 Glacially Induced Stresses

Glacially induced stresses are induced in the lithosphere during a glacial cycle, and they change during the loading and unloading phases as well as after deglaciation (Johnston, 1987). Since glacially induced stresses can vary from a very short time to thousands of years (Wu & Hasegawa, 1996a,b) and tectonic stresses take several million years to change appreciably (Gordon, 1998), the latter are taken as time-independent during a glacial cycle. The assumption of constant tectonic stresses is not valid for areas like Alaska and Patagonia, where the glacially induced stress field interacts with the time-dependent tectonic one (Sauber et al., 2000; Sauber & Ruppert, 2013; Sauber et al., see Chapter 20). In addition, stress changes due to nearby large earthquakes are not considered.

Stress changes that occur during a glacial cycle are not the same everywhere. The glacially induced vertical stress is only determined by the thickness of the ice sheet at the instant when it is formed and is therefore greater than zero beneath the ice sheet, but zero in the surroundings (e.g. in the area of the peripheral bulge). In contrast, the glacially induced horizontal stress depends not only on the size of the ice load but also on the composition of the Earth (e.g. crustal and lithospheric thickness and elasticity, mantle viscosity; see, for example, Wu & Hasegawa, 1996a,b; Klemann & Wolf, 1998; Kaufmann et al., 2005; Lund, 2005; Lund et al., 2009; Steffen et al., 2014b, 2019) and the effect of stress migration from the mantle into the lithosphere. This is explained in some detail in the next paragraph. While both horizontal and vertical stresses respond immediately to surface load, only horizontal stresses show a viscoelastic response as well and change slowly in time (Wu & Hasegawa, 1996a). In addition, the induced horizontal stress varies from compressional beneath the ice sheet to tensional in the peripheral bulge above the so-called neutral plane (Figure 2.2). The maximum tensional stresses are, however, not induced in the bulge (maximum uplift during loading) but between the ice margin and the bulge (Figure 2.2). This might be related to the pull in the peripheral bulge and beneath the ice margin in opposite directions, which stretches the upper lithosphere most in this area and thus induces large tensional stresses. In the lower half of the lithosphere, below the neutral plane, the horizontal stress conditions are reversed with tensional conditions beneath the ice sheet and compression in the peripheral bulge (Figure 2.2). But, the stress magnitudes are not mirrored in the lower lithosphere compared to the upper lithosphere due to the influence of the viscous mantle in the former case. However, since seismicity within continents (apart from plate boundaries) occur only in the seismogenic part of the crust to a depth of up to 40 km (e.g. Watts & Burov, 2003; Handy & Brun, 2004), the stresses in the lower lithosphere are therefore not used here in the analysis of fault stability.

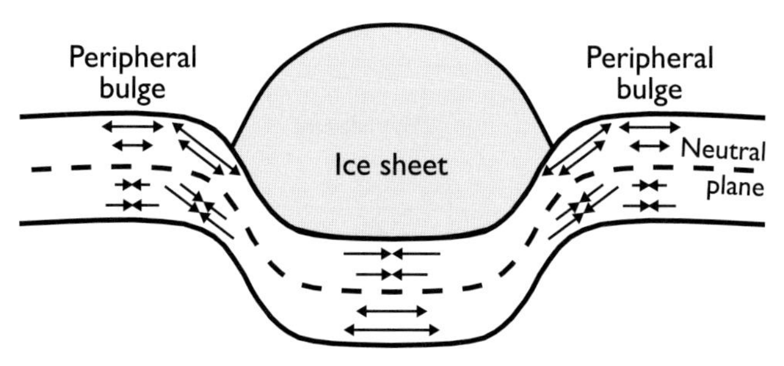

Figure 2.2 Horizontal stress (σ_{xx}) distribution during glacial loading. The stresses are shown for the entire lithosphere, with a so-called neutral plane in the middle. The peripheral bulge is characterized by uplift and surrounds the area covered by the ice sheet. The stresses due to stress migration from the mantle are not shown. The ice sheet and Earth structure are not to scale.

The horizontal stresses are due to flexure of the lithosphere induced by the ice load, as are shown in Figure 2.2 (see also Turcotte & Schubert, 1982). In addition, the thickness of the mechanically strong elastic layer that can support the load is time dependent (Minster & Anderson, 1980), and this leads to stress migration: a load left on the surface of a viscoelastic Earth is supported elastically by the whole mantle at the beginning and this results in an immediate elastic response (Peltier, 1974). However, as time progresses the deeper and thus warmer part of the viscoelastic mantle starts to flow and the thickness of the mechanically strong part of the mantle that can support the load decreases towards the surface (Minster & Anderson, 1980). As mantle temperature increases with depth (e.g. Ranalli, 1995), a much higher viscosity and Maxwell time (ratio of viscosity over shear modulus) exists near the Earth's surface than what is seen deeper below. Thus, elastic behaviour is induced closer to the Earth's surface in the mechanically strong part. Mantle viscosity also increases with depth in the mantle below the mechanically strong part, but on a much smaller scale, only a few orders of magnitude (from 10^{19} Pa·s for a low-viscosity channel in the upper mantle to 10^{22} Pa·s in the lower mantle; Cathles, 1975). For loading times less than the Maxwell time, mantle rocks behave elastically, but when the loading time is longer than the Maxwell time, mantle rocks start to relax and flow like a fluid. As the deeper part of the mantle relaxes, the load becomes supported by a thinner, mechanically strong elastic layer (e.g. the lithosphere), and thus the stress inside the elastic layer will increase. In other words, the stresses will migrate upward from the mantle and become concentrated in the mechanically strong and elastic layer seen by the glacial process. The effect of stress migration is important and so it is critical to consider the entire mantle when modelling glacially induced stresses.

The effect of the glacially induced stresses (horizontal and vertical) on faults in different tectonic stress regimes and at different locations as well as different time steps will be explored in the following sections. Note that when a fault in the peripheral bulge is located parallel to the ice sheet margin and perpendicular to the maximum horizontal stress, its reactivation due to glacial isostatic adjustment (GIA) (for example, at glacial maximum) might be of a thrust-faulting type (Figure 2.3). However, a fault in the peripheral bulge

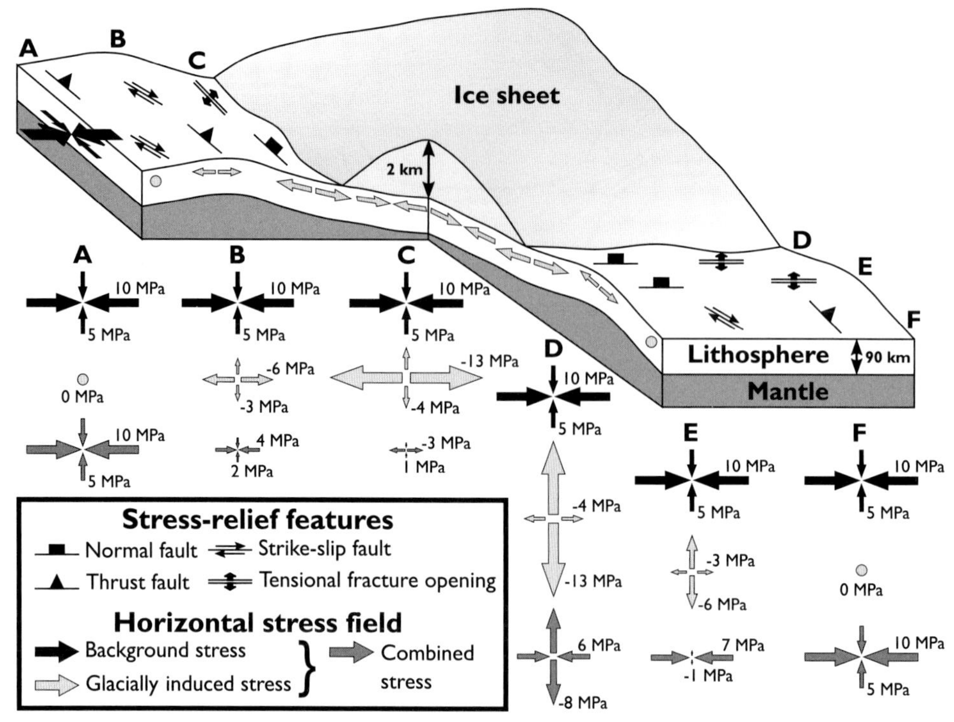

Figure 2.3 Schematic figure showing the effect of the background stress (black arrows) in combination with the glacially induced stress (light grey arrows) on the behaviour of stress-relief features close to the ice margin (locations C and D), in the peripheral bulge (locations B and E) and in the undeformed foreland (locations A and F). The sketch shows the approximate near-surface stresses at glacial maximum for a 2-km thick ice sheet that is located in a compressional stress regime with maximum horizontal stress and minimum horizontal stress equal to 10 MPa and 5 MPa, respectively. Negative values refer to tensional stresses and positive values to compressional stresses. The ice sheet and Earth structure are not to scale. The figure is adapted from Stewart et al. (2000).

located parallel to the ice-sheet margin as well, but also parallel to the maximum horizontal stress direction, would be reactivated as a normal fault (Figure 2.3). Thus, the combination of background stresses and glacially induced stresses are important parameters in finding areas that can be stabilized or destabilized due to ice sheets. However, no stresses are calculated here and only the changes by a load are analyzed with respect to a background stress state. Numerical estimates of glacially induced stresses can be found in Wu et al., in Chapter 22, and Gradmann and Steffen, in Chapter 23.

2.3 Stability of Optimally Orientated Faults in a Thrust-Faulting Stress Regime

Stress changes within a thrust-faulting, compressional stress regime are evaluated at four time points, as seen in Figure 2.4: before glaciation (A), during glaciation (B), during deglaciation (C) and after deglaciation (D); and at three locations: beneath the ice sheet (1), outside the ice margin but before reaching the peripheral bulge (2), and at the peripheral bulge (3). The thrust-faulting stress regime implies that maximum and intermediate stresses are in the horizontal plane, whereas the minimum principal background (BG) stress σ_3^{BG} is vertical (Figure 2.4, left column; Figure 2.1C). For simplicity, we show only the horizontal stress acting in the *x*-direction, which is equivalent to the direction of the maximum horizontal stress σ_1. Background stresses are assumed to vary only with depth, and the fault is close to a critically stressed state before glaciation. The Mohr circle, therefore, is just below the linear failure envelope and completely in the stable zone (row (A) in Figure 2.4). The optimal dip angle of a fault in a thrust-faulting stress regime has a value of $\sim 30°$ for a coefficient of friction of 0.6 (Figure 2.1C). The evaluation of stability during a glacial cycle is carried out under an assumption of zero cohesion (Figure 2.4). Background stress magnitudes do not change during the glacial cycle, and only those additional stresses induced by the ice load change the position and size of the Mohr circle. Stress magnitudes of glacially induced stresses are named σ_x^{GIA} and σ_z^{GIA} for the horizontal and vertical directions, respectively. These combine with their respective background stress magnitudes to create the entire stress field; however, the magnitude of σ_x^{GIA} alone might be smaller than the magnitude of the other horizontal glacially induced stress (σ_y^{GIA}, not shown here) or even σ_z^{GIA}. But, rotation of the entire stress field (background and glacially induced stress field) is not considered here. In addition, if the Mohr circle touches or crosses the linear failure envelope and *CFS* becomes positive, earthquakes may release stresses that alter the stress field as well as reduce the size of the Mohr circle. Such an explicit event is not considered here, and thus *CFS* remains positive in our example. In addition, optimally orientated faults may not be present in a real-world application, and less-optimal faults need to be reactivated, which requires larger stress changes. Thus, the Mohr circle shows the stability of the area rather than a specific fault. An example of a glacially induced fault in a thrust-faulting stress regime is the Pärvie Fault, in northern Sweden (Lagerbäck & Sundh, 2008; Smith et al., Chapter 12).

Loading and unloading during a glacial cycle induces vertical stresses beneath the ice sheet and horizontal stresses in the lithosphere both below the ice and outside the ice margin. For locations beneath the ice sheet (Figure 2.4 (A1–D1)), the increase in glacially induced vertical stress σ_z^{GIA} causes the left endpoint of the Mohr circle to move towards the right, i.e. increased normal stresses. Similarly, the increase in glacially induced horizontal stress causes the magnitude of σ_1 to move towards the right as well (Figure 2.4 (B1)). As the magnitude of the vertical stress is determined by the weight of the ice load but the horizontal stress depends on the dimensions of the ice sheet and the rheological parameters of the lithosphere and mantle, their magnitudes of change can vary, but are assumed to be similar, and so the radius of the Mohr circle and the midpoint of the circle are also changed. In short, for large ice sheets, such as the one in Fennoscandia, the induced GIA stresses move the Mohr circle further away from the linear failure envelope, and stability of the area beneath the ice sheet is increased during glaciation (Figures 2.4 (B1) and 2.5, solid black line). Large differences between the principal stress magnitudes can be obtained for certain ice sheet size and lithospheric thickness combinations due to the effect of stress

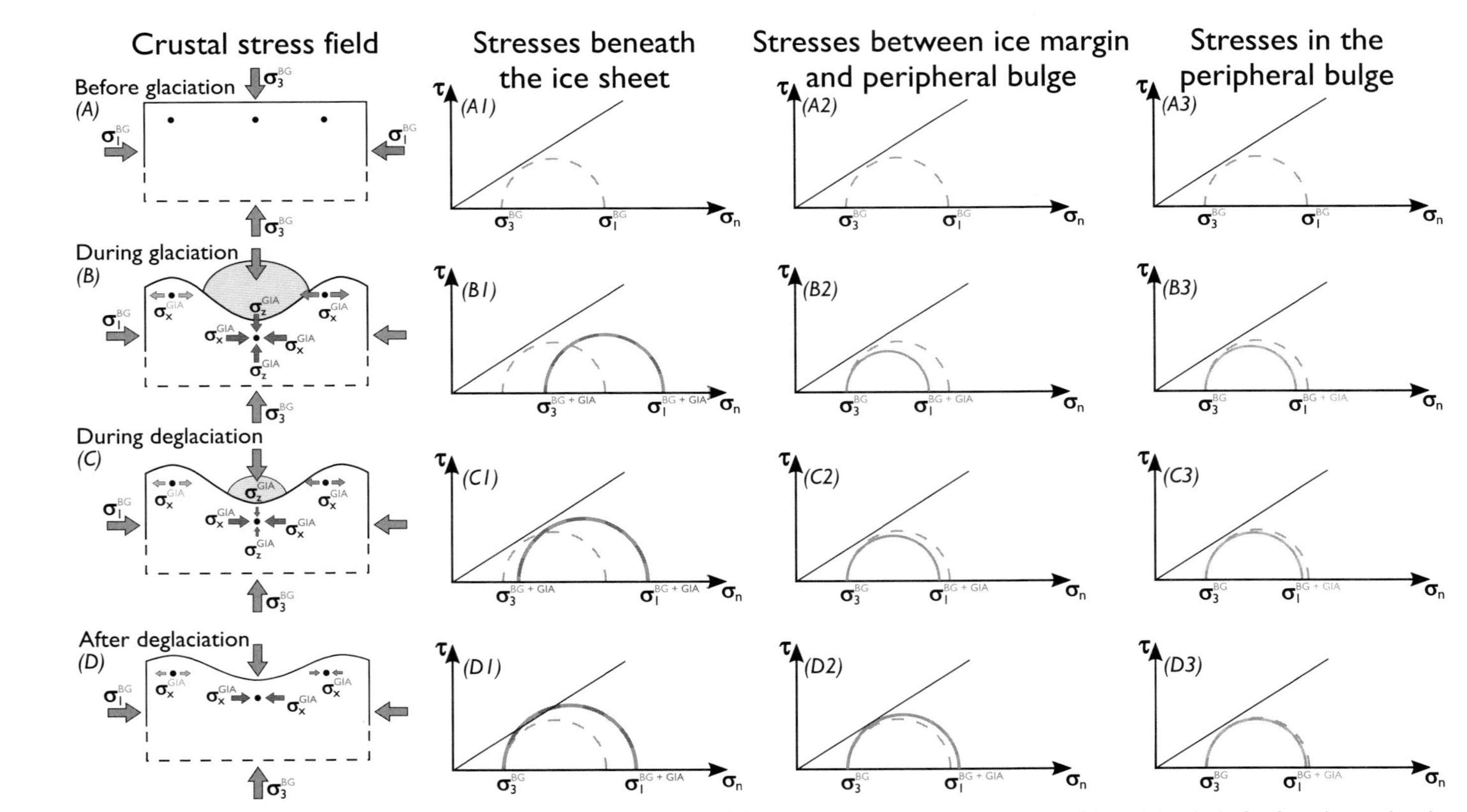

Figure 2.4 Schematic presentation of the stress settings during a glacial cycle and their effect on the size of the Mohr circle for four time points in a thrust-faulting stress regime. The stress settings are shown in the left column (A–D) with background stresses shown as dark grey arrows (BG), GIA stresses beneath the ice sheet as purple arrows, GIA stresses between ice margin and peripheral bulge as red arrows and GIA stresses in the peripheral bulge as orange arrows. The ice load is presented as a blue area in (B) and (C). The surface deformation is not to scale. The second (A1–D1), third (A2–D2) and fourth (A3–D3) columns show the Mohr diagram for a point beneath the ice sheet (purple), a point between ice

amplification (see Wu et al., Chapter 22, for more details). During the deglaciation phase (Figure 2.4 (C1)), the vertical stress due to the ice load decreases, while the decline of horizontal stresses is slower. Thus, the Mohr circle increases in size and moves back closer to the linear failure envelope. As soon as the formerly ice-covered region becomes ice-free, after deglaciation (Figure 2.4 (D1)), the vertical stress returns to the background value. In contrast, horizontal stress σ_x decreases slowly back to its initial state due to the viscoelastic behaviour of the mantle. The combined horizontal stress of background and GIA remains large, and the radius of the circle is increased. The Mohr circle now touches or even crosses the linear failure envelope, allowing certain types of faults to be reactivated. The range of possible faults depends on the intersection points of the Mohr circle and the linear failure envelope, and on the corresponding fault angles. *CFS* becomes positive after deglaciation, and the potential to trigger earthquakes is increased (Figure 2.5). In general, unless there is a release of the stress due to fault reactivation, instability in the area can be maintained for thousands of years after the end of deglaciation. The above results for sites inside the ice margin are consistent with the numerical results found in Wu & Hasegawa (1996a,b) and Lund et al. (2009).

Two locations outside the ice-covered area are considered for the reactivation potential of faults during a glacial cycle. The first is in the peripheral bulge (yellow points in Figure 2.4(A–D) and Mohr circles in Figure 2.4(A3–D3)), and the second is between the ice margin at glacial maximum and the peak of the peripheral bulge (red points in Figure 2.4(A–D) and Mohr circles in Figure 2.4(A2–D2)), where induced tensional stresses reach a maximum during glaciation (Figures 2.2 and 2.3). During the glaciation phase, fault stability at these locations also increases, as (i) there is no change in the magnitude of the vertical stress, thus the left endpoint of the Mohr circle is at the same location as before glaciation started; and (ii) the tensional stress in the upper lithosphere over the peripheral bulge (see Figures 2.2 and 2.5) causes the total horizontal stress to decrease. The radius of the Mohr circle is thus reduced by an amount that depends on the rheology of the Earth and ice-load dimensions. During deglaciation, the tensional stresses decrease at both locations (Figure 2.4(C2,C3)), and the Mohr circle increases in size again, without changing σ_3^{BG} (the left edge of the Mohr circle). However, as soon as the ice has completely melted (after deglaciation, Figure 2.4(D2,D3)), the situation differs between the two locations. For sites between the ice margin and the peripheral bulge (red point in Figure 2.4(A–D)), the glacially induced stress decreases to zero and becomes compressional afterward (Figure 2.4(D2)) due to the collapse of the peripheral bulge and the uplift beneath the

---

Figure 2.4 (*cont.*) margin and peripheral bulge (red), and a point in the peripheral bulge (orange), respectively. The Mohr circle of the background stress is marked with a dashed grey line and is visible in all Mohr diagrams as reference. The combined stress field of background stress $\left(\sigma_{1,3}^{BG}\right)$ and glacially induced stress $\left(\sigma_{x,z}^{GIA}\right)$ is shown in the Mohr diagram. The maximum principal stress is in the horizontal plane and the minimum principal stress in the vertical. For the sake of convenience, the intermediate principal stress σ_2 (minimum horizontal stress) is not shown here as it has no effect on the size of the Mohr circle. (A black and white version of this figure will appear in some formats. For the colour version, please refer to the plate section.)

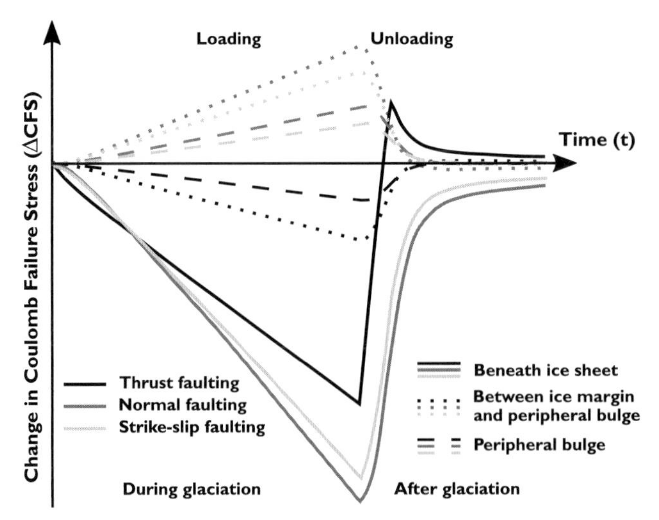

Figure 2.5 Sketch of the change in Coulomb failure stress (ΔCFS, assuming a critically stressed crust before glaciation) for different stress regimes and beneath the ice sheet (solid line), between ice margin and peripheral bulge (dotted line) and in the peripheral bulge (dashed line). Thrust-faulting stress regime: black lines; normal-faulting stress regime: grey lines; strike-slip-faulting stress regime: light grey lines.

formerly ice-covered area. The Mohr circle is therefore enlarged and ΔCFS can become positive (Figure 2.5, dotted black line) if the area was critically stressed before glaciation started. However, for the location at the peak of the peripheral bulge, the tensional stresses simply decrease after deglaciation and the circle grows to its original size, before glaciation. ΔCFS is therefore negative for the entire glacial cycle and slowly approaches zero if the crust was critically stressed before (Figure 2.5, dashed black line). Results for sites outside the ice margin were not considered in Figure 1 of Wu & Hasegawa (1996a), but these regions are important for understanding glacially induced faults outside the ice margin of previously glaciated areas (Brandes et al., 2012, 2015) or in currently glaciated regions. Thus, a seismic hazard might exist surrounding Greenland and Antarctica.

2.4 Stability of Optimally Orientated Faults in a Normal-Faulting Stress Regime

For a normal-faulting stress regime, maximum principal stress σ_1 is in the vertical direction, and intermediate and minimum principal stresses are in the horizontal plane (Figures 2.6 and 2.1C). The optimal dip angle of a fault in a normal-faulting stress regime is $\sim 60^\circ$. For comparison with the previous section we shall apply the same initial conditions as for the thrust-faulting stress regime (e.g. nearly critical stressed crust before glaciation, homogenous stresses). The glacially induced vertical stress is again named σ_z^{GIA}, and the horizontal stress, which is equivalent to the minimum principal stress direction, is σ_x^{GIA}. Again, a rotation of the entire stress field is not considered, and the glacially induced stress components are added to the background stress field without a change in the principal stress direction. The

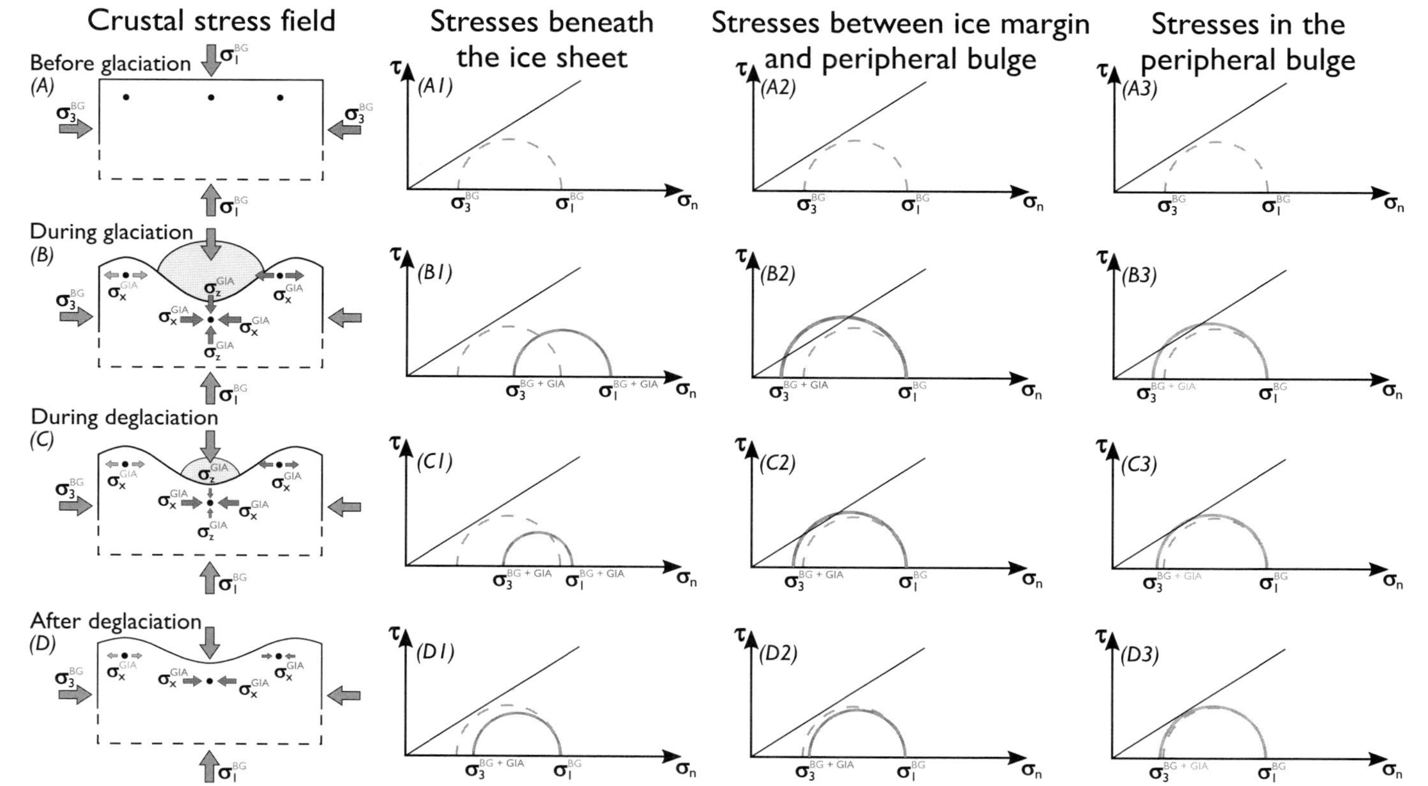

Figure 2.6 Schematic presentation of the stress settings during a glacial cycle and their effect on the size of the Mohr circle for four time points in a normal-faulting stress regime. Colour and description are the same as in Figure 2.4, except for the naming of the horizontal and vertical background stresses (BG) and glacially induced stresses (GIA). The maximum principal stress has a vertical direction, while the minimum principal stress is in the horizontal plane. For the sake of convenience the intermediate principal stress σ_2 (maximum horizontal stress) is not shown in the Mohr diagram as it has no effect on the size of the Mohr circle. (A black and white version of this figure will appear in some formats. For the colour version, please refer to the plate section.)

Kerlingar Fault, in Iceland (Hjartardóttir et al., 2011), is an example of a potential glacially induced fault in a normal-faulting stress regime (see Steffen & Steffen, Chapter 21).

As in the previous section, locations beneath the ice sheet are considered first (Figure 2.6 (A1–D1)). During glaciation, the maximum and minimum principal stresses increase due to the ice load. The change in vertical GIA stress σ_z^{GIA} causes the right edge of the Mohr circle to move to the right towards increased normal stress values. Additionally, the increase in horizontal GIA stress σ_x^{GIA} causes the left endpoint of the Mohr circle to move to the right as well. However, the radius of the circle remains mostly unchanged, since the vertical and horizontal GIA stress magnitudes are close to each other for large ice loads (e.g. Laurentian and Fennoscandian ice sheets) and for typical rheological models of the Earth. Consequently, the Mohr circle moves away from the linear failure envelope and so stable conditions are promoted (Figure 2.5, solid grey line and Figure 2.6(B1)). However, the increase in vertical and horizontal stress depends also on the glaciation rate. As Earth's response is much slower than the rate at which ice builds up, vertical stresses might exceed horizontal stresses in the early stages of the glaciation. This would increase the size of the Mohr circle and might lead to unstable conditions beneath the ice sheet. During deglaciation, the radius of the Mohr circle is reduced as melting of the ice load causes the vertical stress σ_1 to decrease, while the horizontal stresses σ_3 continue to enlarge, since the flexure and stress migration are still progressing. As the normal stress values of the right endpoint of the Mohr circle decrease, more stable conditions during deglaciation are promoted (Figures 2.5 and 2.6(C1)). After deglaciation, there is no change in the vertical stress σ_1, so it has the same value as the background vertical stress. The right edge of the Mohr circle therefore returns to its previous location, before the onset of glaciation. However, the horizontal GIA stress remains large. This leads to an increase in the radius of the Mohr circle when compared to what happens during deglaciation, and the midpoint of the Mohr circle moves towards its initial location before the onset of glaciation. Unlike the result of the thrust-faulting stress regime, where the area that was once covered by ice becomes unstable at the end of deglaciation, stability in the same area is predicted after glaciation for a normal-faulting stress regime. As time progresses and the deformation slowly vanishes, the radius and midpoint of the Mohr circle go back to their original state before the onset of glaciation. Therefore, if the crust has a normal-faulting stress condition, sites within the ice margin will experience stability during most of the glacial cycle with activation phases only in the early stages of the glaciation (Figures 2.5 and 2.6(D1)).

On the other hand, instability of the area is obtained for sites outside the ice sheet during glaciation (Figure 2.6(A2–D2) and (A3–D3)). Since vertical stress is related to ice load alone, no vertical GIA stress is induced outside the ice margin, so the total vertical stress is unchanged. However, tensional stresses (negative horizontal GIA stresses) are induced there, which reduce the minimum principal stress. Therefore, the radius of the Mohr circle becomes larger while the midpoint of the Mohr circle moves leftward. In other words, instability of the area is promoted from just outside the ice sheet towards the peripheral bulge if the stress regime is normal-faulting and the area was almost critically stressed before the onset of glaciation. *CFS* is larger (more positive; Figure 2.5, dotted grey line) for a fault in the upper crust, between the ice margin and peripheral bulge, than at the peak of the peripheral bulge (Figure 2.5, dashed grey line). This is due to the larger tensional stresses at the first location, which results in lower minimum principal stresses and an increased radius of the Mohr circle compared to the Mohr circle in the peripheral bulge. During deglaciation, if the fault in the peripheral bulge is reactivated, then some of the stress may be released by an earthquake, and the fault may become marginally stabilized. If it is not reactivated, then horizontal GIA stresses become less tensional during deglaciation, as the left end of the circle has larger values than during glaciation, though still lower than before the onset of glaciation (Figure 2.6(C2,C3)). After deglaciation, the Mohr circle approaches values from before the onset of glaciation for the site located at the peak of the peripheral bulge. For the site between the ice margin and peripheral bulge, the horizontal stresses also change from tensional conditions (during glaciation and deglaciation) to compressional after deglaciation, leading to negative *CFS* values and therefore more stable conditions (Figures 2.5 and 2.6(D2)).

2.5 Stability of Optimally Orientated Faults in a Strike-Slip-Faulting Stress Regime

For a strike-slip-faulting stress regime, maximum and minimum principal stresses are horizontal (Figure 2.7), but the intermediate principal stress σ_2 is vertical (Figure 2.1C). As σ_2 is not considered in the analysis of the relative location between the Mohr circle and the linear failure envelope, the vertical GIA stress has no effect on the reactivation potential of optimally orientated faults in this stress regime, if stress regime changes are neglected. Optimally orientated faults in a strike-slip-faulting stress regime have strike values $\sim 30°$ to the maximum horizontal stress direction. The optimal dip angle is 90°, which arises from the vertical direction of the intermediate principal stress and σ_2 being parallel to the optimal fault plane. The maximum and minimum principal stresses are parallel to the x- and y-directions, respectively. The same initial conditions as in the previous sections are used in the analysis of fault stability during a glacial cycle (e.g. nearly critical stressed crust before glaciation, background stress variations with depth only). The glacially induced vertical stress is again named σ_z^{GIA}, and the glacially induced horizontal stresses σ_x^{GIA} and σ_y^{GIA}. As before, the entire stress field of background stress and glacially induced stress is assumed to retain its direction and no rotation is considered. A glacially induced fault with a strike-slip motion was found in Denmark, the Børglum Fault (Brandes et al., 2018). However, the stress regime might not be purely strike-slip-faulting in this area (Heidbach et al., 2018).

First, sites within the ice margin are considered. During the glacial phase, bending of the lithosphere by the ice load results in horizontal stresses, which are reasonably similar in magnitude to σ_x^{GIA} and σ_y^{GIA}, unless the lithosphere is highly anisotropic in the horizontal direction or the shape of the ice sheet is highly elliptical. Therefore, the radius of the Mohr circle remains nearly constant, but the midpoint of the Mohr circle is moved to the right along the normal stress axis (Figure 2.7(B1)), resulting in a decrease of *CFS* and thus inducing fault stability for strike-slip faults (Figure 2.5, solid light grey line). After deglaciation, the horizontal stresses of GIA slowly relax and the midpoint of the Mohr circle moves slowly back to its initial position before the onset of glaciation (Figure 2.7 (C1–D1)). Thus, for sites beneath the ice sheet, fault stability is induced throughout the entire glacial cycle (Figure 2.5). However, for strike-slip-faulting stress regimes the results depend more on the form of the ice sheet (eccentricity of the ellipse) than is the case for thrust- and normal-faulting stress regimes.

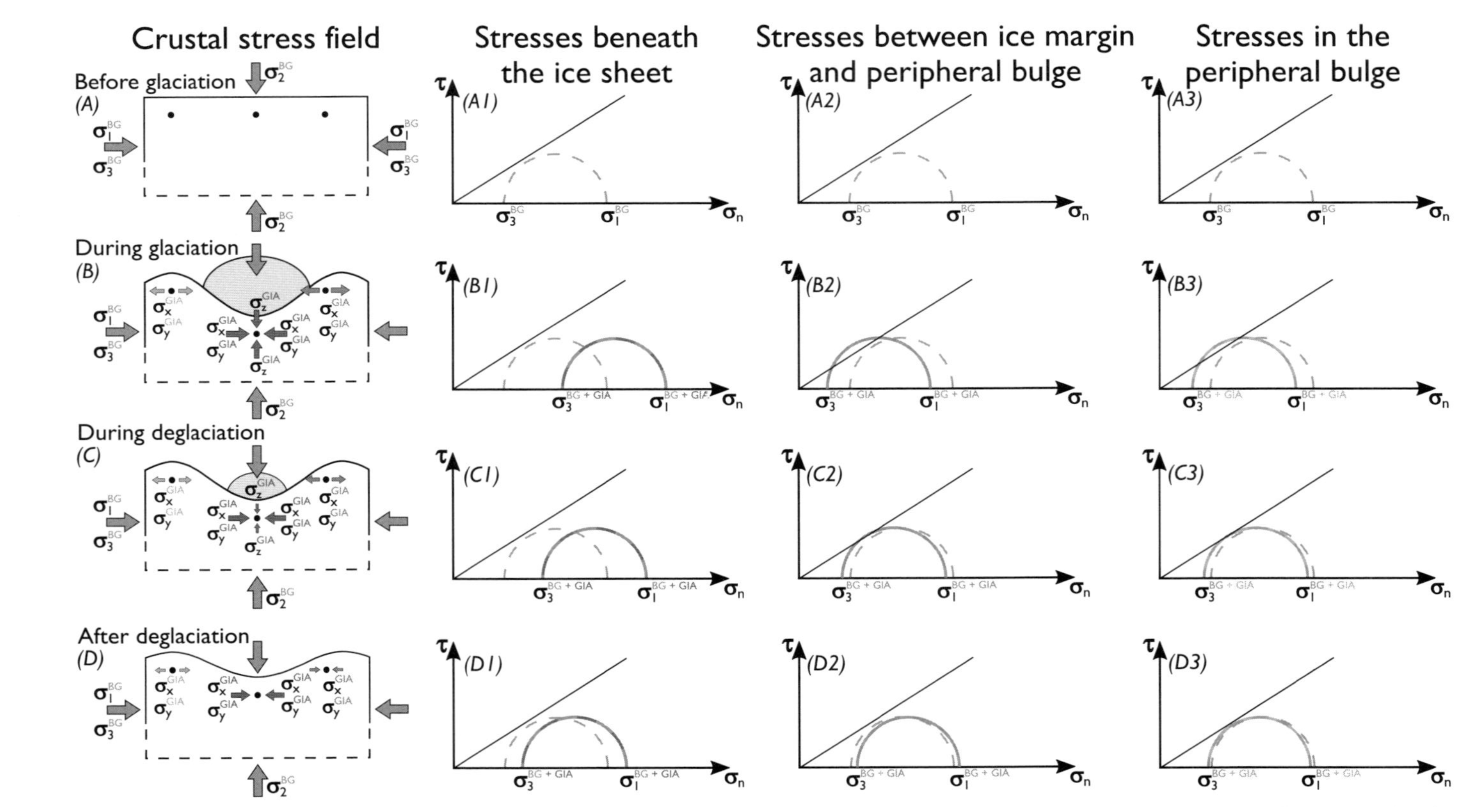

Figure 2.7 Schematic presentation of the stress settings during a glacial cycle and their effect on the size of the Mohr circle for four time points in a strike-slip-faulting stress regime. Colour and description are the same as in Figure 2.4, except for the naming of the horizontal and vertical background stresses (BG) and glacially induced stresses (GIA). Maximum principal stress and minimum principal stress are in the horizontal plane. The intermediate principal stress σ_2 (vertical direction) is not shown in the Mohr diagram, since it has no effect on the size of the Mohr circle. (A black and white version of this figure will appear in some formats. For the colour version, please refer to the plate section.)

For sites at the peripheral bulge, tensional stresses might cause unstable fault conditions during glaciation (Figure 2.7(B3)). However, as soon as deglaciation starts, the deformation in the peripheral bulge decreases and tensional stresses decrease as well (Figure 2.7(C3)). Thus, the midpoint of the Mohr circle moves back to its initial position before the onset of glaciation and away from the linear failure envelope (Figure 2.7(D3)). After deglaciation, the crust experiences more stable conditions.

For sites between the ice margin and the peripheral bulge, *CFS* is similar during glaciation and deglaciation to locations in the peripheral bulge, but larger tensional stresses lead to slightly larger instability of the area during glaciation (Figure 2.5, dotted light grey line). After deglaciation, the induced compressional stresses, due to the uplift beneath the formerly ice-covered area and the collapse of the peripheral bulge, result in more stable conditions than are found in the peripheral bulge (Figure 2.7(D2,D3)). In summary, instability of the area may be promoted during glaciation and deglaciation for sites between the ice margin and the peripheral bulge, but not after deglaciation (Figure 2.5).

2.6 Other Factors Affecting Fault Stability

Increased pore-fluid pressure leads to a decrease in normal stress but an unchanged shear stress. The Mohr circle thus moves towards the origin of the graph and can intersect with the failure envelope. Faults that would have been stable under dry conditions can become active with increasing pore pressure. Thus, pore pressure changes are an additional factor to consider when analyzing stress changes due to a glacial cycle.

A change of the coefficient of friction changes the slope of the linear failure envelope. Lower values than the standard value of 0.6 (Byerlee, 1978) result in a flatter-angle linear failure envelope while retaining the principal stresses. This leads to a larger variety of faults that can be reactivated. In contrast, larger coefficients of friction increase the angle of the linear failure envelope, and additional stresses are required to generate earthquakes. In addition, orientation of the optimally orientated fault also changes slightly for different coefficients of friction. A coefficient of friction of 0.4, for example, changes the optimal dip of a fault in a thrust-faulting stress regime from $\sim 30^\circ$ (for a coefficient of friction of 0.6) to $\sim 34^\circ$. A lower coefficient of friction is a more suitable assumption in active regions like Alaska (e.g. Sauber et al., Chapter 20). Other studies point to larger coefficients of friction (between 0.64 and 0.88) using the orientation of faults and their slip directions (e.g. Yin & Ranalli, 1995). It also must be noted that the coefficient of friction changes during the slipping process. While a static friction exists before fault slip starts, a lower, so-called dynamic, friction is found in experiments during fault movement (Di Toro et al., 2011). The dynamic friction is usually only 10–30 per cent of the static friction (Di Toro et al., 2011).

2.7 Conclusion

The effect of a glacial cycle on the lithosphere is manifested not only in displacement and gravity changes but also in induced stress changes. Those stresses alter the existing stress field, which can lead to reactivation of faults in the glaciated area and its surroundings. However, glacially triggered reactivation of faults depends not only on the location of the fault with respect to the ice sheet but also on the background stress regime (Figure 2.3). While faults beneath the ice sheet can be reactivated only in a thrust-faulting stress regime and only after deglaciation, faults in the peripheral bulge can be unstable during glaciation only in a normal or strike-slip-faulting stress regime. For normal-faulting and strike-slip-faulting stress regimes, this area is still slightly unstable during and shortly after deglaciation, but it reaches stability soon thereafter due to the low amount of tensional stresses existing in the peripheral bulge. The tensional stresses that become compressive in the area between peripheral bulge and ice margin might lead to instability of this area in a thrust-faulting stress regime. Thus, when analyzing fault stability (both qualitative and quantitative), the background stress field needs to be included (Figure 2.3). Additional stress changes (e.g. poroelastic-stress changes) can alter the total stress field, leading to an increase or decrease of stability. Glacially induced stresses are capable of triggering the reactivation of both optimally and non-optimally orientated faults, but when this occurs, analysis using the full stress tensor is required. This would also allow the inclusion of stress rotation (as shown by Lund et al., 2009) as well as a change of stress regime (as shown by Wu & Hasegawa, 1996a). The additional glacially induced stress can thus modify the background stress regime, leading to different conclusions on the stability of faults. In addition, the location and orientation of non-optimally orientated faults with respect to the ice sheet in three dimensions has an important effect on the stability or instability of an area and needs to be considered in future studies.

Acknowledgements

We are thankful to Jeanne Sauber, Christian Brandes and David Tanner for their reviews, as well as Ken McCaffrey for feedback on an earlier version of this manuscript. We would like to thank Pierre-Michel Rouleau (Memorial University of Newfoundland), Holger Steffen (Lantmäteriet) and Lenn Hills (University of Calgary) for fruitful discussions on this topic.

References

Anderson, E. M. (1951). *The Dynamics of Faulting and Dyke Formation with Applications to Britain*, 2nd ed., Oliver & Boyd, Edinburgh.

Brandes, C., Winsemann, J., Roskosch, J. et al. (2012). Activity along the Osning Thrust in Central Europe during the Lateglacial: ice-sheet and lithosphere interactions. *Quaternary Science Reviews*, **38**, 49–62, doi.org/10.1016/j.quascirev.2012.01.021.

Brandes, C., Steffen, H., Steffen, R. and Wu, P. (2015). Intraplate seismicity in northern Central Europe is induced by the last glaciation. *Geology*, **43**, 611–614, doi.org/10.1130/G36710.1.

Brandes, C., Steffen, H., Sandersen, P. E., Wu, P. and Winsemann, J. (2018). Glacially induced faulting along the NW segment of the Sorgenfrei–Tornquist Zone, northern Denmark: implications for neotectonics and Lateglacial fault-bound basin formation.

Quaternary Science Reviews, **189**, 149–168, doi.org/10.1016/j.quascirev.2018.03.036.

Byerlee, J. D. (1978). Friction of rock. *Pure and Applied Geophysics*, **116**, 615–626, doi.org/10.1007/BF00876528.

Cathles, L. M., III (1975). *The Viscosity of the Earth's Mantle*. Princeton University Press, Princeton.

Coulomb, C. A. (1776). Essai sur une application des règles des maximis et minimis à quelquels problemes de statique relatifs, à la architecture. *Mémoires de mathématique et physique presenté à l'Acádemie des sciences par savantes étrangères*, **7**, 343–382.

Di Toro, G., Han, R., Hirose, T. et al. (2011). Fault lubrication during earthquakes. *Nature*, **471**, 494–498, doi.org/10.1038/nature09838.

Gordon, R. G. (1998). The plate tectonic approximation: plate non-rigidity, diffuse plate boundaries, and global plate reconstructions. *Annual Review of Earth and Planetary Sciences*, **26**, 615–642, doi.org/10.1146/annurev.earth.26.1.615.

Handy, M. R. and Brun, J.-P. (2004). Seismicity, structure and strength of the continental lithosphere. *Earth and Planetary Science Letters*, **223**, 427–441, doi.org/10.1016/j.epsl.2004.04.021.

Harris, R. A. (1998). Introduction to special section: stress triggers, stress shadows, and implications for seismic hazard. *Journal of Geophysical Research*, **103**(B10), 24347–24358, doi.org/10.1029/98JB01576.

Heidbach, O., Rajabi, M., Cui, X. et al. (2018). The World Stress Map database release 2016: crustal stress pattern across scales. *Tectonophysics*, **744**, 484–498, doi.org/10.1016/j.tecto.2018.07.007.

Heyman, J. (1972). *Coulomb's Memoir on Statics*. Cambridge University Press, London.

Hjartardóttir, Á. R., Einarsson, P. and Brandsdóttir, B. (2011). The Kerlingar fault, Northeast Iceland: a Holocene normal fault east of the divergent plate boundary. *Jökull*, **60**, 103–116.

Ishibe, T., Shimazaki, K., Tsuruoka, H., Yamanaka, Y. and Satake, K. (2011). Correlation between Coulomb stress changes imparted by large historical strike-slip earthquakes and current seismicity in Japan. *Earth, Planets and Space*, **63**, 12, doi.org/10.5047/eps.2011.01.008.

Jaeger, J. C., Cook, N. G. W. and Zimmerman, R. W. (2007). *Fundamentals of Rock Mechanics*. Blackwell Publishing, Malden, Massachusetts.

Johnson, R. B. and DeGraff, J. V. (1988). *Principles of Engineering Geology*. John Wiley & Sons, New York.

Johnston, A. C. (1987). Suppression of earthquakes by large continental ice sheets. *Nature*, **330**, 467–469, doi.org/10.1038/330467a0.

Kaufmann, G., Wu, P. and Ivins, E. R. (2005). Lateral viscosity variations beneath Antarctica and their implications on regional rebound motions and seismotectonics. *Journal of Geodynamics*, **39**, 165–181, doi.org/10.1016/j.jog.2004.08.009.

King, G. C. P., Stein, R. S. and Lin, J. (1994). Static stress changes and the triggering of earthquakes. *Bulletin of the Seismological Society of America*, **84**(3), 935–953.

Klemann, V. and Wolf, D. (1998). Modelling of stresses in the Fennoscandian lithosphere induced by Pleistocene glaciations. *Tectonophysics*, **294**, 291–303, doi.org/10.1016/S0040-1951(98)00107-3.

Lagerbäck, R. and Sundh, M. (2008). *Early Holocene Faulting and Paleoseismicity in Northern Sweden*. SGU Research Paper, **C386**, 80 pp.

Lund, B. (2005). *Effects of Deglaciation on the Crustal Stress Field and Implications for Endglacial Faulting: A Parametric Study of Simple Earth and Ice Models*. SKB

Technical Report TR-05-04, Swedish Nuclear Fuel and Waste Management Co., Stockholm, 68 pp.

Lund, B., Schmidt, P. and Hieronymus, C. (2009). *Stress Evolution and Fault Stability during the Weichselian Glacial Cycle*. SKB Technical Report TR-09-15, Swedish Nuclear Fuel and Waste Management Co., Stockholm, 106 pp.

Minster, J. B. and Anderson, D. L. (1980). Dislocations and nonelastic processes in the mantle. *Journal of Geophysical Research*, **85**(B11), 6347–6352, doi.org/10.1029/JB085iB11p06347.

Mohr, O. (1914). Abhandlungen aus dem Gebiete der Technische Mechanik. *Treatise on Topics in Engineering Mechanics*, 2nd ed., Ernst und Sohn, Berlin.

Nur, A. and Booker, J. R. (1972). Aftershocks caused by pore fluid flow? *Science*, **175**, 885–887, doi.org/10.1126/science.175.4024.885.

Peltier, W. R. (1974). The impulse response of a Maxwell Earth. *Reviews of Geophysics and Space Physics*, **12**, 649–669, doi.org/10.1029/RG012i004p00649.

Ranalli, G. (1995). *Rheology of the Earth*, 2nd ed., Chapman & Hall, London.

Sauber, J. M. and Ruppert, N. (2013). Rapid ice mass loss: does it have an influence on earthquake occurrence in Southeast Alaska? In J. T. Freymueller, P. J. Haeussler, R. L. Wesson and G. Ekström, eds., *Active Tectonics and Seismic Potential of Alaska*. American Geophysical Union, Geophysical Monograph Series, Vol. 179, doi.org/10.1029/179GM21.

Sauber, J., Plafker, G., Molnia, B. F. and Bryant, M. A. (2000). Crustal deformation associated with glacial fluctuations in the eastern Chugach Mountains, Alaska. *Journal of Geophysical Research*, **105**, 8055–8077, doi.org/10.1029/1999JB900433.

Scholz, C. H. (2019). *The Mechanics of Earthquakes and Faulting*, 3rd ed., Cambridge University Press, doi.org/10.1017/9781316681473.

Sibson, R. H. (1985). A note on fault reactivation. *Journal of Structural Geology*, **7**, 751–754, doi.org/10.1016/0191-8141(85)90150-6.

Steffen, R., Wu, P., Steffen, H. and Eaton, D.W. (2014a). On the implementation of faults in finite-element glacial isostatic adjustment models. *Computers & Geosciences*, **62**, 150–159, doi.org/10.1016/j.cageo.2013.06.012.

Steffen, R., Wu, P., Steffen, H. and Eaton, D. W. (2014b). The effect of earth rheology and ice-sheet size on fault slip and magnitude of postglacial earthquakes. *Earth and Planetary Science Letters*, **388**, 71–80, doi.org/10.1016/j.epsl.2013.11.058.

Steffen, H., Steffen, R. and Tarasov, L. (2019). Modelling of glacially-induced stress changes in Latvia, Lithuania and the Kaliningrad District of Russia. *Baltica*, **32**(1), 78–90, doi.org/10.5200/baltica.2019.1.7.

Stein, R. S. (1999). The role of stress transfer in earthquake occurrence. *Nature*, **402**, 605–609, doi.org/10.1038/45144.

Stewart, I. S., Sauber, J. and Rose, J. (2000). Glacio-seismotectonics: ice sheets, crustal deformation and seismicity. *Quaternary Science Reviews*, **19**, 1367–1389, doi.org/10.1016/S0277-3791(00)00094-9.

Turcotte, D. L. and Schubert, G. (1982). *Geodynamics – Applications of Continuum Physics to Geological Problems*. Wiley, New York.

Twiss, R. J. and Moores, E. M. (2007). *Structural Geology*, 2nd ed., W. H. Freeman, New York.

Walcott, R. I. (1970). Isostatic response to loading of the crust in Canada. *Canadian Journal of Earth Sciences*, **7**, 716–727, doi.org/10.1139/e70-070.

Watts, A. B. and Burov, E. B. (2003). Lithospheric strength and its relationship to the elastic and seismogenic layer thickness. *Earth and Planetary Science Letters*, **213**, 113–131, doi.org/10.1016/S0012-821X(03)00289-9.

Wu, P. and Hasegawa, H. S. (1996a). Induced stresses and fault potential in Eastern Canada due to a disc load: a preliminary analysis. *Geophysical Journal International*, **125**, 415–430, doi.org/10.1111/j.1365-246X.1996.tb00008.x.
Wu, P. and Hasegawa, H. S. (1996b). Induced stresses and fault potential in Eastern Canada due to a realistic load: a preliminary analysis. *Geophysical Journal International*, **127**, 215–229, doi.org/10.1111/j.1365-246X.1996.tb01546.x.
Yin, Z.-M. and Ranalli, G. (1995). Estimation of the frictional strength of faults from inversion of fault-slip data: a new method. *Journal of Structural Geology*, **17**, 1327–1335, doi.org/10.1016/0191-8141(95)00028-C.
Zoback, M. L. (1992). First and second order patterns of stress in the lithosphere: The World Stress Map Project. *Journal of Geophysical Research*, **97**, 11703–11728, doi .org/10.1029/92jb00132.
Zoback, M. D. and Townend, J. (2001). Implications of hydrostatic pore pressures and high crustal strength for the deformation of intraplate lithosphere. *Tectonophysics*, **336**, 19–30, doi.org/10.1016/S0040-1951(01)00091-9.
Zoback, M. L. and Zoback, M. (2015). Lithosphere stress and deformation. *Treatise on Geophysics*, **6**, 255–271, doi.org/10.1016/B978-0-444-53802-4.00115-9.
Zoback, M. L., Zoback, M. D., Adams, J. et al. (1989). Global patterns of tectonic stress. *Nature*, **341**, 291–298, doi.org/10.1038/341291a0.

Part II

Methods and Techniques for Fault Identification and Dating

The following chapters introduce geological, geodetic and geophysical methods and techniques that specifically help in the identification of glacially induced faults. In addition, a summary of methods for dating of fault (re-)activation is presented, and the forthcoming drilling project into the Pärvie Fault is introduced.

3

Earthquake-Induced Landforms in the Context of Ice-Sheet Loading and Unloading

PETER B. E. SANDERSEN AND RAIMO SUTINEN

ABSTRACT

Prominent earthquake-induced landforms have been found at several locations in northern Fennoscandia in areas formerly situated below the centre of the Pleistocene ice sheets, and more subtle examples have been recognized in regions far away from the former ice sheet centre. Stress changes in the subsurface created by loading and unloading of the ice sheets can result in reactivation of deep-seated faults. Glacially induced faulting can happen during glaciation in a proglacial or subglacial setting, in a distal setting away from the ice margin, or in a postglacial setting after the ice sheet has melted away. Thus, the timing and location of the tectonic event is important for the resulting landform creation or landform change. Identification of earthquake-induced landforms can be used in interpretations of palaeoseismic events, for location of previously unrecognized fault zones and in evaluations of likelihood of future seismic events. Interpretations of earthquake-induced landforms in and around former glaciated areas can therefore add important information to interpretations of both the Quaternary geology and the deep structural framework. In this chapter we present examples of earthquake-induced geomorphology ranging from readily visible surface expression to more subtle and complex landforms.

3.1 Introduction

New landforms and alterations to existing landforms created by glacially induced tectonics are gaining attention in countries in and around former glaciated areas (e.g. Palmu et al., 2015; Mikko et al., 2015). Identification of tectonic geomorphology in the present-day terrain can lead to interpretations of palaeoseismic events, recently reactivated fault zones and eventually to an evaluation of the likelihood of future seismic events. This is of high societal relevance when assessing seismic hazards – even in intra-cratonic areas considered tectonically stable (Stewart et al., 2000).

Tectonic geomorphology in and around former glaciated areas, whether the landforms are the result of seismic events from the higher or the lower end of the intensity scale, can add important information to interpretations of both the Quaternary geology and the deeper structural framework (e.g. Brandes et al., 2018). The interplay between the ice sheets and the deeper subsurface thus enables the use of geomorphological features as key to new understandings of the tectonic framework. Consequently, by adding the concept of glacially induced tectonics to the 'Quaternary geology toolbox,' landforms that previously were difficult to explain with classic geological concepts now find new hypotheses and explanations (Johnson et al., 2015; Sutinen et al., 2014a; Sandersen & Jørgensen, 2015).

Earthquake-induced landforms are expected to occur not only in areas formerly covered by ice sheets and close to the former ice sheet margin but also in peripheral bulge areas at large distances from the maximum extent of the ice sheet (e.g. Stewart et al., 2000; Brandes et al., 2015). The timing and location of the tectonic event is important for the resulting landform creation or landform change. Glacially induced faulting can happen during glaciation in a proglacial or subglacial setting, in a setting far away from the ice margin or in a postglacial setting after the ice sheet has melted away. In recent years, a number of investigations point to not just one but multiple episodes of seismic activity along the same fault lines during the Lateglacial and the Holocene (Ojala et al., 2018, 2019a; Mattila et al., 2019; Brandes et al., 2018; van Balen et al., 2019; Smith et al., 2018). Because the basic mechanisms and tectonic framework remain the same, repeated glaciations with loading and unloading of ice sheets may result in repeated reactivation of the same fault zones (e.g. Sandersen & Jørgensen, 2015). This means that creation of earthquake-induced landforms can, hypothetically, recur at the same geographic locations over extended time intervals.

Recession of the Fennoscandian ice sheet released glacial isostatic processes such that, according to modelling by Wu et al. (1999), fault activity started 15 thousand calibrated years before present (ka BP) and culminated 13–10 ka BP. This time frame was concurrent with well constrained climatic warming during the Bølling–Allerød interstadial, 14.7–12.9 ka BP (Rasmussen et al., 2014) and ice streaming towards the Younger Dryas end moraines, 12.8–11.5 ka BP (Muscheler et al., 2008). Seismotectonic events that take place underneath an ice sheet will result in the creation of landforms in a highly dynamic setting constrained by primarily glacial load and subglacial meltwater under high pressure (Sutinen et al., 2018). The resulting landforms can thus be highly complex and will reveal secondary earthquake-induced landforms – and not necessarily any primary landforms. In a sedimentary basin, reactivation of a deep basement fault will likely create fault arrays and folds in the unconsolidated sediments above, where the energy and thus the geomorphological effect will be spread out over larger areas, eventually resulting in a series of more subtle surface expressions (e.g. Sandersen & Jørgensen, 2015; Brandes et al., 2018).

Depending on the timing of the event and the distance to the ice margin, later erosive modifications of the landforms can diminish the imprint of the tectonic event in the present-day topography. However, in some cases the opposite can be the case because fault zones usually are easier to erode.

Thus, the impact of earthquakes on landforms will range from creation of prominent fault scarps, formation of syn- and anticlines and formation of complex subglacial landforms, to landform changes like landslides, slope changes, shoreline displacements, liquefaction features and impact on drainage and erosional patterns. Because the earthquake-induced landforms are not related to active plate margins but to glacial and periglacial terrain, the morphology is typically smooth and without large altitude variations. Added to this, erosion and degradation during and after deglaciation often result in complex and subtle surficial expressions that cannot be interpreted unambiguously as earthquake related. However, with the use of new data types – especially light detection and ranging (LiDAR) – new opportunities for detailed delineation of tectonic geomorphology are at hand (Johnson et al., 2015).

In this chapter, we focus on earthquake-induced landforms in the context of the Pleistocene glaciations in intraplate settings and show examples from Fennoscandia and northern Central Europe.

3.2 The Timing and Preservation of Earthquake-Induced Landforms

Distinguishing between landforms related to glacially induced earthquakes and those that have other causes can be difficult, but nonetheless it is important to do so (e.g. Van Vliet-Lanoë et al., 2016). A distinct and linear scarp in a formerly glaciated landscape, for example, cannot be determined as a fault scarp related to glacially induced tectonics unless a distinct marker with a specific age has been displaced or deformed or unless structural descriptions and dating of sediment samples from trenches or outcrops support the interpretation. Otherwise, it will be difficult to distinguish the landform from periglacial phenomena, glacial erosion, meltwater erosion or the result of a postglacial erosional response to an antecedent subsurface feature.

Well-preserved raised shorelines, surfaces of lake sediments, marine seabeds and glacial outwash plains are often excellent chronological markers because the surfaces represent a specific moment in time (Burbank and Anderson, 2012; Lykke-Andersen et al., 1996; Sandersen & Jørgensen, 2015). Alterations to these types of surfaces will post-date their formation and thereby place the events chronologically.

Earthquake-induced landforms and structures related to ice-sheet loading and unloading can be formed in a glacial setting – either proglacial (e.g. Brandes et al., 2012; Grube, 2019; Pisarska-Jamroży et al., 2018) or subglacial (e.g. Sutinen et al., 2014a), in a deglaciation setting – either proximal or distal (e.g. Brandes et al., 2018; Brandes & Winsemann, 2013), or in a postglacial setting, where the ice sheet is no longer present (e.g. Ojala et al., 2018; Sandersen & Jørgensen, 2015). In the glacial setting, the dynamics of the ice sheet and the meltwater will subject any earthquake-induced landforms to alteration or erosion. Especially, landforms related to earthquake events underneath the ice sheet will be heavily influenced by forces acting in the subglacial environment. In the proglacial setting, sedimentation and erosion by the meltwater combined with oscillations of the ice margin will reduce the likelihood of preservation of earthquake-induced landforms. Generally, any re-advance or movement of the glacier over earthquake-induced landforms will most likely cause these to be removed or altered beyond recognition. In this case, identification of deformations related to the glacially induced tectonic event will most likely be traceable within the underlying sediments only.

In the deglaciation setting the ice no longer has a direct influence on landform creation. In the periglacial environment, landform alterations are mainly the result of the erosional forces of meltwater streams and surface water runoff in combination with freeze-thaw mechanisms, gravitational forces and wind (French, 2017). In addition, landforms in areas inundated by marine transgressions or large lakes can be altered by erosion and draped by blankets of marine or freshwater deposits (e.g. Öhrling et al., 2018). In a postglacial setting, climate amelioration will stimulate the vegetation and reduce the alteration of landforms compared to the glacial and periglacial settings. Thus, distance from the former ice margin and exposure time since formation will be important factors influencing the preservation of earthquake-induced landforms.

3.3 Classification of Earthquake-Induced Landforms in the Context of Ice-Sheet Loading and Unloading

Earthquake-induced landforms in general can attain many shapes and be related to both tectonic deformation and seismic shaking; they can be located close to or far from the fault; and they can be either co-seismic or post-seismic (McCalpin & Nelson, 2009). Primary features such as faults with large offsets rupturing the terrain surface can, in many cases, easily be related to known fault zones, whereas secondary features such as landslides give a hint of a seismic event but do not necessarily provide a link to a specific fault. Glacially induced faults and seismic events related to loading and unloading of ice sheets can be either primary or secondary features as described by McCalpin and Nelson (2009). However, apart from the spectacular surface ruptures in northern Fennoscandia, many of the surficial expressions of glacially induced faulting are subtle and complex primarily due to reworking in the glacial environment and the transient nature of the stress changes related to the growth and decay of the ice sheets (e.g. Sandersen et al., see Chapter 15).

Table 3.1 is a modified version of the hierarchical classification of palaeoseismic features by McCalpin and Nelson (2009) including landforms only. In Table 3.1, the primary landform group includes features directly related to tectonic deformation. Fault scarps where the fault is readily visible at the surface or beneath a thin cover of soil are the most prominent members of this group. Other primary landforms close to the fault can be surface ruptures, folds over the fault, depressions and raised blocks or ridges. Landforms farther away from the fault can be tilted surfaces, uplifted or drowned shorelines and tilted lakes.

The secondary landform group of Table 3.1 comprises features formed as A) a response to seismic shaking and B) landforms created as a response to changes in erosion and sedimentation. The first subgroup includes surface expressions of liquefaction, landslides or subglacial sediment mobilization or modification. The second subgroup contains landforms created or modified as a response to changes in erosion and sedimentation, such as stream deflections, changes in drainage patterns, stream erosion and sedimentation (e.g. Schumm et al., 2002). Included in the subgroup are also surficial sediment wedges or blankets formed in relation to changes in elevation and drainage.

Just as for fault zones in regions of active plate tectonics (e.g. Dramis & Blumetti, 2005; McCalpin & Nelson, 2009; Burbank & Anderson, 2012), it should be expected that landforms created by glacially induced earthquakes can be very diverse. Attention should

Table 3.1. *Landforms related to glacially induced faulting (based on McCalpin and Nelson, 2009).*

Primary earthquake-induced landforms	
Landforms created by tectonic deformation	
On fault/near-field	**Off fault/far-field**
Fault scarps Fissures Folds over faults Pressure ridges Subsided areas Uplifted areas	Tilted surfaces Lake tilt Uplifted shorelines Drowned shorelines
Secondary earthquake-induced landforms	
A) Landforms created by seismic shaking	
On fault/near-field	**Off fault/far-field**
Landslides Subglacial landforms Sand blows/liquefaction Sink holes Subsidence (sediment compaction) Talus shuttering/fractured bedrock	Landslides Subglacial landforms Sand blows/liquefaction Sink holes Subsidence (sediment compaction)
B) Landforms created or modified by changes in erosion and sedimentation	
On fault/near-field	**Off fault/far-field**
Stream deflections Changes in drainage patterns Changes in stream erosion/sedimentation Changes in erosional pattern Colluvial aprons	

therefore be given to searches for all the landforms of Table 3.1 and complex combinations of these.

3.4 Examples of Earthquake-Induced Landforms

In the following, we describe selected examples of earthquake-induced landforms given in Table 3.1; interpretations typically have been based on a combination of LiDAR data, field reconnaissance, geological data and geophysical data.

3.4.1 Primary Landforms Created by Tectonic Deformation

3.4.1.1 Fault Scarps

In northern Fennoscandia, surface expressions of faults can be seen in the terrain as pronounced linear fault scarps. Olesen et al. (2013) list twelve documented glacially induced fault scarps in Finland, Norway and Sweden, occurring within a 400 × 400-km area, where the Precambrian basement lies close to the terrain (see also Olesen et al.; Smith et al.; Sutinen et al., Chapters 11–13). The faults are predominantly NE-SW-trending reverse faults with offsets of up to 30 m and lengths of up to 155 km. Although they may seem simple, these fault scarps have proven to represent complicated and segmented fault systems (Mattila et al., 2019, and references herein). The Pärvie Fault in northern Sweden consists of a series of linear, west-facing fault scarps, forming a 155-km-long fault line with occasional bedrock exposures (Lagerbäck & Sundh, 2008). The fault scarp height is 3–10 m, and occasionally overhanging cliffs indicate steeply dipping reverse fault planes.

The Stuoragurra Fault in northern Norway (Olesen, 1988) occurs along an 80-km SW-NE-trending zone, where fault segments are seen as linear steps of up to 7 m in the till cover.

The Ruokojärvi–Pasmajärvi Fault, in Finland, has a SW-NE orientation and a height of up to 12 m (Figure 3.1, see also Figure 13.2, Sutinen et al., Chapter 13, and Markovaara-Koivisto et al., 2020). Trenching revealed a fault in the bedrock that also cross-cuts the overlying Quaternary sediments (Markovaara-Koivisto et al., 2020; see drillings in Kuivamäki et al., 1998). Detailed mapping based on LiDAR data showed additional, smaller scarps associated with the main fault scarp (Figure 3.1).

In Norway, glacially induced fault scarps are found in the northern part of the country (Olesen et al., 2013; and see Chapter 11). In Finland, confirmed glacially induced fault scarps are found in the northern part, whereas proposed glacially induced fault scarps are described in central and southern Finland (Palmu et al., 2015; Sutinen et al., see Chapter 13). In Sweden, fault scarps have been found in the northern and central parts of the country, but not in the southern parts (Mikko et al., 2015; Smith et al., Chapter 12).

The LiDAR-based image of the Ruokojärvi–Pasmajärvi Fault in Figure 3.1 sums up the overall appearance of the fault scarps of northern Fennoscandia, where the Precambrian bedrock is covered by a thin cover of Quaternary sediments and where the scarps are fairly obvious in the terrain or in LiDAR data. Within the sedimentary basins in the southern parts of Scandinavia and further to the south, creation of earthquake-induced landforms requires that movements on deep bedrock faults deform the overlying sediments all the way to the terrain. As described in e.g. Sandersen et al., Chapter 15, and Müller et al., Chapter 16, the number of unambiguous examples of Lateglacial and glacially induced fault scarps is limited in these areas. However, in the following two examples from Denmark, fault scarps are interpreted to dissect well-defined surfaces.

At Nørre Lyngby in northern Denmark, Lateglacial reactivation of faults in the Sorgenfrei–Tornquist Zone (STZ) is evidenced by faults and soft-sediment deformation structures (SSDS) in exposed Lateglacial marine and freshwater sediments (Brandes et al., 2018). The top of these sediments constitutes the terrain in the area. A subtle, WNW-ESE-oriented

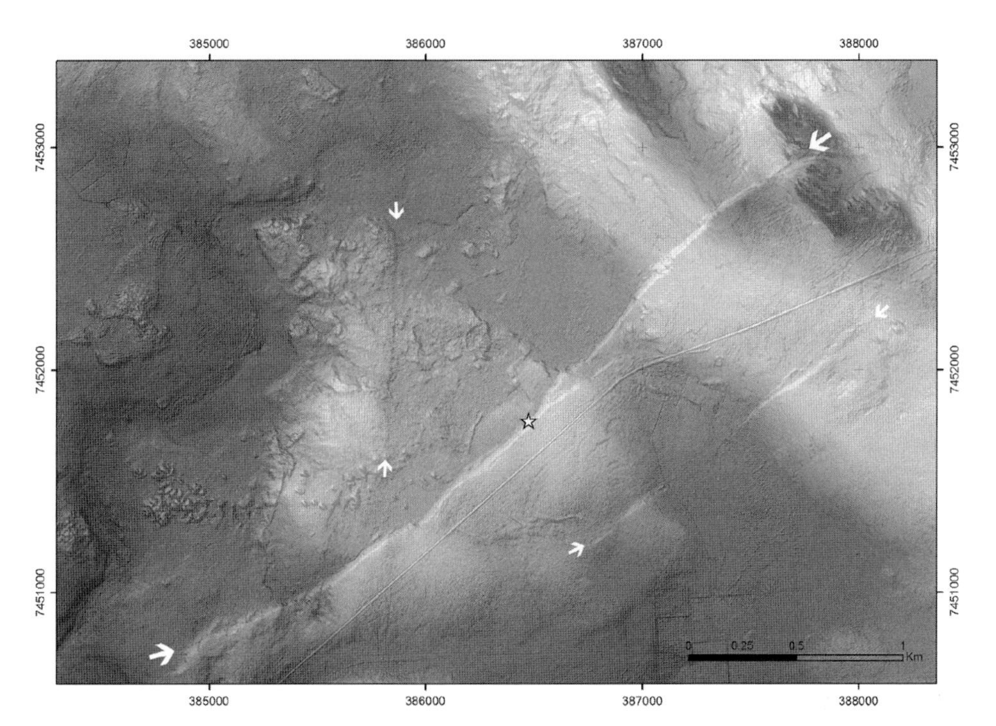

Figure 3.1 A hill-shaded LiDAR digital elevation model (DEM) of the Ruokojärvi–Pasmajärvi glacially induced fault, Finland (from Palmu et al., 2015). Red colours indicate high elevations (up to 235 m above sea level, a.s.l.), blue colours show low elevations (down to 178 m a.s.l.). White arrows mark fault scarps. The yellow star marks trenching described by Markovaara-Koivisto et al. (2020); see Figure 13.2 of Sutinen et al., Chapter 13. (A black and white version of this figure will appear in some formats. For the colour version, please refer to the plate section.)

escarpment with a height of up to 2 m runs across the marine surface over around 8 km (see Figure 15.5 in Sandersen et al., Chapter 15). The escarpment appears as a degraded linear kink in the terrain separating two drainage networks. It has the same orientation as the STZ, and it is interpreted as a surface expression of Lateglacial movements along faults of the STZ (Brandes et al., 2018). However, no direct link to deep bedrock faults has yet been made.

At Tinglev, in southern Denmark, deformations of Late Weichselian outwash plain sediments are interpreted to be the result of early Holocene movements of the faults of the Tønder Graben below (Sandersen & Jørgensen, 2015). The outwash plain constitutes a gently W-SW-sloping surface created in front of the Late Weichselian ice sheet, but parts of the outwash plain deviate from the expected picture. Figure 3.2 shows a LiDAR-based map of the eastern outwash plain topography. Highlighted with a pair of white arrowheads is one out of four lineaments of up to 7 km length, across which elevation changes of the outwash plain of up to 2 m can be seen. The lineaments are sub-parallel and roughly perpendicular to the boundary faults of the graben and are interpreted to represent near-surface faults within the sediments above the reactivated faults (Sandersen & Jørgensen, 2015).

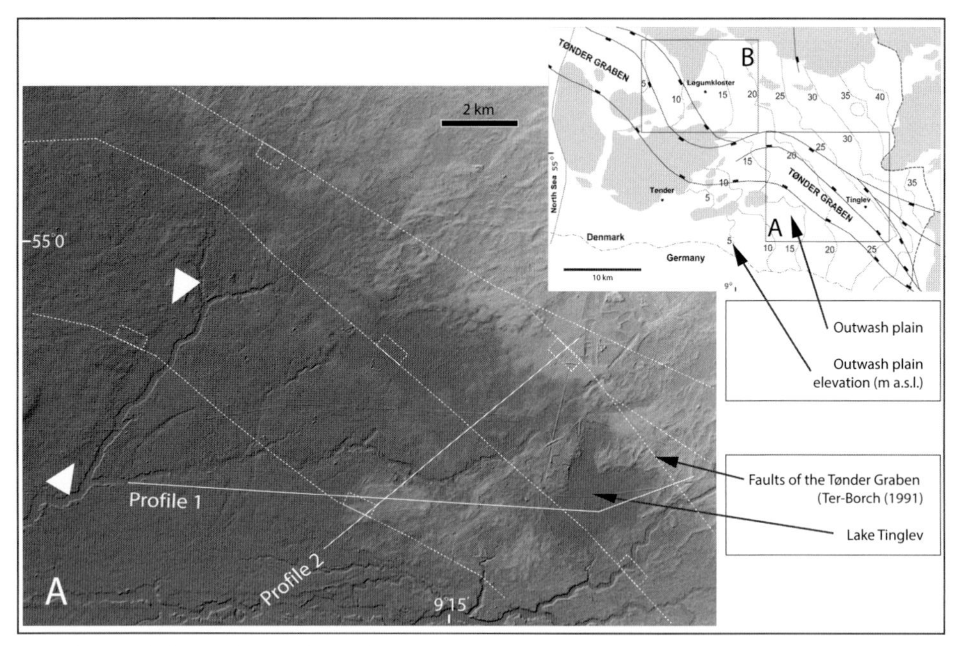

Figure 3.2 Selected area of the Tinglev outwash plain, Denmark. The topography is shown as grey hillshade (LiDAR data). Hatched white lines outline deep-seated faults of the Tønder Graben. The locations of Profiles 1 and 2 (Figure 3.3) are shown as white lines. White arrowheads mark a lineament referred to in the text. Area B on the insert map shows the area covered by Figure 3.4 (based on Sandersen & Jørgensen, 2015).

3.4.1.2 Other Primary Landforms

Subsided Areas

At the Nørre Lyngby site, a W-E-oriented depression, 200–300 m wide and 900 m long is present above faulted Lateglacial sediments (see Figure 15.5 – 'Nørre Lyngby basin', in Sandersen et al., Chapter 15). The depression lies around 5 m below the surrounding areas and is interpreted as a fault-parallel sag pond formed as the result of Lateglacial reactivation of deep faults of the STZ (Brandes et al., 2018). The marine sediments constitute the terrain surface in the area and therefore it is likely that other elongated depressions in the area also represent small fault-bounded basins providing accommodation space for Lateglacial to Holocene sediments (Brandes et al., 2018).

In the Tinglev area mentioned previously, early Holocene reactivation of the Tønder Graben is interpreted to have created a NW-SE, 5 × 8-km depression of the Late Weichselian outwash plain (Sandersen & Jørgensen, 2015). The depression is seen on Figure 3.2 as an irregular NW-SE-oriented, roughly rectangular area from the centre to the lower right part of the map (darker grey shade). The depression roughly follows the outline of the underlying graben structure. Two profiles crossing the graben image the depression between 8,600–13,500 m and 2,200–6,600 m, respectively (Figure 3.3; see location on Figure 3.2). The depression measures up to 8 m, but locally under Lake Tinglev, the outwash plain lies at a depth of 16 m below the expected elevation.

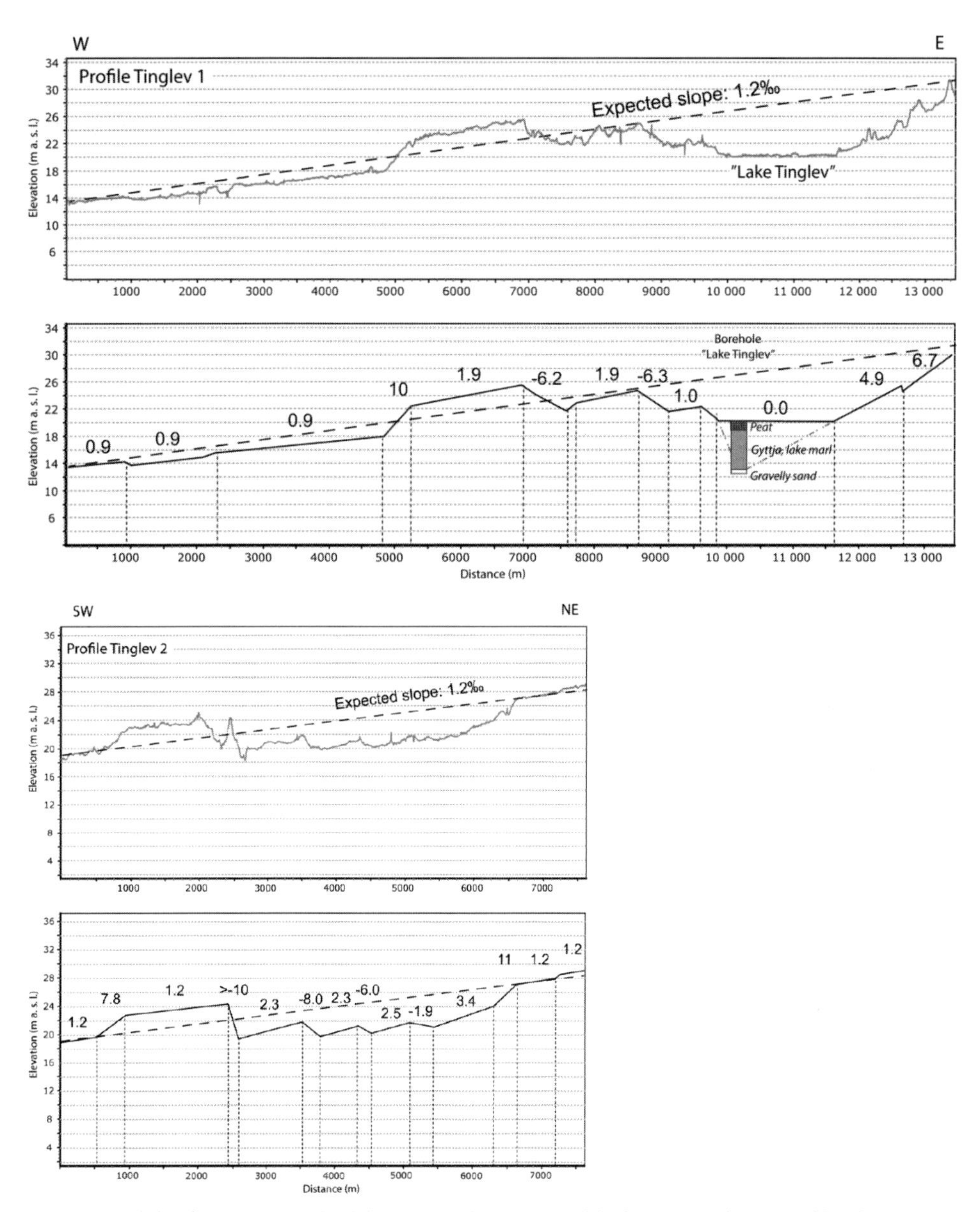

Figure 3.3 Tinglev outwash plain, Denmark. Topographical cross-sections: Profile Tinglev 1 (upper profile) and 2 (lower profile); see location on Figure 3.2. The upper panel of each profile shows the LiDAR-based digital terrain model. The lower panel of each profile shows a simplified topography; numbers above the terrain show local slope. Negative values indicate slopes opposite to the expected SW/W direction. Note: Vertical exaggeration $\times 100$ (from Sandersen & Jørgensen, 2015).

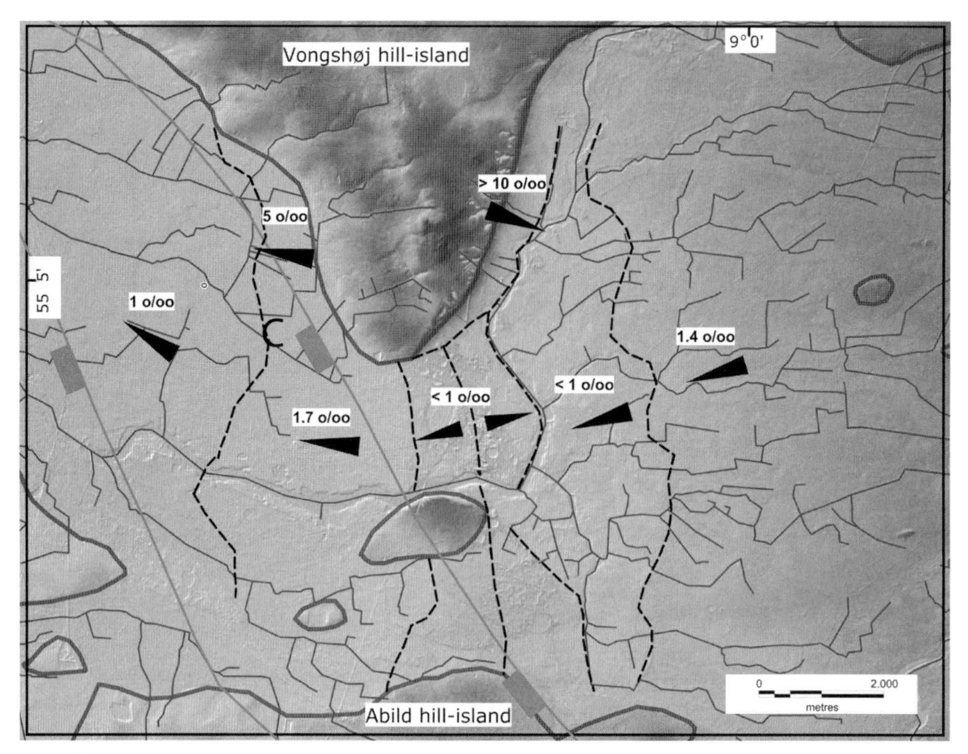

Figure 3.4 A N-S threshold on the Tinglev outwash plain between Vongshøj and Abild hill-islands. For location see Figure 3.2, area B. The figure shows slope and drainage pattern on the outwash plain. The highest elevations on the hill-islands are shown in brown; lowest elevations in pale blue (LiDAR DEM). Streams and brooks are shown as blue lines. Black arrows indicate slope orientation. Faults in the Top Chalk surface about 400 m below sea level shown in red (modified from Sandersen & Jørgensen, 2015) (A black and white version of this figure will appear in some formats. For the colour version, please refer to the plate section.).

In northern Germany, Grim and Sirocko (2012) found a correlation between negative landforms and the flanks of the tectonic Glückstadt Graben. The authors concluded that the negative landforms indirectly indicated tectonic activity at the end of the last glaciation or during the Holocene. Moreover, 29 large depressions (larger than 5 km) were found to be associated, at least partly, with basin subsidence and tectonic activity of the graben flanks.

Folds Over Faults and Other Uplifted Areas

The reactivation of the faults in the Tønder Graben (Figure 3.2) was interpreted to have a strike-slip component, creating areas of both uplift and subsidence as the result of localized compression and extension (Sandersen & Jørgensen, 2015). On Profile 1 (Figure 3.3), an uplifted part of the outwash plain can be seen at 5,000–7,000 m, and at 500–2,500 m on Profile 2. Further to the north-west along the northern graben flank, slope variations of the outwash plain reveal a N-S threshold, where the plain is uplifted 2 m. Figure 3.4 shows slope changes from 1.4‰ to the east to below 1.0‰ in the central part over the threshold and 1.7‰ west of the threshold. This is supported by a seismic line revealing an anticline underneath, which suggests compressive forces along the graben (Sandersen & Jørgensen, 2015).

As opposed to marine shorelines, which are sensitive to both isostatic and eustatic changes, lake shorelines are sensitive only to the level of the outlet and can be direct indicators of differential postglacial uplift (Påsse, 1998). Seppä et al. (2012) studied Lake Pielinen in Finland and found changes in palaeogeography and palaeohydrology as a result of the tilting of eastern Fennoscandia following the deglaciation. The lake is parallel to the uplift, which makes it especially sensitive to tilting. At the time of formation (10.2 ka BP) the lake was 143 km long; because of the tilting, regression has now reduced the length to 99 km.

3.4.2 Secondary Landforms Created by Seismic Shaking

3.4.2.1 Landslides

Northern Fennoscandia has experienced high-magnitude Lateglacial and postglacial earthquakes associated with the release of lithospheric stresses during and after the retreat of the ice sheets (Wu et al., 1999). It has been postulated that landslides (see classification e.g. in Hungr et al., 2014) are, at least in several cases, triggered by earthquakes and therefore can be used as indicators of postglacial seismic activity (Kujansuu, 1972; Lagerbäck & Sundh, 2008; Mikko et al., 2015; Ojala et al., 2018, 2019a; Olesen et al., 2004; Palmu et al., 2015; Sutinen et al., 2009a, 2014c, 2018). The LiDAR mapping e.g. in Finnish Lapland has shown that landslide scars are found within a radius of 35 km from the nearest glacially induced fault (GIF) (Ojala et al., 2019a); yet not all GIFs are surrounded by landslides and not all landslides are located next to LiDAR detectable fault scarps (e.g. Sutinen et al., 2014b; Ojala et al., 2018). One of the critical requirements for sliding is full saturation of the sediments ($>$ 40 per cent volume water). A potential time-envelope, e.g. up to 10 weeks saturation for fine-grained tills, can be achieved during the snowmelt period (e.g. Sutinen et al., 2007).

Based on the size and shape of the landslide scars and using empirical correlations between landslide volume area and earthquake moment magnitude, Ojala et al. (2019a) estimated maximum magnitude $M_w \approx 6.9-7.7$ for the palaeoearthquakes in the Suasselkä, Isovaara–Riikonkumpu, Venejärvi and Vaalajärvi areas in northern Finland. These figures are comparable to $M_w \approx 6.5-7.5$ estimates based on the fault lengths and displacements of the GIFs.

Palaeolandslides tend to be spatially distributed in the areas of GIFs and recent seismic activity (see e.g. Sutinen et al., 2019a), hence it is plausible that GIFs have been reactivated repeatedly during Late- and postglacial times (Mattila et al., 2019; Ojala et al., 2019a), and landslide buried organic materials may provide insight into the timing and recurrence of past earthquakes. Currently, the time span of palaeolandslides in the interior of the former Fennoscandian ice sheet (FIS) range from 1.297 to 11.292 ka BP, with three distinct episodes, 9–11 ka BP, 5–6 ka BP and 1–3 ka BP (Ojala et al., 2018; Figure 3.5; Smith et al., Chapter 8).

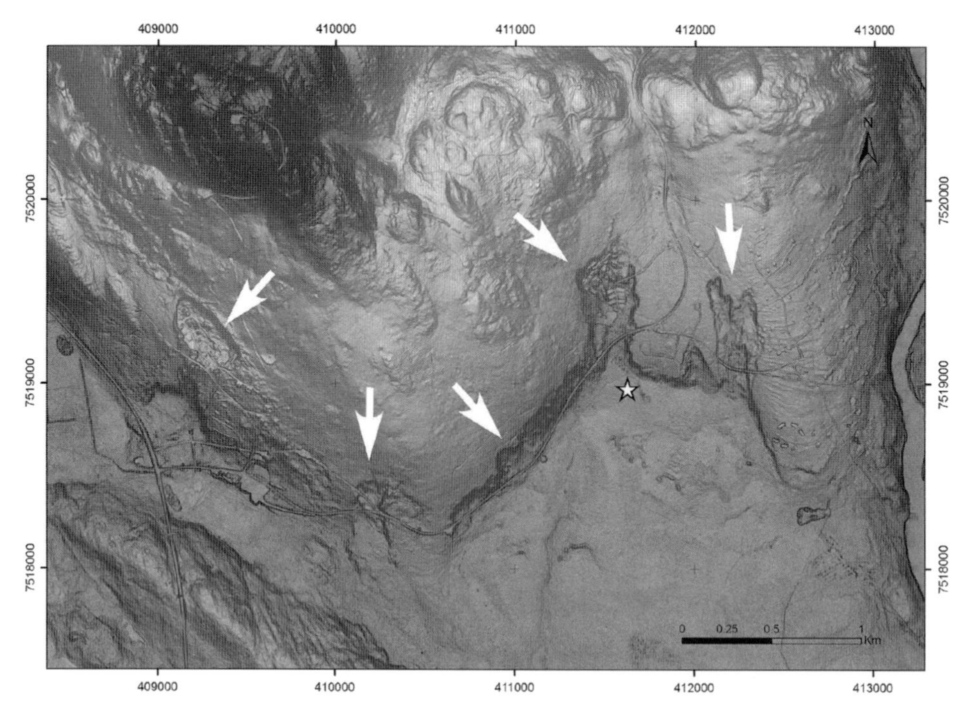

Figure 3.5 Hill-shaded LiDAR DEM of palaeolandslides from Levi Fell in Kittilä, Finland (Palmu et al., 2015). The lowest terrain is at the bottom of the map (approximately 181 m above sea level, a.s.l.), and the highest is at the top of the map (approximately 531 m a.s.l.). White arrows mark the slides. The star indicates dated sediments of the bog beneath the slide (4.965 ka BP; Sutinen et al., 2014c).

3.4.2.2 *Liquefaction Spreads*

Liquefaction spreads, or lateral spreads, are a common landslide type in the areas of recent earthquake terrains (Obermeier, 2009; see also Hungr et al., 2014). Similar deformation features have also been observed on drumlins in the Kuusamo area, SE Finnish Lapland (Sutinen et al., 2018). Generally, they appear on distal slopes (1.3–2.8 per cent) of the streamlined landforms, yet often the original drumlin ridge forms have been totally destroyed by the spreads (Figure 3.6b). Similar to rotational landslides (Sutinen et al., 2014c), the orientation of the liquefaction spreads deviates from the ice flow direction suggesting postdepositional processes (Sutinen et al., 2019b).

3.4.2.3 *Subglacial Landforms*

Model predictions by Wu et al. (1999) suggest that the onset of fault instability started at 15 ka BP and that maximum fault instability was reached at 13–10 ka BP in Fennoscandia (see also Stewart et al., 2000), hence it is postulated that not only subaerial but also subglacial types of deformations may have been associated with past seismic events (e.g. Sutinen et al., 2009b). This is particularly true for the deformation within the ice-steam

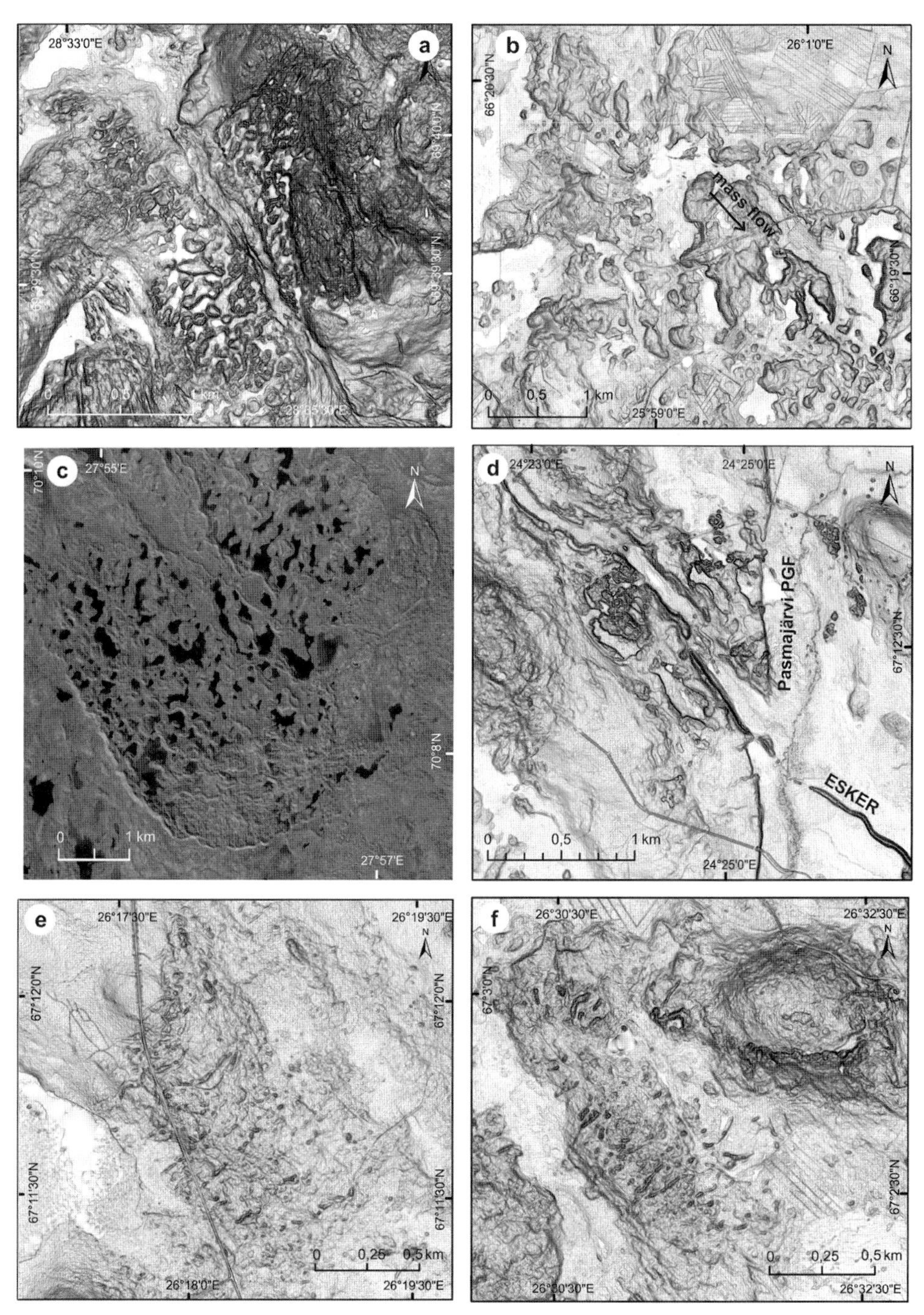

Figure 3.6 a) LiDAR DEM images of the Pulju moraine field in Sevetti–Näätämöjoki area, next to Sevetti GIF, in NE Finnish Lapland (modified from Sutinen et al., 2019a); b) mass flow deposits in central Finnish Lapland (modified from Sutinen et al., 2009b); c) aerial photo (Norkart Geoservice AS and Geovekst) showing Maskevarri Ráhppát in Finnmark, Norway (modified from Sutinen et al., 2014a); d) LiDAR images of esker collapse morphologies next to the Pasmajärvi GIF in western Finnish Lapland (modified from Markovaara-Koivisto et al., 2020); e,f) squeeze-up ridges 20–40 km down-ice from the Vaalajärvi–Risttonmännikkö GIF Complex in central Finnish Lapland (modified from Sutinen et al., 2021).

streamlined landforms heading towards the Younger Dryas end moraines (12.8–11.5 ka BP; Muscheler et al., 2008). Fast ice streaming was probably triggered by subglacial flooding (e.g. Schoof, 2010), hence fully saturated sediments of the glacier bed and in the newly exposed drumlins and flutings were prone to earthquake-induced deformations. Based on LiDAR DEM observations and sediment electromagnetic (EM) anisotropy data on the landforms, such as the rim-ridged/arcuate Pulju moraine (Kujansuu, 1967; Sutinen et al., 2014b), hummocky ridge fields of mass flow sediments (Sutinen et al., 2009b; Middleton et al., 2020b), Ráhppát landforms (Sutinen et al., 2014a) as well as esker collapse systems (Markovaara-Koivisto et al., 2020) and squeeze-up ridge fields (Sutinen et al., 2021) may be linked to subglacial earthquake activity.

Pulju Moraine

Pulju moraine (see morphology in Figure 3.6a), according to the type area near Pulju village, western Finnish Lapland, is a morphological term for clusters of irregular fields of elongated, ring-formed subglacial ridges, whose height is generally less than 2 m in Finnish Lapland (Kujansuu, 1967; Sutinen et al., 2014a). Traditionally, non-oriented and rim-ridged morainic landforms are considered as stagnant or semi-stagnant ice features (Hoppe, 1952; Kujansuu, 1967; Knudsen et al., 2006). A number of studies have argued against a supraglacial concept (see Menzies & Shilts, 2002), favoring instead the squeezing hypothesis with diapirism and squeezing of (clay-rich) till into the subglacial crevasses (Hoppe, 1952). Pulju moraines have been reported particularly on the ice-divide zone of the former FIS and in regions with known postglacial faults (Kujansuu, 1967; Sutinen et al., 2009a, 2018), hence their spatial distribution may be a notable feature to act as a pathfinder for potential GIFs (Middleton et al., 2020a). LiDAR DEM data display no ice streamlining on the surfaces of Pulju moraine ridges (Figure 3.6a; Sutinen et al., 2019a), hence the Pulju moraine appears to be a stagnant ice feature (Sutinen et al., 2018; Middleton et al., 2020a). Electromagnetic (EM) anisotropy data indicate the material (diamicton) eventually was squeezed by the earthquake(s) underneath the ice (Sutinen et al., 2019b). This sedimentary pattern is in accordance with the subglacial squeeze-up hypothesis as presented by Hoppe (1952). Furthermore, the Pulju moraine morphology reveals a transition from rim ridges to arcuate ridges or single esker ridges of sinusoidal forms and finally to anastomosing esker networks, implying that all these landforms began in wet-based conditions beneath the ice (Sutinen et al., 2009a, 2014b). Even though the morphology of the Pulju moraine resembles that of the so-called Veiki moraine in northern Sweden (Hoppe, 1952; Lagerbäck, 1988) or ice-disintegration (hummocky) moraines of the Canadian plains, supraglacial and non-seismic origin of the Veiki moraine cannot be ruled out, particularly as it dates back to Early Weichselian (Lagerbäck, 1988). However, west of the so-called Lainio Arc (Veiki moraine; Hoppe, 1952; Lagerbäck, 1988), and next to Lainio–Suijavaara GIF in Norrbotten, Sweden, subaerial seismic morphologies, such as ground surface depressions (Lagerbäck & Sundh, 2008; called liquefaction bowls in Sutinen et al., 2019b) can be seen on the LiDAR DEM (Figure 3.7c,d).

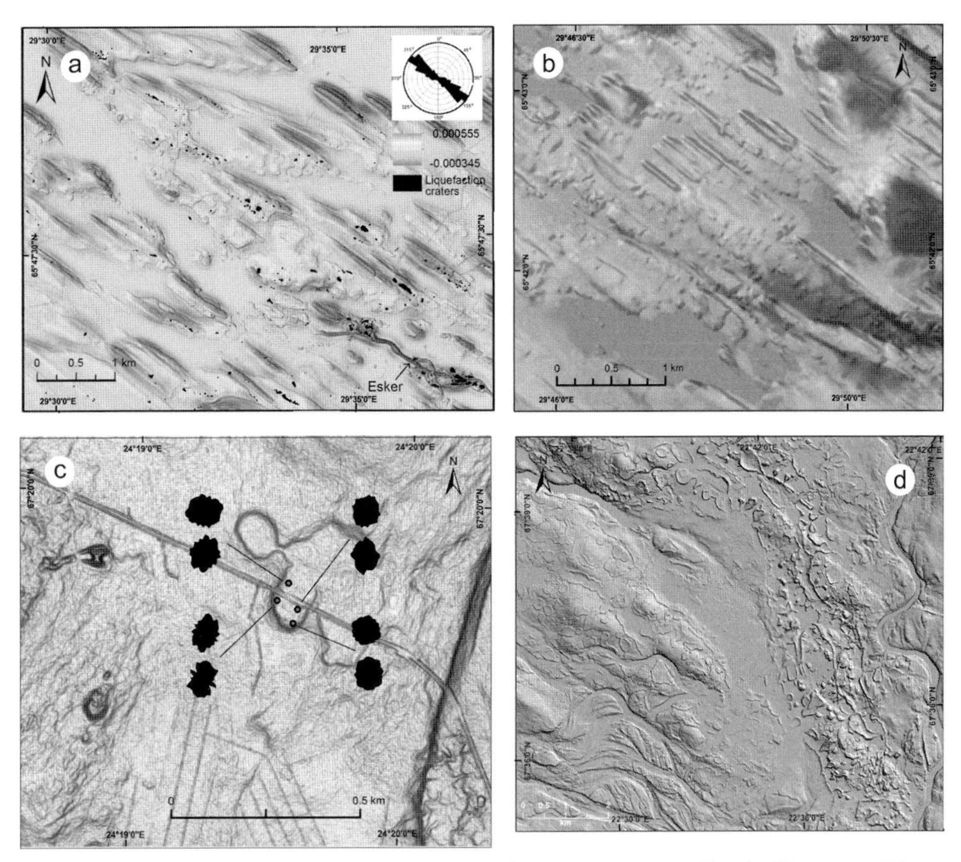

Figure 3.7 a) LiDAR DEM showing liquefaction craters on drumlins in Kuusamo, SE Finnish Lapland (modified from Sutinen et al., 2019a); b) liquefaction spreads deforming the ice-streamlined landforms in Kuusamo, SE Finnish Lapland (modified from Sutinen et al., 2018); c) liquefaction bowls next to Naamivittikko GIF in Kolari, Finnish Lapland (Mattila et al., 2019, modified from Sutinen et al., 2019b); d) LiDAR DEM showing a part of the Lainiobågen (Lainio Arc, right) with several kilometre-long undulating NW-SE-oriented (coarse-textured gravel) ridges north of Kangos in Norrbotten, Sweden. The river Lainioälven is seen to the right; round-shaped liquefaction bowls cover the western part of the image (modified from Sutinen et al., 2019b).

Mass Flow Deposits

Crudely stratified sediments are common constituents of hummocky moraine fields in northern Fennoscandia; one good example is the subglacial mass-flow field in Kemijärvi (Figure 3.6b; Sutinen, 1985; Nordkalott Project, 1986; Sutinen et al., 2019b; Middleton et al., 2020b). On the basis of morphometric analysis (OBIA) of LiDAR DEMs (Middleton et al., 2020b) and EM sedimentary anisotropy records, crudely stratified sediments were transported as flows of high bedload concentration possibly due to outburst(s) of subglacial water bodies (lakes) (Sutinen et al., 2009b). With respect to seismogenic triggering mechanisms for the mass flows in Kemijärvi, one potential source is the NNW-SSE-trending

Vaalajärvi GIF. It is some 60 km up-ice from Kemijärvi exhibiting a 7-km-long fault scarp, and it hosted multiple slip events $M_w \approx 6.7-7.0$ (Ojala et al., 2020).

Ráhppát Landforms

Ráhppát landforms are found in Finnmark, northern Norway, to be built up of a network of stony ridges and mounds (Figure 3.6c). In Maskevarri, the arch-shaped and sinusoidal ridges on fell terraces exhibit EM sedimentary anisotropy parallel-to-ridge trends. The ridges are interconnected to meltwater gullies suggesting generation through short-lived conduit infills (Sutinen et al., 2014a). Maskevarri Ráhppát is characterized by 500 ponds and small lakes on three different terrace elevations descending in an up-ice direction. These scarps may have been generated through Lateglacial earthquake(s) contributing to subglacial deformation at the base of the ice.

Esker Collapses

Esker collapse morphologies are found in northern Finland and may be linked to GIFs. As an example, the pressure gradient required for the dendritic esker systems is depicted as a single-ridged esker on the eastern side of the Pasmajärvi GIF, yet the same glaciofluvial system shows total collapse on the western side of the ramp (Figure 3.6d; Markovaara-Koivisto et al., 2020). The lobe and fan-like morphology suggests collapse of an esker tunnel system. These pressurized outburst esker ridges differ from the conventional time-transgressive evolution of eskers formed in the 'steady state' pressure gradient (see in lower right corner of Figure 3.6d, Sutinen et al., 2009a).

Squeeze-Up Moraines

South-east of the Vaalajärvi–Ristonmännikkö GIF System in central Finnish Lapland, LiDAR DEMs show irregular groupings of straight or curvilinear and transverse-to-ice flow ridges (Figure 3.6e,f; Sutinen et al., 2021). These are distributed along the SE-trending down-ice transect from the fault, and EM anisotropy indicates these moraines are not comparable to annual or ice-marginal end moraines. It is proposed that these squeeze-ups are subglacial (see Hoppe, 1952) and that the driving force for the ice crevassing was an earthquake (see glacial earthquakes in Ekström et al., 2006), which created pathways for the saturated material to squeeze into the fractured ice system. Forty kilometres down-ice from the GIF is a field of arcuate/semi-circular Pulju moraines, associated with transverse-to-ice squeeze-up ridges forming a complex of seismic-induced subglacial landforms.

3.4.2.4 Liquefaction Features

Liquefaction Craters

Postglacial (sub-aerial) deformations are easily detectable e.g. with LiDAR morphometry (Figure 3.7). Since ice-streaming has been shown to be associated with discharge of highly pressurized subglacial meltwater, newly exposed ice-stream landforms are fully water-saturated and prone to earthquake shaking-induced liquefaction processes. Figure 3.7a shows an example of liquefaction craters on drumlins in Kuusamo, SE Finnish Lapland

(Sutinen et al., 2018, 2019a). Generally, these holes are 20–30 m in diameter and less than 5 m in depth and are morphologically rather similar to craters after the 1886 earthquake in Charleston, South Carolina (see Obermeier, 2009). Liquefaction craters are different from cryogenic craters (Buldovicz et al., 2018). The loss of anisotropic character of the sediments at the craters (e.g. Sutinen et al., 2019b) indicates postdepositional reworking, presumably in a manner of vertical grading as described by Lagerbäck and Sundh (2008).

Liquefaction Bowls

Lagerbäck and Sundh (2008, see their Figures 55–58) found normally graded tills to be a common palaeoseismic feature in Norrbotten, Sweden. Within the grading a primary glacial stratigraphy is modified, and maximum grain size increases with depth. This earthquake-induced liquefaction phenomenon is particularly well demonstrated in the Landsjärv area and shown with experimental vibration tests (Lagerbäck & Sundh, 2008, their Figures 55–58). Many times, the morphology is bowl-shaped with rims surrounding the internal bowl depression (Figure 3.7d; depressions in Lagerbäck & Sundh, 2008). Lack of preferred orientation of the sediments, as measured with EM sedimentary anisotropy (Figure 3.7c, Sutinen et al., 2019b) in the liquefaction bowls supports the grading till concept of Lagerbäck and Sundh (2008). The timing of liquefaction features is not well constrained, yet Lagerbäck and Sundh (2008) postulated that the graded tills were created shortly after the deglaciation.

3.4.3 Stream Deflections and Changes in Drainage Patterns

At Tinglev, Denmark, the slope changes created by a 2-m uplift had a pronounced impact on the drainage pattern of the outwash plain sediments (Figure 3.4, Sandersen & Jørgensen, 2015). This threshold on the outwash plain caused the streams to converge just east of the threshold to gain enough force to erode a passage through the obstacle. Obviously, the threshold was formed faster than the stream could erode, thus leading to a reorganization of the drainage pattern.

In Brande, Denmark, a correlation between the location of interpreted faults and the present streams on the Late Weichselian outwash plain led Lykke-Andersen et al. (1996) to conclude that the Quaternary faults created zones in the unconsolidated near-surface sediments that were easier to erode.

The inferred kilometre-long fault scarp found at Nørre Lyngby, mentioned earlier, created a distinct drainage divide. During the Lateglacial, the area was uplifted due to the GIA, and the former seabed drained. This was accompanied by faulting and tilting of the former seabed creating a drainage divide separating two newly formed streams (Brandes et al., 2018).

3.5 Discussion

The examples described have all been interpreted as formations resulting from seismotectonic events caused by the loading and unloading of the Pleistocene ice sheets. The degree of surface deformation revealed at several localities underlines that seismic events from the higher end of the intensity scale were not uncommon around the time of deglaciation. However, even for the largest seismotectonic events, the surficial expressions visible in the topography today will depend on the character and thickness of the unconsolidated sediments above the bedrock as well as the length of time since the event. The examples show that different landforms and surface deformations can be expected in areas affected by seismotectonic events. They also highlight the differences in the resulting landform types depending on whether the event took place in a subglacial, proglacial or postglacial environment. In addition to geomorphological and EM sedimentary anisotropy surveys (e.g. Middleton et al., 2020a, 2020b), trenching for sedimentological and stratigraphic studies (e.g. Ojala et al., 2020) is needed to reveal the genetic constraints of landforms developed in different environments.

Distinguishing between tectonic and nontectonic geomorphology can be difficult – especially for subtle landform types or for landforms formed in direct contact with the glacier ice. The Danish examples of primary landforms show less pronounced topographic features compared to the examples from northern Fennoscandia; how these relate to specific deeper structures has not been verified. The same can be said about the examples of the secondary landforms from northern Fennoscandia. However, even though the origin of some of the landform types can be discussed, the described features represent morphologies that one could expect in tectonically active areas (e.g. Burbank & Anderson, 2012; McCalpin, 2009).

In all cases, it is important to establish an event chronology based on both detailed digital elevation models and data from geophysics, boreholes, trenches and outcrops. Deformations of well-dated shorelines, lake sediments, surfaces of marine sediments or outwash plains will help establish a robust chronology.

LiDAR data show the terrain in exceptional detail and have enabled observations of subtle landforms that before had been unnoticed or subject to misinterpretations (e.g. Johnson et al., 2015; Ojala et al., 2019b; Sandersen & Jørgensen, 2015). However, care should be taken when categorizing landforms as earthquake-induced. For instance, periglacial landforms and sediment structures such as cryoturbation, ice wedges and thermokarst features can resemble seismotectonic deformations (Stewart et al., 2000; Van Vliet-Lanoë, 2004; Van Vliet-Lanoë et al., 2016; Buldovicz et al., 2018), and glacially lineated landscapes and parts of the subglacial landforms called 'murtoos' can resemble fault scarps (Öhrling et al., 2018; Ojala et al., 2019b). It is thus important to combine different types of data and look for several types of seismotectonic indicators within the same area.

The loading and unloading of glacier ice can affect salt structures in the subsurface resulting in salt flow towards zones of lower overburden pressure (e.g. Lang et al., 2014). Studies from Germany find that halokinetic movements can be induced by ice-sheet loading and unloading (Al Hseinat et al., 2016; Lang et al., 2014; Sirocko et al., 2002, 2008). The result can be deformations of the sedimentary cover and eventually landform creation. However, landforms created in this way can be difficult to distinguish from halokinetic movements induced by sediment load only. This will for example concern the sedimentary basins of northern Central Europe, where salt structures are common (e.g. Britze & Japsen, 1991; Lang et al., 2014).

If the individual Pleistocene ice sheets had roughly the same impact on the lithosphere, then stress changes can be expected to follow the same overall pattern during each glacial cycle. This will most likely result in recurrent reactivation of existing structures, with recurrence intervals between the major episodes of stress release matching the timespan of the glacial cycles. Evidence of multiple slip events on postglacial faults seems to support this hypothesis (Mattila et al., 2019; Smith et al., 2018).

3.6 Perspectives

The identification of earthquake-induced landforms is important when assessing seismic hazards in and around areas formerly covered by ice sheets (Middleton et al., 2020a; Öhrling et al., 2018). The magnitude of the palaeoseismic event can tentatively be inferred from the dimensions of the earthquake-induced landforms combined with observations of the deformed rocks and sediments (Michetti et al., 2007). Added to this, determining an age for seismotectonic events that formed or modified a terrain and near-surface succession is important for evaluations of recurrence.

Studying earthquake-induced landforms and landform change in the context of loading and unloading of the Pleistocene ice sheets adds knowledge about the past and present seismicity of former glaciated areas. But in addition to this, valuable new information on the structural setting of the subsurface can be gained, both for the near-surface and deep domains. This is for instance highly important for evaluations of deep geologic repositories, carbon capture storage (CCS), assessments of geothermal potential, assessments of groundwater vulnerability and geotechnical surveys.

Acknowledgements

We are thankful to Christoph Grützner and an anonymous reviewer for their constructive reviews, which helped improve the manuscript.

References

Al Hseinat, M., Hübscher, C., Lang, J., Lüdmann, T., Ott, I., and Polom, U. (2016). Triassic to recent tectonic evolution of a crestal collapse graben above a salt-cored anticline in the Glückstadt Graben/North German Basin. *Tectonophysics*, **680**, 50–66, doi.org/10.1016/j.tecto.2016.05.008.

Brandes, C. and Winsemann, J. (2013). Soft-sediment deformation structures in NW Germany caused by Late Pleistocene seismicity. *International Journal of Earth Sciences*, **102**, 2255–2274, doi.org/10.1007/s00531-013-0914-4.

Brandes, C., Winsemann, J., Roskosch, J. et al. (2012). Activity along the Osning Thrust in Central Europe during the Lateglacial: ice-sheet and lithosphere interactions. *Quaternary Science Reviews*, **38**, 49–62, doi.org/10.1016/j.quascirev.2012.01.021.

Brandes, C., Steffen, H., Steffen, R. and Wu, P. (2015). Intraplate seismicity in northern Central Europe is induced by the last glaciation. *Geology*, **43**, 611–614, doi.org/10.1130/G36710.1.

Brandes, C., Steffen, H., Sandersen, P. B. E., Wu, P. and Winsemann, J. (2018). Glacially induced faulting along the NW segment of the Sorgenfrei-Tornquist Zone, northern Denmark: implications for neotectonics and Lateglacial fault-bound basin formation. *Quaternary Science Reviews*, **189**, 149–168, doi.org/10.1016/j.quascirev.2018.03.036.

Britze, P. and Japsen, P. (1991). Geological map of Denmark 1:400 000: the Danish Basin: «Top Zechstein» and the Triassic (two-way travel time and depth, thickness and interval velocity). *Geological Survey of Denmark, Map Series*, **31**, 1–4.

Buldovicz, S. N., Khilimonyuk, V. Z., Bychkov, A. Y. et al. (2018). Cryovolcanism on the Earth: origin of a spectacular crater in the Yamal Peninsula (Russia). *Scientific Reports*, **8**, 13534, doi.org/10.1038/s41598-018-31858-9.

Burbank, D. W. and Anderson, R. S. (2012). *Tectonic Geomorphology*, 2nd ed., Wiley-Blackwell, Hoboken, New Jersey.

Dramis, F. and Blumetti, A. M. (2005). Some considerations concerning seismic geomorphology and paleoseismology. *Tectonophysics*, **408**, 177–191, doi.org/10.1016/j.tecto.2005.05.032

Ekström, G., Nettles, M. and Tsai, V. C. (2006). Seasonality and increasing frequency of Greenland glacial earthquakes. *Science*, **311**(5768), 1756–1758, doi.org/10.1126/science.1122112.

French, H. M. (2017). *The Periglacial Environment*, 4th ed., John Wiley & Sons Ltd., Hoboken, New Jersey.

Grim, S. and Sirocko, F. (2012). Natural depressions on modern topography in Schleswig-Holstein (Northern Germany) – indicators for recent crustal movements or “only” kettle holes? *Zeitschrift der deutschen Gesellschaft für Geowissenschaften*, **163**(4), 469–481, doi.org/10.1127/1860-1804/2012/0163-000.

Grube, A. (2019). Palaeoseismic structures in Quaternary sediments of Hamburg (NW Germany), earthquake evidence during the younger Weichselian and Holocene. *International Journal of Earth Sciences*, **108**(3), 845–861, doi.org/10.1007/s00531-019-01681-2.

Hoppe, G. (1952). Hummocky moraine regions with special reference to the interior of Norrbotten. *Geografiska Annaler*, **34**, 1–72.

Hungr, O., Leroueil, S. and Picarelli, L. (2014). The Varnes classification of landslide types, an update. *Landslides*, **11**, 167–194, doi.org/10.1007/s10346-013-0436-y.

Johnson, M. D., Fredin, O., Ojala, A. E. K. and Peterson, G. (2015). Unraveling Scandinavian geomorphology: the LiDAR revolution. *GFF*, **137**, 245–251, doi.org/11035897.1111410.

Knudsen, C. G., Larsen, E., Sejrup, H. P. and Stalsberg, K. (2006). Hummocky moraine landscape on Jæren, SW Norway – implications for glacier dynamics during the last glaciations. *Geomorphology*, **77**, 153–168, doi.org/10.1016/j.geomorph.2005.12.011.

Kuivamäki, A., Vuorela, P. and Paananen, M. (1998). *Indications of Postglacial and Recent Bedrock Movements in Finland and Russian Karelia*. Geological Survey of Finland Nuclear Waste Disposal Research Report YST-99, Espoo, Finland, 92 pp.

Kujansuu, R. (1967). On the deglaciation of western Finnish Lapland. *Geological Survey of Finland Bulletin*, **232**.

Kujansuu, R. (1972). On landslides in Finnish Lapland. *Geological Survey of Finland Bulletin*, **256**.

Lagerbäck, R. (1988). The Veiki moraines in northern Sweden – widespread evidence of an Early Weichselian deglaciation. *Boreas* **17**, 469–486, doi.org/10.1111/j.1502-3885.1988.tb00562.x

Lagerbäck, R. and Sundh, M. (2008). Early Holocene faulting and paleoseismicity in northern Sweden. *Geological Survey of Sweden Research Paper C 836*, 80 pp.

Lang, J., Hampel, A., Brandes, C. and Winsemann, J. (2014). Response of salt structures to ice-sheet loading: implications for ice-marginal and subglacial processes. *Quaternary Science Reviews*, **101**, 217–233, doi.org/10.1016/j.quascirev.2014.07.022.

Lykke-Andersen, H., Madirazza, I. and Sandersen, P. B. E. (1996). Tektonik og landskabsdannelse i Midtjylland [Tectonics and landscape formation in Mid-Jutland]. *Geologisk Tidsskrift 1996*, **3**, 1–32.

Markovaara-Koivisto, M., Ojala, A. E. K., Mattila, J. et al. (2020). Geomorphological evidence of paleoseismicity: surficial and underground structures of Pasmajärvi postglacial fault. *Earth Surface Processes and Landforms*, **45**(12), 3011–3024, doi.org/10.1002/esp.4948.

Mattila, J., Ojala, A. E. K., Ruskeeniemi, T. et al. (2019). Evidence of multiple slip events on postglacial faults in northern Fennoscandia. *Quaternary Science Reviews*, **215**, 242–252, doi.org/10.1016/j.quascirev.2019.05.022.

McCalpin, J. P. (2009). *Paleoseismology. International Geophysics Series Vol. 95*, 2nd ed., Elsevier, Amsterdam, doi.org/10.1016/S0074-6142(09)95001-X.

McCalpin, J. P. and Nelson, A. R. (2009). Introduction to paleoseismology. In J. P. McCalpin, ed., *Paleoseismology. International Geophysics Series Vol. 95*, 2nd ed., Elsevier, Amsterdam, pp. 1–27, doi.org/10.1016/S0074-6142(09)95001-X.

Menzies, J. and Shilts, W. W. (2002). Subglacial environments. In J. Menzies, ed., *Modern & Past Glacial Environments*. Butterworth-Heinemann, Oxford, pp. 183–278.

Michetti, A. M., Esposito, E., Guerrieri, L. et al. (2007). Environmental seismic intensity scale – ESI 2007. *Memorie descrittive della carta geologica d'Italia*, **74**.

Middleton, M., Heikkonen, J., Nevalainen, P., Hyvönen, E. and Sutinen, R. (2020a). Machine learning-based mapping of micro-topographic earthquake-induced paleo Pulju moraines and liquefaction spreads. *Geomorphology*, **358**, 107099, doi.org/10.1016/j.geomorph.2020.107099.

Middleton, M., Nevalainen, P., Hyvönen, E., Heikkonen, J. and Sutinen, R. (2020b). Pattern recognition of LiDAR data and sediment anisotropy advocate polygenetic subglacial mass-flow origin of the Kemijärvi hummocky moraine field in northern Finland. *Geomorphology*, **362**, 107212, doi.org/10.1016/j.geomorph.2020.107212.

Mikko, H., Smith, C. A., Lund, B., Ask, M. V. S. and Munier, R. (2015). LiDAR-derived inventory of post-glacial fault scarps in Sweden. *GFF*, **137**, 334–338, doi.org/10.1080/11035897.2015.1036360.

Muscheler, R., Kromer, B., Björk, S. et al. (2008). Tree rings and ice cores reveal ^{14}C calibration uncertainties during the Younger Dryas. *Nature Geoscience*, **1**(4), 263–267, doi.org/10.1038/ngeo128.

Nordkalott Project (1986). Geological Map, Northern Fennoscandia, 1:1 million. Geological Surveys of Finland, Norway and Sweden.

Obermaier, S. F. (2009). Using liquefaction-induced and other soft-sediment features for paleoseismic analysis. *International Geophysics*, **95**, 497–564, doi.org/10.1016/S0074-6142(09)95007-0.

Öhrling, C., Peterson, G. and Mikko, H. (2018). *Detailed Geomorphological Analysis of LiDAR Derived Elevation Data, Forsmark. Searching for Indicatives of Late- and Postglacial Seismic Activity*. SKB Report R-18-10, Swedish Nuclear Fuel and Waste Management Co., Stockholm, 38 pp.

Ojala, A. E. K., Markovaara-Koivisto, M., Middleton, M. et al. (2018). Dating of paleo-landslides in western Finnish Lapland. *Earth Surface Processes and Landforms*, **43**, 2449–2462, doi.org/10.1002/esp.4408.

Ojala, A. E. K., Mattila, J., Markovaara-Koivisto, M. et al. (2019a). Distribution and morphology of landslides in northern Finland: an analysis of postglacial seismic activity. *Geomorphology*, **326**, 190–201, doi.org/10.1016/j.geomorph.2017.08.045.

Ojala A. E. K., Peterson G., Mäkinen J. et al. (2019b). Ice-sheet scale distribution and morphometry of triangular shaped hummocks (murtoos): a subglacial landform produced during rapid retreat of the Scandinavian Ice Sheet. *Annals of Glaciology*, **60**(80), 115–126, doi.org/10.1017/aog.2019.34.

Ojala, A. E. K., Mattila, J., Middleton, M. et al. (2020). Earthquake-induced deformation structures in glacial sediments – evidence on fault reactivation and instability at the Vaalajärvi fault in northern Fennoscandia. *Journal of Seismology*, **24**(3), doi.org/10.1007/s10950-020-09915-6.

Olesen, O. (1988). The Stuoragurra Fault, evidence of neotectonics in the Precambrian of Finnmark, northern Norway. *Norsk Geologisk Tidsskrift*, **68**, 107–118.

Olesen, O., Blikra, L. H., Braathen, A. et al. (2004). Neotectonic deformation in Norway and its implications: a review. *Norwegian Journal of Geology*, **84**, 3–34.

Olesen, O., Bungum, H., Dehls, J. et al. (2013). Neotectonics, seismicity and contemporary stress field in Norway – mechanisms and implications. In L. Olsen, O. Fredin and O. Olesen, eds., *Quaternary Geology of Norway, Geological Survey of Norway Special Publication*, **13**, pp. 145–174.

Palmu, J.-P., Ojala, A. E. K., Ruskeeniemi, T., Sutinen R. and Mattila, J. (2015). LiDAR DEM detection and classification of postglacial faults and seismically-induced landforms in Finland: a paleoseismic database. *GFF*, **137**, 344–352, doi.org/10.1080/11035897.2015.1068370.

Påsse, T. (1998). Lake-tilting, a method for estimation of glacio-isostatic uplift. *Boreas*, **27**, 69–80, doi.org/10.1111/j.1502-3885.1998.tb00868.x.

Pisarska-Jamroży, M., Belzyt, S., Börner, A. et al. (2018). Evidence from seismites for glacio-isostatically induced crustal faulting in front of an advancing land-ice mass (Rügen Island, SW Baltic Sea). *Tectonophysics*, **745**, 338–348, doi.org/10.1016/j.tecto.2018.08.004.

Rasmussen, S. O., Bigler, M., Blockey, S. P. et al. (2014). A stratigraphic framework for abrupt climatic changes during the last Glacial period based on three synchronized Greenland ice-core records: refining and extending the INTIMATE event stratigraphy. *Quaternary Science Reviews*, **106**, 14–28, doi.org/10.1016/j.quascirev.2014.09.007.

Sandersen, P. B. E. and Jørgensen, F. (2015). Neotectonic deformation of a Late Weichselian outwash plain by deglaciation-induced fault reactivation of a deep-seated graben structure. *Boreas*, **44**, 413–431, doi.org/10.1111/bor.12103.

Schoof, C. (2010). Ice-sheet acceleration driven by melt supply variability. *Nature*, **468**(7325), 803–806, doi.org/10.1038/nature09618.

Schumm, S. A., Dumont, J. F. and Holbrook, J. M. (2002). *Active Tectonics and Alluvial Rivers*. Cambridge University Press, Cambridge.

Seppä, H., Tikkanen, M. and Mäkiaho, J.-P. (2012). Tilting of Lake Pielinen, eastern Finland – an example of extreme transgressions and regressions caused by differential post-glacial isostatic uplift. *Estonian Journal of Earth Sciences*, **61**(3), 149–161, doi.org/10.3176/earth.2012.3.02.

Sirocko, F., Reicherter, K., Lehne, R. W. et al. (2008). Glaciation, salt and the present landscape. In R. Littke et al., eds., *Dynamics of Complex Intracontinental Basins: The Central European Basin System*. Springer Verlag, Heidelberg, pp. 234–245.

Sirocko, F., Szeder, T., Seelos, C. et al. (2002). Young tectonic and halokinetic movements in the North-German-Basin: its effect on formation of modern rivers and surface morphology. *Netherlands Journal of Geosciences/Geologie en Mijnbouw*, **81**(3-4), 431–441, doi.org/10.1017/S0016774600022708.

Smith, C. A., Grigull, S. and Mikko, H. (2018). Geomorphic evidence of multiple surface ruptures of the Merasjärvi "postglacial fault", northern Sweden. *GFF*, **140**(4), 318–322, doi.org/10.1080/11035897.2018.1492963.

Stewart, I. S., Sauber, J. and Rose, J. (2000). Glacio-seismotectonics: ice sheets, crustal deformation and seismicity. *Quaternary Science Reviews*, **19**, 1367–1389, doi.org/10.1016/S0277-3791(00)00094-9.

Sutinen, R. (1985). On the subglacial sedimentation of hummocky moraines and eskers in northern Finland. *Striae*, **22**, 21–25.

Sutinen, R., Middleton, M., Hänninen, P. et al. (2007). Dielectric constant time stability of glacial till at a clear-cut site. *Geoderma*, **141**, 311–319, doi.org/10.1016/j.geoderma.2007.06.016.

Sutinen, R., Piekkari, M. and Middleton, M. (2009a). Glacial geomorphology in Utsjoki, Finnish Lapland proposes Younger Dryas fault-instability. *Global and Planetary Change*, **69**, 16–28, doi.org/10.1016/j.gloplacha.2009.07.002.

Sutinen, R., Middleton, M., Liwata, M., Piekkari, M. and Hyvönen, E. (2009b). Sediment anisotropy coincides with moraine ridge trend in south-central Finnish Lapland. *Boreas*, **38**, 638–646, doi.org/10.1111/j.1502-3885.2009.00089.x.

Sutinen, R., Aro, I., Närhi, P., Piekkari, M. and Middleton, M. (2014a). Maskevarri Ráhhpát in Finnmark, northern Norway – is it an earthquake-induced landform complex? *Solid Earth*, **5**, 683–691, doi.org/10.5194/se-5-683-2014.

Sutinen, R., Hyvönen, E., Middleton, M. and Ruskeeniemi, T. (2014b). Airborne LiDAR detection of postglacial faults and Pulju moraine in Palojärvi, Finnish Lapland. *Global and Planetary Change*, **115**, 24–32, doi.org/10.1016/j.gloplacha.2014.01.007.

Sutinen, R., Hyvönen, E. and Kukkonen, I. (2014c). LiDAR detection of paleolandslides in the vicinity of the Suasselkä postglacial fault, Finnish Lapland. *International Journal of Applied Earth Observation and Geoinformation*, **27**, 91–99, doi.org/10.1016/j.jag.2013.05.004.

Sutinen, R., Hyvönen, E., Middleton, M. and Airo, M.-L. (2018). Earthquake-induced deformations on ice-stream landforms in Kuusamo, eastern Finnish Lapland. *Global and Planetary Change*, **160**, 46–60, doi.org/10.1016/j.gloplacha.2017.11.011.

Sutinen, R., Andreani, L. and Middleton, M. (2019a). Post-Younger Dryas fault instability and de-formations on ice lineations in Finnish Lapland. *Geomorphology*, **326**, 202–212, doi.org/10.1016/j.geomorph.2018.08.034.

Sutinen, R., Hyvönen, E., Liwata-Kenttälä, P. et al. (2019b). Electrical-sedimentary anisotropy of landforms adjacent to postglacial faults in Lapland. *Geomorphology*, **326**, 213–224, doi.org/10.1016/j.geomorph.2018.01.008.

Sutinen, R., Sutinen, A. and Middleton, M. (2021). Subglacial squeeze-up moraines adjacent to the Vaalajärvi-Ristonmännikkö glacially-induced fault system, Finnish Lapland. *Geomorphology*, **384**, 107716, doi.org/10.1016/j.geomorph.2021.107716.

Ter-Borch, N. (1991). Geological map of Denmark, 1:500,000. Structural map of the top chalk group. *Geological Survey of Denmark Map Series* 7, 4 pp. Copenhagen.

van Balen, R. T., Bakker, M. A. J., Kasse, C. et al. (2019). A Late Glacial surface rupturing earthquake at the Peel Boundary fault zone, Roer Valley Rift System, the Netherlands. *Quaternary Science Reviews*, **218**, 254–266, doi.org/10.1016/j.quascirev.2019.06.033.

Van Vliet-Lanoë, B., Brulhet, J., Combes, P. et al. (2016). Quaternary thermokarst and thermal erosion features in northern France: origin and palaeoenvironments. *Boreas*, **46** (3), 442–461, doi.org/10.1111/bor.12221.

Van Vliet-Lanoë, B., Maygari, A. and Meilliez, F. (2004). Distinguishing between tectonic and periglacial deformations of quaternary continental deposits in Europe. *Global and Planetary Change*, **43**, 103–127, doi.org/10.1016/j.gloplacha.2004.03.003.

Wu, P., Johnston, P. and Lambeck, K. (1999). Postglacial rebound and fault instability in Fennoscandia. *Geophysical Journal International*, **139**, 657–670, doi.org/10.1046/j.1365-246x.1999.00963.x.

4

The Challenge to Distinguish Soft-Sediment Deformation Structures (SSDS) Formed by Glaciotectonic, Periglacial and Seismic Processes in a Formerly Glaciated Area

A Review and Synthesis

KATHARINA MÜLLER, JUTTA WINSEMANN, MAŁGORZATA (GOSIA) PISARSKA-JAMROŻY, THOMAS LEGE, THOMAS SPIES AND CHRISTIAN BRANDES

ABSTRACT

This chapter gives an overview of the use of soft-sediment deformation structures (SSDS) as palaeoearthquake indicators in formerly glaciated and periglacial areas. We review the most important processes of soft-sediment deformation and the various nomenclature used in scientific communities.

In recent years many studies have focused on SSDS to identify past seismic events. So-called seismites are beds with SSDS that formed as a result of seismic shaking. However, in regions affected by glacial and periglacial processes, the use of SSDS as palaeoearthquake indicator is challenging, and interpretation must be done with care. Earthquakes are only one trigger process of many that can cause liquefaction and/or fluidization of sediments, leading to the formation of SSDS such as load casts, flame structures, ball-and-pillow structures, convolute bedding, sand intrusions, dish-and-pillar structures, clastic dykes, sand volcanoes, craters/bowls and gravity induced mass-flows. Ice-sheet loading, glaciotectonism and freeze and thaw processes in glacial and periglacial environments are also potential trigger processes that can cause the formation of similar types of SSDS, which can easily be mistaken for seismites. Therefore, it is important to use clear criteria to recognize seismites in the field. Characteristic features of seismically induced SSDS are: 1) their occurrence close to major faults; 2) their presence in several outcrops in the same stratigraphic interval; 3) their large lateral extent, although high lateral variabilities of the deformation style, pattern and bed thicknesses are possible, depending on the susceptibility of the sediments to liquefaction and/or fluidization; and 4) the occurrence of deformation bands close to the tip line, where fault displacement goes to zero.

The combination of deformation bands that occur in the vicinity of basement faults with carefully evaluated SSDS is a robust indicator for palaeoearthquakes. The results presented in this chapter are transferable to other comparable, seismically active intraplate regions.

4.1 Introduction

Soft-sediment deformation structures (SSDS) are used as indicators for past seismic events (e.g. Montenat et al., 2007; Obermeier, 2009; Tuttle et al., 2019). However, in regions that were frequently affected by ice-sheet loading/unloading and periglacial processes, the use of SSDS for interpreting seismic events is challenging. In these regions glacial and periglacial processes affected the near-surface sediments and led to the formation of SSDS (e.g. Van Vliet-Lanoë et al., 2004; van Loon, 2009; Brandes & Winsemann, 2013; Gehrmann & Harding, 2018) similar to those caused by earthquakes. Also, depositional loading, gravity-induced sediment failure, storm or flood events in fluvial, lacustrine or shallow marine settings or salt tectonics may lead to the formation of SSDS (e.g. Li et al., 1996; Molina et al., 1998; Fossen, 2010; van Loon et al., 2019; Vandekerkhove et al., 2020) similar to those caused by earthquakes. It may therefore be difficult to distinguish the trigger process of the formation of SSDS. In addition, the use of different nomenclatures for similar SSDS (Figure 4.1) in the various scientific communities (e.g. seismology, sedimentology, Quaternary geology, civil engineering) may lead to confusion.

A detailed analysis of SSDS and the depositional environment in which they occur is thus important, and the use of SSDS as indicator for palaeoearthquakes must be done with care to avoid misinterpretation. In this context, formation and differentiation of SSDS and seismites was discussed in several studies (e.g. Van Vliet-Lanoë et al., 2004; Moretti & Sabato, 2007; Obermeier, 2009; van Loon, 2009; Shanmugam, 2016a; Tuttle et al., 2019).

Seismites are represented by beds or bed sets that contain seismically induced SSDS (Moretti et al., 2016). The term 'seismites' was introduced by Seilacher (1969) and referred to sediment beds that were deformed by earthquake-related shaking. These deformed beds occur sandwiched between undeformed beds (e.g. van Loon et al., 2016). The deformation structures are caused by liquefaction and/or fluidization of the sediment, triggered by seismic waves (e.g. Moretti & Sabato, 2007) if the magnitude of an earthquake is high enough ($M \geq 5.0$) (e.g. Atkinson et al., 1984; Rodríguez-Pascua et al., 2000; Obermeier, 2009). An earthquake with the intensity of V to VI can cause liquefaction and fluidization of sediments that occur at up to 40 km from the epicentre, depending on the nature of the sediments (Galli, 2000).

Pleistocene to Holocene seismites related to glacially induced earthquakes are known from e.g. Belgium, Canada, Denmark, Finland, Germany, Latvia, Lithuania, the Netherlands, Poland, Russia, Scotland and the United States (e.g. Davenport et al., 1989; Ringrose, 1989; Hoffmann & Reicherter, 2012; Brandes & Winsemann, 2013; van Loon & Pisarska-Jamroży, 2014; van Loon et al., 2016; Druzhinina et al., 2017; Brandes et al., 2018a; Brooks, 2018; Ojala et al., 2018; Pisarska-Jamroży et al., 2018, 2019a,b; Woźniak & Pisarska-Jamroży, 2018; Grube, 2019a,b; Pisarska-Jamroży & Woźniak, 2019; van Balen et al., 2019; Belzyt et al., 2021).

The most common SSDS triggered by seismic waves are load casts, flame structures, ball-and-pillow structures, convolute bedding and liquefaction spreads (Figure 4.1), which are mainly related to liquefaction (Rydelek & Tuttle, 2004; Obermeier, 2009; van Loon, 2009; van Loon & Pisarska-Jamroży, 2014; Sutinen et al., 2019a,b; Naik et al., 2020; van Loon et al., 2020). Brittle deformation as well as water-escape structures (e.g. dish-and-pillar structures, clastic dykes, sand volcanoes, craters/bowls and hydrofractures) related to fluidization indicate higher pore-water pressure and often occur closer to the potential seismogenic fault (Brandes & Winsemann, 2013; Brandes et al., 2018a) and are caused by high magnitude earthquakes (e.g. Rydelek & Tuttle, 2004; Obermeier, 2009; Naik et al., 2020).

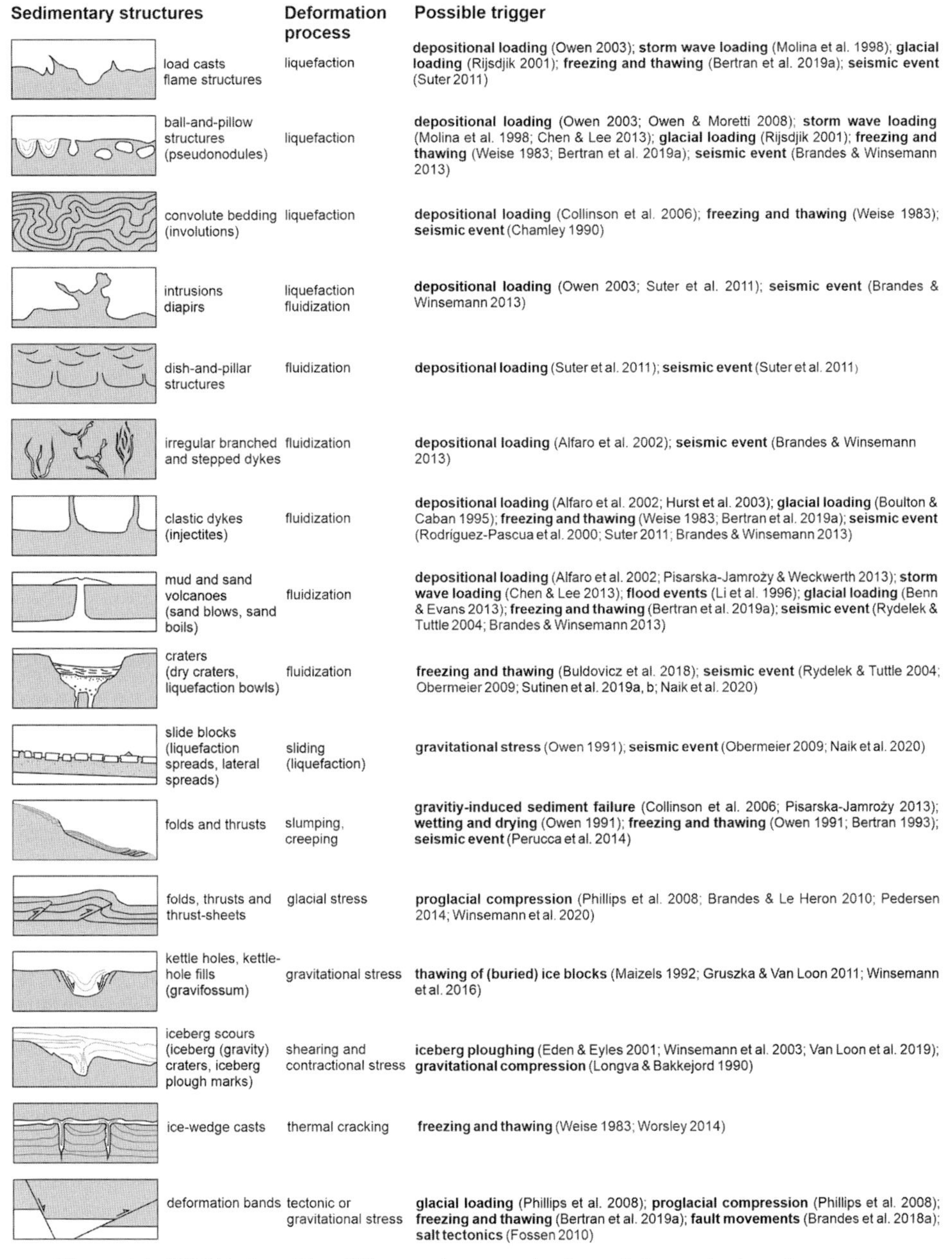

Figure 4.1 SSDS induced by different trigger mechanisms. It has to be kept in mind that SSDS cannot be treated as equivalent to seismites. This figure shows a compilation of common SSDS and does not include all structures and trigger mechanisms that may induce SSDS. The definition of deformation mechanisms (fluidization, liquefaction) is based on Lowe (1975).

This chapter reviews the use of SSDS as palaeoearthquake indicators in periglacial and glacial environments with a focus on northern Central Europe. We describe the most important SSDS, their formation, different nomenclatures and potential trigger processes. Based on this review we deliver criteria to identify a neotectonically active fault based on SSDS. Our synthesis is transferable to other intraplate regions.

4.2 Formation Processes of SSDS

Important factors that control SSDS formation are hydraulic gradient, grain size, permeability, tensile strength and flexural resistance of the sediments (e.g. Mörz et al., 2007; Obermeier, 2009; Giona Bucci et al., 2017, 2019; Pisarska-Jamroży & Woźniak, 2019). Lowe (1975) identified two main processes of fluid escape referred to as liquefaction and fluidization that are responsible for soft-sediment deformation. The main characteristics of these processes are briefly summarized below.

4.2.1 Liquefaction

Liquefaction is the loss of grain-to-grain contact, resulting from increased pore pressure of a static fluid (Frey et al., 2009). During liquefaction a rapid temporary increase in the pore-fluid pressure occurs, leading to a sudden loss of shearing resistance, which is associated with a grain framework collapse and the resulting mobilization and deformation of the liquefied bed. As fluid flow rapidly dissipates upwards, a grain-supported framework is re-established. Reversed density gradients in the liquefied sediments lead to the formation of SSDS (e.g. Lowe, 1975; Rodrígues et al., 2009; Ross et al., 2011). Liquefaction mostly develops in sediments that were buried at less than 5 m (Owen & Moretti, 2011) because higher vertical effective stress caused by the overlying sediment greatly increases the shearing and deformation resistance of the sediment (Obermeier, 2009).

Characteristic SSDS caused by liquefaction are load casts, flame structures, ball-and-pillow structures, convolute bedding (involutions), slides (liquefaction spreads) and slumps (e.g. Collinson et al., 2006; Obermeier, 2009; Sutinen et al., 2019a,b). Several processes can cause liquefaction. These include depositional loading or loading during advancing and/or overriding ice sheets, storm waves, flood events, seismic events and freeze and thaw processes in periglacial environments (e.g. Boulton & Caban, 1995; Li et al., 1996; Alfaro et al., 2002; Van Vliet-Lanoë et al., 2004; Brandes & Winsemann, 2013; van Loon et al., 2016; Bertran et al., 2019a).

4.2.2 Fluidization

Fluidization is a process that requires pore fluid to move upwards with sufficient velocity to suspend or carry individual grains with it. Therefore, fluidization often needs an additional fluid source (Van Vliet-Lanoë et al., 2004; Shanmugam, 2016a).

The increased pore pressure leads to the expansion of the pore spaces until grain interaction is negligible and particles are free to move with the fluid (e.g. Lowe, 1975;

Nichols et al., 1994; Frey et al., 2009). The fluid tends to follow preferred escape paths (e.g. faults, heterogeneities in sediment properties and layer thicknesses), while the surrounding sediment remains largely unfluidized (Mörz et al., 2007; Frey et al., 2009; Ross et al., 2011). The primary lamination is destroyed in fluidized zones, and fluid-escape structures develop that cross-cut the primary bedding, accompanied by the injection of sand or mud into overlying layers (Lowe, 1975; Hurst & Cartwright, 2007; Frey et al., 2009; Hurst et al., 2011).

Characteristic SSDS that form due to fluidization are mud or sand intrusions, dish-and-pillar structures, clastic dykes, sand or mud volcanoes and craters/bowls (e.g. Collinson et al., 2006; Sutinen et al., 2019b). Sand volcanoes and craters in terrestrial sediments are commonly regarded as highly diagnostic features for palaeoearthquakes (e.g. Obermeier, 2009; Sutinen et al., 2019b). However, fluidization can also be caused by different processes such as depositional loading, freeze and thaw cycles in periglacial environments (Lowe, 1975; Brandes & Winsemann, 2013; Vandenberghe, 2013; Bertran et al., 2019a), river floods or artesian groundwater rise (e.g. Deynoux et al., 1990; Li et al., 1996; Obermeier, 2009).

4.3 Common Trigger Processes and Timing of SSDS Formation

As shown above, different processes can cause the formation of SSDS.

The timing of soft-sediment deformation provides an important criterion for the genetic interpretation of SSDS. Based on the style of faulting/folding or the style of truncation of bedding/lamination within the deformed sediment, 1) syndepositional, 2) metadepositional and 3) postdepositional deformation can be distinguished (Figure 4.2). Syndepositional soft-sediment deformation forms during the deposition of the sediment; metadepositional deformation occurs after deposition but before the overlying sediments were deposited; and postdepositional deformation takes place after the sediments have been deposited (e.g. Collinson et al., 2006; van Loon, 2009).

4.3.1 Earthquakes

Earthquakes can induce many different liquefaction and fluidization structures that vary in size, morphology and deformation style, depending on the earthquake magnitude and the strength of the sediment (e.g. Rodríguez-Pascua et al., 2000; Obermeier, 2009; Giona Bucci et al., 2017, 2019). In an ideal case, the formation of liquefaction or fluidization structures can be directly tied to earthquakes, as shown e.g. in the studies of Giona Bucci et al. (2017, 2019) for the 2010–2012 Canterbury earthquake sequence (M_w 5.8 to 7.1) and the 2016 Valentine earthquake (M_w 5.7) in New Zealand.

During earthquakes the applied shear stress and accumulation of shear strain can cause a breakdown of the grain framework and an increase of the pore-water pressure, which results in liquefaction and/or fluidization of unconsolidated sediments (e.g. Obermeier et al., 2005; Obermeier, 2009). The sedimentary microfabric of SSDS is similar for different depositional environments (e.g. Giona Bucci et al., 2019).

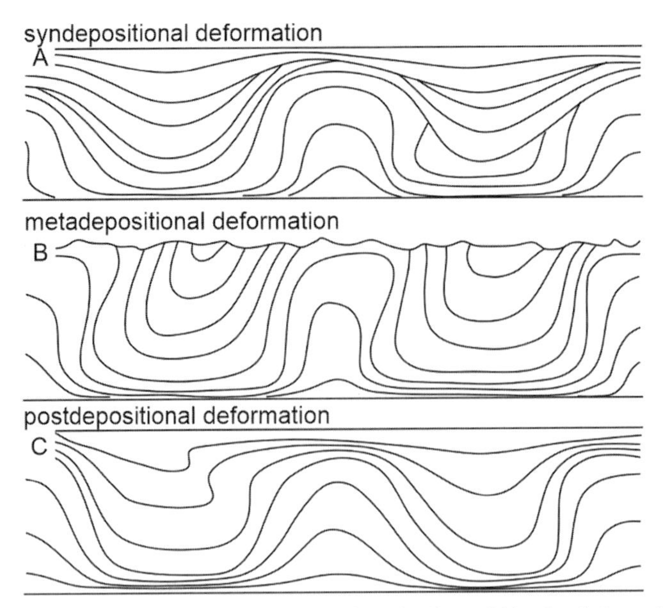

Figure 4.2 Based on the style of truncation of lamination within the deformed sediment (convolute bedding) the timing of soft-sediment deformation can be derived: A) syndepositional deformation occurs during the deposition of the sediment; B) metadepositional deformation occurs after deposition but before the overlying sediment was deposited; C) postdepositional deformation occurs after the sediments have been deposited (modified after Collinson et al., 2006).

Besides the standard SSDS that are described from outcrops and artificial trenches, seismic events can cause mass flows such as slides and slumps that may have a large lateral extent in the range of hundreds to thousands of metres. So-called liquefaction-induced lateral spreading (liquefaction spreads) occurs when unsaturated sediments slide as intact blocks over liquefied sediment. These slide processes can occur on very gentle slopes or even relatively flat terrains (inclined as gently as 0.1–5 per cent) (Obermeier, 2009; Hungr et al., 2014; Sutinen, 2019a,b). The internal structures may include fissures, scarps and grabens. On steeper slopes, slumps and debris flows can develop, which may transition into turbidity currents in subaqueous settings (e.g. Shanmugam, 2016b; Vandekerkhove et al., 2020).

Brittle deformation and fluidization of sediment may occur in settings with higher pore-water pressure and often indicate higher magnitude earthquakes and a closer location to the potential seismogenic fault (Rodríguez-Pascua et al., 2000; Brandes & Winsemann, 2013; Brandes et al., 2018a).

However, these features may not always indicate a higher magnitude or a closer location to the earthquake epicentre and instead may also reflect a higher susceptibility of the sediment for deformation (Giona Bucci et al., 2017; Morsilli et al., 2020).

Longer-lived (glacial) lakes may provide excellent archives for reconstructing palaeoearthquakes, based on well-dated flood or mass-failure events related to earthquakes that occur contemporaneously in different lakes (e.g. Monecke et al., 2006; Strasser et al.,

2006; Vandekerkhove et al., 2020). Holocene earthquakes triggered by the glacial rebound of Fennoscandia are recorded in the form of SSDS in varves that were deposited in a glacial lake in Finland (Ojala et al., 2018).

In recent years SSDS have been used to identify palaeoearthquakes related to glacial isostatic adjustment (GIA). Hoffmann and Reicherter (2012) analyzed SSDS in Late Pleistocene meltwater deposits, which are interpreted to have been induced by earthquakes at the south-west Baltic Sea coast in north-east Germany. Brandes et al. (2018a) reconstructed strong Lateglacial earthquakes along the north-west segment of the Sorgenfrei–Tornquist Zone in northern Denmark. Brandes et al. (2012) and Brandes and Winsemann (2013) found evidence for palaeoearthquakes in Lateglacial sediments near the Osning Thrust, where a large variety of SSDS are developed in mixed alluvial-aeolian sediments (Meinsen et al., 2014). Several studies in the Lower Rhine Rift and the Roer Valley Rift present evidence of Pleistocene to Holocene palaeoearthquakes and higher slip rates at several faults during the Quaternary, and they interpreted glacial rebound as a possible trigger (e.g. Vanneste et al., 1999; Houtgast et al., 2005; Vanneste et al., 2018; van Balen et al., 2019). van Loon and Pisarska-Jamroży (2014), Pisarska-Jamroży and Woźniak (2019) reported evidence for palaeoearthquakes in north-west and northern Poland, and van Loon et al. (2016) and Pisarska-Jamroży et al. (2019a) found evidence for palaeoearthquakes in Latvia and Lithuania. However, more recently Pisarska-Jamroży et al. (2019a) stated that these structures alternatively may have formed by large-scale stick-slip processes during ice-sheet motion.

4.3.2 Depositional Loading

The most common process that creates SSDS is depositional loading in water-saturated sediments, which is commonly related to high sedimentation rates (e.g. Lowe, 1975; Moretti et al., 2001; Oliveira et al., 2009). These sediments may include all styles of liquefaction and fluidization structures (Figure 4.1) if the overburden is thick enough. In (glacio)lacustrine settings high sedimentation rates may cause the formation of synsedimentary faults and liquefaction structures. Flood events, high-magnitude lake-level fluctuations or strong wave action may also generate mass failure and liquefaction structures (Monecke et al., 2006; Vandekerkhove et al., 2020).

4.3.3 Gravity-Driven Sediment Failure

Gravity-driven sediment failure occurs on slopes and can be induced by different processes such as depositional loading, wetting and drying, freezing and thawing (Owen, 1991) or earthquakes (e.g. Monecke et al., 2006; Obermeier, 2009; Sutinen et al., 2019b). Climate, hydrology, vegetation, geology and topography define the rates and types of downslope displacement (Matsuoka, 2001).

Mass-flow processes include creeping, sliding and slumping, and the resulting mass-flow deposits are distinguished by the degree of their internal deformation.

Creep is a granular flow with slow intergranular frictional sliding downslope with quasi-static grain contacts driven by gravity (Owen, 1991). In periglacial environments these downslope movements are referred to as solifluction or gelifluction and are caused mainly by freeze and thaw processes in the so-called active layer (e.g. French, 2017). Typical structures that develop due to these processes are thrusts and fold structures (e.g. Bertran, 1993).

Slides are translational or rotational coherent masses with a commonly planar glide plane and with minor or no internal deformation. Slumps often have concave glide planes and considerable internal deformation such as folds and thrusts (Owen, 1991; Shanmugam, 2016a). Slides and slumps may transform downslope into debris-flows, and in subaqueous settings they may transform further into turbidity currents (Shanmugam, 2016b).

4.3.4 Glaciotectonic Deformation

Glaciotectonic deformation structures are induced by the advance of ice sheets (Figure 4.3A,B). In recent decades, the processes and products of glaciotectonic deformation have been studied in detail (e.g. Van der Wateren et al., 2000; Bennett et al., 2004; Phillips et al., 2008; Pedersen, 2014; Woronko et al., 2018; Winsemann et al., 2020). Glaciotectonic deformation includes: (1) proglacial contractional structures, formed at the margin of an ice sheet and (2) subglacial predominantly extensional and shearing related deformation beneath the ice sheet (Hart & Boulton, 1991). Different deformation structures commonly develop due to stress field variations caused by the advancing ice sheet. At the margin of large ice sheets, SSDS can develop due to high pore-water pressure in front or at the toe of the ice sheet (Boulton & Caban, 1995).

4.3.4.1 Subglacial Deformation

Subglacial deformation is highly variable in style and intensity and can often result in normal faulting and the formation of horst and graben structures (Figure 4.3A), heterogeneous folds and SSDS, which are associated with subglacial shear zones (e.g. Åmark, 1986; Piotrowski et al., 2004; Phillips et al., 2008). The initial water (or ice) content of the deformed sediments controls the pattern of deformation within the shear zone (Lee & Phillips, 2008; Phillips et al., 2008; Szuman et al., 2013; Kowalski et al., 2018). Subglacial deformation can also affect the bedrock (e.g. Kenzler et al., 2010; Gehrmann & Harding, 2018; Winsemann et al., 2020).

4.3.4.2 Proglacial Deformation

Proglacial deformation leads to the formation of large-scale contractional structures such as folds, reverse and thrust faults (Figure 4.3B,C, Aber & Ber, 2007; Phillips et al., 2008; Pedersen, 2014; Gehrmann & Harding, 2018). Depending on the rheology, competence, strain rate and strain history of the sediments and the behaviour of the ice sheet, folding (ductile deformation) or faulting (brittle deformation) may occur (Hart & Boulton, 1991; Brandes & Le Heron, 2010).

Iceberg Scours, Iceberg Gravity Craters and Kettle Holes

Iceberg scours, iceberg gravity craters and kettle holes are common features in proglacial areas. The keel of a floating iceberg creates a curvilinear scour by ploughing through the substrate. The plough mark is usually preserved as a furrow, bounded by normal faults, and a frontal ridge of scoured material on either side (Longva & Bakkejord, 1990; Eden & Eyles, 2001; Winsemann et al., 2003; van Loon et al., 2019). In cross-section the sheared scour fill may resemble sand volcanoes or clastic dykes (cf. Winsemann et al., 2003). Iceberg gravity craters form when icebergs strand and get stuck into the sediment. The resulting structures are semicircular depressions, rimmed by low ridges (e.g. Longva & Bakkejord, 1990).

Kettle holes form by the melt-out of (buried) isolated blocks of glacier ice and commonly consist of near-circular depressions, which may be bounded by steep normal faults (Maizels, 1992). The kettle-hole fill may be characterized by a pronounced downwarping of strata into the central depression, showing small scale deformation structures such as folds and faults (e.g. Gruszka & van Loon, 2011; Winsemann et al., 2016).

4.3.5 Periglacial Processes

The periglacial environment is the marginal zone of an ice sheet that is not directly influenced by the glacier but characterized by permafrost conditions (Figure 4.3C–G; French, 2017). Cryoturbation is a widely used term for different deformation structures that develop in unconsolidated sediments under periglacial conditions. It is related to seasonal freeze and thaw processes in the so-called active layer (e.g. Dobiński, 2011; French, 2017).

Processes that produce cryoturbation structures are shrinkage, frost heave pressure or swelling and differential loading, which cause vertical grain movement or gravity-induced lateral mass movements (Bockheim & Tarnocai, 1998; Van Vliet-Lanoë et al., 2004; Ogino & Matsuoka, 2007). Buldovicz et al. (2018) and Bertran et al. (2019a) showed that SSDS may be caused by the formation of ground ice and a related pore-water overpressure in underlying or overlying sediments if water is available. Therefore, the most important controlling factors for the type and abundance of cryoturbation structures are moisture content, thermal gradient and grain size of the sediments (e.g. Van Vliet-Lanoë et al., 2004).

4.3.5.1 Ice Wedges and Ice-Wedge Casts

Ice wedges are wedge-shaped bodies with their apex pointing downwards; they are composed of foliated or vertically banded ice (Harry, 1988). Ice wedges develop in thermal contraction cracks, in which hoar frost forms and water from melting snow percolates into the near-surface sediments (Weise, 1983).

Repeated annual contraction cracking of the ice in the wedge, followed by freezing of water in the crack, leads to a gradual increase in the width and depth of the wedge and causes vertical banding of the ice mass (Harry, 1988; Collinson et al., 2006). The size of ice wedges typically ranges 1–3 m in width and 2–6 m in depth (Worsley, 2014).

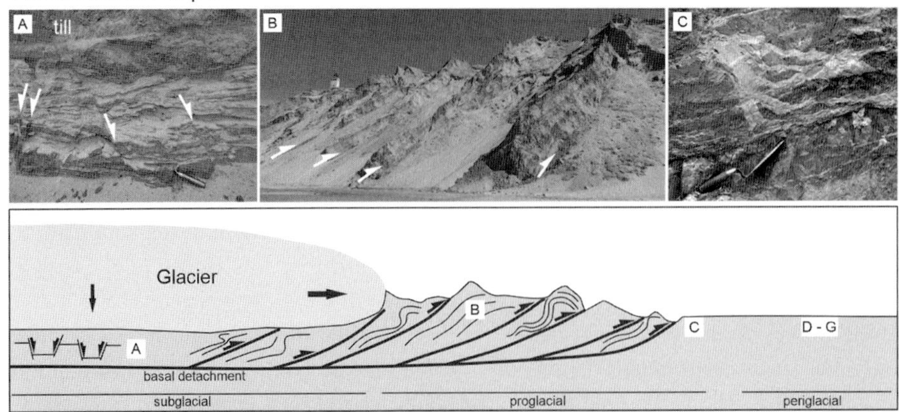

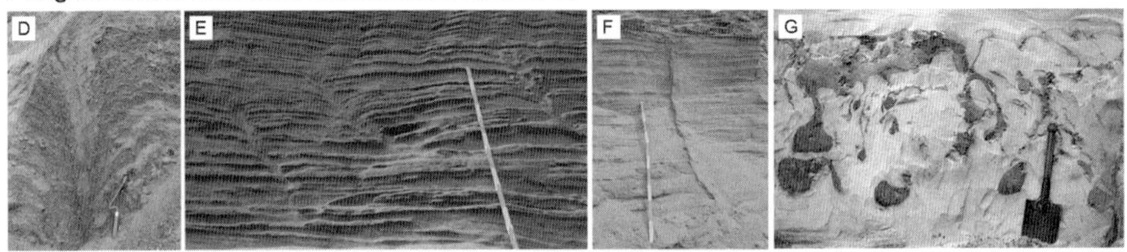

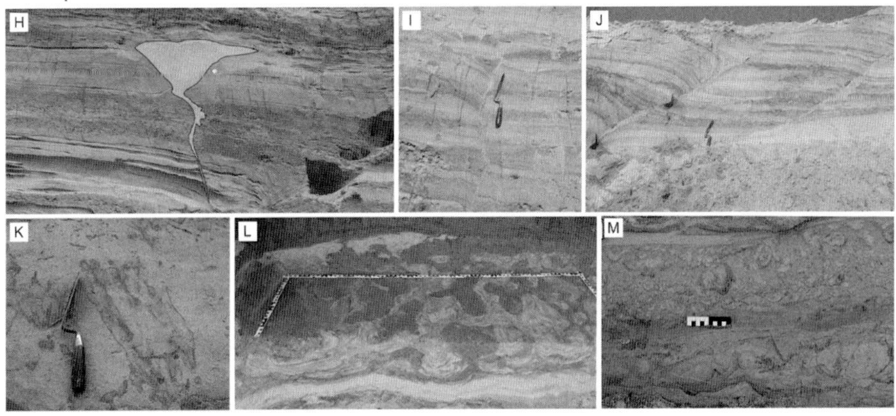

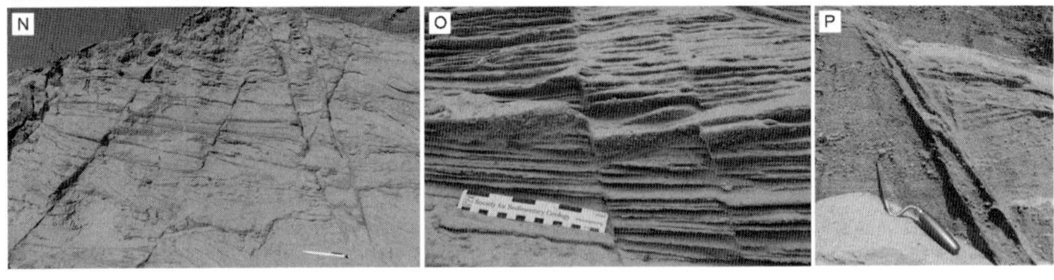

Figure 4.3 Typical SSDS formed by glaciotectonics, periglacial processes and earthquakes. A–C) Deformation structures related to glaciotectonic complexes (based on Aber & Ber, 2007). A) Normal faults in meltwater sand, overlain by till (Gardelegen, Germany). The faults

An ice-wedge cast is the fill in the space formerly occupied by ice (Figure 4.3D–F). Ice-wedge casts are a widespread indicator of past periglacial conditions and permafrost.

4.3.5.2 Periglacial Involutions and Craters

Involutions and convolute bedding describe comparable structures, reported from past and present periglacial environments with permafrost or deep seasonal frost (Figure 4.3G; Ogino & Matsuoka, 2007; Vandenberghe, 2013). These deformation structures include folding of a pre-existing lamination, commonly into upright cuspate forms with sharp anticlines and more gentle synclines (Collinson et al., 2006) or more swirl-like or tear-drop patterns (Bockheim & Tarnocai, 1998; Pisarska-Jamroży & Zieliński, 2012).

In the literature the term 'involution' is connected with cryoturbation. Involutions that occur in seasonal frost regions show smaller vertical amplitudes (0.6 m) than involutions with up to 2-m amplitudes that developed in permafrost regions (Vandenberghe, 2013; French, 2017). However, the size of the involutions can also differ with the lithology and the forming process (Ogino and Matsuoka, 2007).

Craters may form by collapsing pingos, also referred to as cryovolcanism. In a recent study, Buldovicz et al. (2018) showed that pingos in Siberia collapsed under cryogenic hydrostatic pressure build-up in the closed system of a freezing talik. This happened before the freezing was complete, when a core of wet sediment remained unfrozen and stored a large amount of carbon dioxide dissolved in pore water. When gas-phase saturation was reached, the resulting overpressure led to an explosion of the pingos.

Figure 4.3 (*cont.*) were formed by subglacial deformation. B) Thrust-sheets in glaciolacustrine deposits (Lønstrup Klint, Denmark) formed by proglacial deformation. C) Hydrofractures in silt. These structures were most probably caused by high pore-water pressure in front of an ice sheet (Lønstrup Klint, Denmark). D–G) Cryoturbation features that formed by freeze and thaw processes in periglacial settings. D) Ice-wedge cast in glaciolacustrine deposits (Bolsehle, Germany). E) Three ice-wedge casts in alluvial-fan deposits (Dresden, Germany). F) Thin ice-wedge cast in meltwater sand (Groß Eilstorf, Germany). G) Involutions in meltwater deposits and till (Stukenbrock, Germany, courtesy of K. Skupien). H–M) Earthquake-induced SSDS. H) Sand volcano in mixed alluvial-aeolian sediment (Oerlinghausen, Germany; Brandes & Winsemann, 2013). I) Clastic dykes in mixed alluvial-aeolian sediment (Oerlinghausen, Germany; Brandes et al., 2012; Brandes & Winsemann, 2013). J) Inversion structures with a typical harpoon shape in mixed alluvial-aeolian sediments (Oerlinghausen, Germany; Brandes & Winsemann, 2013). K) Irregular branched and stepped dykes in mixed alluvial-aeolian sediment (Augustdorf, Germany; Brandes & Winsemann, 2013). L) Load casts, pseudonodules and flame structures in glaciolacustrine deposits (Dwasieden, Germany; Pisarska-Jamroży et al., 2018). M) Load casts, flame structures and water escape structures in lacustrine deposits (Baltmuiža, Latvia; Belzyt et al., 2018). N–P) Deformation bands. N) Set of conjugate deformation bands with normal displacement in ice-marginal delta slope sediments (Altenhagen, Germany). O) Two thin deformation bands with normal displacement in meltwater sand (Groß Eilstorf, Germany). P) Deformation bands in ice-marginal delta slope sediment (Freden, Germany). (A black and white version of this figure will appear in some formats. For the colour version, please refer to the plate section.)

4.4 Limitation of SSDS to Identify Palaeoearthquakes

As shown above, SSDS can be used to indicate past seismic events, and in an ideal case they can directly indicate palaeoearthquakes.

However, it must be kept in mind that SSDS can also be induced by a set of different non-seismic processes as discussed in Section 4.3. (Figures 4.1 and 4.3A–G), and it is often challenging to derive the related triggers, especially in areas that were affected by glacial and periglacial processes. Moreover, not all earthquakes trigger liquefaction and fluidization processes. SSDS are likely to be formed only in susceptible sediments (Figure 4.3H–M). They often develop in fine- to medium-grained water-saturated sand with a loose grain packing (high porosity). Barriers with low permeability such as mud laminae or matrix-rich impermeable beds that support the increase in pore-water pressure must be present. The susceptibility of the sediment is therefore the most important requirement for the formation of seismically triggered liquefaction and/or fluidization features (Rodríguez-Pascua et al., 2000; Giona Bucci et al., 2017, 2019; Morsilli et al., 2020). Thus, the most distinct liquefaction and fluidization features may not necessarily be located close to the earthquake epicentre. Instead their distribution may reflect the grain size, sorting and cohesion of the sediments and the position of the groundwater table (Giona Bucci et al., 2017; Morsilli et al., 2020).

To ensure a careful application of SSDS in palaeoseismological studies, various criteria and schemes were proposed, which in general are quite similar (e.g. Wheeler 2002; Van Vliet-Lanoë et al., 2004, Obermeier, 2009; Owen & Moretti, 2011; van Loon et al., 2016; Morsilli et al., 2020).

To identify seismically induced SSDS, Wheeler (2002) introduced a test scheme that involves the following criteria: 1) sudden formation, 2) synchroneity, 3) zoned distribution over several outcrops, 4) size, 5) tectonic setting and 6) depositional setting.

Owen and Moretti (2011) defined six criteria to ensure identification of seismites in the field: 1) large areal extent, 2) lateral continuity, 3) vertical repetition of beds with SSDS, 4) comparable morphology of SSDS with deformation structures described from earthquakes, 5) proximity to active faults and 6) complexity and frequency of SSDS as dependent on the distance to the triggering fault.

According to van Loon et al. (2016) the most important criterion for recognizing seismites is the vertical repetition of beds with SSDS. However, Morsilli et al. (2020) reported that the number of deformed beds and their main physical character (size of deformation, morphologies and thicknesses of the deformed intervals) may change laterally.

So, many applied criteria are not diagnostic and do not rule out other trigger mechanisms (Owen & Moretti, 2011; Brandes & Winsemann, 2013). A vertical repetition of beds with SSDS in (glacial) lakes (e.g. Monecke et al., 2006), often regarded as diagnostic for seismic events, can also be the result of major (meltwater) flood events or retrogressive slides leading to the rapid deposition of thicker beds. In periglacial environments, for example, seasonal freeze and thaw cycles can lead to the formation of vertically stacked beds with SSDS, which are produced by loading or diapirism (Vandenberghe,

2013). In non-glacigenic settings, fast depositional loading of water-saturated sediments may lead to the formation of all styles of liquefaction and fluidization structures (Figure 4.1) if the overburden and pore pressure is high enough (Owen, 2003; Rodrígues et al., 2009). These SSDS may also appear in different stratigraphic levels but are not the result of an earthquake. However, many of the criteria described above are difficult to evaluate. It remains a challenge to identify a large areal extent of seismites because outcrops are often isolated or restricted to sand and gravel pits.

The limitations in the use of SSDS as palaeoearthquake indicators have been recognized by several authors. Montenat et al. (2007) pointed out that SSDS are not always univocal and their interpretation has to be placed in the geological context. Shanmugam (2016a) summarized that earthquakes are only one of 21 mechanisms that can cause liquefaction and/or fluidization. Consequently, the use of SSDS as indicators for palaeoseismicity is controversial (e.g. Bertran et al., 2019b).

4.5 Deformation Bands as Indicators for Neotectonic Fault Activity

The inconclusive nature of SSDS requires a more robust indicator for neotectonic activity. The work of Cashman et al. (2007), Brandes and Tanner (2012), Shipton et al. (2017) and Brandes et al. (2018a,b) showed that near-surface deformation bands in unconsolidated sediments are such an indicator for neotectonic activity at basement faults. Deformation bands are tabular zones of local deformation, which can occur in unconsolidated sandy sediments and sedimentary rocks (Ballas et al., 2015). They form in porous material (Figure 4.3N–P) and represent equivalents to faults that usually form in nonporous rocks. Recent studies show that near-surface deformation bands, which formed in unconsolidated sediments, can develop due to tectonic activity along large faults (Cashman et al., 2007; Brandes et al., 2018a,b). Especially, shear-deformation bands are a relevant tool in palaeo-seismological and neotectonic studies and, if developed in young, unconsolidated sediments, can serve as a strong indicator for recently active faults. Deformation bands that formed in a glaciotectonic environment can show a spread in strike direction, whereas those formed by neotectonics follow the strike of the regional basement faults (Brandes et al., 2018a,b). In the latter case the deformation processes were caused by fault propagation and reflect the fault-related deformation above the tip line of reactivated basement faults (Brandes et al., 2018b). Deformation bands might also be formed by cryoturbation processes. In such a case they reflect the local near-surface extension and contraction of the sediments related to freeze and thaw processes (Bertran et al., 2019a).

Cataclastic deformation bands are an indicator for fault rupture processes (Cashman et al., 2007). The non-cataclastic deformation bands shown by Brandes and Tanner (2012) and Brandes et al. (2018b) are not indicators for palaeoearthquakes. Nevertheless, their orientation follows the strike of underlying basement faults, thus indicating a close relation. We argue that deformation bands in combination with (carefully evaluated) SSDS are the most reliable indicator for palaeoearthquakes if the deformation bands follow the strike of the nearby fault and the SSDS match the criteria of Wheeler et al. (2002) and Owen and

Moretti (2011). The deformation bands reflect the fault activity, and the SSDS indicate the passage of seismic waves.

4.6 Conclusions

Different processes that include depositional loading, flood and storm events, ice loading and proglacial compression, iceberg ploughing, freeze and thaw processes, mass failure and salt tectonics can produce SSDS that may be mistaken for earthquake-induced deformation structures. Therefore, the use of SSDS as indicators for palaeoearthquakes in intraplate areas (characterized by rather low magnitudes and long earthquake recurrence intervals) that were affected by glaciotectonic deformation and periglacial processes is difficult and has to be done with care to ensure a correct application.

Nevertheless, SSDS can deliver valuable hints to palaeoseismology if they 1) are developed close to major faults; 2) are exposed in several outcrops in the same stratigraphic interval; 3) show a large lateral extent, although high lateral variabilities of the deformation style, pattern and bed thicknesses are possible, depending on the susceptibility of the sediments for liquefaction and/or fluidization; and 4) are associated with deformation bands that have the same strike direction as the regional faults.

Acknowledgements

We thank Gösta Hoffmann and Ronald van Balen for their constructive reviews, which helped improve the manuscript. The work has been financially supported by the project Paläoseismische Untersuchung Norddeutschlands from the BGR (No. 201-10079313) and GREBAL from the National Science Centre Poland (No. 2015/19/B/ST10/00661).

References

Aber, J. S. and Ber, A. (2007). *Glaciotectonism. Developments in Quaternary Science, Vol. 6*. Elsevier, Amsterdam.

Alfaro, P., Delgado, J., Estévez, A. et al. (2002). Liquefaction and fluidization structures in Messinian storm deposits (Bajo Segura Basin, Betic Cordillera, southern Spain). *International Journal of Earth Sciences*, **91**, 505–513, doi.org/10.1007/s00531-001-0241-z.

Åmark, M. (1986). Clastic dikes formed beneath an active glacier. *Geologiska Foereningen i Stockholm Förhandlingar*, **108**, 13–20, doi.org/10.1080/11035898609453740.

Atkinson, G. M., Finn, W. L. and Charlwood, R. G. (1984). Simple computation of liquefaction probability for seismic hazard applications. *Earthquake Spectra*, **1**, 107–123, doi.org/10.1193/1.1585259.

Ballas, G., Fossen, H. and Soliva, R. (2015). Factors controlling permeability of cataclastic deformation bands and faults in porous sandstone reservoirs. *Journal of Structural Geology*, **76**, 1–21, doi.org/10.1016/j.jsg.2015.03.013.

Belzyt, S., Nartišs, M., Pisarska-Jamroży, M., Woronko, B. and Bitinas, A. (2018). Large-scale glaciotectonically-deformed Pleistocene sediments with deformed layers sandwiched between undeformed layers, Baltmuiža site, Western Latvia. In M. Pisarska-Jamroży and A. Bitinas, eds., *Soft-Sediment Deformation Structures and Palaeoseismic Phenomena in the South-Eastern Baltic Region. Excursion Guide of International Palaeoseismological Field Workshop, 17–21st September 2018, Vilnius, Lithuania*. Lithuanian Geological Survey, Lithuanian Geological Society, pp. 38–42.

Belzyt, S., Pisarska-Jamroży, M., Bitinas, A. et al. (2021). Repetitive Late Pleistocene soft-sediment deformation by seismicity-induced liquefaction in north-western Lithuania. *Sedimentology*, doi.org/10.1111/sed.12883.

Benn, D. I. and Evans, D. J. A. (2013). *Glaciers and Glaciation*, 2nd ed. Routledge, New York.

Bennett, M. R., Huddart, D., Waller, R. I. et al. (2004). Styles of ice-marginal deformation at Hagafellsjökull–Eystri, Iceland during the 1998/99 winter-spring surge. *Boreas*, **33**, 97–107, doi.org/10.1111/j.1502-3885.2004.tb01132.x.

Bertran, P. (1993). Deformation-induced microstructures in soils affected by mass movements. *Earth Surface Processes and Landforms*, **18**, 645–660, doi.org/10.1002/esp.3290180707.

Bertran, P., Font, M., Giret, A., Manchuel, K. and Sicilia, D. (2019a). Experimental soft-sediment deformation caused by fluidization and intrusive ice melt in sand. *Sedimentology*, **66**(3), 1102–1117, doi.org/10.1111/sed.12537.

Bertran, P., Manchuel, K. and Sicilia, D. (2019b). Discussion on 'Palaeoseismic structures in Quaternary sediments, related to an assumed fault zone north of the Permian Peissen–Gnutz salt structure (NW Germany) – neotectonic activity and earthquakes from the Saalian to the Holocene' (Grube, 2019). *Geomorphology*, **365**, 106704, doi.org/10.1016/j.geomorph.2019.03.010.

Bockheim, J. G. and Tarnocai, C. (1998). Recognition of cryoturbation for classifying permafrost-affected soils. *Geoderma*, **81**, 281–293, doi.org/10.1016/S0016-7061(97)00115-8.

Boulton, G. S. and Caban, P. (1995). Groundwater flow beneath ice sheets: part II – its impact on glacier tectonic structures and moraine formation. *Quaternary Science Reviews*, **14**, 563–587, doi.org/10.1016/0277-3791(95)00058-W.

Brandes, C. and Le Heron, D. P. (2010). The glaciotectonic deformation of Quaternary sediments by fault-propagation folding. *Proceedings of the Geologists' Association*, **121**, 270–280, doi.org/10.1016/j.pgeola.2010.03.001.

Brandes, C. and Tanner, D. C. (2012). Three-dimensional geometry and fabric of shear deformation-bands in unconsolidated Pleistocene sediments. *Tectonophysics*, **518–521**, 84–92, doi.org/10.1016/j.tecto.2011.11.012.

Brandes, C. and Winsemann, J. (2013). Soft-sediment deformation structures in NW Germany caused by Late Pleistocene seismicity. *International Journal of Earth Sciences*, **102**, 2255–2274, doi.org/10.1007/s00531-013-0914-4.

Brandes, C., Winsemann, J., Roskosch, J. et al. (2012). Activity along the Osning Thrust in Central Europe during the Lateglacial: ice-sheet and lithosphere interactions. *Quaternary Science Reviews*, **38**, 49–62, doi.org/10.1016/j.quascirev.2012.01.021.

Brandes, C., Steffen, H., Sandersen, P. B. E., Wu, P. and Winsemann, J. (2018a). Glacially induced faulting along the NW segment of the Sorgenfrei-Tornquist Zone, northern Denmark: implications for neotectonics and Lateglacial fault-bound basin formation. *Quaternary Science Reviews*, **189**, 149–168, doi.org/10.1016/j.quascirev.2018.03.036.

Brandes, C., Igel, J., Loewer, M. et al. (2018b). Visualisation and analysis of shear-deformation bands in unconsolidated Pleistocene sand using ground-penetrating radar: implications for paleoseismological studies. *Sedimentary Geology*, **367**, 135–145, doi.org/10.1016/j.sedgeo.2018.02.005.

Brooks, G. R. (2018). Deglacial record of palaeoearthquakes interpreted from mass transport deposits at three lakes near Rouyn–Noranda, north-western Quebec, Canada. *Sedimentology*, **65**, 2439–2467, doi.org/10.1111/sed.12473.

Buldovicz, S. N., Khilimonyuk, V. Z., Bychkov, A. Y. et al. (2018). Cryovolcanism on the Earth: origin of a spectacular crater in the Yamal Peninsula (Russia). *Scientific Reports*, **8**, 13534, doi.org/10.1038/s41598-018-31858-9.

Cashman, S. M., Baldwin, J. N., Cashman, K. V., Swanson, K. and Crawford, R. (2007). Microstructures developed by coseismic and aseismic faulting in near-surface sediments, San Andreas Fault, California. *Geology*, **35**, 611–614, doi.org/10.1130/G23545A.1.

Chamley, H. (1990). *Sedimentology*. Springer, Berlin/Heidelberg.

Chen, J. and Lee, H. S. (2013). Soft-sediment deformation structures in Cambrian siliciclastic and carbonate storm deposits (Shandong Province, China): differential liquefaction and fluidization triggered by storm-wave loading. *Sedimentary Geology*, **288**, 81–94, doi.org/10.1016/j.sedgeo.2013.02.001.

Collinson, J. D., Mountney, N. P. and Thompson, D. B. (2006). *Sedimentary Structures*, 3rd ed., Terra Publishing, England.

Davenport, C. A., Ringrose, P. S., Becker, A., Hancock, P. and Fenton, C. (1989). Geological investigations of late and post glacial earthquake activity in Scotland. In S. Gregersen and P. W. Basham, eds., *Earthquakes at North-Atlantic Passive Margins: Neotectonics and Postglacial Rebound. NATO ASI Series 266*. Springer, Dordrecht, pp. 175–194.

Deynoux, M., Proust, J. N., Durand, J. and Merino, E. (1990). Water-transfer cylindrical structures in the Late Proterozoic eolian sandstones in the Taoudeni Basin, West Africa. *Sedimentary Geology*, **66**, 227–242, doi.org/10.1016/0037-0738(90)90061-W.

Dobiński, W. (2011). Permafrost. *Earth-Science Reviews*, **108**, 158–169, doi.org/10.1016/j.earscirev.2011.06.007.

Druzhinina, O., Bitinas, A., Molodkov, A. and Kolesnik, T. (2017). Palaeoseismic deformations in the Eastern Baltic region (Kaliningrad District of Russia). *Estonian Journal of Earth Sciences*, **66**, 119–129, doi.org/10.3176/earth.2017.09.

Eden, D. J. and Eyles, N. (2001). Description and numerical model of Pleistocene iceberg scours and ice-keel turbated facies at Toronto, Canada. *Sedimentology*, **48**, 1079–1102, doi.org/10.1046/j.1365-3091.2001.00409.x.

Fossen, H. (2010). Deformation bands formed during soft-sediment deformation: observations from SE Utah. *Marine and Petroleum Geology*, **27**, 215–222, doi.org/10.1016/j.marpetgeo.2009.06.005.

French, H. M. (2017). *The Periglacial Environment*, 4th ed., John Wiley and Sons, Chichester.

Frey, S. E., Gingras, M. K. and Dashtgard, S. E. (2009). Experimental studies of gas-escape and water-escape structures: mechanisms and morphologies. *Journal of Sedimentological Research*, **79**(11), 808–816, doi.org/10.2110/jsr.2009.087.

Galli, P. (2000). New empirical relationships between magnitude and distance for liquefaction. *Tectonophysics*, **324**, 169–187, doi.org/10.1016/S0040–1951(00)00118-9.

Gehrmann, A. and Harding, C. (2018). Geomorphological mapping and spatial analyses of an Upper Weichselian glacitectonic complex based on LiDAR data, Jasmund Peninsula (NE Rügen), Germany. *Geosciences*, **8**(6), 208, doi.org/10.3390/geosciences8060208.

Giona Bucci, M., Almond, P., Villamor, P. et al. (2017). When the earth blisters: exploring recurrent liquefaction features in the coastal system of Christchurch, New Zealand. *Terra Nova*, **29**, 162–172, doi.org/10.1111/ter.12259.

Giona Bucci, M., Smith, C. M., Almond, P. C., Villamor, P. and Tuttle, M. P. (2019). Micromorphological analysis of liquefaction features in alluvial and coastal environments of Christchurch, New Zealand. *Sedimentology*, **66**, 963–982, doi.org/10.1111/sed.12526.

Grube, A. (2019a). Palaeoseismic structures in Quaternary sediments of Hamburg (NW Germany), earthquakes evidence during the younger Weichselian and Holocene. *International Journal of Earth Sciences*, **108**, 845–861, doi.org/10.1007/s00531-019-01681-2.

Grube, A. (2019b). Palaeoseismic structures in Quaternary sediments, related to an assumed fault zone north of the Permian Peissen-Gnutz salt structure (NW Germany) – neotectonic activity and earthquakes from the Saalian to the Holocene. *Geomorphology*, **328**, 15–27, doi.org/10.1016/j.geomorph.2018.12.004.

Gruszka, B. and van Loon, A. J. (2011). Genesis of a giant gravity-induced depression (gravifossum) in the Enköping esker, S. Sweden. *Sedimentary Geology*, **235**, 304–313, doi.org/10.1016/j.sedgeo.2010.10.004.

Harry, D. G. (1988). Ground ice and permafrost. In M. J. Clark, ed., *Advances in Periglacial Geomorphology*, Wiley, Chichester, pp. 113–149.

Hart, J. K. and Boulton, G. S. (1991). The interrelation of glaciotectonic and glaciodepositional processes within the glacial environment. *Quaternary Science Reviews*, **10**, 335–350, doi.org/10.1016/0277-3791(91)90035-S.

Hoffmann, G. and Reicherter, K. (2012). Soft-sediment deformation of Late Pleistocene sediments along the southwestern coast of the Baltic Sea (NE Germany). *International Journal of Earth Sciences*, **101**, 351–363, doi.org/10.1007/s00531-010-0633-z.

Houtgast, R. F., van Balen, R. T. and Kasse, C. (2005). Late Quaternary evolution of the Feldbiss Fault (Roer Valley Rift System, the Netherlands) based on trenching, and its potential relation to glacial unloading. *Quaternary Science Reviews*, **24**, 489–508, doi.org/10.1016/j.quascirev.2004.01.012.

Hungr, O., Leroueil, S. and Picarelli, L. (2014). The Varnes classification of landslide types, an update. *Landslides*, **11**, 167–194, doi.org/10.1007/s10346-013-0436-y.

Hurst, A. and Cartwright, J. (2007). Relevance of sand injectites to hydrocarbon exploration and production. In A. Hurst and J. Cartwright, eds., *Sand Injectites: Implications for Hydrocarbon Exploration and Production*, AAPG Memoir **87**, Tulsa, pp. 1–19, doi.org/10.1306/1209846M871546.

Hurst, A., Cartwright, J. and Duranti, D. (2003). Fluidization structures produced by upward injection of sand through a sealing lithology. In P. Van Rensbergen, R. R. Hillis, A. J. Maltman and C. K. Morley, eds., *Subsurface Sediment Mobilization*. Geological Society, London, Special Publication, Vol. 216, pp. 123–138, doi.org/10.1144/GSL.SP.2003.216.01.09.

Hurst, A., Scott, A. and Vigorito, M. (2011). Physical characteristics of sand injectites. *Earth-Science Reviews*, **106**, 215–246, doi.org/10.1016/j.earscirev.2011.02.004.

Kenzler, M., Obst, K., Hüneke, H. and Schütze, K. (2010). Glazitektonische Deformation der kretazischen und pleistozänen Sedimente an der Steilküste von Jasmund nördlich des Königsstuhls (Rügen) [Glaciotectonic deformation of the Cretaceous and Pleistocene sediments on the steep coast of Jasmund north of the Königsstuhl (Rügen Island)]. *Brandenburger Geowissenschaftliche Beiträge*, **17**, 107–122.

Kowalski, A., Makoś, M. and Pitura, M. (2018). New insights into the glacial history of southwestern Poland based on large-scale glaciotectonic deformations – a case study from the Czaple II Gravel Pit (Western Sudetes). *Annales Societatis Geologorum Poloniae*, **88**, 341–359, doi.org/10.14241/asgp.2018.022.

Lee, J. R. and Phillips, E. R. (2008). Progressive soft sediment deformation within a subglacial shear zone – a hybrid mosaic-pervasive deformation model for Middle Pleistocene glaciotectonised sediments from eastern England. *Quaternary Science Reviews*, **27**, 1350–1362, doi.org/10.1016/j.quascirev.2008.03.009.

Li, Y., Craven, J., Schweig, E. S. and Obermeier, S. F. (1996). Sand boils induced by the 1993 Mississippi River flood: could they one day be misinterpreted as earthquake-induced liquefaction? *Geology*, **24**, 171–174, doi.org/10.1130/0091-7613(1996)024<0171:SBIBTM>2.3.CO;2.

Longva, O. and Bakkejord, K. J. (1990). Iceberg deformation and erosion in soft sediments, southeast Norway. *Marine Geology*, **92**, 87–104, doi.org/10.1016/0025-3227(90)90028 I.

Lowe, D. R. (1975). Water escape structures in coarse-grained sediments. *Sedimentology*, **22**, 157–204, doi.org/10.1111/j.1365-3091.1975.tb00290.x.

Maizels, J. K. (1992). Boulder ring structures produced during jökulhaups flows: origin and hydraulic significance. *Geografiska Annaler*, **74A**, 21–33, doi.org/10.1080/04353676.1992.11880346.

Matsuoka, N. (2001). Solifluction rates, processes and landforms: a global review. *Earth-Science Reviews*, **55**, 107–134, doi.org/10.1016/S0012-8252(01)00057-5.

Meinsen, J., Winsemann, J., Roskosch, J. et al. (2014). Climate control on the evolution of Late Pleistocene alluvial-fan and aeolian sand-sheet systems in NW Germany. *Boreas*, **43**, 42–66, doi.org/10.1111/bor.12021.

Molina, J. M., Alfaro, P., Moretti, M. and Soria, J. M. (1998). Soft-sediment deformation structures induced by cyclic stress of storm waves in tempestites (Miocene, Guadalquivir Basin, Spain). *Terra Nova*, **10**, 145–150, doi.org/10.1046/j.1365-3121.1998.00183.x.

Monecke, K., Anselmetti, F. S., Becker, A. et al. (2006). Earthquake-induced deformation structures in lake deposits: a Late Pleistocene to Holocene paleoseismic record for Central Switzerland. *Eclogae Geologicae Helvetiae*, **99**, 343–362, doi.org/10.1007/s00015-006-1193-x.

Montenat, C., Barrier, P., d'Estevou, P. O. and Hibsch, C. (2007). Seismites: an attempt at critical analysis and classification. *Sedimentary Geology*, **196**, 5–30, doi.org/10.1016/j.sedgeo.2006.08.004.

Moretti, M. and Sabato, L. (2007). Recognition of trigger mechanisms for soft-sediment deformation in the Pleistocene lacustrine deposits of the Sant'Arcangelo Basin (Southern Italy): seismic shock vs. overloading. *Sedimentary Geology*, **196**, 31–45, doi.org/10.1016/j.sedgeo.2006.05.012.

Moretti, M., Miguel, J., Alfaro, P. and Walsh, N. (2001). Asymmetrical soft-sediment deformation structures triggered by rapid sedimentation in turbiditic deposits (Late Miocene, Guadix Basin, Southern Spain). *Facies*, **44**, 283–294, doi.org/10.1007/BF02668179.

Moretti, M., Alfaro, P. and Owen, G. (2016). The environmental significance of soft-sediment deformation structures: key signatures for sedimentary and tectonic processes. *Sedimentary Geology*, **344**, 1–4, doi.org/10.1016/j.sedgeo.2016.10.002.

Morsilli, M., Giona Bucci, M., Gliozzi, E., Lisco, S. and Moretti, M. (2020). Sedimentary features influencing the occurrence and spatial variability of seismites (late Messinian, Gargano Promontory, southern Italy). *Sedimentary Geology*, **401**, 105628, doi.org/10.1016/j.sedgeo.2020.105628.

Mörz, T., Karlik, E. A., Kreiter, S. and Kopf, A. (2007). An experiment setup for fluid venting in unconsolidated sediments: new insights to fluid mechanics and structures. *Sedimentary Geology*, **196**, 251–267, doi.org/10.1016/j.sedgeo.2006.07.006.

Naik, S. P., Mohanty, A., Porfido, S. et al. (2020). Intensity estimation for the 2001 Bhuj earthquake, India on ESI-07 scale and comparison with historical 16th June 1819 Allah Bund earthquake: a test of ESI-07 application for intraplate earthquakes. *Quaternary International*, **536**, 127–143, doi.org/10.1016/j.quaint.2019.12.024.

Nichols, R. J., Sparks, R. S. J. and Wilson, C. J. N. (1994). Experimental studies of fluidization of layered sediments and the formation of fluid escape structures. *Sedimentology*, **41**, 233–253, doi.org/10.1111/j.1365-3091.1994.tb01403.x.

Obermeier, S. F. (2009) Using liquefaction-induced and other soft-sediment features for paleoseismic analysis. In J. P. McCalpin, ed., *Paleoseismology, Vol. 95*. International Geophysics Series, Elsevier, Amsterdam, pp. 497–564, doi.org/10.1016/S0074-6142(09)95007-0.

Obermeier, S. F., Olson, S. M. and Green, R. A. (2005). Field occurrences of liquefaction-induced features: a primer for engineering geologic analysis of paleoseismic shaking. *Engineering Geology*, **76**, 209–234, doi.org/10.1016/j.enggeo.2004.07.009.

Ogino, Y. and Matsuoka, N. (2007). Involutions resulting from annual freeze – thaw cycles: a laboratory simulation based on observations in northeastern Japan. *Permafrost and Periglacial Processes*, **18**, 323–335, doi.org/10.1002/ppp.597.

Ojala, A. E. K., Mattila, J., Virtasalo, J., Kuva, J. and Luoto, T. P. (2018). Seismic deformation of varved sediments in southern Fennoscandia at 7400 cal BP. *Tectonophysics*, **744**, 58–71, doi.org/10.1016/j.tecto.2018.06.015.

Oliveira, C. M. M, Hodgson, D. M. and Flint, S. S. (2009). Aseismic controls on in situ soft-sediment deformation processes and products in submarine slope deposits of the Karoo Basin, South Africa. *Sedimentology*, **56**, 1201–1225, doi.org/10.1111/j.1365-3091.2008.01029.x.

Owen, L. A. (1991). Mass movement deposits in the Karakoram Mountains: their sedimentary characteristics, recognition and role in Karakoram landform evolution. *Zeitschrift für Geomorphologie*, **35**, 401–424.

Owen, G. (2003). Load structures: gravity-driven sediment mobilization in the shallow subsurface. In F. Van Rensbergen, R. R. Hillis, A. J. Maltman and C. K. Morley, eds., *Subsurface Sediment Mobilization*. Geological Society, London, Special Publication, Vol. 216, pp. 21–34, doi.org/10.1144/GSL.SP.2003.216.01.03.

Owen, G. and Moretti, M. (2008). Determining the origin of soft-sediment deformation structures: a case study from Upper Carboniferous delta deposits in south-west Wales, UK. *Terra Nova*, **20**, 237–245, doi.org/10.1111/j.1365-3121.2008.00807.x.

Owen, G. and Moretti, M. (2011). Identifying triggers for liquefaction-induced soft-sediment deformation in sands. *Sedimentary Geology*, **235**(3–4), 141–147, doi.org/10.1016/j.sedgeo.2010.10.003.

Pedersen, S. A. S. (2014). Architecture of glaciotectonic complexes. *Geosciences*, **4**(4), 269–296, doi.org/10.3390/geosciences4040269.

Perucca, L. P., Godoy, E. and Pantano, A. (2014). Late Pleistocene–Holocene earthquake-induced slumps and soft-sediment deformation structures in the Acequion River Valley, Central Precordillera, Argentina. *Geologos*, **20**(2), 147–156, doi.org/10.2478/logos-2014-0007.

Phillips, E., Lee, J. R. and Burke, H. (2008). Progressive proglacial to subglacial deformation and syntectonic sedimentation at the margins of the Mid-Pleistocene British Ice Sheet: evidence from north Norfolk, UK. *Quaternary Science Reviews*, **27**, 1848–1871, doi.org/10.1016/j.quascirev.2008.06.011.

Piotrowski, J. A., Larsen, N. K. and Junge, F. W. (2004). Reflections on soft subglacial beds as a mosaic of deforming and stable spots. *Quaternary Science Reviews*, **23**, 993–1000, doi.org/10.1016/j.quascirev.2004.01.006.

Pisarska-Jamroży, M. (2013). Varves and megavarves in the Eberswalde Valley (NE Germany) – a key for the interpretation of glaciolimnic processes. *Sedimentary Geology*, **291**, 84–96, doi.org/10.1016/j.sedgeo.2013.03.018.
Pisarska-Jamroży, M. and Woźniak, P. P. (2019). Debris-flow and glacio isostatic-induced soft-sediment deformation structures in a Pleistocene glaciolacustrine fan: the southern Baltic Sea coast, Poland. *Geomorphology*, **326**, 225–238, doi.org/10.1016/j.geomorph.2018.01.015.
Pisarska-Jamroży, M. and Zieliński, T. (2012). Specific erosional and depositional processes in a Pleistocene subglacial tunnel in the Wielkopolska region, Poland. *Geografiska Annaler*, **94A**, 429–443, doi.org/10.1111/j.1468-0459.2012.00466.x.
Pisarska-Jamroży, M., Belzyt, S., Börner, A. et al. (2018). Evidence from seismites for glacio-isostatically induced crustal faulting in front of an advancing land-ice mass (Rügen Island, SW Baltic Sea). *Tectonophysics*, **745**, 338–348, doi.org/10.1016/j.tecto.2018.08.004.
Pisarska-Jamroży, M., Belzyt, S., Bitinas, A., Jusienė, A. and Woronko, B. (2019a). Seismic shocks, periglacial conditions and glaciotectonics as causes of the deformation of a Pleistocene meandering river succession in central Lithuania. *Baltica*, **32**, 63–77, doi.org/10.5200/baltica.2019.1.6.
Pisarska-Jamroży, M., Belzyt, S, Börner, A. et al. (2019b). The sea cliff at Dwasieden: soft-sediment deformation structures triggered by glacial isostatic adjustment in front of the advancing Scandinavian Ice Sheet. *DEUQUA Special Publications*, **2**, 61–67, doi.org/10.5194/deuquasp-2-61-2019.
Pisarska-Jamroży, M. and Weckwerth, P. (2013). Soft-sediment deformation structures in a Pleistocene glaciolacustrine delta and their implications for the recognition of sub-environments in delta deposits. *Sedimentology*, **60**, 637–665, doi.org/10.1111/j.1365-3091.2012.01354.x.
Rijsdijk, K. F. (2001). Density-driven deformation structures in glacigenic consolidated diamicts: examples from Traeth Y Mwnt, Cardiganshire, Wales, UK. *Journal of Sedimentary Research*, **71**, 122–135, doi.org/10.1306/042900710122.
Ringrose, P. S. (1989). Palaeoseismic (?) liquefaction event in late Quaternary lake sediment at Glen Roy, Scotland. *Terra Nova*, **1**, 57–62, doi.org/10.1111/j.1365-3121.1989.tb00326.x.
Rodrígues, N., Cobbold, P. R. and Løseth, H. (2009). Physical modeling of sand injectites. *Tectonophysics*, **474**, 610–632, doi.org/10.1016/j.tecto.2009.04.032.
Rodríguez-Pascua, M. A., Calvo, J. P., De Vicente, G. and Gómez-Gras, D. (2000). Soft-sediment deformation structures interpreted as seismites in lacustrine sediments of the Prebetic Zone, SE Spain, and their potential use as indicators of earthquake magnitudes during the Late Miocene. *Sedimentary Geology*, **135**, 117–135, doi.org/10.1016/S0037-0738(00)00067-1.
Ross, J. A., Peakall, J. and Keevil, G. M. (2011). An integrated model of extrusive sand injectites in cohesionless sediments. *Sedimentology*, **58**, 1693–1715, doi.org/10.1111/j.1365-3091.2011.01230.x.
Rydelek, P. A. and Tuttle, M. (2004). Explosive craters and soil liquefaction. *Nature*, **427**, 115–116, doi.org/10.1038/427115a.
Seilacher, A. (1969). Fault-graded beds interpreted as seismites. *Sedimentology*, **13**, 155–159, doi.org/10.1111/j.1365-3091.1969.tb01125.x.
Shanmugam, G. (2016a). The seismite problem. *Journal of Palaeogeography*, **5**, 318–362, doi.org/10.1016/j.jop.2016.06.002.
Shanmugam, G. (2016b). Submarine fans: A critical retrospective (1950–2015). *Journal of Palaeogeography*, **5**, 110–184, doi.org/10.1016/j.jop.2015.08.011.

Shipton, Z. K., Meghraoui, M. and Monro, L. (2017). Seismic slip on the west flank of the Upper Rhine Graben (France–Germany): evidence from tectonic morphology and cataclastic deformation bands. In A. Landgraf, S. Kuebler, E. Hintersberger and S. Stein, eds., *Seismicity, Fault Rupture and Earthquake Hazards in Slowly Deforming Regions*. Geological Society, London, Special Publication, Vol. 432, pp. 147–161, doi .org/10.1144/SP432.12.

Strasser, M., Anselmetti, F. S., Fäh, D., Giardini, D. and Schnellmann, M. (2006). Magnitudes and source areas of large prehistoric northern Alpine earthquakes revealed by slope failures in lakes. *Geology*, **34**, 1005–1008, doi.org/10.1130/G22784A.1.

Suter, F., Martínez, J. I. and Vélez, M. I. (2011). Holocene soft-sediment deformation of the Santa Fe–Sopetrán Basin, northern Colombian Andes: evidence for pre-Hispanic seismic activity? *Sedimentary Geology*, **235**, 188–199, doi.org/10.1016/j.sedgeo .2010.09.018.

Sutinen, R., Andreani, L. and Middleton, M. (2019a). Post-Younger Dryas fault instability and deformations on ice lineations in Finnish Lapland. *Geomorphology*, **326**, 202–212 doi.org/10.1016/j.geomorph.2018.08.034.

Sutinen, R., Hyvönen, E., Liwata-Kenttälä, P. et al. (2019b). Electrical-sedimentary anisotropy of landforms adjacent to postglacial faults in Lapland. *Geomorphology*, **326**, 213–224, doi.org/10.1016/j.geomorph.2018.01.008.

Szuman, I., Ewertowski, M. and Kasprzak, L. (2013). Thermo-mechanical facies representative of fast and slow flowing ice sheets: the Weichselian ice sheet, a central west Poland case study. *Proceedings of the Geologists' Association*, **124**, 818–833, doi .org/10.1016/j.pgeola.2012.09.003.

Tuttle, M. P., Hartleb, R., Wolf, L. and Mayne, P. W. (2019). Paleoliquefaction studies and the evaluation of seismic hazard. *Geosciences*, **9**, 311, doi.org/10.3390/ geosciences9070311.

van Balen, R. T., Bakker, M. A. J., Kasse, C., Wallinga, J. and Woolderink, H. A. G. (2019). A Late Glacial surface rupturing earthquake at the Peel Boundary fault zone, Roer Valley Rift System, the Netherlands. *Quaternary Science Reviews*, **218**, 254–266, doi.org/10.1016/j.quascirev.2019.06.033.

Van der Wateren, F. M., Kluiving, S. J. and Bartek, L. R. (2000). Kinematic indicators of subglacial shearing. In A. J. Maltman, B. Hubbard and M. J. Hambrey, eds., *Deformation of Glacial Materials*. Geological Society, London, Special Publication, Vol. 176, pp. 259–278, doi.org/10.1144/GSL.SP.2000.176.01.20.

van Loon, A. J. T. (2009). Soft-sediment deformation structures in siliciclastic sediments: an overview. *Geologos*, **15**, 3–55.

van Loon, A. J. T. and Pisarska-Jamroży, M. (2014). Sedimentological evidence of Pleistocene earthquakes in NW Poland induced by glacio-isostatic rebound. *Sedimentary Geology*, **300**, 1–10, doi.org/10.1016/j.sedgeo.2013.11.006.

van Loon, A. J. T., Pisarska-Jamroży, M., Nartišs, M., Krievāns, M. and Soms, J. (2016). Seismites resulting from high-frequency, high-magnitude earthquakes in Latvia caused by Late Glacial glacio-isostatic uplift. *Journal of Palaeogeography*, **5**, 363–380, doi.org/10.1016/j.jop.2016.05.002.

van Loon, A. J. T., Soms, J., Nartišs M., Krievāns, M. and Pisarska-Jamroży, M. (2019). Sedimentological traces of ice-raft grounding in a Weichselian glacial lake near Dukuli (NE Latvia). *Baltica*, **32**, 170–181, doi.org/10.5200/baltica.2019.2.4.

van Loon, A. J. T., Pisarska-Jamroży, M. and Woronko, B. (2020). Sedimentological distinction in glacigenic sediments between load casts induced by periglacial processes from those induced by seismic shocks. *Geological Quarterly*, **64**, 626–640, doi .org/10.7306/gq.1546.

Van Vliet-Lanoë, B., Magyar, I. A. and Meilliez, F. (2004). Distinguishing between tectonic and periglacial deformations of quaternary continental deposits in Europe. *Global and Planetary Change*, **43**, 103–127, doi.org/10.1016/j.gloplacha.2004.03.003.

Vandenberghe, J. (2013). Cryoturbation structures. In S. A. Elias and C. J. Mock, eds., *The Encyclopedia of Quaternary Science*, 2nd ed., Elsevier, Amsterdam, pp. 430–435, doi.org/10.1016/B978-0-444-53643-3.00096-0.

Vandekerkhove, E., Van Daele, M., Praet, N. et al. (2020). Flood-triggered versus earthquake-triggered turbidites: a sedimentological study in clastic lake sediments (Eklutna Lake, Alaska). *Sedimentology*, **67**(1), 364–389, doi.org/10.1111/sed.12646.

Vanneste, K., Meghraoui, M. and Camelbeeck, T. (1999). Late Quaternary earthquake-related soft-sediment deformation along the Belgian portion of the Feldbiss Fault, Lower Rhine Graben system. *Tectonophysics*, **309**, 57–79, doi.org/10.1016/S0040-1951(99)00132-8.

Vanneste, K., Camelbeeck, T., Verbeeck, K. and Demoulin, A. (2018). Morphotectonics and past large earthquakes in Eastern Belgium. In A. Demoulin, ed., *Landscapes and Landforms of Belgium and Luxembourg, World Geomorphological Landscapes*. Springer, Cham, pp. 215–236, doi.org/10.1007/978-3-319-58239-9_13.

Wheeler, R. L. (2002). Distinguishing seismic from nonseismic soft-sediment structures: criteria from seismic-hazard analysis. In F. R. Ettensohn, N. Rast and C. E. Brett, eds., *Ancient Seismites*. Geological Society of America, Special Paper 359, pp. 1–11, doi.org/10.1130/0-8137-2359-0.1.

Weise, O. R. (1983). *Das Periglazial. Geomorphologie und Klima in gletscherfreien, kalten Regionen* [The Periglacial. Geomorphology and Climate in Glacier-Free, Cold Regions], Gebrüder Borntraeger, Stuttgart.

Winsemann, J., Asprion, U., Meyer, T., Schultz, H. and Victor, P. (2003). Evidence of iceberg ploughing in a subaqueous ice-contact fan, glacial Lake Rinteln, NW Germany. *Boreas*, **32**, 386–398, doi.org/10.1111/j.1502-3885.2003.tb01092.x.

Winsemann, J, Alho, P., Laamanen, L. et al. (2016). Flow dynamics, sedimentation and erosion of glacial lake outburst floods along the Middle Pleistocene Scandinavian ice sheet (northern Central Europe). *Boreas*, **45**, 260–283, doi.org/10.1111/bor.12146.

Winsemann, J., Koopmann, H., Tanner, D. et al. (2020). Seismic interpretation and structural restoration of the Heligoland glaciotectonic thrust-fault complex: implications for multiple deformation during (pre-)Elsterian to Warthian ice advances into the southern North Sea Basin. *Quaternary Science Reviews*, **227**, 106068, doi.org/10.1016/j.quascirev.2019.106068.

Woronko, B., Belzyt, S., Bujak, Ł. and Pisarska-Jamroży, M. (2018). Glaciotectonically deformed glaciofluvial sediments with ruptured pebbles (the Koczery study site, E Poland). *Bulletin of the Geological Society of Finland*, **90**, 145–159.

Worsley, P. (2014). Ice-wedge growth and casting in a Late Pleistocene periglacial, fluvial succession at Baston, Lincolnshire. *Mercian Geologist*, **18**, 159–170.

Woźniak, P. P. and Pisarska-Jamroży, M. (2018). Debris flows with soft-sediment clasts in a Pleistocene glaciolacustrine fan (Gdańsk Bay, Poland). *Catena*, **165**, 178–191, doi.org/10.1016/j.catena.2018.01.022.

5

Glacially Induced Fault Identification with LiDAR, Based on Examples from Finland

JUKKA-PEKKA PALMU, ANTTI E. K. OJALA, JUSSI MATTILA,
MIRA MARKOVAARA-KOIVISTO, TIMO RUSKEENIEMI, RAIMO SUTINEN,
TOBIAS BAUER AND MARIE KEIDING

ABSTRACT

The use of information gained from airborne light detection and ranging (LiDAR) data has produced a breakthrough in identification of postglacial faults and earthquake-deformed Quaternary deposits. LiDAR digital elevation models (DEMs) also improve the collection of detailed information on spatial distribution, characteristics and geometry, and they provide guidance for more costly and time-consuming field studies. In areas of weak glacial erosion, younger and older (i.e. Pre-Late Weichselian) ruptures have been discovered superimposed on the same or adjacent postglacial fault (PGF) segments, and the same have been identified when DEM information is combined with test pit sedimentological studies. We discuss Finnish examples and identify advantages, disadvantages and limitations. Advantages include verification of previously known fault scarps, detection of new PGF segments, systems and entire complexes and the ability to measure the dimensions (lengths and offset) of fault scarps obtained from the LiDAR DEM data. Disadvantages include that an inventory of the sub- and postglacial fault scarps is only possible when linear scarps cross-cut glacial and postglacial sediments.

5.1 Introduction

The utilization of information from airborne LiDAR (light detection and ranging) data has provided a breakthrough in detection of postglacial faults (PGFs) and earthquake-deformed Quaternary deposits (e.g. Sutinen et al., 2014a, 2014b; Mikko et al., 2015; Palmu et al., 2015; Ojala et al., 2019a). LiDAR-based digital elevation models (DEMs) provide an accurate and relatively low-cost methodology, which has advanced screening, mapping and DEM analysis in the Scandinavian Shield area (e.g. Mikko et al., 2015; Palmu et al., 2015; Ojala et al., 2017; Keiding et al., 2018; Smith et al., 2014; Mattila et al., 2019). LiDAR DEMs enable detection and mapping of PGFs and landslides in areas where they have not previously been recognized. It also enhances the collection of detailed information on their spatial distribution and characteristics and provides guidelines for the more costly and labour-intensive field studies.

Precise knowledge about the lengths and possible offsets of PGFs is necessary to estimate the potential maximum moment magnitudes of past earthquakes (e.g. Wells &

Coppersmith, 1994). Landslides are presumably related to seismic activity when they occur in the vicinity of the known postglacial fault scarps (Sutinen, 2005; Sutinen et al., 2007; Sutinen et al., 2014a; Ojala et al., 2018; Ojala et al., 2019b), and the area of the landslide region may be used to estimate the earthquake magnitude (Keefer, 1984; Malamud et al., 2004).

5.2 Mapping Postglacial Faults

During the LiDAR screening process, mostly conducted by Nordic geological surveys (Geological Survey of Finland, GTK; Geological Survey of Sweden, SGU; Geological Survey of Norway, NGU), the investigated areas are visually inspected by researchers using geographic information system (GIS) software ArcGIS (© ESRI). The screening scale is typically set to 1:10,000, but areas of interest are additionally studied at a scale of 1:2,000 to 1:5,000. The vertical accuracy (RMSE) of the LiDAR DEM is 0.3 m (Maanmittauslaitos, 2016); this error margin in practice limits the ability to identify fault scarp offset features to a minimum height of 0.2–0.3 m. Regarding the horizontal minimum length of fault scarps, typically, only rupture segments of 50 m or more in length were identified, except in cases where they clearly exhibited orientation and offset along longer PGF systems (e.g. Mattila et al., 2019).

High-resolution DEM is based on point cloud data produced in country-wide programmes by the National Land Surveys of Finland, Sweden and Norway. By the autumn of 2019, the LiDAR datasets covered approximately 90 per cent of these countries (Figure 5.1). LiDAR cloud point density in Finland and Sweden is 0.5 point per m^2 and in Norway the density ranges from 2 to 5 points per m^2. In Finland and Sweden, the standardized visualization includes a 2-m grid DEM; in Norway, the grid resolution is a variable, partly 1-m grid DEM.

For the mapping of PGFs and the associated seismically induced features, DEM raster and hillshading were optimized to allow visualization of the geomorphology. We consider testing and optimizing of DEM derivatives an essential aspect of the method, which often must deal with site-specific features. In research conducted in Finland, two DEM visualization derivatives were used: (i) MDOW data, which incorporates a multi-directional, oblique-weighted (MDOW) hillshade (Jenness, 2013) and (ii) the slope theme accessed dynamically from the DEM data, an additional visualization component in ArcGIS version 10.6. For the MDOW, a primary illumination direction of 315° was used, with a vertical exaggeration factor of 4. In this procedure the MDOW and slope data layers were incorporated with transparency settings generally of 50 per cent for both themes and DEM height colour classification scaling selected for the display extent. An alternative for MDOW is to use a series of normal hillshade visualizations, for example from 45°. This approach has been successfully used by SGU. The use of hillshading, slope or other visualization methods must also take into account human visual perception. There is general consensus that the illumination should come from above the horizon, and recent studies show that the best virtual light direction (if using ordinary hillshading) is from 337.5° (Biland & Çöltekin,

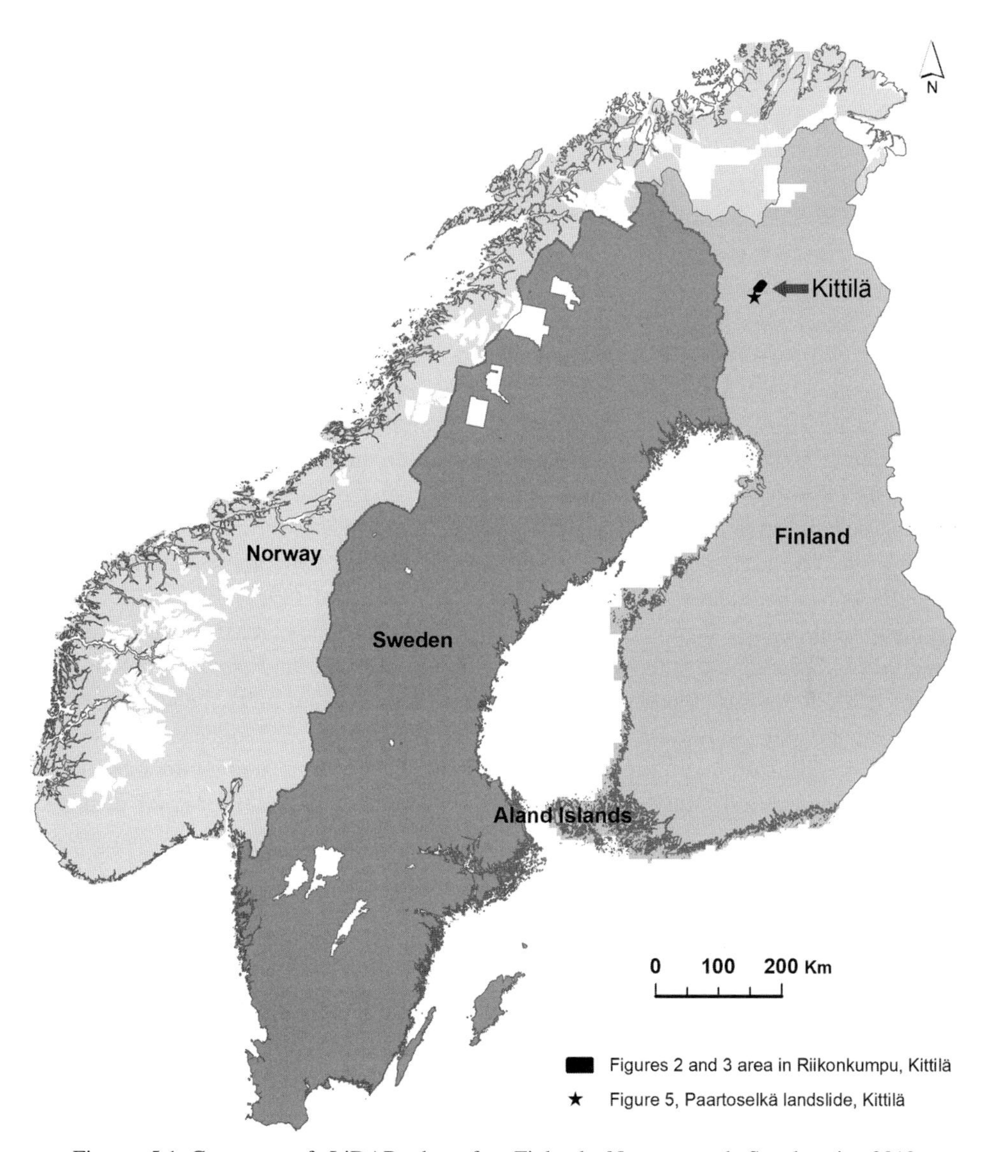

Figure 5.1 Coverage of LiDAR data for Finland, Norway and Sweden in 2019 (© Maanmittauslaitos, 2019, © Kartverket, 2019, © Lantmäteriet, 2020). Locations from Kittilä, Riikonkumpu area (Figures 5.2 and 5.4) and Paartoselkä landslide location (Figure 5.5) are shown on the map.

2016). Hillshading parameters, especially the stretching parameters, are also important, and various histogram classification options should be tested and used to get the best out of LiDAR DEM data.

In the remote sensing of fault scarps based on LiDAR data, vector or raster data and interfaces provided by the National Land Surveys have also been routinely used for additional information on areal characteristics. This has included topographical maps and orthophotos (including false colour image data) for the screened areas. In addition, the GIS-data interface services of the geological surveys – for instance, maps of Quaternary deposits and bedrock and geophysical data – were frequently accessed during the screening process. These various datasets were used to verify or exclude the features under study. As an example, both topographic maps and orthophotos were inspected to determine whether a particular linear feature was generated by a road or some other anthropogenic structures or if it was a true geological feature.

We note that in Finland the screening of PGFs and landslides began with examining the LiDAR characteristics of the previously known fault scarps in the Palojärvi region, Enontekiö and Ruokojärvi–Venejärvi region in Kolari (Kujansuu, 1964, 1972; Kukkonen & Kuivamäki, 1985; Kuivamäki, 1986; Kuivamäki et al., 1988; Kuivamäki et al., 1998; Sutinen et al., 2014b). Based on these experiences and the detection of visual characteristics, Palmu et al. (2015) started to systematically screen the entire country.

The excellence of the LiDAR data compared with the data previously available is perhaps best illustrated by showing the various data types from an example location (Figure 5.2). The example is from Riikonkumpu, part of the Isovaara–Riikonkumpu PGF Complex in Kittilä, Finland, that has been studied by Ojala et al. (2017). Figure 5.2A shows the topographical map, Figure 5.2B shows the previously applied DEM10 provided by the National Land Survey of Finland with 10 × 10-m grid resolution based on contour data, Figure 5.2C shows the orthophoto of Riikonkumpu, and Figure 5.2D shows the LiDAR DEM. Figure 5.2E shows LiDAR data in an oblique view (2.5 dimensions), and in Figure 5.2F a detailed map shows a trenching site, with the associated topography shown in profile (Figure 5.2G), easily processed from the LiDAR DEM data. The superb quality of LiDAR DEM for mapping detailed features is obvious.

5.3 Geometrical Analysis of Fault Scarps and Landslides

In addition to the mapping of PGFs and associated landslides described above, the LiDAR DEM can be further used for the analysis of fault geometries and landslide volumes. A detailed description of methodologies for such analysis developed by GTK is described in the following.

Fault slip profiles were obtained from the LiDAR data to determine the variability in scarp spatial offset and length of surface ruptures in isolated segments, systems and the entire PGF complex. In this approach (Mattila et al., 2016; Ojala et al., 2017), lines were first digitized at the upthrow and downthrow sides of fault scarps and then the lines were draped onto the LiDAR DEM to honour the true topography. After this, one of the generated lines, either on the upthrow and downthrow side, was divided into a set of points, spaced one metre apart. For each of those points, the closest point on the remaining line was computed and next the vertical distance between the points (i.e. throw) was computed. The vertical slip profiles for the fault scarps were then acquired by extracting the height differences as a function of distance along the fault segment. Leapfrog 3D geological modelling software (© Seequent Limited) was used to run the DEM data treatment and Rhinoceros 3D software using its algorithmic modelling plugin

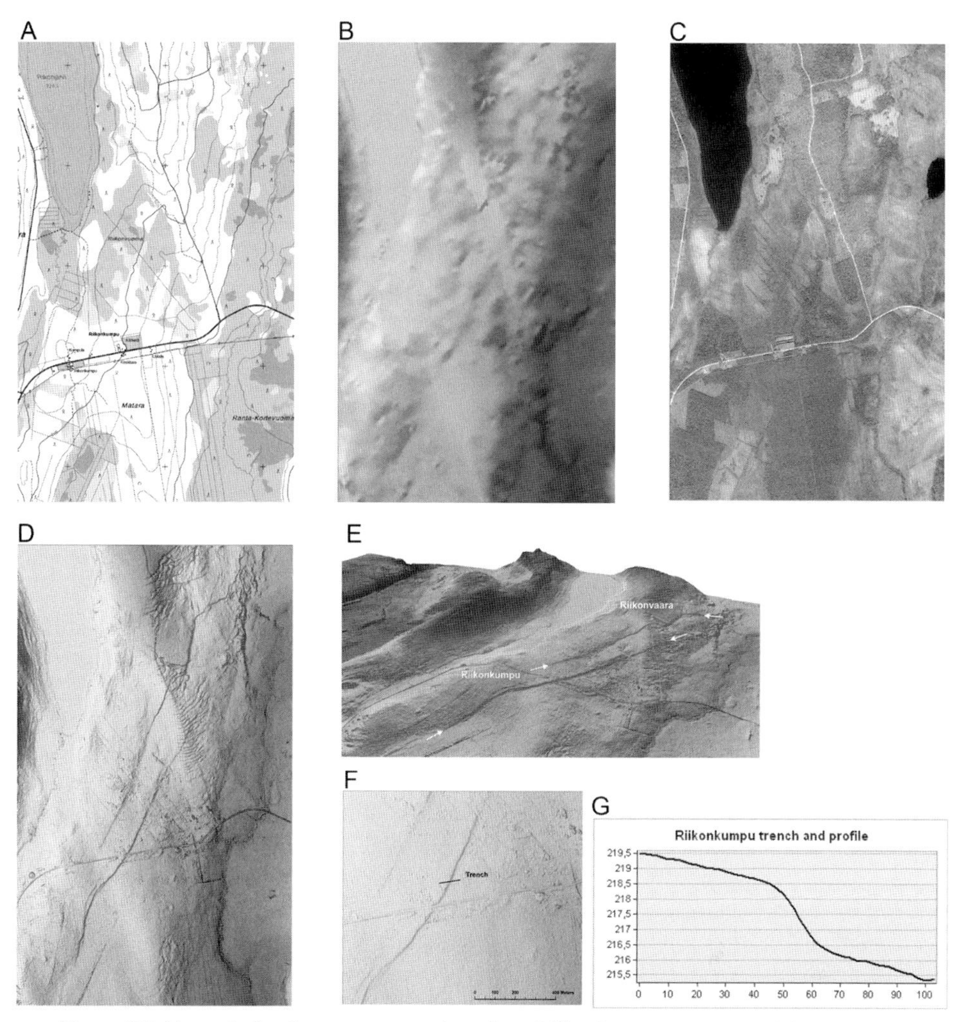

Figure 5.2 Example for data type comparison from Riikonkumpu area, part of the Isovaara–Riikonkumpu PGF Complex in Kittilä, Finland (see Figure 5.3): A) topographical map; B) previously applied DEM10 with 10 × 10 m grid resolution based on contour data; C) orthophoto of Riikonkumpu; D) LiDAR DEM; E) LiDAR DEM data in oblique view towards NNW, visualization with ArcScene (© ESRI); F) detailed map from Riikonkumpu showing the trenching site; and G) the associated topography also in a profile presentation. Topographic map, orthophoto, DEM10 and LiDAR Point Cloud data © Maanmittauslaitos (National Land Survey of Finland). LiDAR DEM processing and visualization by GTK. (A black and white version of this figure will appear in some formats. For the colour version, please refer to the plate section.)

Grasshopper in the generation of the height data of surface ruptures. To reduce noise in the data, a fast Fourier transformation was employed for the calculated scarp heights and reduction of amplitudes corresponding to the error in the topographic data. Details of the applied methodology are presented in Figures 5.3 and 5.4. The offset measured with the presented methodology yields only vertical offsets and therefore does not account for the

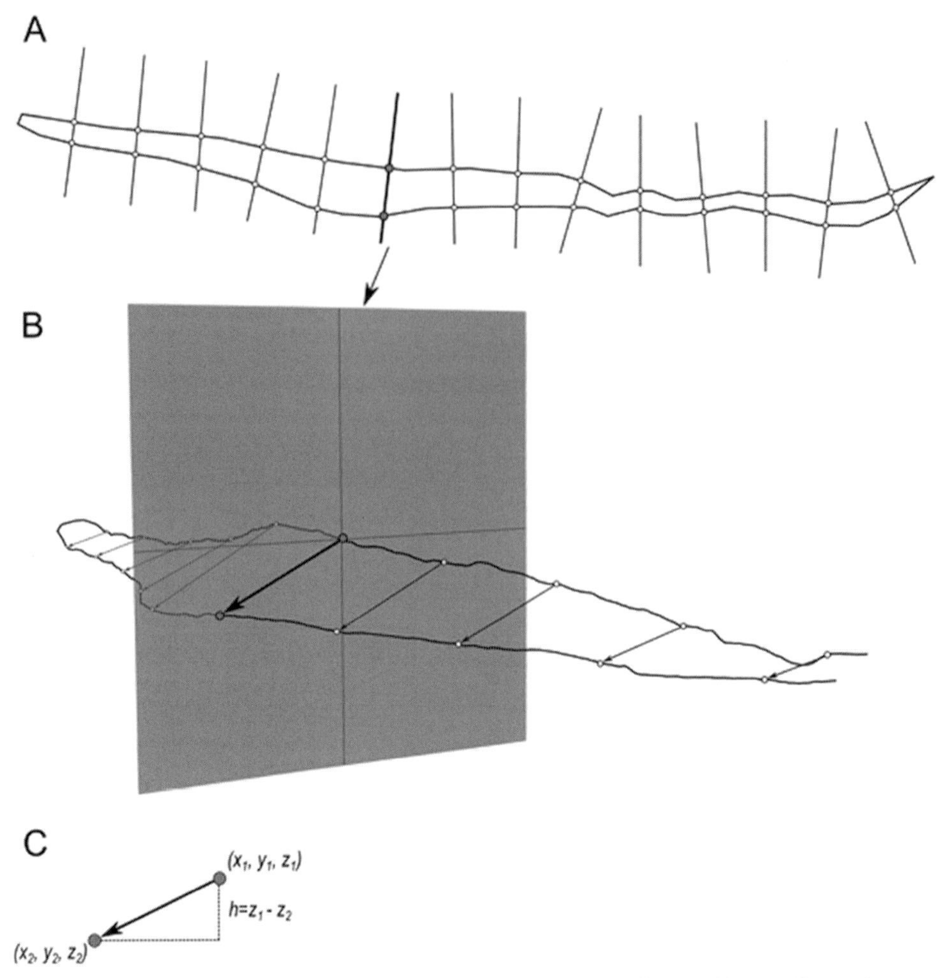

Figure 5.3 Methodology for extraction of continuous vertical offset profiles. A) Generation of a sequence of vertical planes on fault segment. In the example here, the distances are larger than in the actual extraction (1-m interval) for clarity. B) Formation of vectors along the upthrow and downthrow hinge lines and C) calculation of throw (modified from Mattila et al., 2016). For the description of the methodology see supplementary material in Ojala et al. (2017).

dip of the fault plane. In general, apart from vertical faults the measured vertical offset is thus smaller than the real offset. Furthermore, the method does not account for the geometrical error caused by scarps located on sloping terrain, in which case the measured vertical offset is larger than the offset measured perpendicular to the slope. However, such cases are rare, and the error caused by the effect is considered small. For a detailed description of the methodology see also Ojala et al. (2017), supplementary material.

In a study by Ojala et al. (2019b), landslides were first manually delineated in ArcMap (© ESRI) as polygon features during the systematic screening of LiDAR DEMs. Scarp

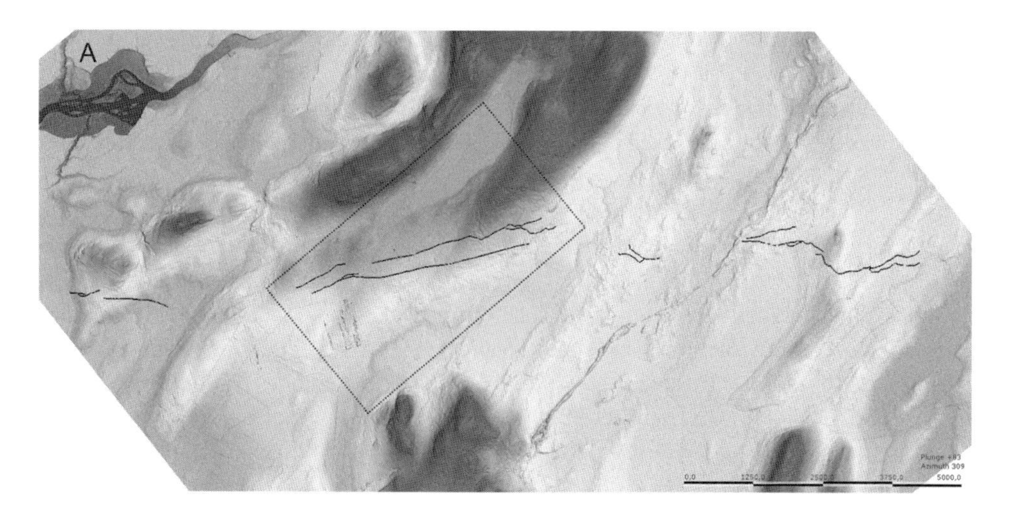

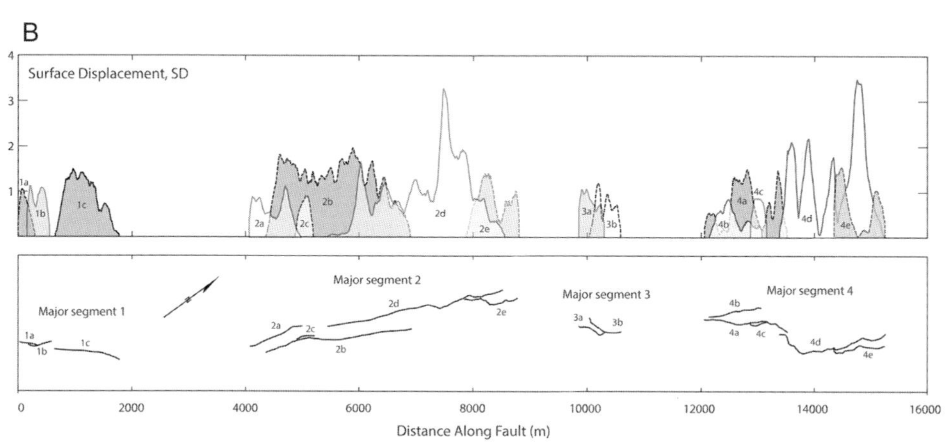

Figure 5.4 A) Riikonkumpu Fault System (up is towards NW (309°). B) Calculated surface displacements of the segments. Orientation of B is the same as for A; see also the N arrow on the lower part of B. Map area of Figures 5.2A–D is shown with a black box on the central area of A. LiDAR data © Maanmittauslaitos, 2019.

edges were then fine-tuned utilizing the Boolean AND algorithm to find cells in a ±40-m buffer zone around the polygon features where the slope was e.g. >20° and the tilt derivative >70°. The tilt derivative (TDR), which is an arctangent of the ratio between vertical and horizontal derivatives, was used to delineate landslide features because it equalizes the amplitude of DEM anomalies and enhances morphometric changes (e.g. Verduzco et al., 2004; Smith & Clark, 2005). The scarp and slide masses were manually delineated within each landslide. The slide geometry was then calculated using GIS software as follows. ArcMap (© ESRI) workflow started by converting the LiDAR DEM raster into points with a 200-m buffer area around the landslide polygon. Points within the landslide polygon were removed, and the base terrain was interpolated from the remaining points with the natural neighbour method. Then, the original LiDAR DEM was reduced from the interpolated surface, and the subtracted DEM provided a morphological three-dimensional model for each landslide. The boundary between the scarp and slide masses was then outlined based on positive (scarp) and negative (mass) DEM values (Figure 5.5).

Detailed geometry of both the fault scarps and landslides has been further used in the estimation of the moment magnitudes of the earthquakes related to these features (Ojala et al., 2019a,b; Mattila et al., 2019). Though there is a lack of instrumental data on earthquake magnitudes, these can be estimated from scaling laws that link the magnitudes to maximum surface rupture displacements and lengths (Wells & Coppersmith, 1994; Leonard, 2010; Moss & Ross, 2011). Similarly, moment magnitudes can be derived from the sizes of landslides (Malamud et al., 2004).

5.4 Conclusions

For mapping PGFs and associated seismically induced features, LiDAR DEM and hillshading visualizations were optimized in order to see the terrain geomorphology. Testing and optimizing the use of DEM derivatives is an essential part of the approach needed to study different sites with varied local geomorphology. LiDAR DEM data allow reliable and precise verification of previously known fault scarps and detection and identification of new PGF segments, systems and entire complexes. Thus, the various elements of the faults can now be discerned. There is also the capacity to measure the dimensions of the fault scarps from the LiDAR DEM data.

Based on experiences of systematic screening of LiDAR DEMs for the detection and characterization of PGFs in the previously glaciated terrain of the Fennoscandian Shield, the following advantages, disadvantages, limitations and unexpected results were observed:

- Advantages: Reliable and precise verification of previously known fault scarps, detection and identification of new PGF segments, systems and entire complexes; various elements of the faults can now be discerned (branch faults, etc.). Also allows measurement of the dimensions (lengths and offset) of the fault scarps from the LiDAR DEM data.
- Disadvantages: Inventory of the sub- and postglacial fault scarps is only possible when linear scarps cross-cut glacial and postglacial sediments. For strike-slips there is only limited possibility of identification.
- Limitations: Lots of linear features are observed in areas of outcropping bedrock, and it is impossible to assign an age to these from LiDAR DEM. Another challenge is created by the masking effect of postglacial sediments (clay, peat) deposited after the seismic events. Also, there is a range of linear or close to linear artefacts that have some of the same characteristics as PGFs, both natural features (glacial features, iceberg keel marks, etc.) and manmade structures (roads, dikes, land-use features such as stone fences, etc.).
- Unexpected: In areas of weak glacial erosion, younger and older (i.e. pre-Late Weichselian) ruptures can be superimposed on the same or adjacent PGF segments, as has been identified when combining the DEM information and test pit sedimentological

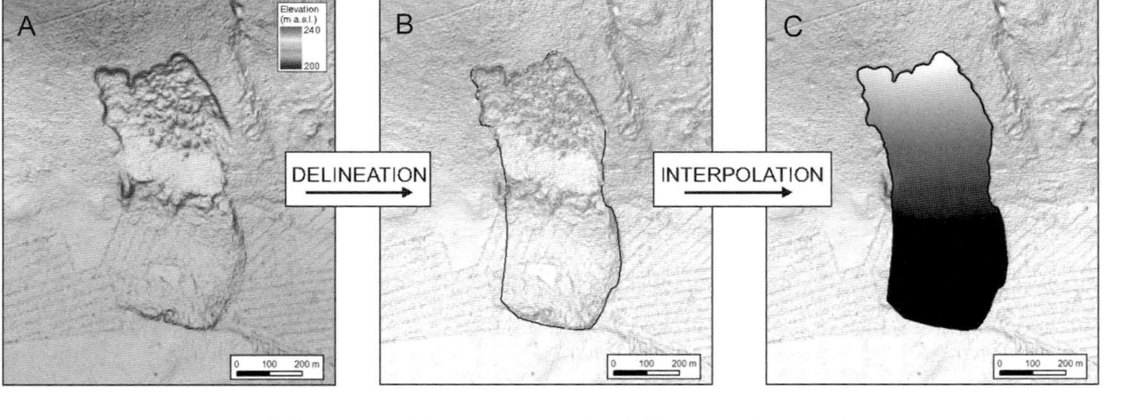

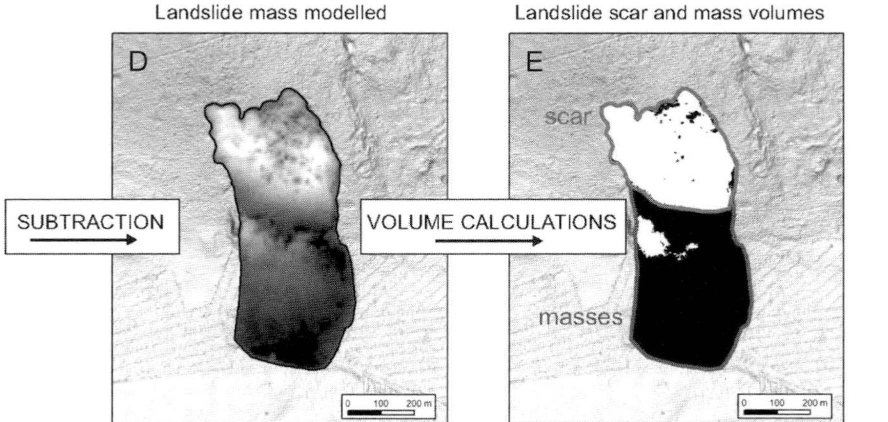

Figure 5.5 Landslide detection from LiDAR DEMs. Their outer boundaries (A) were first delineated manually (B) and then refined utilizing slope angle and tilt derivative. Base terrain (C) was interpolated from LiDAR topography after removing area affected by the landslide. Height difference between base terrain and current topography indicates where mass was removed (scar) and where it was deposited (masses) as well as mass volumes. In the upper central map that shows determined landslide boundaries, orange and red colours show the slopes greater than 35° steepness. In the lower right map that shows landslide scar and mass volumes, white is negative and black is positive, i.e. mass loss/mass addition. This example is from Paartoselkä, Kittilä, SE of Isovaara PGF (Ojala et al., 2019b; Ojala et al., 2018). LiDAR data © Maanmittauslaitos, 2019. (A black and white version of this figure will appear in some formats. For the colour version, please refer to the plate section.)

studies (e.g. Ojala et al., 2019a, and the references therein). An example of this is the Suaspalo PGF System (Ojala et al., 2019c). So, at least in Finnish Central Lapland we have old scarp elements with almost no signs of glacial modification.

References

Biland, J. and Çöltekin, A. (2017). An empirical assessment of the impact of the light direction on the relief inversion effect in shaded relief maps: NNW is better than NW. *Cartography and Geographic Information Science*, **44**(4), 358–372, doi.org/10.1080/15230406.2016.1185647.

Jenness, J. (2013). DEM surface tools. Jenness Enterprises. www.jennessent.com/arcgis/surface_area.htm.

Keefer, D. K. (1984). Landslides caused by earthquakes. *Geological Society of America Bulletin*, **95**(4), 406–421.

Keiding, M., Olesen, O. and Dehls, J. (2018). Neotectonic map of Norway and adjacent areas. Geological Survey of Norway, doi.org/10.13140/RG.2.2.32996.48005.

Kuivamäki, A. (1986). *Havaintoja Venejärven ja Ruostejärven postglasiaalisista siirroksista*. [English abstract: Observations of the Venejärvi and Ruostejärvi Postglacial Faults]. Geological Survey of Finland Espoo, Report YST-52, Espoo, Finland, 20 pp.

Kuivamäki, A., Paananen, M. and Kukkonen, I. (1988). The Pasmajärvi postglacial Fault – a reactivated pre-existing fracture zone. Nordiske geologiske vintermøde: abstracts af foredrag anmeldt til det 18. Nordiske Geologiske Vintermøde 12–14 January 1988, 240, København.

Kuivamäki, A., Vuorela, P. and Paananen, M. (1998). *Indications of Postglacial and Recent Bedrock Movements in Finland and Russian Karelia*. Geological Survey of Finland Nuclear Waste Disposal Research Report, YST-99, Espoo, Finland, 92 pp.

Kujansuu, R. (1964). Nuorista siirroksista Lapissa. [English summary: Recent faults in Lapland]. *Geologi*, **6**, 30–36.

Kujansuu, R. (1972). On landslides in Finnish Lapland. *Bulletin de la Commission Géologique de Finlande*, **256**, 22 pp.

Kukkonen, I. and Kuivamäki, A. (1985). *Geologisia ja geofysikaalisia havaintoja Pasmajärven ja Suasselän postglasiaalisista siirroksista*. [English abstract: Geological and Geophysical Observations of the Pasmajärvi and Suasselkä Postglacial Faults]. Geological Survey of Finland Report YST-46, Espoo, Finland, 14 pp.

Leonard, M. (2010). Earthquake fault scaling: self-consistent relating of rupture length, width, average displacement, and moment release. *Bulletin of the Seismological Society of America*, **100**, 1971–1988, doi.org/10.1785/0120090189.

Maanmittauslaitos (2016). Kansallisen maastotietokannan laatumalli, korkeusmallit, 1.3, biturl.top/yUBjQv.

Malamud B. D., Turcotte D. L., Guzzetti F. and Reichenbach P. (2004). Landslide inventories and their statistical properties. *Earth Surface Processes and Landforms*, **29**, 687–711, doi.org/10.1002/esp.1064.

Mattila, J., Ojala, A., Sutinen, R., Palmu, J.-P. and Ruskeeniemi, T. (2016). Digging deeper with LiDAR: vertical slip profiles of post-glacial faults. In *LITHOSPHERE 2016 Symposium, November 9–11, 2016, Espoo, Finland*, pp. 87–90.

Mattila, J., Ojala, A. E. K., Ruskeeniemi, T. et al. (2019). Evidence of multiple slip events on postglacial faults in northern Fennoscandia. *Quaternary Science Reviews*, **215**, 242–252, doi.org/10.1016/j.quascirev.2019.05.022.

Mikko, H., Smith, C. A., Lund, B., Ask, M. V. S. and Munier, R. (2015). LiDAR-derived inventory of post-glacial fault scarps in Sweden. *GFF*, **137**, 334–338, doi.org/10.1080/11035897.2015.1036360.

Moss, E. S. and Ross, Z. E. (2011). Probabilistic fault displacement hazard analysis for reverse faults. *Bulletin of the Seismological Society of America*, **101**, 1542–1553, doi.org/10.1785/0120100248.

Ojala, A. E. K., Mattila, J., Ruskeeniemi, T. et al. (2017). Postglacial seismic activity along the Isovaara-Riikonkumpu fault complex. *Global and Planetary Change*, **157**, 59–72, doi.org/10.1016/j.gloplacha.2017.08.015.

Ojala, A. E. K., Markovaara-Koivisto, M., Middleton, M. et al. (2018). Dating of seismically-induced paleolandslides in western Finnish Lapland. *Earth Surface Processes and. Landforms*, **43**, 2449–2462, doi.org/10.1002/esp.4408.

Ojala, A. E. K., Mattila, J., Ruskeeniemi, T. et al. (2019a). *Postglacial Faults in Finland – A Review of PGSdyn Project Results*. Posiva Report 2019-01, 118 pp., Posiva Oy, Eurajoki.

Ojala, A. E. K., Mattila, J., Markovaara-Koivisto, M. et al. (2019b). Distribution and morphology of landslides in northern Finland: an analysis of postglacial seismic activity. *Geomorphology*, **326**, 190–201, doi.org/10.1016/j.geomorph.2017.08.045.

Ojala, A. E. K., Mattila, J., Ruskeeniemi, T. et al. (2019c). Postglacial reactivation of the Suasselkä PGF complex in SW Finnish Lapland. *International Journal of Earth Sciences*, **108**, 1049–1065, doi.org/10.1007/s00531-019-01695-w.

Palmu, J.-P., Ojala, A. E. K., Ruskeeniemi, T., Sutinen, R. and Mattila, J. (2015). LiDAR DEM detection and classification of postglacial faults and seismically-induced landforms in Finland: a paleoseismic database. *GFF*, **137**, 344–352, doi.org/10.1080/11035897.2015.1068370.

Smith, M. J. and Clark, C. D. (2005). Methods for the visualization of digital elevation models for landform mapping. *Earth Surface Processes and Landforms*, **30**, 885–900, doi.org/10.1002/esp.1210.

Smith, C. A., Sundh, M. and Mikko, H. (2014). Surficial geology indicates early Holocene faulting and seismicity, central Sweden. *International Journal of Earth Sciences*, **103**(6), 1711–1724, doi.org/10.1007/s00531-014-1025-6.

Sutinen, R. (2005). Timing of early Holocene landslides in Kittilä, Finnish Lapland. *Geological Survey of Finland, Special Paper*, **40**, 53–58.

Sutinen, R., Piekkari, M. and Liwata, P. (2007). Time-transgressive evolution of landslides possibly induced by seismotectonic events in Lapland. Applied Quaternary research in the central part of glaciated terrain. *Geological Survey of Finland, Special Paper*, **46**, 121–128.

Sutinen, R., Hyvönen, E. and Kukkonen, I. (2014a). LiDAR detection of paleolandslides in the vicinity of the Suasselkä postglacial fault, Finnish Lapland. *International Journal of Applied Earth Observation and Geoinformation*, **27**, 91–99, doi.org/10.1016/j.jag.2013.05.004.

Sutinen, R., Hyvönen, E., Middleton, M. and Ruskeeniemi, T. (2014b). Airborne LiDAR detection of postglacial faults and Pulju moraine in Palojärvi, Finnish Lapland. *Global and Planetary Change*, **115**, 24–32, doi.org/10.1016/j.gloplacha.2014.01.007.

Verduzco, B., Fairhead, J. D., Green, C. M. and MacKenzie, C. (2004). New insights into magnetic derivatives for structural mapping. *The Leading Edge*, **23**(2), 116–119, doi.org/10.1190/1.1651454.

Wells, D. L. and Coppersmith, K. J. (1994). New empirical relationships among magnitude, rupture length, rupture width, rupture area, and surface displacement. *Bulletin of the Seismological Society of America*, **84**, 974–1002.

6
Fault Identification from Seismology

NICOLAI GESTERMANN AND THOMAS PLENEFISCH

ABSTRACT

Regions affected by glacial isostatic adjustment (GIA) experience stress changes. Stress is released either by slow aseismic movements along faults or by sudden stress release in the form of earthquakes. The location and source mechanism of these earthquakes can play a major role in providing an understanding of past and ongoing geodynamic processes in a GIA-affected region. On the one hand, alignments of earthquake hypocentres may act as an indicator of active faults that might not have been known from geology before. On the other hand, calculation and interpretation of earthquake focal mechanisms represents a key to stress and stress changes. We present an overview of seismological methods and tools to retrieve fault geometry and motion.

6.1 Introduction

Tectonic earthquakes are the expression of instantaneous stress release in Earth's crust or mantle. They are caused by a sudden slip on a portion of a fault surface (co-seismic slip). A small part of the elastic deformation energy is converted into seismic energy. Seismic waves (compressional (P) and shear (S) waves) are released by the co-seismic process and travel through the Earth at different velocities, depending on the elastic properties of the medium they pass through and on the type of wave.

The occurrence of seismic waves without deformation at the surface is clear evidence of active faults in the ground. Recording and analysis of these waves can be used to deduce information about the rupture process along the fault below the Earth's surface. The degree of detail that can be achieved depends on many factors, such as the number and the distribution of recording stations and the quality of the recordings. Locating the precise hypocentre, the point origin, is crucial for associating an earthquake with the location of a particular fault.

One of the challenges in seismology is to relate the observed seismic waves to the parameters describing the source (Udías & Buforn, 2017). This requires defining the seismic source in terms of a mechanical model that represents the physical fracture. The fracturing process can be approached in two different ways, as a kinematic or as a dynamic source model. Kinematic models consider the slip on the fault without relating it to the stress that causes it. Dynamic source models consider the complete fracture process relating the fault slip to stresses/forces. Source models require a physical understanding of the rupture and knowledge of the material properties at the source region. In seismology, it is difficult to distinguish between effects from travel paths that occur near the source structure and those near the receiver structure. In the following, we describe how to derive earthquake source parameters from ground motion measurements.

6.2 Measurements and Analysis of Seismic Events

6.2.1 Ground Motion Measurement and Instrumentation

Ground motion, caused by seismic waves arriving at the surface, is measured by seismometers. These transform the ground motion into electric power, which in turn is converted into digital values. These values are stored at a recording device or transmitted to seismological observatories for further investigations. Communication networks for real time data transmission are readily available in almost all regions, ensuring data availability and quality control in real time and timely data processing. Ground motion can be described mathematically as displacement, velocity or acceleration. The transformation from one physical quantity to the other requires the precise knowledge of the seismometer transfer function.

Seismological recordings usually use instruments measuring the translational movement, with a vertical component and two horizontal components (usually north-south and east-west). Some measurements require only the vertical component, which usually is less noisy. The vertical component in particular is important for the arrival of P waves, while the later-arriving S waves are more evident on the horizontal components.

For several years, portable rotation motion sensors with a high sensitivity in a broad range of frequencies have been available. These sensors measure the three rotational components of the seismic wavefield. The combination of observed translational and rotational motion enables the processing of a single station that would usually require a network or an array of conventional three-component measurements. The measurement of the rotational component of the seismic wavefield enables direct identification of the S wave constituents and provides direct information about the velocities below the seismometer location from the dispersion relation, which can be helpful for the hypocentre determination of seismic events (e.g. Schmelzbach et al., 2018).

Besides the sensor for recording the ground motion the seismometers must be equipped with a timing device. Precise timing is crucial for synchronization of seismograms (time series of ground motion recording, Figure 6.1) recorded at various locations. This enables the exchange of waveform data between different seismological observatories.

6.2.2 Seismogram Analysis

Routine seismogram analysis starts with identification of the arrival time of the seismic phases in the recordings. An appropriate frequency range has to be selected by digital filtering, which

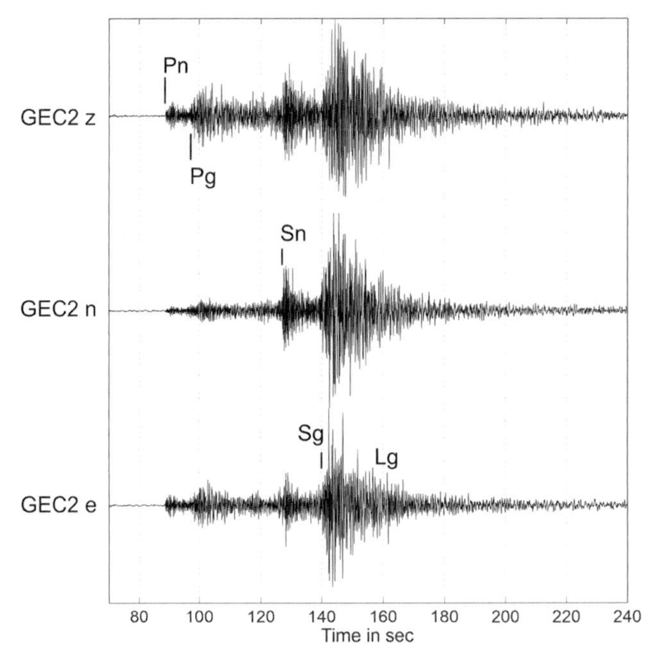

Figure 6.1 Recording of an earthquake close to Constance, South Germany (47.74°N, 9.11°E, magnitude 3.7 M_L) at a depth of about 3 km on 29 July 2019, with vertical component (z) and two horizontal components, north-south (n) and east-west (e). The recording station GEC2 at a distance of about 362 km is the key element of the GERESS Array, north of Passau (south-east Germany). Arrival times of the main crustal phases Pn, Pg, Sn and Sg are marked. The Lg wavetrain, which follows the arrival of the Sg phase, has normally the highest amplitudes (ground motion). The recording is bandpass-filtered in the frequency range from 0.6 to 10.0 Hz.

depends on several criteria, e.g. the source type, distance between source and seismic station, and the seismic phase to be investigated. A frequency range from 0.8 Hz to 15.0 Hz is generally useful for recordings within a local distance range (up to about 150 km).

Seismic waves travel along various paths and velocities (P- and S-phases) between source and receiver. A sketch for a layer above a half space, simplifying Earth's crust with velocities $v1_P$, $v2_P$, $v1_S$ and $v2_S$, is shown in Figure 6.2 for a local distance range and receivers at three arbitrary distances. The ray paths of the dominant seismic phases, their representatives in the seismograms and the corresponding travel times in the distance-time plot are displayed. The time difference between arrival times of seismic phases can directly calculate the source-receiver distance if the Earth velocity model is known. In the case of networks with many stations and large time windows, automatic algorithms for phase identification and arrival time estimation are crucial (Di Stefano et al., 2006).

Seismic wave amplitude decays with distance r to the source. Propagating seismic waves lose energy due to geometrical spreading, intrinsic attenuation and scattering attenuation. The amount of decay is frequency-dependent. Decay amplitude is proportional to $1/r^2$ for body waves like P and S waves. Seismic surface waves travel along the Earth's

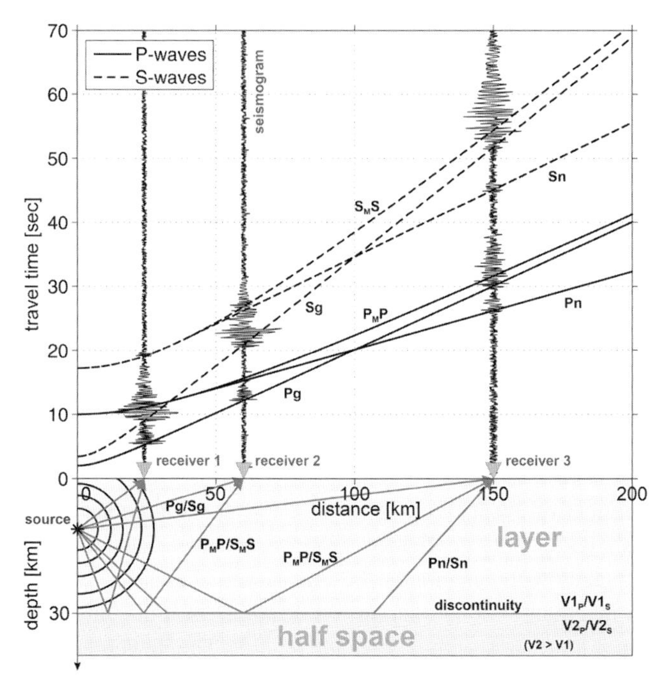

Figure 6.2 Theoretical travel time curves for P and S waves and corresponding seismogram sketches at three distances (upper part) for a simple crust model with a layer over half space (lower part). Grey lines with arrows mark ray paths of the waves in the lower part. Pg/Sg phases are the direct wave in the layer (crust). P_MP/S_MS are the reflected phases at the discontinuity between layer and half space. Pn/Sn travel along the discontinuity with the velocity of the half space.

surface (Rayleigh and Love waves) with an amplitude decay proportional to $1/r$. These waves are detectable at larger distances.

6.2.3 Seismic Networks

The combination of seismic stations is called a seismic network. In most cases a seismic network is necessary for a reliable and precise hypocentre determination. The stations of the network should at least cover the entire area under investigation where earthquakes are expected. The design of a station network – meaning the number of stations, interstation distances and network aperture, which is defined by the largest horizontal distance between two stations – depends on the specific tasks that need to be solved. Other site-specific parameters like the coherency of the relevant waves or the required detection threshold have to be considered as well (Joswig, 1992).

Detection capability of a network is one of the most important key parameters in seismological monitoring. Various methods exist to improve the detection capability of an investigated area. The most trivial and well-tried approach is to increase the density of the network, but this has the disadvantage of increased costs and maintenance effort.

Careful characterization and selection of seismometer sites and adapted network designs (Kraft et al., 2013) are appropriate items to consider for improvements.

The main motivation for improving the detection capability of a network is to increase the number of events that can be detected and localized. In a region with low seismic activity (seismicity rate), it might take a long time to detect and locate a sufficient number of earthquakes to get an image of an active fault. With a high detection capability more events can be detected and located. Consequently, a better image and characterization of active faults is obtained.

The Gutenberg–Richter empirical law describes the magnitude-frequency distribution of earthquakes measured for any specific region. It turns out that there is a relation between the number of earthquakes with low magnitude to those with higher magnitude. The Gutenberg–Richter law is defined as

$$\log_{10} N = a - bM,$$

where N is the number of events with a magnitude $\geq M$ within a specific time rate and a and b are constants (Gutenberg & Richter, 1956). A b-value of 1.0 means that for a given number of earthquakes with magnitude 3.0 and larger, there will be 10 times as many earthquakes with magnitude 2.0 and larger. An example of a magnitude-frequency distribution is seen in Figure 6.3 with a b-value of 1.2. The fewer earthquakes at lower magnitudes in the lower part of Figure 6.3 is an effect of the limiting properties of the monitoring network and demonstrates the consequence of the detection capability.

A special case of seismic monitoring is the application of a so-called seismic array. It is an ensemble of stations deployed in an appropriate configuration that works like an antenna. They allow the calculation of apparent velocity and azimuth of the crossing seismic waves (Harjes, 1990; Rost & Thomas, 2002; Schweitzer et al., 2012). Both parameters are useful for the data analysis and localization procedure. The summation of the recordings at the individual array elements improve the signal-to-noise ratio and decrease the detection threshold for the area of investigation. In some cases, the use of seismic arrays instead of networks or a combination of both is more suitable and leads to significantly better results (Sick et al., 2014; Duncan, 2005, Olesen et al., Chapter 11).

A seismic array differs from a local network of seismic stations mainly with regard to the techniques used for the data analysis. The application of array methods requires a sufficient coherence of the seismic waves under investigation in the appropriate frequency range. The distance between source location and array centre must be larger than the array aperture. The aperture depends on the apparent velocity and frequency of the investigated waves. Data from seismic events in the local distance range (less than 150 km) require smaller aperture arrays with smaller interstation distances compared to events from teleseismic distances (more than 3,000 km).

6.2.4 Signal Processing

Digital ground motion data from seismological stations can be processed by a wide range of manual or automatic methods with the aim to enhance and identify weak signals from

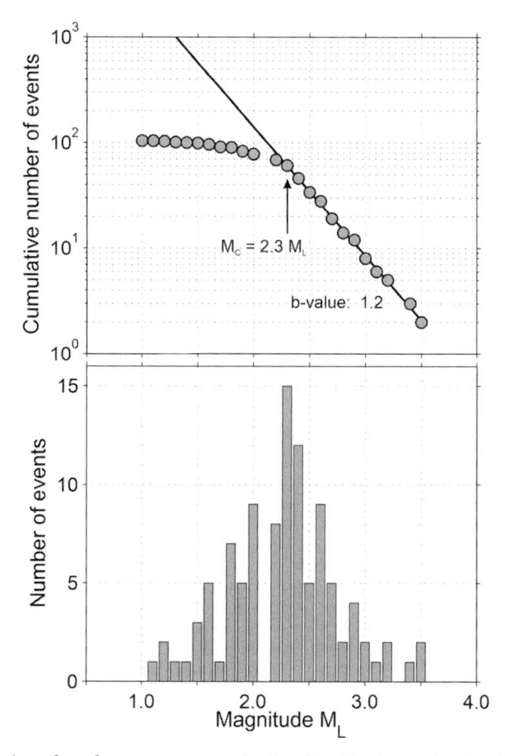

Figure 6.3 Example of a frequency-magnitude distribution of seismic events within a defined region with a *b*-value of 1.2. The solid line in the upper part is derived from the linear regression of the observed events for magnitudes above the magnitude of completeness (M_c). It is expected that all seismic events with a magnitude above M_c were detected by the seismological network. The gradient of the solid line is the *b*-value of the Gutenberg–Richter relationship.

earthquakes, to discriminate between noise and seismic signals or to characterize various wave types (Havskov & Ottemöller, 2010). They help to improve the quality and completeness of seismic catalogs for a region under investigation.

Filtering is one of the most commonly used steps; it highlights relevant features within an adapted frequency band. Many analysis methods are based on spectral analysis (transformation from time to frequency domain) of the recorded signals, for example, source spectral analysis for the source characterization or the surface wave analysis.

Correlation methods are applied to quantify the similarities of different signals in seismograms from various seismic stations or from different time intervals. They are helpful for automatic identification of seismic events or for source characterization and are applied at some waveform-based location methods, as is done with cross-correlation stacking (Schuster et al., 2004).

New processing techniques are based on ideas from artificial intelligence like machine learning. Meier et al. (2019) use machine-learning classifiers on deep neural network architectures for rapid and reliable signal/noise discrimination. Riggelsen and Ohrnberger

(2014) applied a machine-learning method for the automatic classification of seismic waveforms to improve the detection capabilities at three-component seismic stations. Modern approaches to pattern recognition use automatic classification methods to identify different types of seismic sources (Provost et al., 2017), which can aid understanding of source mechanisms.

In the case of data from seismic arrays, procedures like frequency-wavenumber analysis can be applied to calculate the apparent velocity and back azimuth of the selected seismic waves (Schweitzer et al., 2012). Apparent velocity is helpful for phase type identification. The back azimuth is additional information to improve epicentre location accuracy.

6.3 Hypocentre Determination

Source location of earthquakes plays one of the most fundamental roles in the analysis of seismological data. It is the key information concerning the identification of faults.

For hypocentre determination, the source of an earthquake is usually approximated by a point source. A complex time-dependent dynamic source process with an extended rupture surface is omitted. In point source approximation, the calculated epicentre coordinates represent the spatial start point (nucleus) of the rupture. The origin time is the temporal start of the fault rupture.

The earthquake location process is an optimization task for four unknowns: Cartesian coordinates *Xe* and *Ye*, the depth location *Ze* of the hypocentre, and origin time T_0 of the earthquake. The inputs are waveform data from seismic stations at a given configuration.

The standard location procedures for routine work are based on the use of the travel times of various P and S waves propagating between source and different seismological stations such as Pg, Sg, Pn and Sn according to the wave paths in Figure 6.2. They require phase identification and time picking, automatic or manual, of visible phases in the seismograms. A velocity depth model or empirically determined travel time curves are necessary. A simple model for the Earth's crust is seen in Figure 6.2.

The location procedure is based on the search for a minimum misfit between observed onset times and theoretical travel times. Various methods exist with differences in the misfit function definition and the applied inversion strategy. One challenge is to discriminate between relative minima and the relevant absolute minimum of the misfit function.

The earthquake location process with travel times is non-linear in the sense that there is no linear relationship between the observed arrival times and the desired spatial and temporal parameters of the source, which can be solved analytically.

The first method for solving the location problem was developed by Ludwig Geiger (1910, 1912). His method is based on an iteratively linearized search process to minimize the least squares error between observed and predicted arrival times at the stations. A minimum of three arrival times, one from each of three different seismic stations, or data from a single seismic array is required.

Modern approaches are based on non-linear inversion techniques (e.g. Lomax et al., 2000; Schweitzer, 2001). They minimize the differences between the parameters of the observed seismic waves (e.g. arrival times, time differences and slowness vectors) and the theoretical parameter calculated for the velocity depth model.

Besides the conventional travel time-based methods that are still in use, a new category of waveform-based methods has emerged; unlike the travel time methods, these do not need phase identification and picking. They are particularly suitable for small events with low signal-to-noise ratio. The common principle of waveform-based methods is that the source is located by focusing or reconstructing the source energy into a discrete grid of points with a certain migration or imaging operator, which uses travel-time information (Li et al., 2020). A good overview of the advances and challenges associated with waveform-based location methods is given by Li et al. (2020).

The third category of location methods comprises relative location techniques. These location methods are applied to an ensemble of earthquakes, positioning the events relative to one another. The double-difference method (Waldhauser & Ellsworth, 2000) minimizes the residuals between observed and theoretical travel-time differences (double-differences) for pairs of earthquakes at each station, while linking together all observed event station pairs. The relative coordinates are transferred into absolute coordinates with the coordinates of a reference earthquake that has the smallest hypocentre uncertainty. With relative location techniques a significantly higher hypocentre accuracy can be reached. This enables a more precise image of the fault structure activated by earthquakes.

For many projects, a big uncertainty in source location occurs with the velocity depth model. Obviously, the more accurate the velocities the more precise the source location. The applied model should reflect as close as possible the real geological situation. Velocity models can vary from a simple half space with constant velocities to a laterally homogeneous-layered model or a complex three-dimensional model, which requires detailed knowledge about the geological structures and rock properties along the travel path of the seismic waves.

Some waveform-based location methods such as the tomographic techniques iteratively update the velocity structure of the model in the location process, opposite to travel time-based methods with a priori defined velocities.

The accuracy of the calculated earthquake hypocentre strongly depends on the quality of the determined onset times in the case of travel time-based location methods. A high effort is required for especially weak events with low signal-to-noise relation.

For later interpretations, it is necessary to quantify the location uncertainty and the location accuracy. The accuracy of epicentre coordinates is generally higher and less difficult to calculate than the source depth. A precise source depth requires a good velocity model with reliable vp/vs relation. The location uncertainty is calculated during the localization process and is quantified commonly as the error ellipse of the hypocentre. The location inaccuracy is the distance between the calculated hypocentre and the ground truth of the event. The latter can only be estimated with plausible assumptions. Calibration events with ground truth information are helpful to quantify the hypocentre accuracy.

Alternative methods such as the jackknife algorithm (Efron, 1980) can estimate a more realistic confidence area. For a jackknife analysis, several subsets are selected from the entire set of onset-times of a seismic event. The epicentre is calculated for each subset. The variation of the hypocentre between the various subsets and their confidence region defines the new uncertainty of the hypocentre. The jackknife process can, for example, identify effects from incorrect phase identification or timing problems at single stations.

The design of a monitoring network and the selection of the applied location method depend on many factors. In most cases, it is a compromise between what is technically and scientifically feasible and other financial and logistic constraints. A careful and detailed planning with a focus on the main topics is crucial.

6.4 From Hypocentre Location to Fault Location and Orientation

In the case of a 'classical' tectonic region – that is, not an anthropogenic or volcano-tectonic environment – the existence of a single earthquake clearly indicates that an active fault exists at the calculated hypocentre. There are three principal ways to associate an earthquake or earthquake sequence with the location and orientation of a fault:

- spatial and temporal distributions of earthquakes monitored over a longer time period;
- location of a single earthquake in combination with the kinematic properties of the source, such as a fault plane solution; and
- spatial distribution of aftershocks that follow a main earthquake.

One way to image the fault zone structure in a specific area is by observing the seismicity over a long period; by looking at the spatial-temporal distribution of earthquakes it is often possible to retrace faults and their geometries (Gregersen et al., see Chapter 10; Olesen et al., see Chapter 11). A prerequisite for this approach is a precise hypocentre determination. Source depth calculation is more challenging than epicentre calculation, requiring a dense network in the epicentre area (Brandes et al., 2019). Another prerequisite is a large enough hypocentre dataset. The monitoring time required to obtain enough events can be reduced with the network design as described before, which offers an advantage to the detection of even small events.

Analysis of the kinematic properties of the source yields the orientation of the fault plane solution of the rupture process, which in turn makes possible an estimation of the fault's orientation. The rough size and offset of the rupture can be estimated from the earthquake magnitude (e.g. Wells & Coppersmith, 1994).

Magnitude scale values quantify the seismic energy release of an earthquake, describing its overall strength. There are several different magnitude scales (Bormann & Dewey, 2014); their application depends mainly on the distance between the recording stations (network) and the epicentre location as well as the type of seismic waves (different body and surface waves). Magnitude is usually calculated from the amplitude of the specific seismic phase.

Most large earthquakes are followed by additional earthquakes, called aftershocks or aftershock sequence when taken as an ensemble. Aftershocks steadily decrease in frequency and magnitude over time, according to Omori's law (Omori, 1894), which is an empirical relation. The absolute number of events and their magnitudes depend on many factors, such as the geological properties or the existence of other faults. The parameters of a specific earthquake (the decay parameter and the productivity of aftershock sequence) are related to the monitoring station network and its detection threshold.

Aftershocks are assumed to be caused by induced stress changes produced by the main shock. They can occur along the fault plane, which was activated by the main shock or along fault branches in the environment. The temporal and spatial distribution of aftershocks therefore provides important information about the rupture process and the geometric shape of the fault plane of the main shock (e.g. Kilb & Rubin, 2002).

6.5 Styles of Faulting

In order to describe the earthquake process from a geometrical point of view, the orientation and movement of the blocks involved are depicted. In seismology the most common description is given by three parameters: strike, dip and rake (Figure 6.4). The strike and the dip define the spatial orientation of the blocks and the fault, respectively. The rake is related to the kinematic part of the earthquake process, namely, the direction of the movement of one block against the other. As defined by Aki and Richards (1980) the strike φ is measured clockwise from the north ($0^\circ \leq \varphi \leq 360^\circ$), the dip δ is measured from the horizontal ($0^\circ \leq \delta \leq 90^\circ$) and the rake λ is defined as the angle ($-180^\circ \leq \lambda \leq 180^\circ$) between the fault strike and the slip vector, which is the direction the hanging wall moves with respect to the footwall during rupture (Figure 6.4).

The rake is the crucial parameter to define the style of faulting that takes place during an earthquake. A positive rake specifies an upward movement of the hanging wall and it is termed a ‘reverse’ or ‘thrust faulting’. Contrary, a downward movement, indicated by a negative rake, is called a normal faulting (Figure 6.5). If the rake is close or equal to 0° or 180°, respectively, the corresponding faulting type is called a left-lateral or right-lateral strike-slip. That is, if an observer is standing on one block and sees the other block move to the right, it is termed ‘right-lateral strike-slip motion’ and vice versa for ‘left-lateral

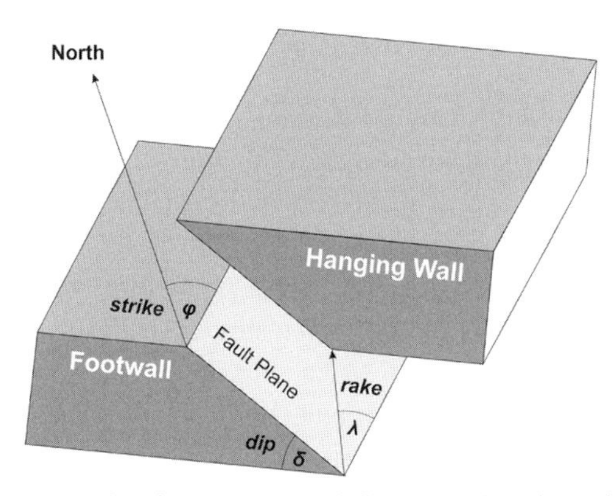

Figure 6.4 Block diagram showing movement during an earthquake and depiction of strike, dip and rake.

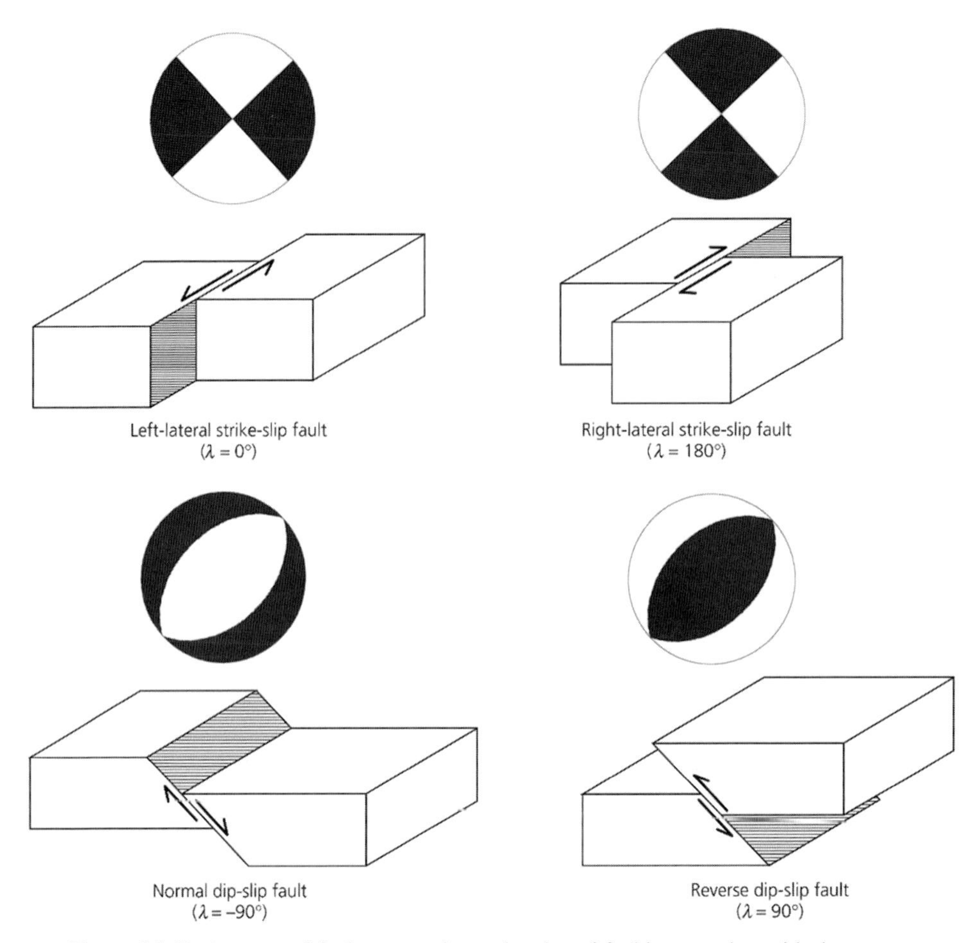

Figure 6.5 Basic types of fault geometries and styles of faulting together with the corresponding beach ball representation, plotted on a stereographic projection of the lower hemisphere (in map view). Black and white areas denote compressional and dilatational quadrants, respectively (modified after Stein & Wysession, 2003).

strike-slip motion'. In nature, there is often a mixture of two faulting types, e.g. an earthquake can show a strike-slip mechanism with a significant normal faulting component.

Most earthquakes take place at plate boundaries and are therefore called interplate earthquakes. At major subduction zones, such as at the Sunda Arc or the Chilean subduction zone, earthquakes with predominant reverse or normal faulting mechanisms occur. Prominent examples of earthquakes with strike-slip mechanisms can be observed, e.g. at the North Anatolian Fault (right-lateral strike-slip), at the Xianshuihe Fault System (left-lateral strike-slip) or at the San Andreas Fault (right-lateral strike-slip). Regions governed by glacial isostatic adjustment (GIA) are situated in intraplate regions rather than at plate boundaries. Typical earthquake focal mechanisms in those regions are normal and thrust faulting events (Gregersen et al., Chapter 10).

Contrary to surface geology features, where striations in outcrops indicate the fault movement at the surface, earthquakes take place inside the Earth (crust and mantle). Except in the case of larger events, earthquakes show no dislocation at the surface that may be interpreted in terms of the faulting geometry. In order to still reveal information on the shear faulting process at the earthquake hypocentre the technique of the so-called fault plane solution is applied, which is also called focal mechanism solution. The fault plane solution method takes advantage of the strong lateral variations of amplitudes and polarities at the Earth's surface. The complexity of the source radiation for P and S waves of a double couple, as described below, even allows inversion and determination of the fault geometry and the relative movement of the fault blocks.

6.6 Determination of the Earthquake Focal Mechanism and Moment Tensor

Tectonic earthquakes are mainly caused by shear faulting. The shearing mostly takes place at already existing zones of weakness; seldom is it connected to a new fracture. Shearing can be modelled as displacement on an internal surface in an elastic media (e.g. Shearer, 2009). The simplest physical model that explains the shearing that occurs during an earthquake and the associated energy radiation involves a pair of force couples with no net torque and termed a 'double couple' (DC). It is a point force model for the earthquake wave radiation. Energy radiation is highly non-uniform, for both P and S waves (Figure 6.6). The radiation pattern can be described as so-called amplitude-lobes, which means there are strong variations in the radiated energy depending on the azimuth and take-off angle of the considered seismic ray or the hypocentre-station geometry. It turns out that the assumption of a double source is justified because it explains the observed radiation of most tectonic earthquakes. Here, we explain the radiation pattern in a qualitative way; for the derivation of the mathematical formulations see Aki and Richards (1980). The qualitative description below is essentially based on a summary given in Chapter 8 of Tanner and Brandes (2019).

The P wave energy is radiated into four quadrants, which are separated by so-called nodal planes, where the displacement is zero (Figure 6.6). One nodal plane is equivalent to the fault itself, and the other nodal plane – the auxiliary plane – is perpendicular to the first one. When approaching the nodal plane, the P wave amplitude is decreasing, and it increases again when striding away from the nodal plane into the neighbouring quadrant. Opposite quadrants have the same polarity, which may have positive or negative initial motion, whereas neighbouring quadrants have different polarities. The positive quadrant is associated with the tensional axis (T-axis). It leads to upward or compressional motion of the P wave at the surface receiver (seismometer). Vice versa, the dilatational quadrant has the compressional axis (P-axis) associated with downward motion at the receiver. The S wave radiation pattern does not have nodal planes. S waves are radiated with maximum amplitudes along the nodal planes and are perpendicular to the P wave nodal planes. They have zero amplitude along the P-, T- and B-axes. S wave motion converges towards the T-axis, diverges from the P-axis and is zero at the so-called nodal point, where both nodal planes intersect (B-axis).

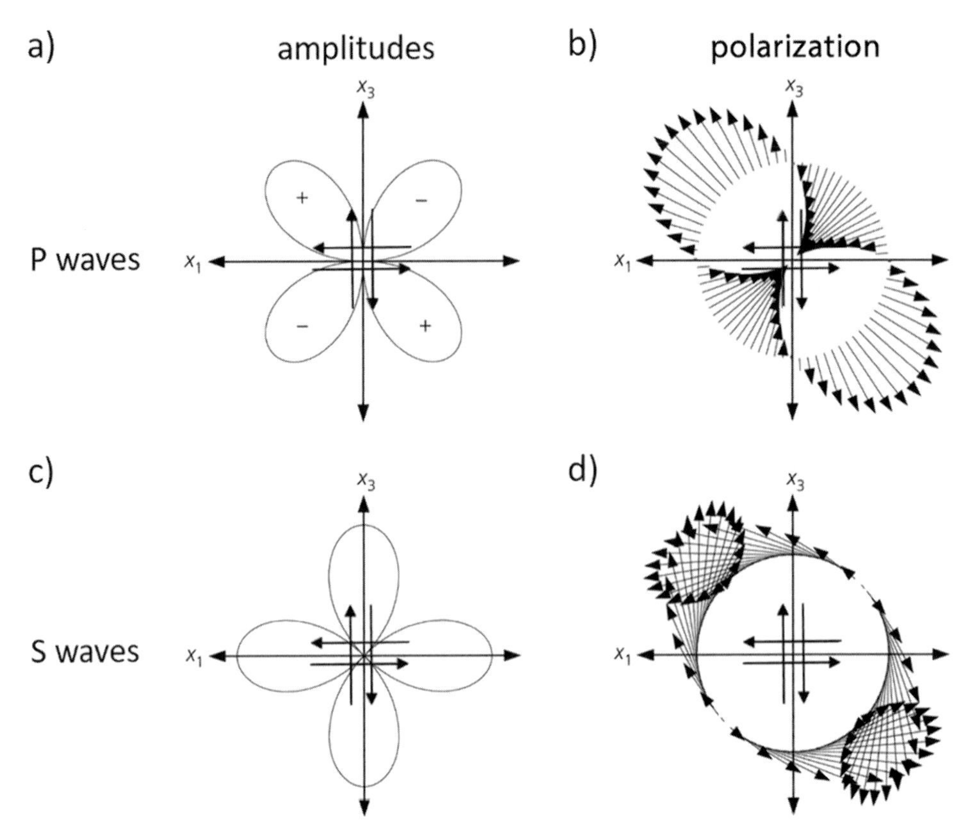

Figure 6.6 Representation of the double couple (DC) in the x_1x_3-plane (a), corresponding radiation pattern of P waves (b) and S waves (c and d) in the same plane (modified after Stein & Wysession, 2003). + and − indicate compressional and dilatational onsets, respectively. Arrows symbolize the polarization of the wave.

The concept of the double couple does not allow distinguishing which of the nodal planes is the actual fault (rupture) plane. In order to solve the ambiguity in the nodal planes, other techniques need to be considered such as the distribution of aftershocks, directivity effects or geological information. These parameters together with the determined fault plane solution often allow the identification of the active fault plane.

The input parameters for determination of an earthquake focal mechanism are polarities, amplitudes and waveforms measured at seismometers of seismic stations at the Earth's surface. The seismic stations need to be well distributed around the seismic event to properly constrain the mechanism (Figure 6.7). In the simplest inversion scheme only polarities of P waves are used. The distinct polarities for the stations and their corresponding take-off angles (seismic rays leave the earthquake focus under the so-called take-off angle measured off the vertical) are plotted in a stereographic projection of the focal sphere. Ideally, contiguous regions of different polarities should appear in the stereographic projection. Then, usually a simple grid search is performed to find those orthogonal (nodal)

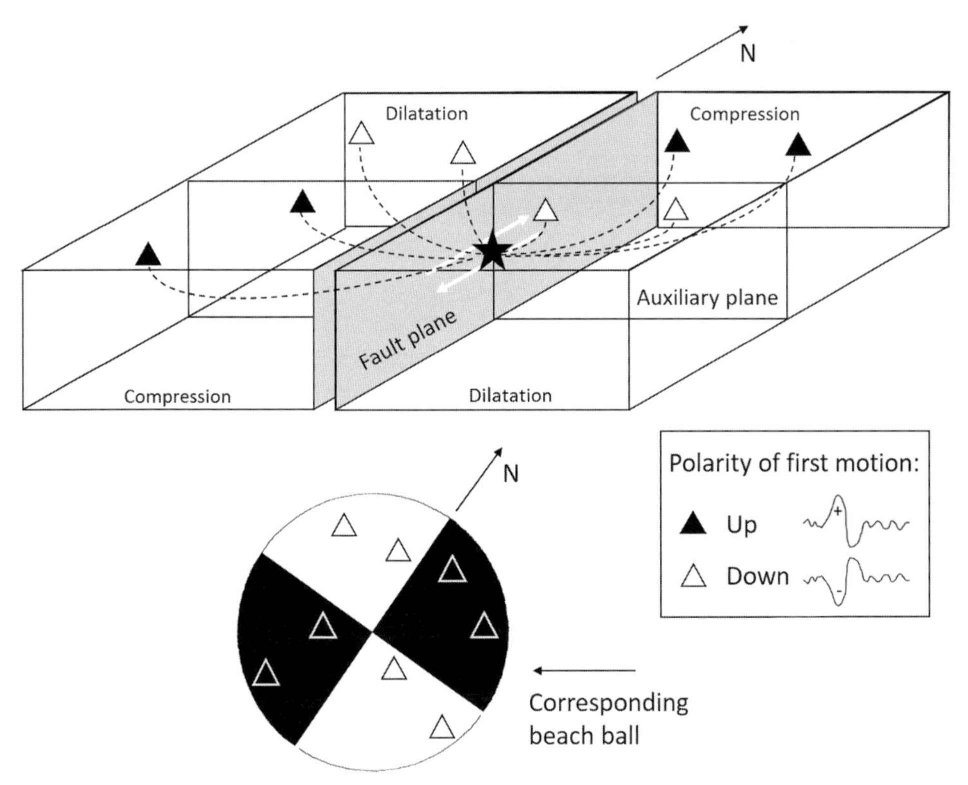

Figure 6.7 Fault and hypocentre (star) together with the radiation of the double couple into the different quadrants of compressional and dilatational motion. Exemplarily, ray paths of the P wave towards eight seismometer stations are shown, which exhibit opposite first motion of the P wave (up and down). A sketch of the 'beach ball' that belongs to the shear failure as well as the rough positions of the stations are given at the bottom. The mechanism represents a left-lateral strike-slip fault on an N-S-oriented fault plane.

planes that separate individual observations into compressional (upward first motion) and dilatational (downward first motion) quadrants. Since there is often a bundle of acceptable solutions, amplitudes and/or amplitude ratios can be additionally used in the inversion to further constrain the final focal mechanism. This is done by minimizing the misfit between observed and calculated amplitudes. An example of a focal mechanism is given in Figure 6.8 for the event at Constance (29 July 2019, 23:17 UT, $M_L = 3.7$), for which the seismogram of Figure 6.1 was also used in the inversion. In order to consider complex sources, the concept of the seismic moment tensor was developed (e.g. Aki & Richards, 1980; Jost & Hermann, 1989). It is a mathematical representation of the internal sources at the hypocentre and based on the assumption of a point source model. The moment tensor is a 3×3 matrix; it is symmetrical due to the conservation of angular momentum and consists of nine single-force couples. The concept of the moment tensor is quite general and comprises not only earthquake sources of classical shear faulting type; it is also applicable for explosions, implosions, landslides, rock falls, fluid injection-induced events, phase

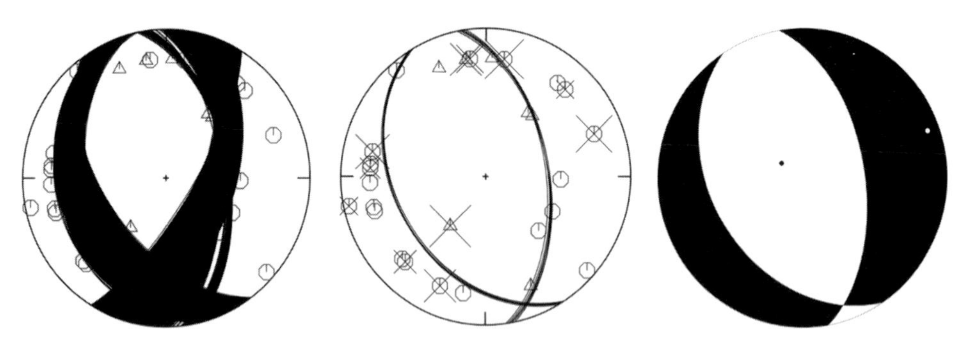

Figure 6.8 Example of focal mechanism determination from observations of P polarities and P/S amplitude ratios. The mechanism is determined for the strongest event of the earthquake sequence at Lake Constance, Germany (29 July 2019, 23:17 UT, M_L = 3.7) (Barth & Plenefisch, 2020). Left: The input parameters (octagon = compression, triangle = dilatation) as well as fitting nodal planes calculated with the program FOCMEC (Snoke, 2003). Middle: Reduction of nodal planes by incorporating amplitude ratios (cross = additional amplitude ratio). Right: The corresponding 'beach ball' with the preferred fault plane.

changes, complex fault plane geometry and so on. The full moment tensor consists of an isotropic part and so-called deviatoric moment tensor.

Decomposition of the moment tensor is often useful to find the most appropriate components that represent geological and physical source processes and support interpretations (e.g. Dahm & Krüger, 2014). There are various ways to decompose the moment tensor and the deviatoric component into different elementary sources (e.g. Jost & Herrmann, 1989). One example is decomposition into a double couple and a so-called compensated linear vector dipole (CLVD). The exact mathematical derivation of the moment tensor and its different decompositions are beyond the scope of this book; we thus refer to Jost and Herrmann (1989) and Dahm and Krüger (2014).

Several seismological agencies (e.g. USGS, GFZ) routinely provide moment tensors partly based on automatic analysis, mostly for larger earthquakes and at teleseismic distances (more than 1,000 km epicentral distance). In general, complete seismograms – consisting of body and surface waves and filtered for lower frequencies – are inverted to reveal the moment tensor. It turns out that the shear faulting component (the double couple) of the moment tensor dominates, with a ratio of more than 90 per cent. Other components are usually low and of minor importance.

For lower magnitude earthquakes, it is more difficult to invert for the full moment tensor and to resolve non-double-couple components because higher frequencies must be considered. Also, more detailed 3D-velocity models are required, which are often not available. Therefore, for small magnitude earthquakes, the double couple approach, with the classical consideration of phase polarities, is still an adequate and often used method.

Thanks to increasing computer power, inversions for extended source models and full moment tensor inversions for smaller earthquake magnitudes are advancing fast. The newest development in the field of source determination is the measure and use of rotational ground motion recordings in addition to the classical translational ground motion. This enables the determination of the moment tensor, even in the case of sparse seismic networks (e.g. Donner et al., 2018).

6.7 Conclusions

We have shown how a digitally recorded ground motion signal can be converted into fault parameters. The focal mechanism contains the information about the geometry of the fault (its strike and dip – which might be pre-existing and not related to a contemporary stress field) and the direction of the slip (rake). The rake is the only decisive parameter of the focal mechanism that puts constraints on the actual stress field. Only the inversion of an ensemble of focal mechanisms allows for a determination of the principal stress axes' orientation and of a relative stress magnitude.

In the case of GIA, the stress field is mostly a superposition of the regional stress field and the stress related to the GIA itself (e.g. Keiding et al., 2015; Steffen et al.; Gradmann & Steffen, see Chapters 2 and 22). This must be considered when interpreting the stress field inferred by the inversion of focal mechanisms. More detailed investigation on focal mechanisms in GIA-affected areas is needed to aid the understanding of contemporary stress conditions in those areas. Combining this knowledge with stress indicators from geology and new data from geodesy can provide further insight into the seismicity of glaciation and deglaciation cycles.

Acknowledgements

We kindly acknowledge the constructive comments of Thomas Braun and the remarks of an anonymous reviewer. The paper also benefitted from discussions with Stefanie Donner.

References

Aki, K. and Richards, P. G. (1980). *Quantitative Seismology: Theory and Methods*. W. H. Freeman, San Francisco.

Barth, L. and Plenefisch, T. (2020). Focal mechanisms for small to intermediate earthquakes in the northern part of the Alps and their seismotectonic interpretation. EGU, Vienna, doi.org/10.5194/egusphere-egu2020-12066.

Bormann, P. and Dewey, J. W. (2014). IS 3.3: The new IASPEI standards for determining magnitudes from digital data and their relation to classical magnitudes. In P. Bormann, ed., *New Manual of Seismological Observatory Practice (NMSOP-2)*. Deutsches GeoForschungsZentrum GFZ, Potsdam, pp. 1–44, doi.org/10.2312/GFZ.NMSOP-2_IS_3.3.

Brandes, C., Plenefisch, T., Tanner, D. C., Gestermann, N. and Steffen, H. (2019). Evaluation of deep crustal earthquakes in northern Germany – Possible tectonic causes. *Terra Nova*, **31**, 83–93, doi.org/10.1111/ter.12372.

Dahm, T. and Krüger, F. (2014). IS 3.9: Moment tensor inversion and moment tensor interpretation. In P. Bormann, ed., *New Manual of Seismological Observatory*

Practice 2 (NMSOP-2). Deutsches GeoForschungsZentrum GFZ, Potsdam, pp. 1–37, doi.org/10.2312/GFZ.NMSOP-2_IS_3.9.

Di Stefano, R., Aldersons, F., Kissling, E. et al. (2006). Automatic seismic phase picking and consistent observation error assessment: application to the Italian seismicity. *Geophysical Journal International*, **165**(1), 121–134, doi.org/10.1111/j.1365-246X.2005.02799.x.

Donner, S., Igel, H., Hadziioannou, C. and the Romy group (2018). Retrieval of the seismic moment tensor from joint measurements of translational and rotational ground motions: sparse networks and single stations. In S. D'Amico, ed., *Moment Tensor Solutions*. Springer, Cham, pp. 263–280, doi.org/10.1007/978-3-319-77359-9_12.

Duncan, P. M. (2005). Is there a future for passive seismic? *First Break*, **23**(6), 111–115.

Efron, B. (1980). *The Jackknife, the Bootstrap, and Other Resampling Plans*. Stanford University, Department of Statistics, Technical Report, NSF 163, 135 pp.

Geiger, L. (1910). Herdbestimmung bei Erdbeben aus den Ankunftszeiten. *Nachrichten von der Gesellschaft der Wissenschaften zu Göttingen, Mathematisch-Physikalische Klasse*, June 1910, 331–349; trans. (1912) Probability method for the determination of earthquake epicentres from the arrival time only. *Bulletin of St. Louis University*, **8**(1), 56–71, eudml.org/doc/58769.

Gutenberg, B. and Richter, C. F. (1956). Magnitude and energy of earthquakes. *Annali di Geofisica*, **9**, 1–15.

Harjes, H.-P. (1990). Design and siting of a new regional array in Central Europe. *Bulletin of the Seismological Society of America*, **80**(6), 1801–1817.

Havskov, J. and Ottemöller, L. (2010). *Routine Data Processing in Earthquake Seismology*. Springer, Dordrecht, doi.org/10.1007/978-90-481-8697-6.

Jost, M. and Herrmann, R. (1989). A student's guide to and review of moment tensor. *Seismological Research Letters*, **60**(2), 37–57, doi.org/10.1785/gssrl.60.2.37.

Joswig, M. (1992). System architecture of seismic networks and its implications to network automatization. *Cahiers Centre Européen de Géodynamique et de Séismologie*, **5**, 75–84.

Keiding, M., Kreemer, C., Lindholm, C. et al. (2015). A comparison of strain rates and seismicity for Fennoscandia: depth dependency of deformation from glacial isostatic adjustment. *Geophysical Journal International*, **202**, 1021–1028, doi.org/10.1093/gji/ggv207.

Kilb, D. and Rubin, A. M. (2002). Implications of diverse fault orientations imaged in relocated aftershocks of the Mount Lewis, ML 5.7, California, earthquake. *Journal of Geophysical Research*, **107**(B11), 2294, doi.org/10.1029/2001JB000149.

Kraft, T., Mignan, A. and Giardini, D. (2013). Optimization of a large-scale microseismic monitoring network in northern Switzerland. *Geophysical Journal International*, **195**(1), 474–490, doi.org/10.1093/gji/ggt225.

Li, L., Tan, J., Schwarz, B. et al. (2020). Recent advances and challenges of waveform-based seismic location methods at multiple scales. *Reviews of Geophysics*, **58**, e2019RG000667, doi.org/10.1029/2019RG000667.

Lomax, A., Virieux, J., Volant, P. and Berge, C. (2000). Probabilistic earthquake location in 3D and layered models: introduction of a Metropolis–Gibbs method and comparison with linear locations. In C. H. Thurber and N. Rabinowitz, eds., *Advances in Seismic Event Location*. Kluwer, Amsterdam, pp. 101–134, doi.org/10.1007/978-94-015-9536-0_5.

Meier, M.-A., Ross, Z. E., Ramachandran, A. et al. (2019). Reliable real-time seismic signal/noise discrimination with machine learning. *Journal of Geophysical Research: Solid Earth*, **124**(1), 788–800, doi.org/10.1029/2018JB016661.

Omori, F. (1894). On after-shocks of earthquakes. *The Journal of the College of Science, Imperial University of Tokyo, Japan*, **7**, 111–200.

Provost, F., Hibert, C. and Malet, J.-P. (2017). Automatic classification of endogenous landslide seismicity using the Random Forest supervised classifier. *Geophysical Research Letters*, **44**, 113–120, doi.org/10.1002/2016GL070709.

Riggelsen, C. and Ohrnberger, M. (2014). A machine learning approach for improving the detection capabilities at 3C seismic stations. *Pure and Applied Geophysics*, **171**, 395–411, doi.org/10.1007/s00024–012-0592-3.

Rost, S. and Thomas, C. (2002). Array seismology: methods and applications. *Reviews of Geophysics*, **40**(3), 2-1–2-27, doi.org/10.1029/2000RG000100.

Schmelzbach, C., Donner, S., Igel, H. et al. (2018). Advances in 6C seismology: applications of combined translational and rotational motion measurements in global and exploration seismology. *Geophysics*, **83**(3), doi.org/10.1190/geo2017-0492.1.

Schuster, G. T., Yu, J. and Sheng, J. (2004). Interferometric/daylight seismic imaging. *Geophysical Journal International*, **157**(2), 838–852, doi.org/10.1111/j.1365-246X .2004.02251.x.

Schweitzer, J. (2001). HYPOSAT – an enhanced routine to locate seismic events. *Pure and Applied Geophysics*, **158**, 277–289, doi.org/10.1007/978-3-0348-8250-7_17.

Schweitzer, J., Fyen, J., Mykkeltveit, S. et al. (2012 online): seismic arrays. In P. Bormann, ed., *New Manual of Seismological Observatory Practice 2 (NMSOP-2)*. Deutsches GeoForschungsZentrum GFZ, Potsdam, pp. 1–80, doi.org/10.2312/GFZ.NMSOP-2_ ch9.

Shearer, P. M. (2009). *Introduction to Seismology*. Cambridge University Press, Cambridge.

Sick, B., Walter, M. and Joswig, M. (2014). Visual event screening of continuous seismic data by supersonograms. *Pure and Applied Geophysics*, **171**, 549–559, doi.org/10 .1007/s00024-012-0618-x

Snoke, J. A. (2003). FOCMEC: FOCal MEChanism determinations. *International Handbook of Earthquake and Engineering Seismology*, pp. 1629–1630.

Stein, S. and Wysession, M. (2003). *An Introduction to Seismology, Earthquakes, and Earth Structure*. Blackwell Publishing, Malden, Massachusetts.

Tanner, D. and Brandes, C. (2019). *Understanding Faults: Detecting, Dating, and Modeling*. Elsevier, Amsterdam, doi.org/10.1016/B978-0-12-815985-9.00001-1.

Udías, A. and Buforn, E. (2017). *Principles of Seismology*. Cambridge University Press, Cambridge.

Waldhauser, F. and Ellsworth, W. E. (2000). A double-difference earthquake location algorithm: method and application to the Northern Hayward Fault, California. *Bulletin of the Seismological Society of America*, **90**(6), 1353–1368, doi.org/10 .1785/0120000006.

Wells, D. L. and Coppersmith, K. J. (1994). New empirical relationships among magnitude, rupture length, rupture width, rupture area, and surface displacement. *Bulletin of the Seismological Society of America*, **84**, 974–1002.

7

Imaging and Characterization of Glacially Induced Faults Using Applied Geophysics

RUTH BECKEL, CHRISTOPHER JUHLIN, ALIREZA MALEHMIR AND OMID AHMADI

ABSTRACT

Geophysical methods have the potential to delineate and map the geometry of glacially induced faults (GIFs) in the hard rock environment of the Baltic Shield. Relevant geophysical methods include seismics, geoelectrics, electromagnetics, magnetics and gravity. However, seismic methods have the greatest potential for determining the geometry at depth due to their higher resolving power. Seismic methods have even been used to identify a previously unknown GIF within the Pärvie Fault System. The other geophysical methods are usually employed to image the near-surface structure of GIFs. In this chapter, we provide a brief review of geophysical principles and how they apply to imaging of GIFs in the hard rock environment. The advantages and challenges associated with various geophysical methods are discussed through several case histories. Results to date show that it is possible to map GIFs dipping at 35–65° from the near surface down to depths of 7–8 km. It is not clear if the limiting factor in their mapping at depth is the nature of the faults or limitations in seismic acquisition parameters, since mapping capacity is highly dependent upon the acquisition geometry and source type used.

7.1 Introduction

The scarps of glacially induced faults (GIFs) are conspicuous features, and most major scarps in Scandinavia were discovered during systematic analyses of aerial photographs (Kujansuu, 1964; Lagerbäck, 1978). Subsequent studies to verify the Late- to postglacial nature of the faults and to reveal their deeper structure have been methodologically diverse. Though applied geophysical methods entail indirect imaging, they have always played an important role in these studies by enabling the mapping of geological features at depth instead of just extrapolating surface observations or interpolating between a small number of drill holes. Apart from this mapping capability, geophysical methods can provide additional information on the characteristics of fault zones and their surroundings. This can help to establish the tectonic setting in which the faults developed and to locate optimal sites for more direct methods like drilling and excavations. However, for

GIFs to be imaged or detected with geophysical methods, certain prerequisites must be fulfilled.

First, geological structures must feature a sufficiently large contrast in physical properties to be identifiable. Furthermore, the geophysical methods used have to provide a sufficiently high-resolution capacity. Drilling results from Norway and Finland suggest that GIF fault gouges are rather narrow features (e.g. Olesen et al., 1992; Roberts et al., 1997; Kuivamäki et al., 1998), which makes imaging the deeper parts challenging. Moreover, GIFs are commonly situated in older deformation zones with complex, three-dimensional geology, complicating interpretation of the results.

Despite all these challenges, several case studies have demonstrated the usefulness of geophysical methods in the study of GIFs. In the following, we will briefly review how different geophysical methods can be used to map GIFs and summarize the major findings of geophysical site surveys at the Lansjärv, Stuoragurra, Pärvie, Burträsk, Suasselkä and Bollnäs faults (Figure 7.1).

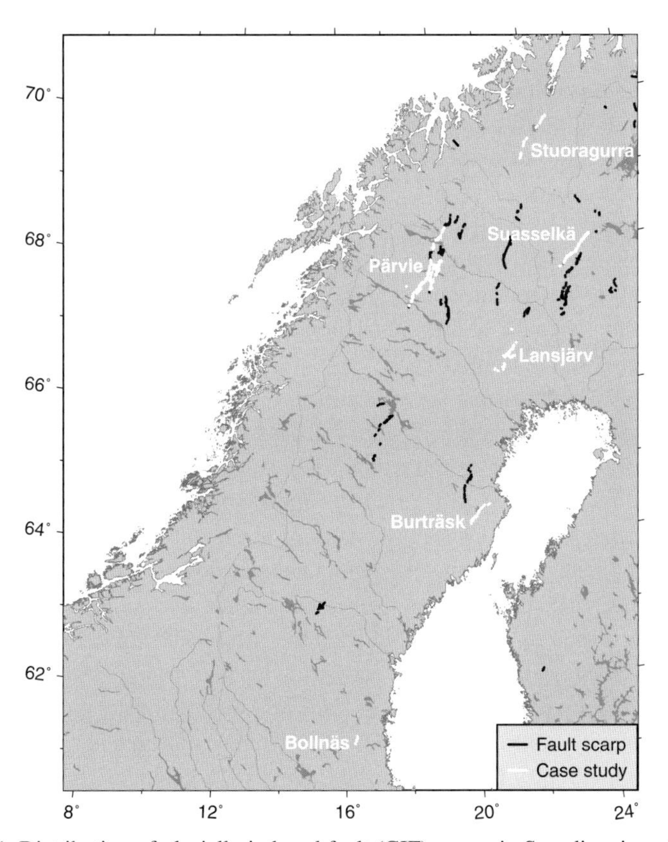

Figure 7.1 Distribution of glacially induced fault (GIF) scarps in Scandinavia and location of the case studies. The faults discussed here are shown in white.

7.2 Geophysical Methods

7.2.1 Seismic Methods

Seismology includes the study of the propagation of elastic waves in the subsurface. The main two ways of analyzing seismic data in exploration geophysics are the seismic reflection and refraction methods, the former being sensitive to relative changes in seismic velocity and density and the latter providing a more direct measurement of seismic velocity.

When a seismic wave encounters an interface where the product of seismic velocity and density – the seismic impedance – changes, a portion of the energy is reflected. The *seismic reflection method* aims at using reflected waves from such interfaces to illuminate the subsurface structure. After processing, impedance contrasts are displayed as reflections delineating the geometry of geological features such as lithological contacts, intrusions and deformation zones. In hard rock environments, faults and fracture zones are usually imaged as elongated linear (2D) or planar (3D) reflections (Figure 7.2) due to intense alteration and re-mineralization of the rocks in the deformation zones (e.g. Mair & Green, 1981; Juhlin, 1995). If the alteration is not strong enough to make the faults reflective, the faults might still be detected by offsets in the background reflectivity, as is common in sedimentary environments (e.g. Sheriff & Geldart, 1995). A number of factors determine which features can be imaged with seismic waves. A structure that is too small compared to the seismic wavelength will not produce a significant reflection and will thus not be detected. Furthermore, the amount of detail that seismic images can resolve depends on the wavelength, with shorter waves delineating much finer details. On the other hand, penetration depth depends on the frequency of the seismic waves, with lower frequencies penetrating much deeper (Figure 7.2). Therefore, it is always necessary to balance resolution and penetration to optimize a seismic survey. Nevertheless, the seismic reflection method is the geophysical method with the highest resolution at depth and is thus the main method for mapping the deeper structure of GIFs.

The path along which seismic waves travel through the subsurface is governed by the velocity structure of the subsurface. A velocity change leads to a change in the propagation direction of the waves, known as refraction. The *seismic refraction method* uses the travel times of such refracted waves to invert for a model of the subsurface velocities. In the simplest case, such a model consists of a series of layers with vertically varying velocities (Figure 7.3a), but with the advance of computational power, it has become possible to retrieve 2D, or even 3D, distributions using tomographic inversions (Figure 7.3b). In the hard rock environment of the Baltic Shield, the sediment/bedrock interface is usually characterized by a distinct increase in velocity with depth, allowing estimates of the bedrock depth from the velocity models. Fracturing leads to a loss of stiffness, manifesting itself as localized low-velocity zones in the otherwise relatively uniform bedrock (e.g. Moos & Zoback, 1983) (Figure 7.3). Depending on the acquisition geometry, the seismic refraction method can reach a high resolution, but the penetration into the bedrock is usually limited in exploration geophysical studies due to the relatively short profile lengths employed. Therefore, the common application of the method is mapping the sediment-bedrock interface and locating potential fracture zones in the shallower parts of the bedrock.

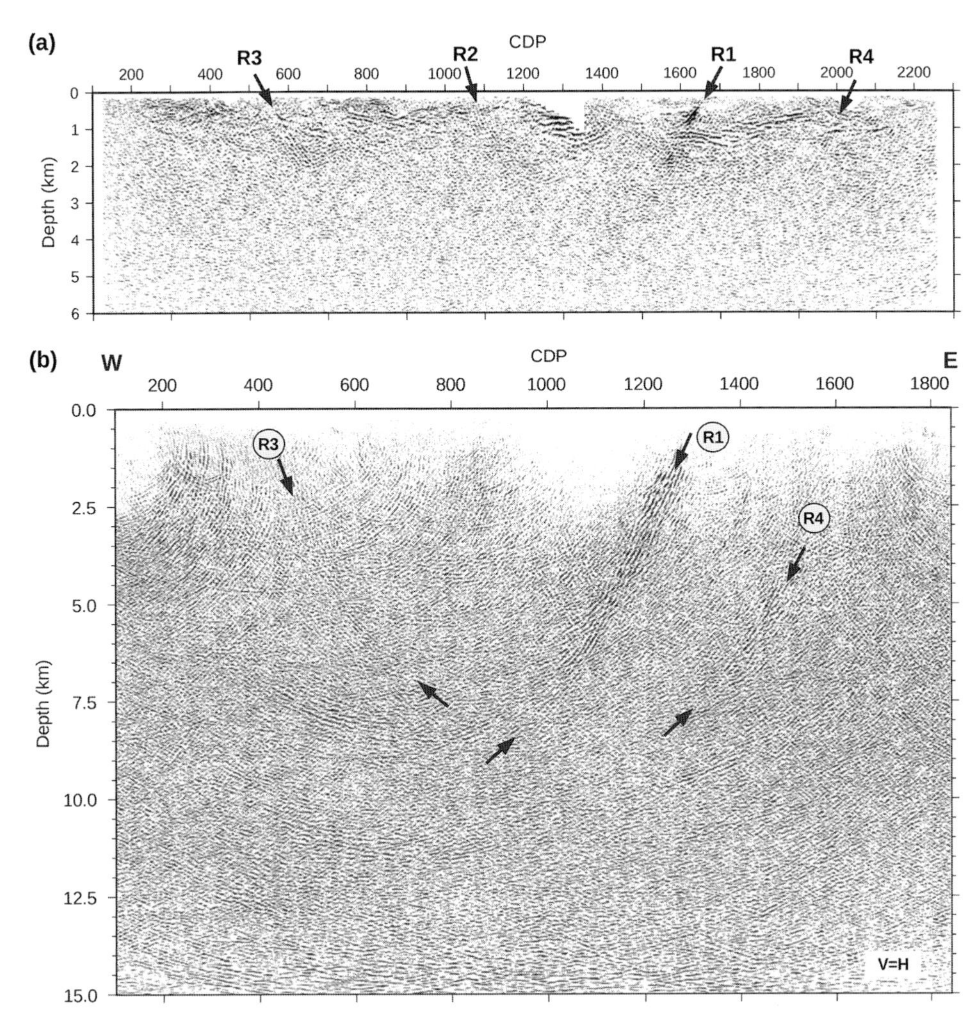

Figure 7.2 Reflection seismic images of the Pärvie Fault. a) First survey with a very dense source spacing enabling imaging the upper 2–3 km in high resolution (modified from Juhlin et al., 2010); b) second survey with a sparser source spacing, but with an explosive source, enabling imaging down to 7–8 km (modified from Ahmadi et al., 2015).

7.2.2 Ground-Penetrating Radar

The principle of ground-penetrating radar (GPR) measurements is similar to the reflection seismic method but using electromagnetic waves for the investigations instead of elastic waves. Recording reflected electromagnetic waves from interfaces with changes in the dielectric constant of the underground allows detailed imaging of subsurface structures (Figure 7.4). Compared to seismic sources, GPR antennas operate with much higher frequencies/shorter wavelengths (e.g. 50 MHz in Mauring et al., 1997; Olsen et al., 2018). Therefore, they provide much higher resolution images but much shallower depth penetration. Thus, GPR is mainly used to map the bedrock surface, shallow fracture zones

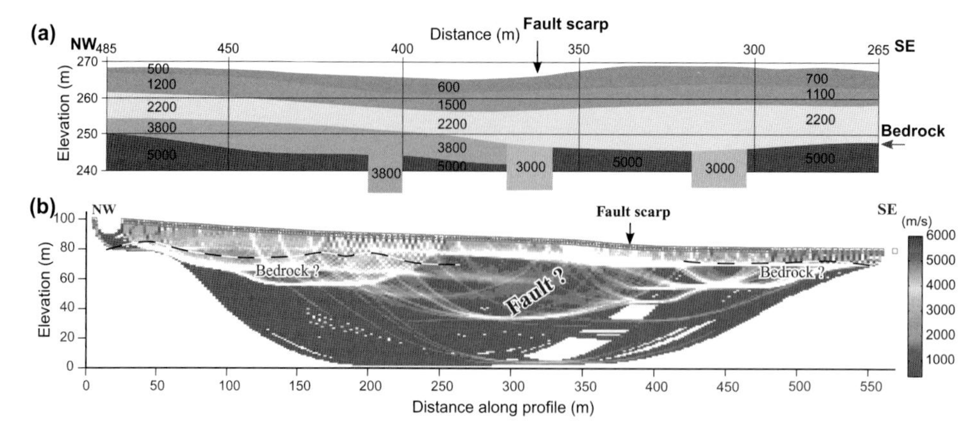

Figure 7.3 Refraction seismic imaging of GIFs. a) Layer-based refraction analysis of a profile across the Lansjärv Fault (redrawn from Henkel et al., 1983); b) travel time tomography of a refraction seismic profile across the Bollnäs Fault (modified from Malehmir et al., 2016). The numbers in (a) denote the velocity in m/s. (A black and white version of this figure will appear in some formats. For the colour version, please refer to the plate section.)

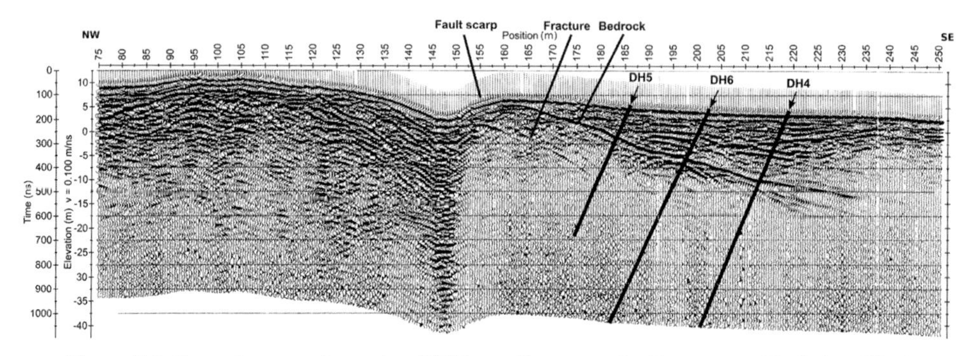

Figure 7.4 Ground-penetrating radar (GPR) profile across the Stuoragurra Fault (modified from Mauring et al., 1997).

or distinct sedimentary layers. Successful employment of GPR methods is highly dependent upon the amount of clay present and its degree of water saturation. Even small amounts of highly conductive clay will attenuate the radar waves very quickly so hardly any penetration into the subsurface occurs.

7.2.3 Geoelectrics

Direct current geoelectric measurements rely on injecting a current in the underground and measuring the resulting voltage distribution to investigate the resistivity structure of the subsurface. Depending on the electrode configuration, it is possible to map the mean resistivity of the upper metres of the subsurface or infer 1D resistivity profiles (Figure 7.5a). Modern instrumentation and processing techniques make it even possible to obtain 2D or 3D resistivity distributions (Figure 7.5b). Since electrical currents are mainly conducted

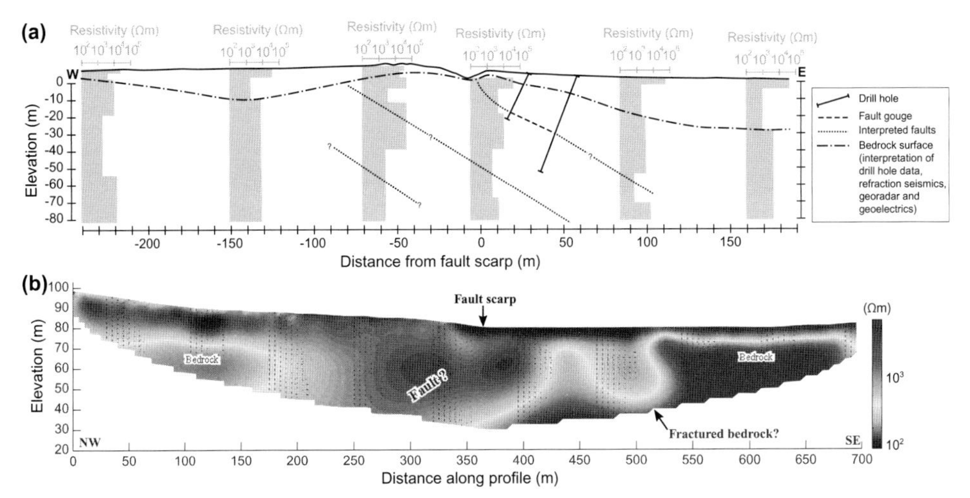

Figure 7.5 Geoelectric imaging of GIFs. a) Two-dimensional interpretation of six 1D resistivity profiles at the Stuoragurra Fault (redrawn from Olesen et al., 1992); b) 2D resistivity section across the Bollnäs Fault (modified from Malehmir et al., 2016) overlapping with the profile in Figure 7.3b. (A black and white version of this figure will appear in some formats. For the colour version, please refer to the plate section.)

through the pore space, typical low-porosity crystalline rocks feature very high resistivity values (Schön, 2011). Glacial sediments, on the other hand, are usually characterized by low resistivities due to higher porosities and a high clay content (Schön, 2011), making the sediment-bedrock contact quite suitable for mapping with geoelectrics. Fracture zones, brecciated and weathered bedrock are typically imaged as low-resistivity zones in the bedrock due to the presence of water, clay-minerals or the fractures themselves (e.g. McDowell, 1979). Depth penetration depends to a large extent on the acquisition geometry but is typically limited to the high tens to a few hundreds of metres. Therefore, geoelectrics are mostly used to map the bedrock surface and locate potential fracture zones close to it. When the data are inverted automatically, geoelectrics are referred to as electrical resistivity tomography (ERT).

7.2.4 Electromagnetic Methods

Electromagnetic measurements provide another way of determining the resistivity of the subsurface. Several different methods using different sources exist, but all rely on the same fundamental principle: a primary electromagnetic field induces secondary fields in the underground that interfere with the primary field and cause both changes in the amplitude and phase of the original signal. In the simplest case, these changes can be directly interpreted to estimate the mean resistivity in the shallow subsurface, but nowadays it is more common to use multi-frequency measurements to invert for a 2D resistivity distribution of the subsurface (Figure 7.6). Electromagnetic methods are especially sensitive to conductive structures (whereas geoelectric methods are more sensitive to resistive structures) and are, therefore, often used to identify potential fracture zones in the bedrock.

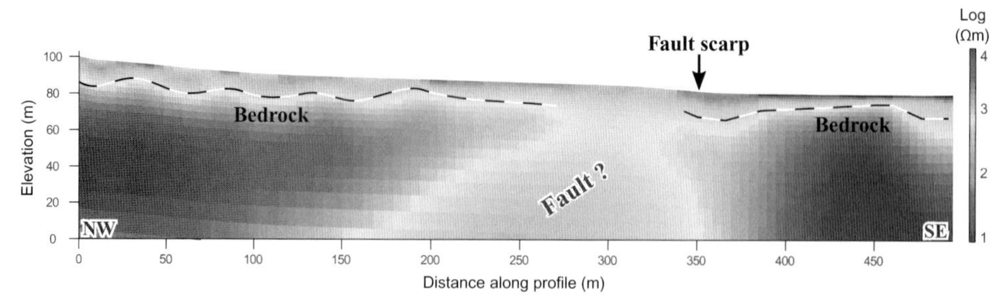

Figure 7.6 2D resistivity section from electromagnetic measurements across the Bollnäs Fault (modified from Malehmir et al., 2016) overlapping with the profile in Figure 7.3b. (A black and white version of this figure will appear in some formats. For the colour version, please refer to the plate section.)

The resolution of electromagnetic measurements decreases quickly with depth, but a good conductor can be detected at several kilometres of depth. There are numerous abbreviations for various electromagnetic measurement configurations. When both the electric and magnetic fields are measured and analyzed in the frequency domain the term 'magnetotellurics' (MT) is used if the sources are natural variations in the fields, and 'controlled source magnetotellurics' (CSMT) if an active source is employed. The very low frequency (VLF) method uses signals in the 15–30-kHz range from transmitters used for communication with submarines. Transient electromagnetics (TEM) involves the measurement of electric and magnetic fields in the time domain using an active loop transmitter.

7.2.5 Magnetics

The strength of Earth's magnetic field is locally influenced by the magnetic susceptibility of the rocks in the subsurface. Susceptibility is mainly controlled by the content of magnetic minerals like magnetite and ilmenite (Schön, 2011), which can vary significantly between different rock units. Aeromagnetic measurements covering large areas make it possible to analyze anomalies in the magnetic field on a regional scale to delineate rock units with distinctive magnetic signatures and to track magnetic anomalies and lineaments. Anomalies can be due to juxtaposition of rocks with different magnetic characteristics or elongated magnetic minima arising from the oxidization of magnetic minerals (e.g. magnetite to hematite) in fault zones (Henkel & Guzmán, 1977). The shape of such minima depends on the geometry of the fault zones, making it possible to estimate their dip. Thus, aeromagnetic measurements can be used to establish a tectonic framework of the area where GIFs occur (Figure 7.7a). Complementary high-resolution ground magnetic measurements may reveal further detailed information on the magnetic properties of GIFs not observable in airborne data (Figure 7.7b).

7.2.6 Gravity

Similar to the magnetic field, the strength of the gravity field depends on the density distribution in the subsurface. It is possible to derive certain information about density

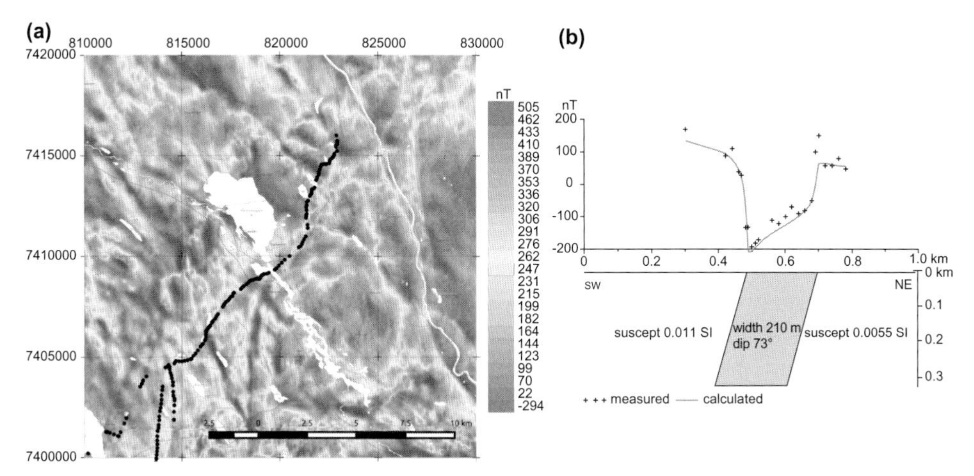

Figure 7.7 a) Aeromagnetic map over the Lansjärv area with the fault scarp shown with black dots (magnetic data © Geological Survey of Sweden); b) forward modelling of the magnetic signature (ground magnetic data) of a fault zone in the Lansjärv area (redrawn from Henkel, 1987). (A black and white version of this figure will appear in some formats. For the colour version, please refer to the plate section.)

variations from gravity measurements, but because such information is ambiguous, gravity data are usually analyzed in combination with other geophysical data to model larger structures like crustal-scale fracture zones. Complementary high-resolution ground gravity measurements have the potential to detect buried bedrock scarps if the cover is thin enough and the bedrock step large enough.

7.3 Case Studies

7.3.1 Near-Surface Studies: Lansjärv, Stuoragurra and Bollnäs

When the first major geophysical investigations of GIFs were conducted at the Lansjärv and Stuoragurra faults[1], acquisition and interpretation possibilities were still quite limited. The investigations at the **Lansjärv Fault** were part of a larger investigation aimed at understanding the regional tectonic framework of glacially induced faulting and locating the optimum site for drilling and trenching. In the first part of the study, Henkel et al. (1983) and Henkel (1987) detected potential regional scale fault zones by joint interpretation of aeromagnetic, gravity and topography data and estimated the dip of the fault zones based on forward modelling of magnetic and gravity anomalies (see Figure 7.7b for modelling of a magnetic anomaly). The Lansjärv Fault follows a gently dipping, NNE-SSW-striking set of fracture zones, suggesting that most of the fault movements occurred as a result of reactivation of older structures (Henkel 1987). Complementary to the regional investigations, Henkel et al. (1983) and Henkel (1987) collected an extensive ground

[1] There were also a number of smaller geophysical surveys across different fault scarps in Finland that were methodologically very similar to the Stuoragurra and Lansjärv surveys. They are not included in this review, but a summary can be found in Kuivamäki et al. (1998).

geophysical dataset over the fault scarp. The fault manifested itself as an asymmetric conductive anomaly in the electromagnetic data, indicating a high water content and a steep easterly dip of the fault. In the seismic refraction data, the fault coincided with localized low-velocity zones in the bedrock and was associated with either a positive or negative step in the bedrock (Figure 7.3a). Henkel (1987) interpreted this as an indication of one steeply and one gently dipping fault segment. Although the geophysical data provided important constraints on the characteristics and orientation of the fault, Henkel et al. (1983) and Henkel (1987) could not directly image the fault at depth.

In a similar study of the **Stuoragurra Fault**, Olesen et al. (1992) first analyzed regional scale aeromagnetic, gravity and topography data to establish the tectonic framework of the area. They found the fault scarp to lie within the regional scale, NE-SW-trending Mierojávri–Sværholt shear zone (MSSZ). To characterize the fault itself, Olesen et al. (1992) carried out a detailed near-surface geophysical survey. Both geoelectric and electromagnetic measurements yielded lower resistivity values in the hanging wall, indicating a higher degree of fracturing than in the footwall. To estimate the 2D resistivity variations across the fault, they combined several 1D resistivity profiles for 2D interpretation (Figure 7.5a). From the complex resistivity pattern, they inferred that the Stuoragurra Fault is situated within a 200-m-wide fault zone belonging to the regional MSSZ. In contrast to the study at the Lansjärv Fault, seismic refraction profiles acquired across the fault did not show localized low velocity zones coinciding with the fault, and bedrock depth estimations did not match results from percussion drilling. The bedrock could, however, be mapped successfully on GPR profiles, where it appeared as a distinct reflector. Reflections from the fault itself were absent, but an additional profile acquired with a more powerful transmitter delineated a dipping reflector parallel to the dip of the Stuoragurra Fault (Mauring et al., 1997) (Figure 7.4). From the combination of the different geophysical results, Olesen et al. (1992) inferred that the Stuoragurra Fault is situated in a several hundred-meter-wide fault zone. They concluded that geophysical methods could characterize the environment where the fault occurred but could not detect or map the fault itself. Recent studies using MT and ERT methods are consistent with previous interpretations of the Stuoragurra Fault (Dalsegg & Olesen, 2014; Mrope et al., 2019).

Since these first surveys, both geophysical instrumentation and analysis methods have developed significantly. In 2014, Malehmir et al. (2016) conducted a multidisciplinary geophysical survey across the newly discovered **Bollnäs Fault** scarp (Smith et al., 2014; Mikko et al., 2015) to corroborate the existence of a fault zone in the bedrock. While the results of the early surveys were mostly limited to one-dimensional or simple, layer-based models, Malehmir et al. (2016) could retrieve two-dimensional sections of the physical parameters (Figures 7.3b, 7.5b and 7.6). Contrary to expectations, gravity and magnetic measurements indicated an even or slightly shallower bedrock level in the footwall of the fault. To obtain more detailed information on bedrock levels, Malehmir et al. (2016) acquired seismic, electromagnetic and geoelectric profiles across the scarp. Both the seismic velocity distribution obtained from tomography based on the seismic travel times and the resistivity sections inverted from the electromagnetic and geoelectric data confirm similar bedrock levels on both sides of the fault and feature a westward dipping low-velocity/high-conductivity zone in the bedrock below, indicating a major fracture zone (Figures 7.3b, 7.5b and 7.6). Thus, they were able to map both the bedrock surface and the uppermost part of the fault. They explain their results with a multi-phase deformation of an older deformation zone that initially acted as a normal fault and that was reactivated as a reverse GIF, leading to the formation of the fault scarp and levelling of the bedrock.

7.3.2 Deeper Studies: Pärvie, Burträsk, Suasselkä

In an effort to understand the processes triggering the formation of GIFs by mapping their geometry at depth, several major faults were investigated with reflection seismics. At the **Pärvie Fault**, Juhlin et al. (2010) acquired a 23-km-long profile with dense source spacing to obtain a high-resolution image of the upper 5–6 km of the crust. After processing, the final image shows a set of four steeply dipping reflections imaged to depths of 1–3 km (Figure 7.2a). Three of these reflections project to the surface very close to the main scarp (R3) and to subsidiary scarps (R2 and R1). Therefore, Juhlin et al. (2010) interpreted them as reflections from different branches of the Pärvie Fault and inferred a dip of approximately 50° east for the main branch. The fourth reflection (R4) could not be associated with any known fault scarp. However, Mikko et al. (2015) later discovered a previously unknown fault scarp on LiDAR data where the reflection projects to the surface.

To reveal the deeper structure of the Pärvie Fault, Ahmadi et al. (2015) conducted a second reflection seismic survey along the same profile using an explosive source, but with a less dense source spacing. This resulted in a sparse dataset with a good signal-to-noise ratio in the deeper part but poorer resolution in the upper 1–2 km. In the final image, the reflection from the main fault is relatively weak but can be traced to a greater depth (R3 in Figure 7.2b). The strongest reflection (R1) is linked to a west-dipping subsidiary fault zone and can now be traced to a depth of 8 km. A third, weaker reflection (R4) can be traced from 3.5 to 7 km and corresponds to the fourth reflection of Juhlin et al. (2010). Using a more accurate 3D velocity model from local earthquake tomography by Lindblom et al. (2015), Ahmadi et al. (2015) estimated a slightly steeper dip of 65° for the main fault. Hence, changing the acquisition scheme made it possible to map the fault system to a depth of 7–8 km, but the continuation below this depth remains uncertain.

Another major seismic survey was conducted across the **Burträsk Fault** using a very similar acquisition geometry as in the first Pärvie survey. The 22-km-long profile followed local roads and intersected the fault close to a sharp bend in the fault scarp (inset in Figure 7.8a). In contrast to the Pärvie survey, the final image does not show any clear reflection linked to the fault scarp (Figure 7.8a). Juhlin and Lund (2011) attributed this to the complicated geometry of both the Burträsk Fault and acquisition line. However, they observed a clear south-east dipping reflection 1 km to the south-east (B2 in Figure 7.8a), which they interpreted to originate from a segment of the scarp to the south-west of the profile. From this reflection, they inferred a dip of 55° for the Burträsk Fault. Furthermore, the image features a second, stronger south-east dipping reflection that can be traced to

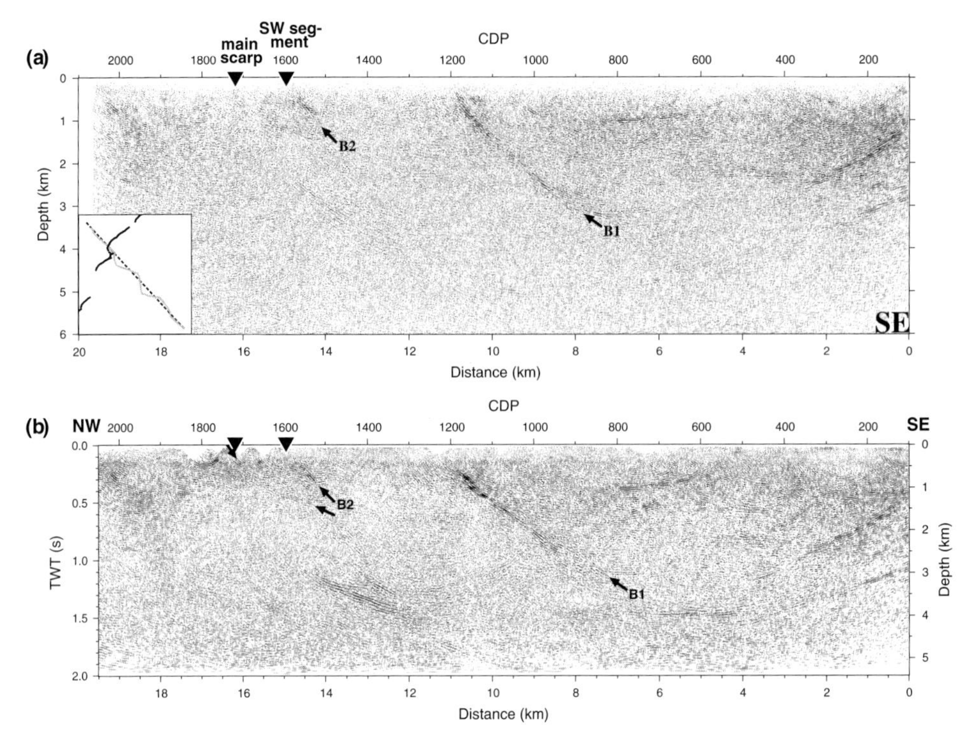

Figure 7.8 Reflection seismic image of the Burträsk Fault. a) Original image (modified from Juhlin & Lund, 2011); b) reprocessed image (modified from Beckel & Juhlin, 2019). The inset shows the fault scarp (black), the acquisition line (light grey) and the processing line (dashed grey). Reflection B2 can be interpreted as being generated by the GIF, whereas B1 cannot be associated with any known fault.

3-km depth (B1 in Figure 7.8a). They interpreted this reflection as a fault zone but found no indications that it was active during the last deglaciation.

Beckel and Juhlin (2019) reprocessed the Burträsk dataset, applying a new and modified cross-dip correction to account for imaging problems caused by the crookedness of the profile. They managed to increase the clarity of the previously imaged reflections significantly (Figure 7.8b), but they also were not able to observe any reflection directly connected to the fault scarp. They explained this as the result of significantly decreased signal quality and fold of the seismic data close to the fault scarp due to the close proximity of residential housing.

In an effort to map the geometry of the **Suasselkä Fault**, Abdi et al. (2015) reprocessed three seismic profiles acquired during the HIRE project – a project designed for delineating the upper crustal structures of a nearby gold mining district (Kukkonen et al., 2009). With a specially adapted processing flow Abdi et al. (2015) were able to obtain a much clearer image of the upper kilometres of the crust, including reflections from the previously poorly imaged Suasselkä Fault. They discovered two south-east dipping reflections extending to a depth of 3 km that could be directly associated with segments of the fault scarp (S1 and S2

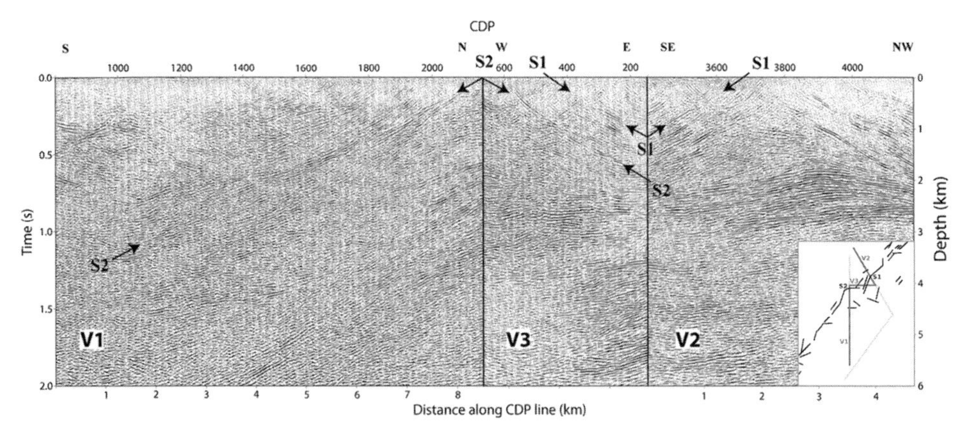

Figure 7.9 Merged reflection seismic image from three profiles across the Suasselkä Fault (modified from Abdi et al., 2015). The inset shows the fault scarp (black) and the processing lines (grey).

in Figure 7.9) and estimated their dips to be 35° and 45°. Since the seismic profiles feature several other reflections with a similar orientation, Abdi et al. (2015) suggested the Suasselkä Fault to be associated to several fault planes.

7.4 Discussion

The case studies presented here illustrate the large potential of applied geophysical methods for identifying and characterizing GIFs, but also the differing fields of application of the methods. While aeromagnetic and gravity data are mostly used in tectonic studies for detecting regional scale fault zones, refraction seismics, geoelectrics, electromagnetics and GPR are often employed for characterizing the near surface of GIFs. The comparison of the early Lansjärv and Stuoragurra surveys and the recent Bollnäs survey demonstrates clearly how near-surface geophysical methods have advanced in past decades. However, the loss of resolution at depth is inherent to the methods and cannot be overcome by more advanced acquisition and processing schemes. Therefore, mapping the deeper structure of the faults is mainly dependent upon using the reflection seismic method.

The reflection seismic method has been successfully applied at the Pärvie, Burträsk and Suasselkä faults, and even led to the detection of a previously unknown subsidiary fault within the Pärvie Fault System. However, the comparison of the two Pärvie surveys also demonstrates how sensitive the method is to the acquisition geometry and the source type. While the use of a more powerful source allows mapping fault zones to much greater depths, it usually entails a sparser shot spacing because of higher operational costs and increased danger of damaging nearby infrastructure. However, good resolution of the shallow subsurface can only be achieved by a very dense shot and receiver spacing. Another challenge is that lower frequencies penetrate deeper, but GIFs seem to be rather narrow features (Olesen et al., 1992; Roberts et al., 1997; Kuivamäki et al., 1998; Ahmadi et al., 2015). Consequently, they might fall below the detection limit below a certain depth, since the higher frequencies become attenuated with depth. The maximum depth of 7–8 km to which the Pärvie Fault has been mapped may, therefore, be either the maximum depth extent of the fault; or a change in the fault's physical parameters due to the closing of micro-fractures at this depth; or just the detection limit of the fault for the chosen acquisition parameters. In addition, the improvements achieved by reprocessing the Burträsk and Suasselkä datasets illustrate how critical the choice of processing methods is. Thus, careful survey design and adapted processing flows are essential for imaging GIFs.

Last but not least, all geophysical methods have in common that their results need to be interpreted to be linked to geological features and that these interpretations are in most cases ambiguous. It is, for example, usually not possible to distinguish between glacially induced and other kinds of fault zones. Therefore, complementary information from e.g. geological or tectonic studies and borehole data is essential to make use of the immense potential of geophysical mapping.

7.5 Conclusions

In summary, applied geophysical methods are valuable tools for imaging GIFs in the subsurface. Reflection seismic methods have shown good capability to delineate the geometry of GIFs in the upper crust and have been used to map faults dipping at 35–65° down to a maximum depth of 7–8 km. Whether this depth represents the true depth extent of the faults or a lower limit of detectability is still not known since reflection seismic imaging is very sensitive to the chosen acquisition and processing parameters. Other methods, including seismic refraction, GPR, geoelectrics, electromagnetics and ground magnetics, are mostly used for characterizing the properties of GIFs in the shallow subsurface and mapping the bedrock surface, since they either do not penetrate deep enough or their resolution at depth is not sufficient for imaging the deeper parts of narrow fault zones. As the interpretation of all geophysical results depends on complementary information, it is most beneficial to combine geophysical investigations with geological, trenching and/or drilling studies.

References

Abdi, A., Heinonen, S., Juhlin, C. and Karinen, T. (2015). Constraints on the geometry of the Suasselkä post-glacial fault, northern Finland, based on reflection seismic imaging. *Tectonophysics*, **649**, 130–138, doi.org/10.1016/j.tecto.2015.03.004.

Ahmadi, O., Juhlin, C., Ask, M. V. S. and Lund, B. (2015). Revealing the deeper structure of the endglacial Pärvie fault system in northern Sweden by seismic reflection profiling. *Solid Earth*, **6**, 621–632 doi.org/10.5194/se-6-621-2015.

Beckel, R. A. and Juhlin, C. (2019). The cross-dip correction as a tool to improve imaging of crooked-line seismic data: a case study from the post-glacial Burträsk fault, Sweden. *Solid Earth*, **10**, 581–598, doi.org/10.5194/se-10-581-2019.

Dalsegg, E. and Olesen, O. (2014). *Resistivitetsmålinger ved Masi, Fiednajohka og Riednajavre og implikasjoner for malmleting, Kautokeino kommune, Finnmark.* Report 2014.021, Geological Survey of Norway, Trondheim, Norway.

Henkel, H. (1987). *Tectonic Studies in the Lansjärv Region*. SKB Technical Report TR 88-07, Swedish Nuclear Fuel and Waste Management Co., Stockholm, 80 pp.

Henkel, H. and Guzmán, M. (1977). Magnetic features of fracture zones. *Geoexploration*, **15**(3), 173–181.

Henkel, H., Hult, K., Eriksson, L. and Johansson, L. (1983). *Neotectonics in Northern Sweden – Geophysical Investigations*. SKB Technical Report TR 83-57, Swedish Nuclear Fuel and Waste Management Co., Stockholm, Sweden.

Juhlin, C. (1995). Imaging of fracture zones in the Finnsjön area, central Sweden, using the seismic reflection method. *Geophysics*, **60**(1), 66–75, doi.org/10.1190/1.1443764.

Juhlin, C. and Lund, B. (2011). Reflection seismic studies over the end-glacial Burträsk fault, Skellefteå, Sweden. *Solid Earth*, **2**, 9–16, doi.org/10.5194/se-2-9-2011.

Juhlin, C., Dehghannejad, M., Lund, B., Malehmir, A. and Pratt, G. (2010). Reflection seismic imaging of the end-glacial Pärvie Fault system, northern Sweden. *Journal of Applied Geophysics*, **70**, 307–316, doi.org/10.1016/j.jappgeo.2009.06.004.

Kuivamäki, A., Vuorela, P. and Paananen, M. (1998). *Indications of Postglacial and Recent Bedrock Movements in Finland and Russian Karelia*. Geological Survey of Finland, Nuclear Waste Disposal Research Report YST-99, Espoo, Finland, 92 pp.

Kujansuu, R. (1964). Nuorista sirroksista Lapissa [English summary: Recent faults in Lapland]. *Geologi*, **16**, 30–36 (in Finnish).

Kukkonen, I., Lahti, I., Heikkinen, P. et al. (2009). *HIRE Seismic Reflection Survey in the Suurikuusikko Gold Mining and Exploration Area, North Finland*. Report Q 23/2009/28, Geological Survey of Finland.

Lagerbäck, R. (1978). Neotectonic structures in northern Sweden. *Geologiska Föreningen i Stockholm Förhandlingar*, **100**(3), 263–269, doi.org/10.1080/11035897809452533.

Lindblom, E., Lund, B., Tryggvason, A. et al. (2015). Microearthquakes illuminate the deep structure of the endglacial Pärvie fault, northern Sweden. *Geophysical Journal International*, **201**, 1704–1716, doi.org/10.1093/gji/ggv112.

Mair, J. A. and Green, A. G. (1981). High-resolution seismic reflection profiles reveal fracture zones within a 'homogeneous' granite batholith. *Nature*, **294**, 439–442, doi.org/10.1038/294439a0.

Malehmir, A., Andersson, M., Mehta, S. et al. (2016). Post-glacial reactivation of the Bollnäs fault, central Sweden – a multidisciplinary geophysical investigation. *Solid Earth*, **7**, 509–527, doi.org/10.5194/se-7-509-2016.

Mauring, E., Olesen, O., Rønning, J. S. and Tønnesen, J. F. (1997). *Ground-Penetrating Radar Profiles across Postglacial Fault at Kåfjord, Troms and Fidnajohka, Finnmark*. Report 97.174, Geological Survey of Norway, Trondheim, Norway.

McDowell, P. W. (1979). Geophysical mapping of water-filled fracture zones in rocks. *Bulletin of the International Association of Engineering Geology*, **19**(1), 258–264, doi.org/10.1007/BF02600485.

Mikko, H., Smith, C. A., Lund, B., Ask, M. V. S. and Munier, R. (2015). LiDAR-derived inventory of post-glacial fault scarps in Sweden. *GFF*, **137**(4), 334–338, doi.org/10.1080/11035897.2015.1036360.

Moos, D. and Zoback, M. D. (1983). In situ studies of velocity in fractured crystalline rocks. *Journal of Geophysical Research*, **88**(B3), 2345–2358, doi.org/10.1029/JB088iB03p02345.

Mrope, F. M., Becken, M., Ruud, B. O. et al. (2019). *Magnetotelluric 2D Inversion and Joint Interpretation of MT, Seismic, Magnetic and Gravity Data from Masi, Kautokeino Municipality, Finnmark*. Report 2019.009, Geological Survey of Norway, Trondheim, Norway.

Olesen, O., Henkel, H., Lile, O. B., Mauring, E. and Rønning, J. S. (1992). Geophysical investigations of the Stuoragurra postglacial fault, Finnmark, northern Norway. *Journal of Applied Geophysics*, **29**, 95–118, doi.org/10.1016/0926-9851(92)90001-2.

Olsen, L., Olesen, O., Dehls, J. and Tassis, G. (2018). Late-/postglacial age and tectonic origin of the Nordmannvikdalen Fault, northern Norway. *Norwegian Journal of Geology*, **98**, 483–500, doi.org/10.17850/njg98-3-09.

Roberts, D., Olesen, O. and Karpuz, M. R. (1997). Seismo- and neotectonics in Finnmark, Kola Peninsula and the southern Barents Sea. Part 1: geological and neotectonic framework. *Tectonophysics*, **270**, 1–13, doi.org/10.1016/S0040-1951(96)00173-4.

Schön, J. H. (2011). *Physical Properties of Rocks, Vol. VIII of Handbook of Petroleum Exploration and Production*. Elsevier, Oxford.

Sheriff, R. E. and Geldart, L. P. (1995). *Exploration Seismology*, 2nd ed., Cambridge University Press, Cambridge, doi.org/10.1017/CBO9781139168359.

Smith, C. A., Sundh, M. and Mikko, H. (2014). Surficial geology indicates early Holocene faulting and seismicity, central Sweden. *International Journal of Earth Sciences*, **103**, 1711–1724, doi.org/10.1007/s00531-014-1025-6.

8

Dating of Postglacial Faults in Fennoscandia

COLBY A. SMITH, ANTTI E. K. OJALA, SUSANNE GRIGULL AND HENRIK MIKKO

ABSTRACT

Numerous methods have been applied to dating postglacial faults in Fennoscandia. Traditionally, these range from determining relative ages based on cross-cutting relationships to determining absolute ages based on stratigraphy and radiocarbon dates. More recently, however, direct dating of fault scarps using terrestrial cosmogenic nuclide dating has been attempted.

The benefits and limitations of these methods are described citing examples from recent literature. Subsequently, the dates themselves are discussed in the context of the longstanding hypothesis that postglacial faults in Fennoscandia ruptured only once during or shortly after deglaciation. While each of the studies reviewed applies only to the investigated faults, collectively recent literature indicates a longer lasting and more complex spatial and temporal history of postglacial faulting in the Fennoscandian Shield area.

8.1 Introduction

Postglacial faults (PGFs), also termed 'glacially induced faults' (GIFs), were first recognized by geomorphic relationships in places where abrupt scarps cross-cut glacial deposits or landforms (Kujansuu, 1964; Lundqvist & Lagerbäck, 1976) (Figure 8.1). The early mapping carried out using aerial photographs as well as the more recent efforts using high-resolution digital elevation models (DEM) rely on this principle (Lagerbäck, 1978; Mikko et al., 2015; Palmu et al., 2015). However, glacial landforms in large parts of northern Fennoscandia do not date to the Late Weichselian. Rather, many glacial sediments and landscapes in the Fennoscandian Shield area have been preserved beneath multiple Late Quaternary glaciations with cold-based ice sheets that were characterized by insignificant glacial erosion (Lagerbäck, 1988; Lagerbäck & Robertsson, 1988). Thus, further investigations are needed to confirm the presence of fault scarps and their postglacial ages, as well as the potential reactivation of the same fault zones during and after subsequent deglaciations of the Late Quaternary. Towards this end, several studies have examined surface ruptures with LiDAR-based remote sensing and excavated trenches across postglacial fault scarps in order to examine their stratigraphy in the Fennoscandian Shield area

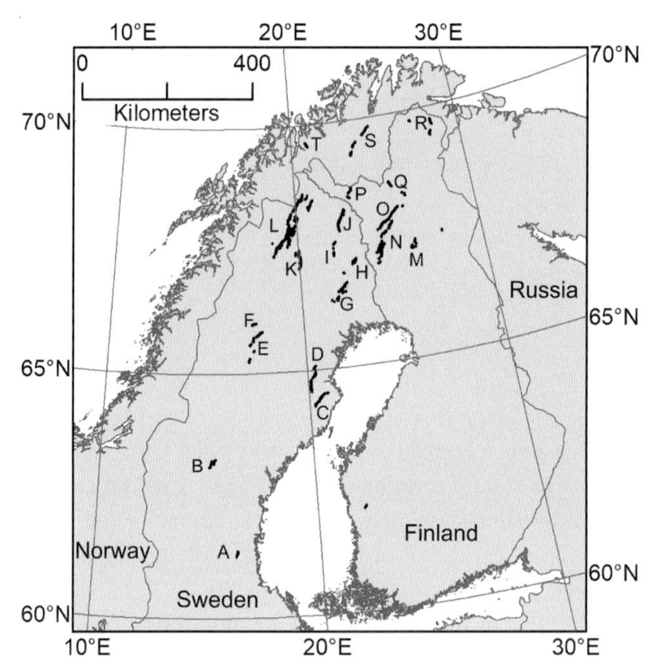

Figure 8.1 The map of northern Fennoscandia shows the locations of known postglacial fault systems. A – Bollnäs; B – Lillsjöhögen/Ismunden; C – Burträsk; D – Röjnoret; E – Sorsele; F – Laisvall; G – Lansjärv; H – Suorsapakka; I – Merasjärvi; J – Lainio; K – Sjaunja; L – Pärvie; M – Vaalajärvi; N – Isovaara-Riikonkumpu; O – Suasselkä; P – Palojärvi; Q – Palo-Peuravaara; R – Sevettijärvi; S – Stuoragurra; T – Nordmannvikdalen.

(Lagerbäck, 1990, 1992; Lagerbäck & Sundh, 2008; Smith et al., 2014; Ojala et al., 2017). Both methods have limitations that depend on the local geology and at best provide relative ages of fault ruptures. More recent studies have sought to employ both indirect and direct dating methods to refine the chronological control of fault ruptures (Ojala et al., 2018a,b; Ojala et al., 2019d).

This chapter reviews the literature and summarizes the different dating methods applied to postglacial faults using illustrative examples of each method. The goal is not a complete list of ages for postglacial faults in Fennoscandia but rather a discussion of dating methods and the latest revelations derived from them.

8.2 Geomorphic Dating

Geomorphic dating of postglacial faults in Fennoscandia relates predominantly to two suites of landforms: glacial/deglacial features and palaeoshorelines. The widespread availability of high-resolution LiDAR DEM in Fennoscandia has revolutionized this method over the past decade. Such tools are now standard (Mikko et al., 2015; Palmu et al., 2015).

8.2.1 Geomorphic Dating with Glacial Landforms

The Pärvie Fault System has long been known to have scarp segments that are both older and younger than the Late Weichselian deglaciation (Lundqvist & Lagerbäck, 1976; Lagerbäck & Sundh, 2008). The most obvious evidence of fault rupture occurring both pre- and postglacially comes from the relationships between eskers and fault scarps (Figure 8.2). In some places, undisturbed eskers overlie fault scarps indicating that fault rupture predates the deposition of the esker beneath an ice sheet. In other places, the fault cross-cuts the esker indicating that fault rupture occurred after deposition of the esker and likely after the retreat of the ice sheet. The preservation of fault scarps beneath ice has been interpreted to suggest that fault rupture occurred shortly before the late Weichselian deglaciation (Lundqvist & Lagerbäck, 1976; Lagerbäck & Sundh, 2008). However, given the widespread existence of preserved Early Weichselian landscapes in northern

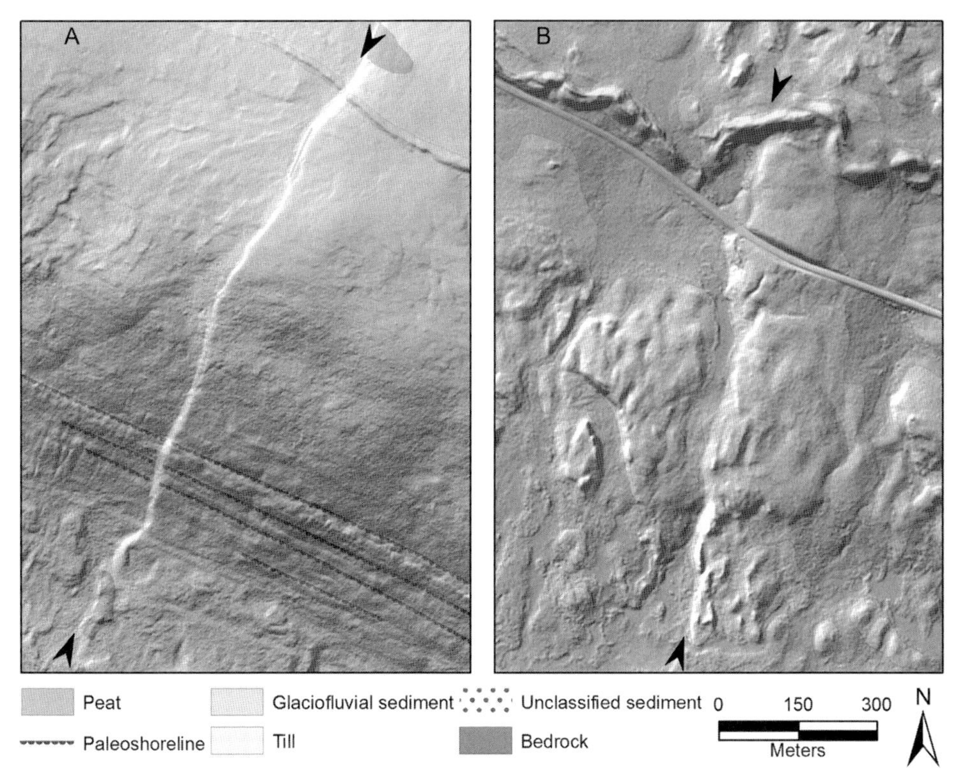

Figure 8.2 A) The north-west facing fault scarp marked by the black arrows cross-cuts not only the esker, which formed subglacially, but also the palaeoshorelines of an ice-dammed lake, which formed subaerially during deglaciation. Thus, faulting occurred after formation of the shorelines. The centre of the figure is at 67.4282°N, 18.7068°E. B) The less distinct west facing fault scarp marked by the black arrows does not cross-cut the esker. Thus, faulting occurred prior to deposition of the esker which occurred beneath an ice sheet. The centre of the figure is at 67.8558°N, 19.4593°E. (A black and white version of this figure will appear in some formats. For the colour version, please refer to the plate section.)

Fennoscandia, a more conservative interpretation is that fault rupture may have occurred both during prior deglaciations and during the Late Weichselian deglaciation.

Numerous other examples of geomorphic dating based on glacial landforms exist across Fennoscandia. A portion of the Lainio Fault scarp constrains a meltwater channel interpreted to be Late Weichselian. Thus, at least part of the scarp height predates deglaciation in order for it to define the course of a meltwater channel (Lagerbäck & Sundh, 2008).

At the Suaspalo and Retu study sites, in northern Finland, the Late Weichselian lateral drainage channels and a proglacial ice lake spillway are cross-cut by the fault segments of the Suasselkä GIF Complex, thus indicating their late- to postglacial rupturing (Ojala et al., 2019d) (Figure 8.3A).

The situation is somewhat different in the Vaalajärvi area, where Late- and postglacial drainage channels cross-cut the Vaalajärvi GIF Segment indicating that the faulting predates the erosional features (Ojala et al., 2019c) (Figure 8.3B). In the Ristonmännikkö GIF Segment of the same Vaalajärvi Complex, the fault offset is not seen at the bottom of a braided subglacial to proglacial meltwater channel that is cutting through the fault rupture (Figure 8.3C, Ojala et al., 2019c). In this case, the fault ramp has either been eroded by the meltwater activity, and subsequently the remaining depression was filled and levelled by postglacial sand and peat, or the meltwater activity really postdates the rupturing event.

In another example, Ojala et al. (2019c) provide an intriguing case of relative dating based on geomorphic evidence in the Ruokojärvi Fault Segment of the Pasmajärvi GIF Complex (Figure 8.3D). They identified a total collapse of an esker drainage tunnel system at the fault location, which indicates that a subglacial earthquake occurred during deglaciation. The eastern side of the fault ramp at the Ruokojärvi site is characterized by a distinct and traditional triangle-shaped esker ridge, whereas on the western side of the ramp the entire glaciofluvial system has collapsed and produced chaotic ridges trending in multiple directions. Their interpretation was that glaciofluvial material was squeezed in crevasses and other ice-fracturing patterns during earth-shaking and sudden ice collapse.

Earlier, Sutinen et al. (2014) described how the Paatsikkajoki Fault Segment of the Palojärvi GIF Complex cross-cuts the Late Weichselian active-ice streamlining lineaments and ice-marginal braided meltwater channels. This evidence indicates that at least one of the faulting events occurred subaerially and after the continental ice sheet had retreated from the area (Figure 8.3E), thus being truly postglacial in nature.

8.2.2 Geomorphic Dating with Palaeoshorelines

The second suite of landforms used to date postglacial fault ruptures are palaeoshorelines. Due to the existence of radiocarbon dated shoreline displacement curves of the Baltic Sea basin in Fennoscandia, cross-cutting relationships between postglacial faults and palaeoshorelines can be dated with decent accuracy provided that such shoreline displacement curves are locally available.

In south-western Finland, Palmu et al. (2015) discovered based on LiDAR DEMs a previously unknown postglacial fault in the Lauhavuori National Park area. The fault

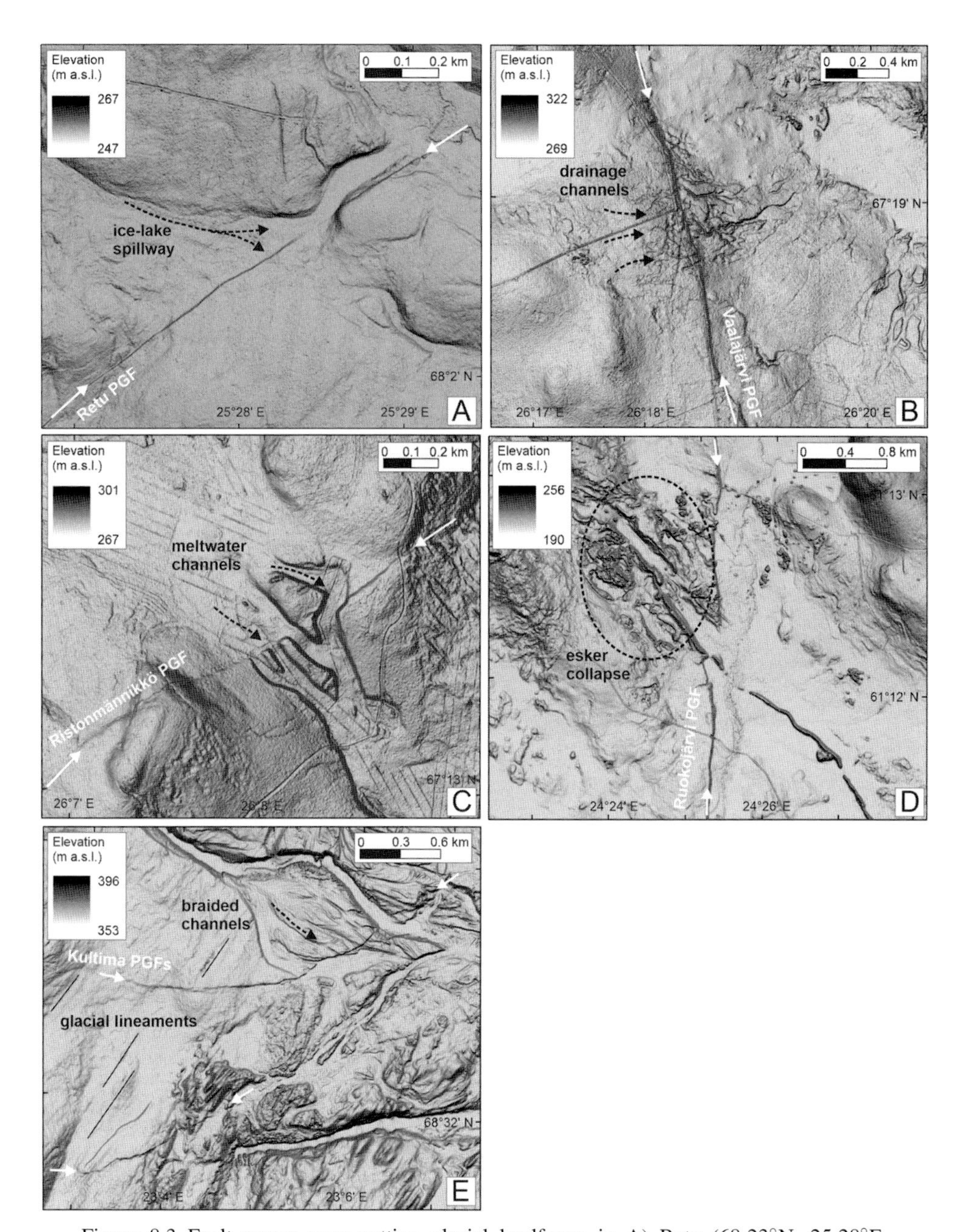

Figure 8.3 Fault scarps cross-cutting glacial landforms in A) Retu (68.23°N, 25.28°E, Suasselkä GIF Complex), B) Vaalajärvi (67.19°N, 26.18°E, Vaalajärvi GIF Complex), C) Ristonmännikkö (67.13°N, 26.82°E, Vaalajärvi GIF Complex); D) Ruokojärvi (61.12°N, 24.25°E, Pasmajärvi GIF Complex) and E) Kultima (68.32°N, 23.45°E, Palojärvi GIF Complex), northern Finland (see Ojala et al., 2019c).

system is 6 km long in total and composed of six GIF segments that rupture to the surface and cross-cut postglacial sediments. The maximum height of the fault segments is about 1.5 m and the mean vertical offset is only 0.6 m. Based on a very linear portrait of the fault, Ojala et al. (2019c) suspected that the orientation of the Lauhavuori GIF must be relatively vertical and striking SW-NE. The importance of the finding is that the LiDAR-detected surface rupture cuts through palaeoshorelines only above approximately 140 m above sea level (a.s.l.), although its continuation was also identified below this elevation with ground-penetrating radar (Ojala et al., 2019c) (Figure 8.4). The interpretation is that surface rupturing of the Lauhavuori GIF occurred about 1,500 years after deglaciation, when the ancient shoreline of the Baltic Sea basin was at 140 m a.s.l. in the Lauhavuori area, which according to the shoreline displacement information of Salomaa (1982) was 9,500–9,600 calibrated years before present (cal BP).

A Swedish example of faulted shorelines lies along an eastern segment of the Lansjärv Fault (Figure 8.5) (see Smith et al., Chapter 12). A flight of palaeoshorelines is cross-cut by a

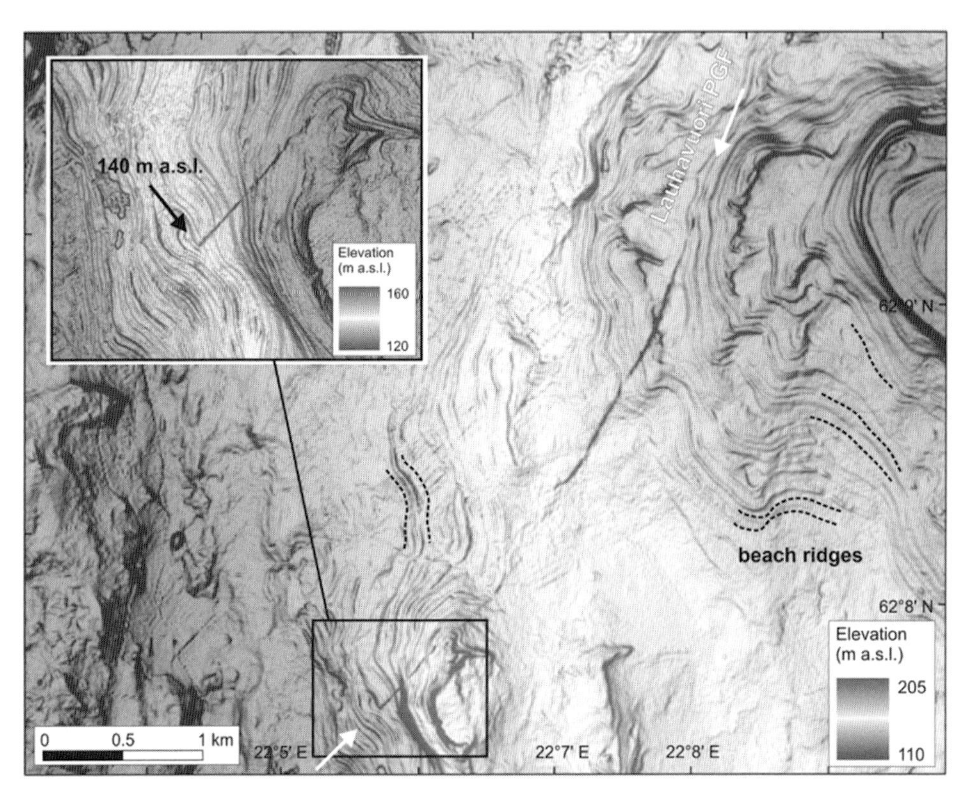

Figure 8.4 The Lauhavuori GIF (62.85°N, 22.65°E) cross-cuts palaeoshorelines only above 140 m above sea level (a.s.l.). The interpretation is that surface rupturing of the Lauhavuori GIF occurred about 1,500 years after deglaciation when the ancient shoreline of Baltic Sea basin was at 140 m a.s.l. in the Lauhavuori area, which, according to the shoreline displacement information of Salomaa (1982) was at 9,500–9,600 calibrated years before present. (A black and white version of this figure will appear in some formats. For the colour version, please refer to the plate section.)

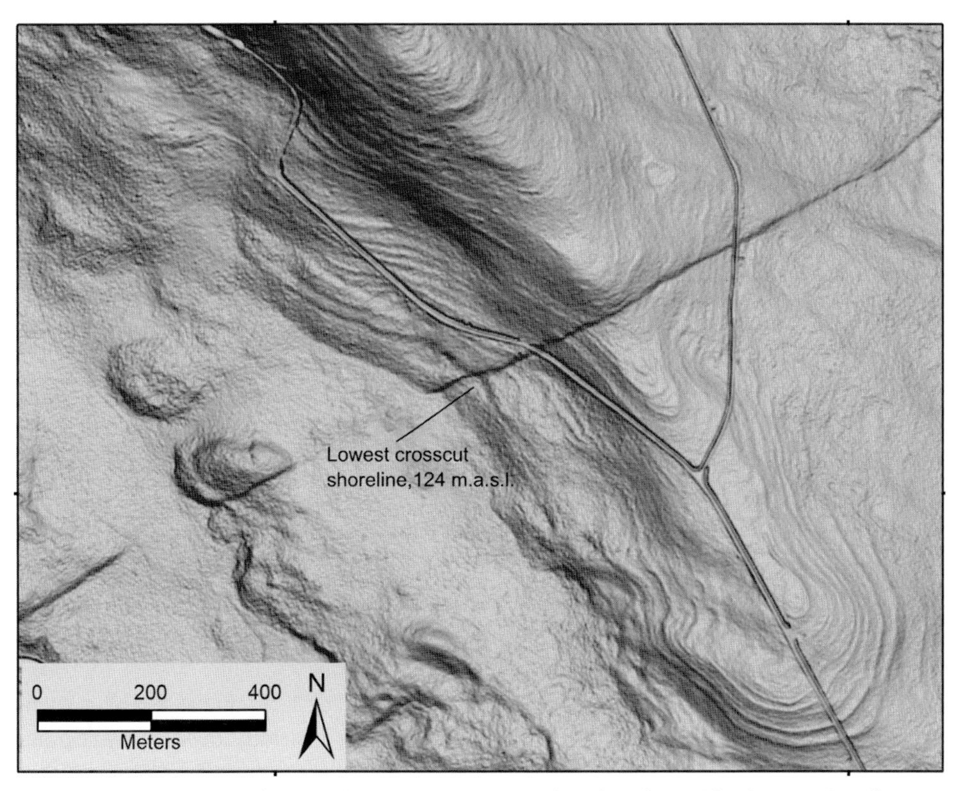

Figure 8.5 The Lansjärv Fault scarp cross-cuts palaeoshorelines. The lowest shoreline cross-cut by the south-east facing fault scarp lies at 124 m a.s.l. Thus, faulting occurred after this elevation rose above the waters of the Baltic basin. This occurred about 9,100 years ago (Lindén et al., 2006). The centre of the figure is at 66.5359°N, 22.2232°E.

south-east facing fault scarp. The lowest elevation shoreline that is definitively cut lies at 124 m a.s.l., an elevation that rose above the waters of the Baltic basin about 9,100 years ago (Lindén et al., 2006). Thus, fault rupture occurred after this time. Interestingly, the more western segments of the Lansjärv Fault System ruptured about a thousand years earlier based on the stratigraphy in numerous trenches excavated across the fault (Lagerbäck, 1990, 1992).

8.2.3 Advantages, Disadvantages and Recent Revelations of Geomorphic Dating

The greatest advantage of geomorphic studies is that the availability of high-resolution DEMs has made the examination of fault zone geomorphology a rapid and inexpensive desktop study. The methodology, however, has limitations. With the Lauhavuori scarp being the exception, geomorphic studies provide only limiting relative ages, generally postglacial or after shore displacement above a certain elevation. Additionally, preserved glacial landscapes in northern Fennoscandia complicate simple age interpretations of cross-cutting relationships because not all landforms date to the Late Weichselian glaciation. For example, cross-cutting relationships between fault scarps and eskers provide useful relative ages, but sometimes the ages of the eskers are uncertain. A final drawback is that dynamic geomorphology is required to record changes, and the mid-to-late Holocene landscape is a slow-changing one relative to the deglacial environment of the early Holocene.

The most interesting recent results derived from relative dating are those of the palaeoshorelines. In both examples discussed above, either the fault or fault segment was discovered using DEMs, and the dates derived from cross-cutting relationships indicate that faulting occurred about 1,000–1,500 years after deglaciation. This is at odds with the longstanding hypothesis that postglacial fault rupture occurs during deglaciation (Lagerbäck & Sundh, 2008).

8.3 Stratigraphic Dating

Fennoscandia can be divided into three environments based upon the regional Late Weichselian and Holocene stratigraphy. First, higher elevation areas, which lie above the highest postglacial shoreline, were not inundated by marine or Baltic waters following deglaciation (Björck, 1995; Ojala et al., 2013). In this environment, the sediments are largely till but often overlain by localized fluvial and glaciolacustrine deposits as well as postglacial peat in the topographic lowlands. Second, lower elevation areas that lie below the highest postglacial shoreline were inundated by marine or Baltic waters following deglaciation and have subsequently been uplifted above water level due to glacial isostatic adjustment. In this environment, the glacial sediments are generally overlain by regional scale glaciomarine or glaciolacustrine sediments, which are in turn overlain by postglacial and littoral sediments. Third, some areas still lie below water, either in marine or lacustrine realms. In this environment, the stratigraphy is similar to that found in terrestrial environments below the highest postglacial shoreline, but a more complete and generally more organic rich Holocene sequence overlies the glaciomarine and glaciolacustrine sediments. Each of these environments provides different opportunities for identification and dating of postglacial fault ruptures as well as different types of soft-sediment deformation structures (SSDS) (Sims, 1973; Owen et al., 2011).

8.3.1 Stratigraphic Dating above the Highest Postglacial Shoreline

Excavation at the Riikonkumpu site in northern Finland provides an example of where sedimentological investigations of glacial and postglacial sediments reveal information on the timing and possible reoccurrence of rupturing event(s) in the Isovaara–Riikonkumpu GIF Complex (Ojala et al., 2017). The study was based on detailed description of lithostratigraphy and sediment structures in a 50–100-m-long trench that was excavated perpendicular to the LiDAR-detected fault line. The trench was several metres deep and reached the surface of the bedrock. Sedimentological logs at the Riikonkumpu site were supplemented with studies by ground-penetrating radar (GPR), 3D photogrammetry of vertical sections of the trench and diamond drilling through the fault zone, which was subsequently instrumented for monitoring purposes (Ojala et al., 2017).

Figure 8.6 The sediment stratigraphy and structures at the Riikonkumpu site (67.73°N, 25.07°E) indicate that the surface rupture is of postglacial origin and was probably formed soon after the deglaciation (modified from Ojala et al., 2017).

The sediment stratigraphy and structures at the Riikonkumpu site indicate that the surface rupture is of postglacial origin and was probably formed soon after the deglaciation at around 11,300 cal BP (Ojala et al., 2017) (Figure 8.6). This interpretation is based on the clear deformation structures in the Late Weichselian till at the scarp location. In addition, the folded phyllitic bedrock was partially intruded into the Late Weichselian till where the folding structure of weathered bedrock was considered as a fault-propagated feature representing relatively recent propagation. Ojala et al. (2017) also noted that the Riikonkumpu fault ramp cuts through the Late Weichselian lateral drainage channels on the SE slope of the Riikonvaara fell, thus supporting the postglacial nature of the rupturing event.

In Finland the same methodology has been applied to investigate the characteristics and timing of the surface rupture(s) of the Suasselkä (Ojala et al., 2019d), Venejärvi–Jauhojärvi (Mattila et al., 2019), Pasmajärvi and Vaalajärvi (Ojala et al., 2019c) GIF complexes.

While the examples cited above use large trenches through the entire Quaternary stratigraphy to the bedrock in order to examine stratigraphy associated with fault rupture, Olesen et al. (see Chapter 11) employ an alternative approach. Short, shallow trenches were excavated through fault scarps and into adjacent peat bogs that existed prior to fault rupture. This allowed for dating organic material overlain by clastic sediments interpreted to have been deposited by mass movement during fault rupture (Olesen et al., Chapter 11). Such an approach may prove useful in obtaining absolute ages of fault rupture(s) elsewhere in Fennoscandia.

8.3.2 Stratigraphic Dating below the Highest Postglacial Shoreline

Working below the highest postglacial shoreline allows for the possibility of bracketing stratigraphic ages. Such is the case along the main strand of the Lansjärv Fault, which lies just below the highest postglacial shoreline. Multiple trenches excavated across the fault reveal the same stratigraphies. Glacial sediments are faulted, but the overlying littoral sediments are undisturbed (Lagerbäck, 1990, 1992). A recent shore displacement curve from Norrbotten indicates that deglaciation occurred about 10,500 years ago and the elevation of the trenches emerged above the waters of the Baltic about 10,200 years ago (Lindén et al., 2006). Thus, fault rupture is bracketed by these two dates.

Although the absolute age is not as tightly constrained, similar stratigraphies showing faulted glacial sediments and undisturbed postglacial sediments were found along the Bollnäs Fault in central Sweden. Here, the stratigraphy was examined both in excavations on land (Smith et al., 2014) and in a hydroacoustical survey beneath Lake Voxsjön (Smith et al., 2018). Based on the stratigraphy and existing varve (Strömberg, 1989) and shoreline displacement (Berglund, 2012) data, Smith et al. (2014) estimated fault rupture at about 10,400 calendar years before present.

8.3.3 Advantages, Disadvantages and Recent Revelations of Stratigraphic Dating

Stratigraphic dating, particularly trenching across scarps, provides solid evidence of faulting. Additionally, stratigraphic dating methods can provide both maximum and minimum limiting ages. The resolution of limiting ages varies dramatically depending on geologic setting. For example, above the highest postglacial shoreline, limiting ages are constrained by tills (Ojala et al., 2019d), often of unknown age. Below the highest postglacial shoreline, ages can often be bracketed to between deglaciation and emergence above water level (Lagerbäck, 1990, 1992; Smith et al., 2014). The primary disadvantage of stratigraphic dating methods is that they are difficult to carry out in remote locations.

Recent trench-based investigations of faults in Finland have indicated a far more complex history than was previously acknowledged. The excavations of glacial strata have revealed that a single rupturing hypothesis of the postglacial fault complexes in the Fennoscandian Shield area is overly conservative (see summary in Ojala et al., 2019c). Based on lithostratigraphical evidence, it is more likely that many of the GIF complexes and separate fault segments within them ruptured independently and several times, both pre- and postdating the Late Weichselian.

8.4 Indirect Dating

Stratigraphical studies of co-seismic landforms, such as palaeolandslides and soft-sediment deformation structures (SSDS) in adjacent basins, are important because they can provide indirect evidence of the age of palaeoseismicity and can often be dated with better accuracy and precision than PGF scarps.

For example, Ojala et al. (2018b) described a distinct turbidite layer and SSDS preserved in clastic-biogenic varves of the Lake Nurmijärvi sequence in central-southern Finland, which they connected to an earthquake at 7,400 cal BP with moment magnitude of $M_W \approx 5.1–6.9$. The type of millimetre-scale varves they studied in Lake Nurmijärvi is similar in character and composition to those described by Tiljander et al. (2003) and by Ojala and Alenius (2005), among others, with comparable formation processes and lake catchment settings.

Ojala et al. (2018b) noticed that in the Nurmijärvi sequence, the varves are perfectly horizontal and undisturbed between the sediment surface and a depth of 490 cm. These deposits are separated from the underlying deformed varves by a 5-cm-thick, light-coloured turbidite, formed when seismic shaking brought sediment particles into suspension, which then flowed under its own gravity to the central lake basin (Figure 8.7). Below the turbidite

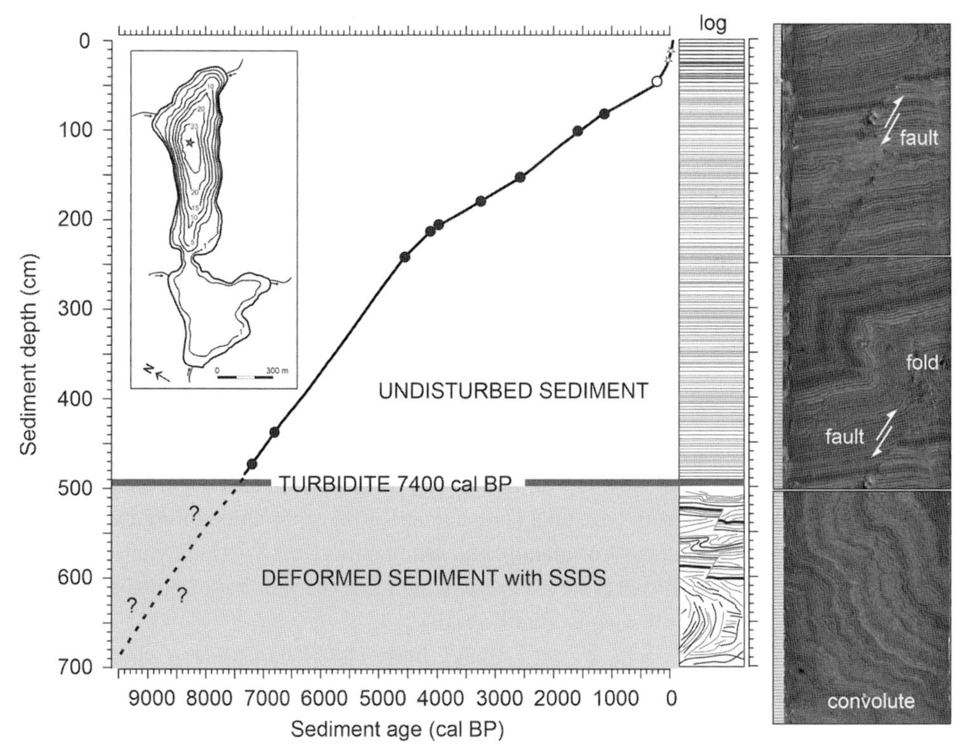

Figure 8.7 Faulted varves and a turbidite layer at Nurmijärvi, Finland (61.58°N, 25.92°E) (modified from Ojala et al., 2018b).

layer, sediment deposits contain SSDS, including centimetre-scale folding and faulting structures and larger convoluted bedding that can be easily detected and measured because of the appearance of a varved structure.

Landslides are considered as sudden or creeping downslope movements of earth material under the influence of gravity (Varnes, 1978; Cruden & Varnes, 1996; Hungr et al., 2014). Different types of landslides are typically triggered by rapid snow melt and intense rainfall, volcanic or anthropogenic activity or earthquakes, and they involve flowing, sliding, toppling, falling, spreading or a combination of these types (Cruden & Varnes, 1996). Interpreting an earthquake origin of a landslide for relative dating of palaeoseismicity requires several viewpoints be considered, as summarized by Jibson and Keefer (1989, 1994), Crozier (1992) and Jibson (1996). These are that: (i) landslides are clustered across a discrete area; (ii) in the presence of ongoing or historical seismicity in the area; (iii) landsides are at close proximity to active fault or seismic zones or in the area with known PGFs; (iv) landslides are of considerable size and (v) have liquefaction and/or SSDS characteristics; and (vi) synchronous ages for several landslides in a discrete area are preferred.

Typically, the age of a landslide is based on radiocarbon dating of either (i) organic deposits that were buried by the landslide masses or (ii) large tree trunks or other macrofossils that were taken down and buried within the earth material during the sliding event or (iii) basal peat deposits in the landslide scar, which only provide the minimum age of the landslide.

In a recent paper, Ojala et al. (2019b) first investigated the distribution of landslides in Finnish Lapland in relation to known PGFs. They then described the morphological characteristics of the landslides and classified their locations and types according to morphometric appearance in order to assess their potential as being seismically triggered. Ojala et al. (2019b) revealed considerable variation in size, morphology and scarp depth characteristics and they demonstrated the complexity of the interaction between the different factors affecting slope stability and landslide evolution. They concluded that fulfillment of the seismic-triggering criteria by Jibson and Keefer (1989, 1994), Crozier (1992) and Jibson (1996) was strongest for the debris slide type of landslides, which are located within 35 km of the nearest known PGF.

In another paper by Ojala et al. (2018a) selected potentially seismically induced landslides, and basal peat sequences in landslide scars were drilled to recover organic material and to constrain the ages of landslides in the vicinity of the known PGFs. An example of a lithological composite log from the Sotka landslide in Kittilä is presented in Figure 8.8. The studied sections generally consist of glaciogenic sedimentary strata with diamicton, gravel and sand lithofacies, and thin organic deposits, which are overlain by an up to 5-m-thick layer of debris representing the landslide material.

Overall, the results of Ojala et al. (2018a) from eleven landslides, which buried organic remains, and basal peat in northern Finland, suggested non-stationary postglacial seismicity throughout the Holocene with three episodes of increased slope instability and formation of landslides: 9,000–11,000 cal BP, 5,000–6,000 cal BP and 1,000–3,000 cal BP.

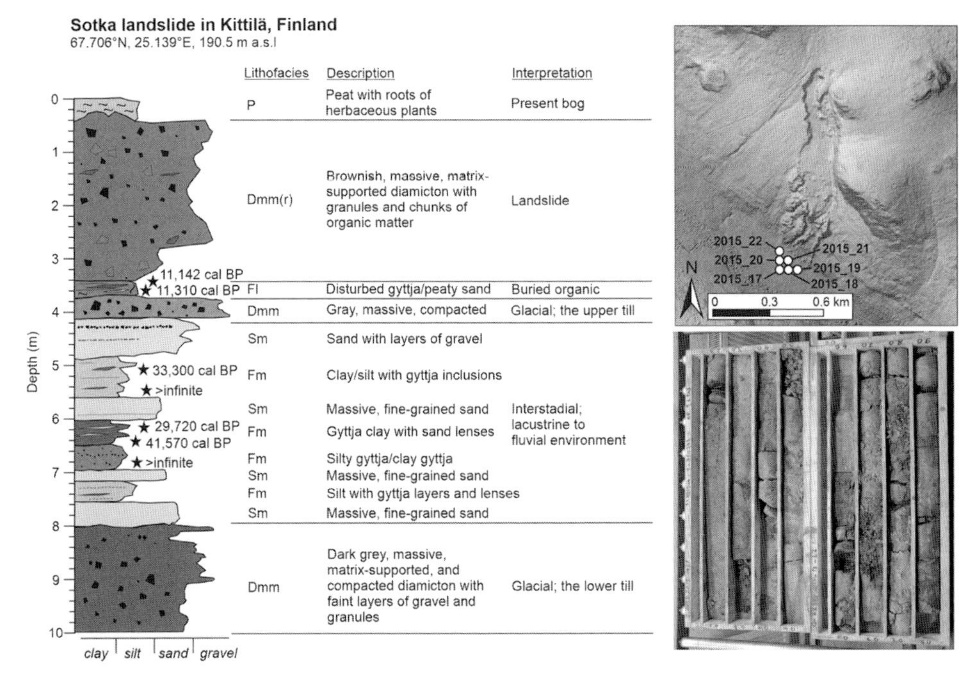

Figure 8.8 Lithological composite log from the Sotka landslide in Kittilä, Finland (67.71°N, 24.15°E) (modified from Ojala et al., 2019b).

8.4.1 Advantages, Disadvantages and Recent Revelations of Indirect Dating

The primary advantage of indirect dating is that organic material for radiocarbon dating is more easily obtained from landslides than from fault scarps. Additionally, dates from landslides can provide both maximum and minimum absolute ages. There is, however, always doubt as to whether the landslides are co-seismic or not.

In Finland, recent drilling of landslides interpreted to be co-seismic (Ojala et al., 2018a) has revealed startlingly young ages. In addition to the deglacial age seismicity, Ojala et al. (2018a) and Ojala et al. (2019a) suggest periods of seismicity between 5,000–6,000 cal BP, and 1,000–3,000 cal BP. Again, this challenges the hypothesis that postglacial faulting and seismicity is limited to occurring during deglaciation or immediately afterwards (Lagerbäck & Sundh, 2008).

8.5 Direct Dating

8.5.1 Surface Exposure Dating

Surface exposure dating is a direct dating method that is based on the in situ production of cosmogenic nuclides in certain rock-forming minerals at the rock surface as a result of the bombardment by cosmic rays (Gosse & Phillips, 2001). One such example is the production of the radioactive isotope ^{10}Be in quartz. For various faults at different locations around the world, recent studies have used terrestrial cosmogenic nuclide dating (TCN) to determine faulting history (Benedetti et al., 2013; Berthet et al., 2014; Benavente et al., 2017; Yang et al., 2018). However, until now no attempt to directly date the PGFs of the Fennoscandian Shield has been reported.

A recent pilot study by the Geological Survey of Sweden (SGU) applying cosmogenic ^{10}Be exposure dating on a segment of the Pärvie Fault north-east of Lake Torneträsk (Figure 8.9) produced ambiguous results. The samples were taken along near-vertical profiles at locations where quartz-bearing bedrock was exposed along the fault scarp. Since the Pärvie Fault is a reverse fault, the actual fault plane is generally not exposed in bedrock due to collapse of the hanging wall. Therefore, SGU's working hypothesis was that, with the

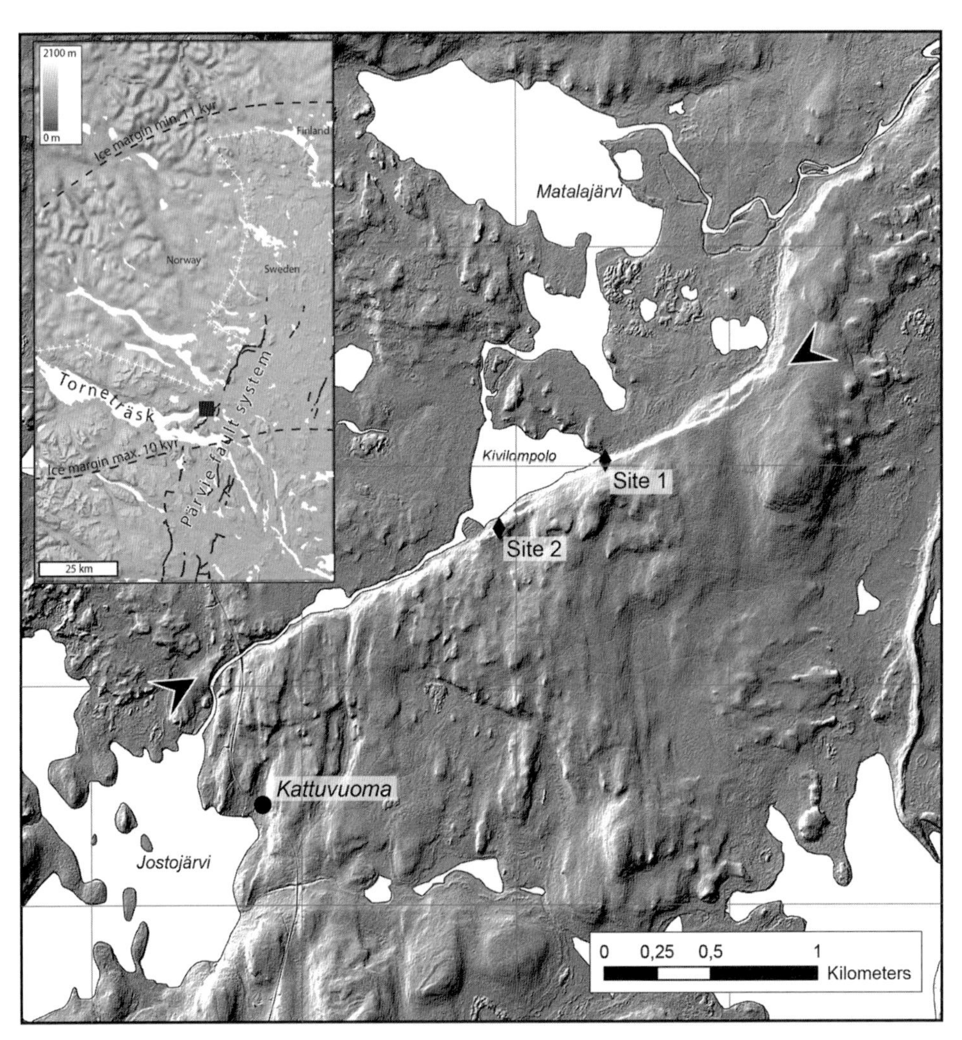

Figure 8.9 Sampling sites where direct dating of the Pärvie Fault was attempted by the Geological Survey of Sweden (SGU). In the inset map, the minimum and maximum isochrons for the ice margin were taken from Hughes et al. (2016). The centre of the figure is at 68.3032°N, 19.9241°E.

help of TCN dating, it may be possible to determine whether the collapsed scarp was exposed prior to or after the latest deglaciation. Assuming that the area around lake Torneträsk was completely free from ice at about ~10 ka (Hughes et al., 2016; Stroeven et al., 2016; Figure 8.9), it was hypothesized that surface exposure dates greater than 10 ka would indicate that at least parts of the fault scarp existed before the latest deglaciation.

When sampling along near-vertical profiles, a certain amount of the measured ^{10}Be is expected to have been produced in the rock due to inheritance prior to fault scarp exposure, especially in the upper 1–2 m (Benavente et al., 2017; Tikhomirov et al., 2019, Tesson & Benedetti, 2019). SGU used a reference sample from a horizontal bedrock surface close to the fault scarp to calculate an inheritance-depth profile for the sampling site. The inheritance at a certain depth was then subtracted from the total measured ^{10}Be concentrations at the respective sampling depths, and the remaining concentrations were assumed to have been produced due to exposure of the fault scarp.

SGU suggests different inheritance scenarios and depth profile modelling. Depending on what inheritance scenario is chosen, the exposure ages range from 1.8 to 18.3 ka, and no clear age/depth relationship could be established. The dates for the most reliable samples lie around 10–11 ka, which is too close to the end of the latest deglaciation in the area to determine with certainty whether the fault scarp was exposed preglacially; SGU recommends a larger sampling set, even from other locations along the Pärvie Fault, for a better understanding of the scarp exposure history.

In summary, fault surface exposure dating can, theoretically, be used to calculate apparent exposure ages to determine if PGF bedrock scarps predate deglaciation. However, locations for sampling need to be chosen with care, and it is recommended that many samples be taken. Also, TCN dating is quite sensitive to shielding. For example, a few degrees difference in the dip of the sample surface can create a difference of several hundreds to thousands of years in the calculation of exposure ages. It is therefore important to determine shielding factors with care. Having additional information about the deglaciation history of the sampling location is also recommended.

8.6 Discussion and Conclusions

While each of the dating methods described here can provide useful information, none of them is a stand-alone solution. Stratigraphic studies must be situated only after consideration of the geomorphology, and indirect dating becomes much more credible if stratigraphic data indicate postglacial fault rupture in the area. Composite studies that make use of multiple dating methods minimize the shortcomings of any single approach. As is often the case in geology, multiple methods yielding consistent results provide the most compelling evidence.

The multiple dating methods described here provide evidence of a far more complex history of postglacial faults than the paradigm that postglacial faults ruptured once during the Late Weichselian deglaciation. In places, the geomorphology shows fault scarps cutting palaeoshorelines that did not form until a thousand years after deglaciation (Palmu et al.,

2015, this study). Stratigraphic investigations across faults in Finland indicate fault rupture both before and after the Late Weichselian (Ojala et al., 2019c). Dating of landslides, an indirect method of dating palaeoseismicity, suggests relatively high-magnitude seismicity in both the Middle and Late Holocene (Ojala et al., 2018a; Ojala et al., 2019a). While each of these results applies only to the investigated faults, collectively they point to a longer and more complex history of postglacial faulting.

References

Benavente, C., Zerathe, S., Audin, L. et al. (2017). Active transpressional tectonics in the Andean forearc of southern Peru quantified by 10Be surface exposure dating of an active fault scarp. *Tectonics*, **36**(9), 1662–1678, doi.org/10.1002/2017TC004523.

Benedetti, L., Manighetti, I., Gaudemer, Y. et al. (2013). Earthquake synchrony and clustering on Fucino faults (Central Italy) as revealed from in situ 36Cl exposure dating. *Journal of Geophysical Research: Solid Earth*, **118**(9), 4948–4974, doi.org/10.1002/jgrb.50299.

Berglund, M. (2012). Early Holocene in Gästrikland, east central Sweden: shore displacement and isostatic recovery. *Boreas*, **41**(2), 263–276, doi.org/10.1111/j.1502-3885.2011.00228.x.

Berthet, T., Ritz, J.-F., Ferry, M. et al. (2014). Active tectonics of the eastern Himalaya: new constraints from the first tectonic geomorphology study in southern Bhutan. *Geology*, **42**(5), 427–430, doi.org/10.1130/G35162.1.

Björck, S. (1995). A review of the history of the Baltic Sea, 13.0–8.0 ka BP. *Quaternary International*, **27**(94), 19–40, doi.org/10.1016/1040-6182(94)00057-C.

Crozier, M. J. (1992). Determination of paleoseismicity from landslides. In D.H. Bell, ed., *Landslides (Glissements de terrain).* Proceedings of the 6th International Symposium, Christchurch, New Zealand, A. A. Balkema, Rotterdam, pp. 1173–1180.

Cruden, D. M. and Varnes, D. J. (1996). Landslide types and processes. In A. K. Turner and R. L. Schuster, eds., *Landslides: Investigation and Mitigation.* Transportation Research Board, US National Research Council Special Report 247, Washington, DC, pp. 36–75.

Gosse, J. C. and Phillips, F. M. (2001). Terrestrial in situ cosmogenic nuclides: theory and application. *Quaternary Science Reviews*, **20**, 1475–1560, doi.org/10.1016/S0277-3791(00)00171-2.

Hughes, A. L. C., Gyllencreutz, R., Lohne, Ø. S., Mangerud, J. and Svendsen, J. I. (2016). The last Eurasian ice sheets – a chronological database and time-slice reconstruction, DATED-1. *Boreas*, **45**, 1–45, doi.org/10.1111/bor.12142.

Hungr, O., Leroueil, S. and Picarelli, L. (2014). The Varnes classification of landslide types, an update. *Landslides*, **11**, 167–194, doi.org/10.1007/s10346–013-0436-y.

Jibson, R. W. (1996). Use of landslides for paleoseismic analysis. *Engineering Geology*, **43**, 291–323, doi.org/10.1016/S0013–7952(96)00039-7.

Jibson, R. W. and Keefer, D. K. (1989). Statistical analysis of factors affecting landslide distribution in the New Madrid seismic zone, Tennessee and Kentucky. *Engineering Geology*, **27**, 509–542, doi.org/10.1016/0013-7952(89)90044-6.

Jibson, R. W. and Keefer, D. K. (1993). Analysis of the seismic orgin of landslides—examples from the New Madrid seismic zone. *Geological Society of America Bulletin*, **105**, 421–436.

Kujansuu, R. (1964). Nuorista siirroksista Lapissa [English summary: Recent faults in Lapland]. *Geologi*, **16**, 30–36 (in Finnish).

Lagerbäck, R. (1978). Neotectonic structures in northern Sweden. *GFF*, **100**(3), 263–269, doi.org/10.1080/11035897809452533.

Lagerbäck, R. (1988). The Veiki moraines in northern Sweden – widespread evidence of an Early Weichselian deglaciation. *Boreas*, **17**, 469–486, doi.org/10.1111/j.1502-3885.1988.tb00562.x.

Lagerbäck, R. (1990). Late Quaternary faulting and paleoseismicity in northern Fennoscandia with particular reference to the Lansjärv area, Northern Sweden. *GFF*, **112**, 333–354, doi.org/10.1080/11035899009452733.

Lagerbäck, R. (1992). Dating of Late Quaternary faulting in northern Sweden. *Journal of the Geological Society*, **149**(2), 285–291, doi.org/10.1144/gsjgs.149.2.0285.

Lagerbäck, R. and Robertsson, A.-M. (1988). Kettle holes – stratigraphical archives for Weichselian geology and palaeoenvironment in northernmost Sweden. *Boreas*, **17**, 439–468, doi.org/10.1111/j.1502-3885.1988.tb00561.x.

Lagerbäck, R. and Sundh, M. (2008). *Early Holocene Faulting and Paleoseismicity in Northern Sweden: Research Paper C 836*. Geological Survey of Sweden pp.

Lindén, M., Möller, P., Björck, S. and Sandgren, P. (2006). Holocene shore displacement and deglaciation chronology in Norrbotten, Sweden. *Boreas*, **35**(1), 1–22, doi.org/10.1111/j.1502-3885.2006.tb01109.x.

Lundqvist, J. and Lagerbäck, R. (1976). The Pärve Fault: a late-glacial fault in the Precambrian of Swedish Lapland. *Geologiska Föreningens i Stockholm Förhandlingar*, **98**, 45–51, doi.org/10.1080/11035897609454337.

Mattila, J., Ojala, A. E. K., Ruskeeniemi, T. et al. (2019). Evidence of multiple slip events on postglacial faults in northern Fennoscandia. *Quaternary Science Reviews*, **215**, 242–252, doi.org/10.1016/j.quascirev.2019.05.022.

Mikko, H., Smith, C. A., Lund, B., Ask, M. V. S. and Munier, R. (2015). LiDAR-derived inventory of post-glacial fault scarps in Sweden. *GFF*, **137**, 334–338, doi.org/10.1080/11035897.2015.1036360.

Ojala, A. E. K. and Alenius, T. (2005). 10000 years of interannual sedimentation recorded in the Lake Nautajärvi (Finland) clastic–organic varves. *Palaeogeography, Palaeoclimatology, Palaeoecology*, **219**(3), 285–302, doi.org/10.1016/j.palaeo.2005.01.002.

Ojala, A. E. K., Palmu, J.-P., Åberg, A., Åberg, S. and Virkki, H. (2013). Development of an ancient shoreline database to reconstruct the Litorina Sea maximum extension and the highest shoreline of the Baltic Sea basin in Finland. *Bulletin of the Geological Society of Finland*, **85**(PART 2), 127–144, doi.org/10.17741/bgsf/85.2.002.

Ojala, A. E. K., Mattila, J., Ruskeeniemi, T. et al. (2017). Postglacial seismic activity along the Isovaara–Riikonkumpu fault complex. *Global and Planetary Change*, **157**(January), 59–72, doi.org/10.1016/j.gloplacha.2017.08.015.

Ojala, A. E. K., Markovaara-Koivisto, M., Middleton M. et al. (2018a). Dating of paleo-landslides in western Finnish Lapland. *Earth Surface Processes and Landforms*, **43**, 2449–2462, doi.org/10.1002/esp.4408.

Ojala, A. E. K., Mattila, J., Virtasalo, J. Kuva, J. and Luoto, T.P. (2018b). Seismic deformation of varved sediments in southern Fennoscandia at 7400 cal BP. *Tectonophysics*, **744**, 58–71, doi.org/10.1016/j.tecto.2018.06.015.

Ojala, A. E. K., Mattila, J., Hämäläinen, J. and Sutinen, R. (2019a). Lake sediment evidence of paleoseismicity: timing and spatial occurrence of late- and postglacial earthquakes in Finland. *Tectonophysics*, **771**, 228227, doi.org/10.1016/j.tecto.2019.228227.

Ojala, A. E. K., Mattila, J., Markovaara-Koivisto, M. et al. (2019b). Distribution and morphology of landslides in northern Finland: an analysis of postglacial seismic activity. *Geomorphology*, **326**, 190–201, doi.org/10.1016/j.geomorph.2017.08.045.

Ojala, A. E. K., Mattila, J., Ruskeeniemi, T. et al. (2019c). *Postglacial Faults in Finland – A Review of PGSdyn – Project Results*, Posiva Report 2019-1, 118 pp., Posiva Oy, Eurajoki.

Ojala, A. E. K., Mattila, J., Ruskeeniemi, T. et al. (2019d). Postglacial reactivation of the Suasselkä PGF complex in SW Finnish Lapland. *International Journal of Earth Sciences*, **108**(3), 1049–1065, doi.org/10.1007/s00531-019-01695-w.

Owen, G., Moretti, M. and Alfaro, P. (2011). Recognising triggers for soft-sediment deformation: current understanding and future directions. *Sedimentary Geology*, **235**, 133–140, doi.org/10.1016/j.sedgeo.2010.12.010.

Palmu, J. P., Ojala, A. E. K., Ruskeeniemi, T., Sutinen, R. and Mattila, J. (2015). LiDAR DEM detection and classification of postglacial faults and seismically-induced landforms in Finland: a paleoseismic database. *GFF*, **137**(4), 344–352, doi.org/10.1080/11035897.2015.1068370.

Salomaa, R. (1982). Post-glacial shoreline displacement in the Lauhanvuori area, Western Finland. *Annales Academiæ Scientiarum Fennicæ A III*, **134**, 81–97.

Sims, J. D. (1973). Earthquake-Induced Structures in Sediments of Van Norman Lake, San Fernando, California. *Science*, **182**, 161–163, doi.org/10.1126/science.182.4108.161.

Smith, C. A., Sundh, M. and Mikko, H. (2014). Surficial geology indicates early Holocene faulting and seismicity, central Sweden. *International Journal of Earth Sciences*, **103**(6), 1711–1724, doi.org/10.1007/s00531-014-1025-6.

Smith, C. A., Nyberg, J. and Bergman, B. (2018). Comparison between hydroacoustical and terrestrial evidence of glacially induced faulting, Lake Voxsjön, central Sweden. *International Journal of Earth Sciences*, **107**, 169–175, doi.org/10.1007/s00531-017-1479-4.

Stroeven, A. P., Hättestrand, C., Kleman, J., et al. (2016). Deglaciation of Fennoscandia. *Quaternary Science Reviews*, **147**, 91–121, doi.org/10.1016/j.quascirev.2015.09.016.

Strömberg, B. (1989) *Late Weichselian Deglaciation and Clay Varve Chronology in East-Central Sweden*, Geological Survey of Sweden (SGU) Series Ca. 73, 70 pp.

Sutinen, R., Hyvönen, E., Middleton, M. and Ruskeeniemi, T. (2014). Airborne LiDAR detection of postglacial faults and Pulju moraine in Palojärvi, Finnish Lapland. *Global and Planetary Change*, **115**, 24–32, doi.org/10.1016/j.gloplacha.2014.01.007.

Tesson, J. and Benedetti, L. (2019). Seismic history from in situ 36Cl cosmogenic nuclide data on limestone fault scarps using Bayesian reversible jump Markov chain Monte Carlo. *Quaternary Geochronology*, **52**, 1–20, doi.org/10.1016/j.quageo.2019.02.004.

Tikhomirov, D., Amiri, N. M., Ivy-Ochs, S. et al. (2019). Fault Scarp Dating Tool – a MATLAB code for fault scarp dating using in-situ chlorine-36 supplemented with datasets of Yavansu and Kalafat faults. *Data in Brief*, **26**, 104476, doi.org/10.1016/j.dib.2019.104476.

Tiljander, M., Saarnisto, M., Ojala, A. E. K. and Saarinen, T. (2003). A 3000-year palaeoenvironmental record from annually laminated sediment of Lake Korttajärvi, central Finland. *Boreas*, **32**(4), 566–577, doi.org/10.1111/j.1502-3885.2003.tb01236.x.

Varnes, D. J. (1978). Slope movement types and processes. In R. L. Schuster and R. J. Krizek, eds., *Landslides: Analysis and Control*, Special Report 176, Transportation Research Board, National Academy of Sciences, Washington, DC., pp. 11–33.

Yang, H., Yang, X., Huang, X. et al. (2018). New constraints on slip rates of the Fodongmiao–Hongyazi fault in the Northern Qilian Shan, NE Tibet, from the 10Be exposure dating of offset terraces. *Journal of Asian Earth Sciences*, **151**, 131–147, doi.org/10.1016/j.jseaes.2017.10.034.

9

Proposed Drilling into Postglacial Faults

The Pärvie Fault System

MARIA ASK, ILMO KUKKONEN, ODLEIV OLESEN, BJÖRN LUND, ÅKE FAGERENG, JONNY RUTQVIST, JAN-ERIK ROSBERG AND HENNING LORENZ

ABSTRACT

The existence of postglacial faults (PGFs) in northern Fennoscandia was first documented in Finland in the last century. Subsequently, over a dozen large PGFs have been observed in Finland, Sweden and Norway. PGFs have been investigated through geophysical methods, trenching and mapping of brittle deformation structures. No direct measurements of the age of faulting exist, although relative ages with respect to glacial (sedimentary) deposits have been established. Very little is known about PGFs through direct measurements. A few short – up to 500 m deep – boreholes exist. Plans for a scientific drilling program were initiated in 2010, but data coverage was relatively unevenly distributed and no PGF was identified as a drilling target. Subsequently, additional data have been collected, and the drilling target has been identified; the Pärvie Fault System is the longest known PGF in the world and has been proposed to have hosted an M8 earthquake near the end or just after the last glaciation. Further, this fault system is still microseismically active. Drill sites are north of the Arctic Circle, in a sparsely populated area. Existing site survey data, established logistics and societal relevance viz the fault's proximity to mining and energy operations make this fault system an appropriate target. The International Continental Scientific Drilling Program (ICDP) approved a full drilling proposal in October 2019. This chapter presents an abbreviated version of the approved proposal.

9.1 Introduction

Intraplate earthquakes contribute significantly to global seismic hazard, destruction and loss of life (e.g. England & Jackson, 2011). Their occurrence challenges one basic assumption of the plate tectonics model: that Earth's outer shell comprises rigid plates that move with respect to each other but do not deform internally. We know that potential slip accumulates relatively quickly at plate margins compared to plate interiors (e.g. Kreemer et al., 2014), leading to earthquake repeat times on the order of tens to hundreds of years; because of slower elastic strain accumulation rates in plate interiors, these may be thousands of years or longer (e.g. England & Jackson, 2011), and aftershock sequences may persist for hundreds of years (Stein & Liu, 2009). Local variations in crustal strength and inherited structures may guide intraplate seismicity more than motion at plate boundaries

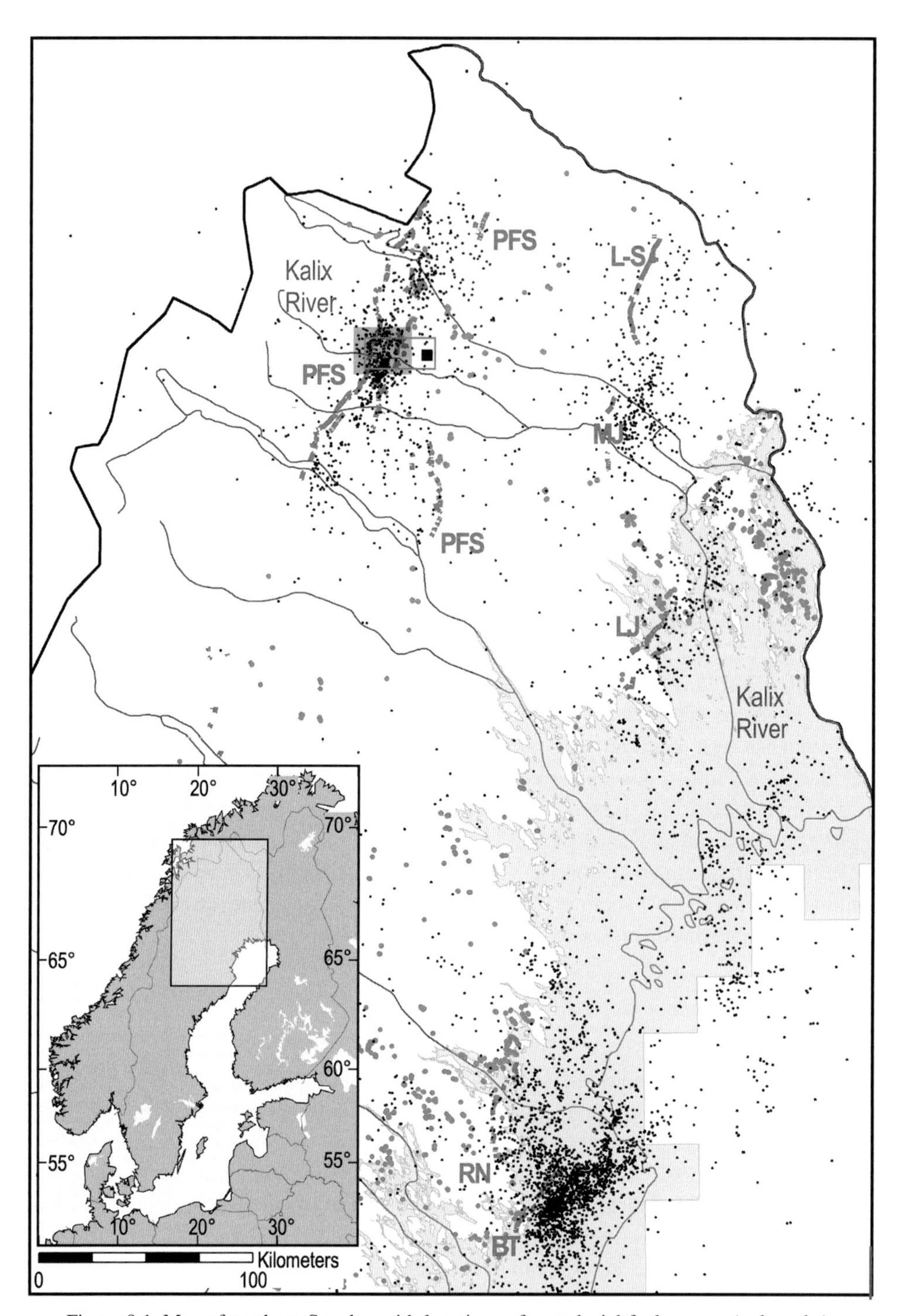

Figure 9.1 Map of northern Sweden with locations of postglacial fault scarps (red marks), palaeolandslides in till (green dots), the highest coastline (light-blue-coloured area) and microseismicity recorded from 2003–2018 by the Swedish National Seismic Network

(e.g. Campell, 1978; Page & Hough, 2014; Smith et al., 2014). The spatial and temporal patterns of intraplate earthquakes appear more complex than those of plate boundary earthquakes (Stein et al., 2009; Li et al., 2009; Calais et al., 2016).

Postglacial fault (PGF) is an all-encompassing term for faults that also have been referred to as glacially triggered, glacially induced, postglacial, Lateglacial or end-glacial faults. These faults represent a particular type of intraplate seismicity, one which was partly triggered during the final stage of melting of a large ice sheet, through a combination of tectonic and isostatic driving stresses and changes in effective stress on the faults (e.g. Lagerbäck, 1978; Muir Wood, 1989; Wu & Hasegawa, 1996a,b; Sutinen et al., 2014; Lund, 2015; Olesen et al., see Chapter 11). This subset of intraplate seismicity is still poorly understood but has great relevance to seismology in general and may pose high risks for society. PGFs were first identified in the field in the Eastern United States (Mather, 1843) and Canada (Matthew, 1894). Since then, PGFs have been observed in several areas, as discussed by several authors in this book (see Chapters 11–21). Whether PGFs have properties distinct from other major faults in low-strain-rate environments is unknown. However, as with intraplate faults in general, earthquakes on PGFs appear clustered in time and triggered by local stress or strength changes, rather than occurring as a result of a gradual accumulation of elastic strain (Calais et al., 2016). In the particular case of PGFs, possibilities for triggering of specific faults include rheological or hydrological properties that promote rupture during glacial retreat.

Knowledge of surface and near-surface structures and properties, and of the geophysical structure of the crust (e.g. from seismicity and reflection seismology) of the Fennoscandian PGFs, is good relative to other areas of PGF activity, and most of the faults are seismically active at present. There is, however, a gap in knowledge that would come from direct observations and measurements of the fault structures at depth. The International Continental Scientific Drilling Program (ICDP) project 'Drilling Active Faults in Northern Europe' (DAFNE) was developed to improve knowledge of the deeper structures of PGFs. The central aim of DAFNE is to advance the theoretical and applied understanding of intraplate seismicity in general and postglacial intraplate seismicity in particular.

The Pärvie Fault System (PFS) is the drilling target of the DAFNE project (Figure 9.1). 'Pärvie' roughly translates into 'like a breaking wave' from the Sami language. We have identified two drill sites in the central part of the PFS where it has been proposed that the system reflects a flower structure (e.g. Riad, 1990; Juhlin et al., 2010). To understand this complex structure, multiple boreholes, logging, in situ testing and monitoring are required. Here, we present an abbreviated summary of our ICDP application for establishing a scientific drilling infrastructure at the drilling target. We first present a brief

---

Figure 9.1 (*cont.*) (black dots). The major PGFs are: PFS – Pärvie Fault System, L-S – Lainio-Suijavara, MJ – Merasjärvi, LJ – Lansjärv, RN – Röjnoret, BT – Burträsk. The location of the city of Kiruna is marked by the black square. The grey-filled rectangle shows the location of Figure 9.2, and the purple unfilled rectangle shows the location of Figure 9.5. The map with data was provided by Henrik Mikko, the Geological Survey of Sweden, on 11 January 2019. (A black and white version of this figure will appear in some formats. For the colour version, please refer to the plate section.)

background, followed by the motivation for drilling, the strategies to address the objectives and concluding remarks.

9.2 Background

9.2.1 Geological Setting

The PFS is located at the eastern margin of the Northern Scandes mountain chain, east of the Caledonian deformation front in the north-western part of the Fennoscandian Precambrian Shield. The area is currently undergoing an uplift of 4 to 5 mm/yr (Vestøl, 2006). Surface Global Positioning System (GPS) data suggest an uplift gradient of $\sim 25 \times 10^{-9}$/yr and an areal strain rate of $\sim 1 \times 10^{-9}$/yr (Kierulf et al., 2014; Keiding et al., 2015). The greatest principal strain rate at the surface from GPS velocity data is $\sim 2 \times 10^{-9}$/yr towards the WNW-ESE, which compares with the NW-SE greatest compressional stress determined from focal mechanisms (Keiding et al., 2015).

The crustal thickness in the region has been estimated at 32.5–35 km from Airy density (England & Ebbing, 2012) and at 40–42.5 km from seismic experiments (Grad et al., 2009). The thickness of the lithosphere has also been constrained from magnetotelluric studies by Korja (2007) and from seismic studies by Plomerova and Babuska (2010); these studies estimate the base of the lithosphere to be at depths of 175–200 km and 120–150 km, respectively.

Along most of its length, the PFS cuts Palaeoproterozoic metasedimentary, metavolcanic and plutonic rocks of about 1.80–1.96 Ga, and in the northernmost part also Archaean, dominantly plutonic rocks ($>$ 2.68 Ga; e.g. Lagerbäck & Sundh, 2008). Figure 9.2 shows the bedrock geology of the central part of the PFS. Exploration drilling (Gerdin, 1979; Carlson & Lundqvist, 1984) revealed Palaeoproterozoic intermediate and andesitic volcanic rocks with a modest greenschist/amphibolite facies metamorphic overprint as well as granite, syenite, diabase and graphitic schist near the planned drill sites.

The basement rocks record a deformation sequence including two regional scale events, D1, a pre-2.1 Ga phase of continental rifting and magmatism, followed by D2–D3, a 1.9–1.8-Ga sequence of convergent-accretionary-collisional deformation and magmatism in the Svecokarelian orogeny (e.g. Wanke & Melezhik, 2005; Lahtinen et al., 2005; Sarlus et al., 2018; Andersson, 2019). Crustal scale deformation zones developed during these regional deformation events and are typically localized in supracrustal rocks (Bergman et al., 2001; Bauer et al., 2018). D1 structures are typically approximately N-S striking, with local E-W-striking linking structures, whereas D2 and D3 are related to an initial N-S-directed shortening, followed by E-W greatest compression (Lahtinen et al., 2005; Bauer et al., 2018). The D2 and D3 events at least partially reactivated D1 structures such that several of the prominent NW- to NE-striking regional deformation zones seen today have experienced Palaeoproterozoic reactivation and represent long-lived weaknesses (Witschard, 1984; Gaál & Gorbatschev, 1987).

Fennoscandian PGFs represent both reactivation of pre-existing structures (e.g. Olesen et al., 1992; Kuivamäki et al., 1998; Lagerbäck & Sundh, 2008) and new fault segments that cross-cut pre-existing structures (Riad, 1990). Brittle deformation structures were measured

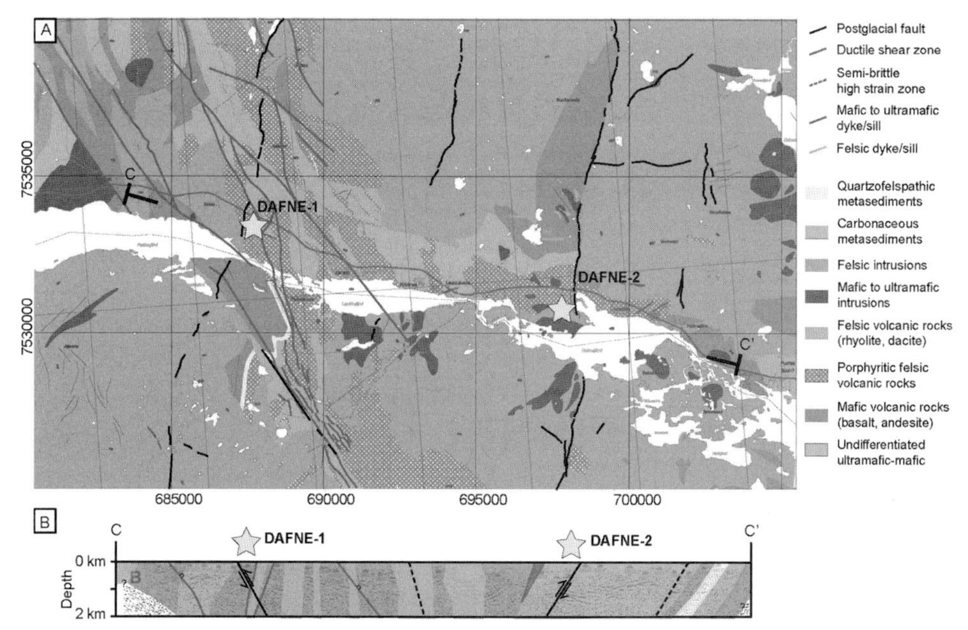

Figure 9.2 A) The bedrock geological map. The location of the map is indicated in Figure 9.1. B) The cross-section is overlain and informed by the reflection surveys of Juhlin et al. (2010) and Ahmadi et al. (2015). Note the cross-section is generalized and underestimates the lithological and structural variability that is likely to be present. The location of the cross-section is shown by C-C' in (A). The maps are based on Geological Survey of Sweden (SGU database) with updates from Andersson (2019). (A black and white version of this figure will appear in some formats. For the colour version, please refer to the plate section.)

by Bäckström et al. (2013) along the Pärvie Fault at Corruvagge. Fracture orientations are consistent with E-W to SW-NE greatest tension, which most likely represents cooling joints in the granites and other fractures parallel to the overall Precambrian structural trend. NE-SW-striking joints, parallel to the strike of the Pärvie Fault, are locally present near the scarp, consistent with a relatively narrow, postglacial fault damage zone.

9.2.2 Geophysical Observations

The establishment of denser and more sensitive seismic networks in Fennoscandia has added much valuable information about certain characteristics of PGFs in the last four decades (Figure 9.1). Clearly, most of the PGFs are seismically active despite the length of time that has passed since deglaciation and the lack of a present ice load, as these faults host most of the current regional microseismicity (e.g. Arvidson, 1996; Lindblom et al., 2015). In northern Sweden, north of latitude 66° N, most of the seismicity recorded in 2000–2013 is located in the neighbourhood of PGFs: 71 per cent of events are within 30 km to the SE and 10 km to the NE of the surface scarps (Lindblom et al., 2015). The analyses also indicate that earthquakes occur to depths down to approximately 35 km (Arvidsson, 1996;

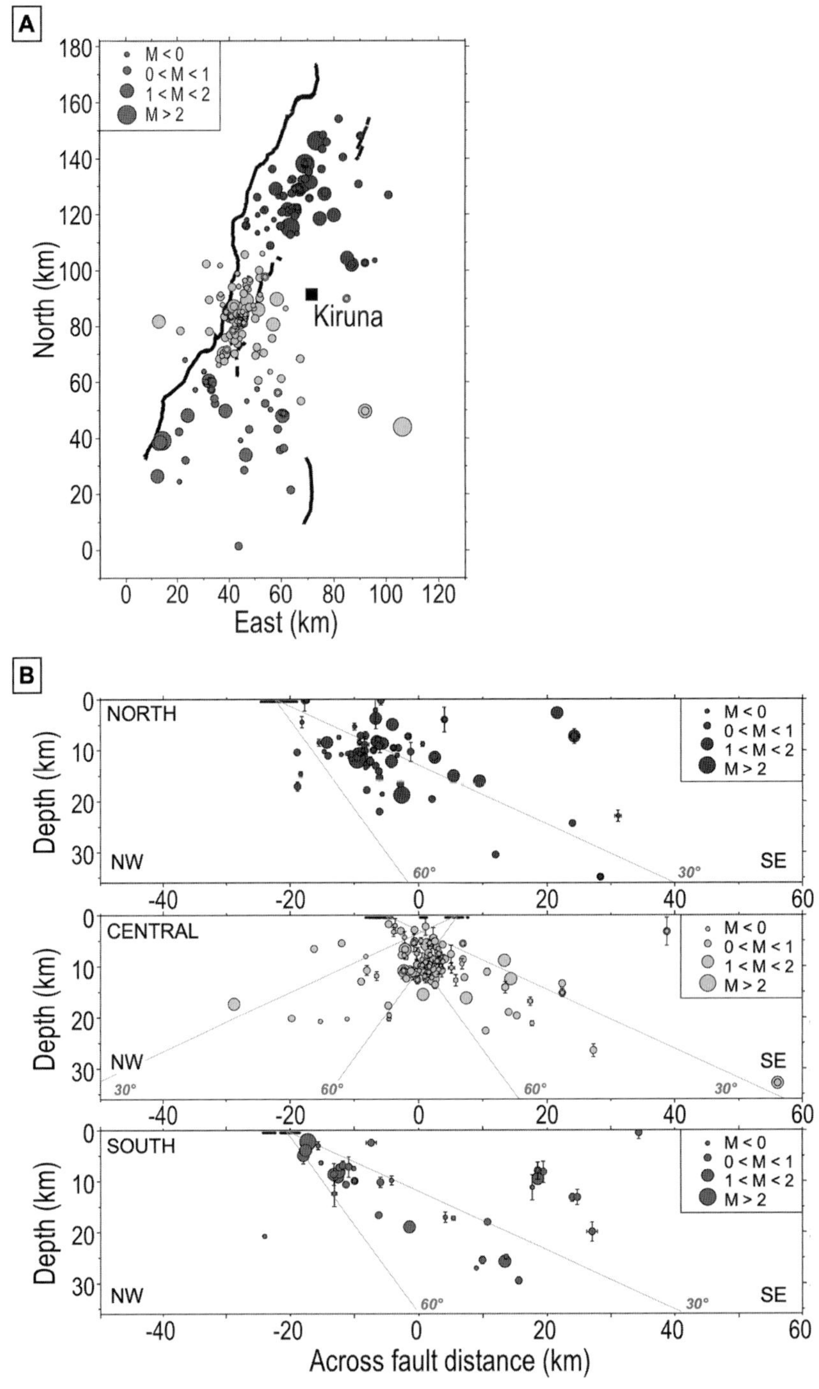

Figure 9.3 A) Epicentral map of the 235 best located earthquakes on the Pärvie Fault System 2007–2010. Blue, yellow and red circles indicate events on the northern, central and southern segments, respectively, and circle size is proportional to magnitude. Black lines mark the PFS. B) Depth distribution of the events for the three segments of the PFS.

Lindblom et al., 2015), implying the faults are crustal-scale structures. This suggests that present-day seismicity is probably occurring along the zones of weakness, allowing reactivation of the PGFs and extending into the lower crust.

The PFS is one of the more active PGFs. A total of 1,921 earthquakes ($M_L \leq 3.7$) were recorded on the PFS between 2003 and 2019. The two largest instrumentally recorded earthquakes along the PFS had $M_L = 3.7$ and occurred on the northern segment of the PFS, one in 1967 and one in 2015. The seismic activity varies along strike, and the seismogenic zone reaches about 35 km depth (Lindblom et al., 2015) (Figure 9.3A). The dip of the zone of earthquakes varies between 30 and 60° (Figure 9.3B). The central section of the PFS is the most seismically active, with the highest density of events occurring south of the Kalix River, an area without road access.

The central part of the PFS consists of the main fault and several subsidiary faults (Figure 9.2). Juhlin et al. (2010) and Ahmadi et al. (2015) explored the geometry at depth and show that a subsidiary fault produced the strongest reflector (R1) compared with the weaker reflector of the main fault (R3). Juhlin et al. (2010) proposed a 50° W dip of the main Pärvie Fault, whereas higher dip values were proposed for the main and subsidiary faults (65° W and 60° E, respectively) by Ahmadi et al. (2015). It has been suggested that the central part of the PFS reflects a flower structure (e.g. Riad, 1990; Juhlin et al., 2010), with seismicity clustered near the projected intersection of the two reflectors R3 and R1 (Ahmadi et al., 2015; Figure 9.4).

9.2.3 Age and Scale of Postglacial Faulting

The postglacial age of the PGFs is based on many observations of the relationship between scarps and various geomorphological features that were unambiguously formed by the latest ice sheet or its meltwater. Deglaciation of the study area occurred at about 9.8–9.9 ka BP and is located NE of the most recently deglaciated area in Fennoscandia (e.g. Stroeven et al., 2016).

Numerous landslide scars have been observed near PGF scarps in Finland (e.g. Kujansuu, 1972; Sutinen et al., 2014; Ojala et al., 2019), Norway (Olesen et al., see Chapter 11) and Sweden (e.g. Lagerbäck, 1992; Lagerbäck & Sundh, 2008; Mikko et al., 2015; Figure 9.1). The spatial relationship between PGF and palaeolandslide in glacial till and in relatively flat terrains suggests earthquake-triggering of the palaeolandslides (e.g. Lagerbäck & Sundh,

---

Figure 9.3 (*cont.*) Cross-sections viewed from the south, perpendicular to N30°E for the northern and southern segments and N15°E for the central segment. Depth uncertainties greater than ±0.3 km are marked by thin black bars. Note that the error cannot be seen assessed for most events. Thin grey lines show dips of 30° and 60°, respectively, and they are centred at the median *x*, *y*-coordinates of the PFS for the respective segment. For the central section we also show 30° and 60° dipping lines from the westerly dipping subsidiary fault located east of the main fault. Black thick lines at the surface show the locations of the PFS traces in each segment (modified from Lindblom et al., 2015). (A black and white version of this figure will appear in some formats. For the colour version, please refer to the plate section.)

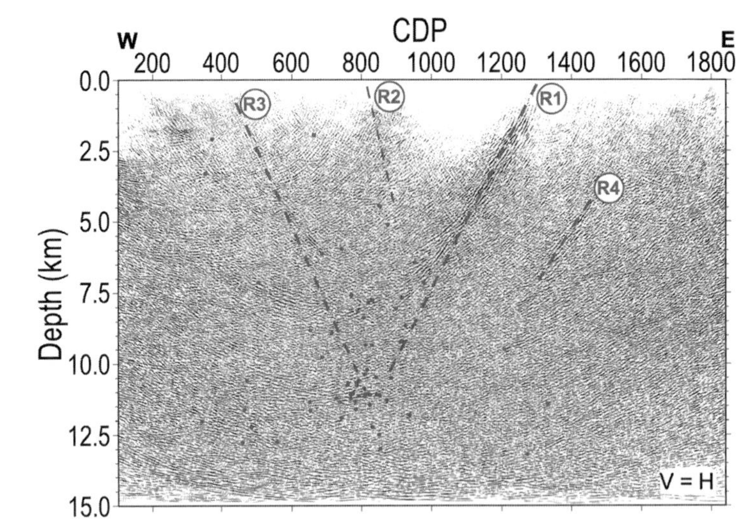

Figure 9.4 The migrated section with earthquake locations from events located within 10 km perpendicular distance from the profile in north and south. The seismicity within 10 km north and south of the seismic line seems to be focused in the area near the intersection of the two reflectors R3 and R1 (Ahmadi et al., 2015). The listric appearance of the R1 and R3 reflections at depth is most likely due to the migration process and not a feature of the faults. We have interpreted the faults to be planar based on the unmigrated sections. The location of the seismic profile is shown in Figure 9.5 (modified from Ahmadi et al., 2015).

2008; Ojala et al., 2019). Radiocarbon dating methods were used to determine the age of the buried organic material under the collapsed land masses. Results from Finland indicate three periods of palaeolandslide age: 11,300–9,480 cal BP, 5,895–5,050 cal BP; 1,774–1,230 cal BP (e.g. Sutinen, 2005; Ojala et al., 2017). Ojala et al. (2018) present mostly similar ages of palaeolandslides (11,000–9,000 cal BP, 6,000–5,000 cal BP and 3,000–1,000 cal BP). Results from Norway by Olesen et al. (see Chapter 11) suggest that strong faulting motion influenced surface sediments. These results suggest that larger earthquakes occurred long after the end of deglaciation, even as recently as <600 cal BP.

No trenching has been carried out on the PFS, and no palaeolandslides near the PFS have been dated. However, individual fault scarps can be dated relative to different morphological features deriving from the most recent deglaciation (Lundqvist & Lagerbäck, 1976; Lagerbäck & Witschard, 1983; Smith et al., see Chapter 8). Presuming that the segments of the main fault and the subsidiary faults ruptured at a similar time, the rupturing of the PFS can be closely related, and dated, to the deglaciation phase of the area (Lagerbäck & Sundh, 2008). The scarps of the PFS can be followed continuously for ~155 km and are generally 3–10 m high, displacing both till and bedrock (Lundqvist & Lagerbäck, 1976; Lagerbäck & Witschard, 1983).

Assuming the whole length of the Pärvie Fault can rupture in a single event, that the depth of the seismogenic zone is 35 ± 5 km and its shear modulus 44 ± 10 GPa, the scarp-forming event had a moment magnitude of 8.0 ± 0.4 (Lindblom et al., 2015). Ruptures of parts of the fault would have created smaller earthquakes; however, in comparison with scaling relationships between fault length and displacement (Leonard, 2010), the relatively large surface displacement is more consistent with a single event along the whole fault (e.g. Muir Wood, 1989; Lindblom et al., 2015).

9.2.4 Deep Biosphere

Hydrogen (H_2) is one of the most important energy sources in the deep subsurface. It has been hypothesized in high velocity friction experiments that considerable amounts of H_2 can be generated even in low magnitude earthquakes, certainly enough to sustain a chemolithoautotrophic community (Hirose et al., 2011). To our knowledge, no studies exist on the deep biosphere in PGFs, but it is likely that the seismic activity of the PFS may produce higher H_2 concentrations and a completely different microbial ecosystem than measured elsewhere in Fennoscandia and seismically inactive Precambrian sites.

9.3 Motivation for Drilling

The DAFNE project is motivated by the geological and geophysical relevance of PGFs. It is especially interesting that faults related to major seismic activity at the end of the last glaciation are still active in the current stress regime, without the ice load, and what this activity may mean for seismic hazard in glaciated and postglacial regions.

There are four scientific objectives:

(1) Make fundamental in situ observations, including documentation of the internal lithological and hydrogeological structure of two major strands of the Pärvie Fault System, measurements of the pressure, temperature and of stress conditions at upper- to mid-crustal depths, and borehole and laboratory characterization of rheological parameters;
(2) Elucidate the accumulation and release of strain along a postglacial intraplate fault zone, including how large intraplate earthquakes are triggered and why seismicity persists for very long time periods in areas of previous large ruptures;
(3) Document and understand the thermal and hydrogeological regimes and the deep biosphere within a PGF and in the surrounding bedrock, as well as reveal thermal and hydrological consequences of seismic slip and concomitant stress change; and
(4) Use the local data to calibrate and test regional models to advance the understanding of glacial isostatic adjustment and study the distal effect of the passive margin development along the western Fennoscandian Shield.

There are three objectives of societal relevance:

(1) Assess seismic risk for mines, tailing dams and hydropower dams near PGFs;
(2) Predict the future behaviour of bedrock, including exposure of more PGFs, during forthcoming glaciations/deglaciations at high latitudes worldwide, in particular for safe disposal of toxic waste in bedrock (e.g. spent nuclear fuel); and
(3) Support the development of sustainable energy production from enhanced geothermal systems in crystalline basement.

Drilling is the only way to obtain the direct observations required to answer the open geological, geophysical and deep biosphere questions surrounding PGFs. We need to collect a wide range of relevant data by conducting direct measurements and observations on drill cores, logging and in situ testing in the borehole and monitoring a subset of parameters within borehole observatories (BHO).

We reiterate that the PFS is an ideal location to study PGFs by scientific drilling. The fault system is clearly related to a glacially triggered, major earthquake event, which created a surface scarp that is available to study for context. Present-day microseismicity is well characterized: it shows the fault is currently active and outlines the seismogenic zone. That the fault is located in a currently active mining area simplifies logistics and creates societal relevance. Existence of roads and infrastructure minimizes the environmental footprint of drilling in the sensitive Arctic environment.

Our scientific team consists of over 40 scientists from 11 countries, who will use the data to address the science objectives as well as the objectives of societal importance. Some of the team members are also involved in the ICDP project, Collisional Orogeny in the Scandinavian Caledonides (COSC), which will ensure integration of data from both projects and address the fourth scientific objective on regional geological history.

9.4 Strategies to Address the Objectives

The central PFS has a complex structure, which must be investigated by multiple boreholes. It is crucial to capture temporal variations of fluid properties and seismic properties. As a result, the drilling targets are the two major reflectors: R3 (main fault branch) and R1 (major subsidiary fault) (Figures 9.5 and 9.6). The two reflectors appear to intersect at about 10 km depth, in a zone of elevated seismic activity (Ahmadi et al., 2015)

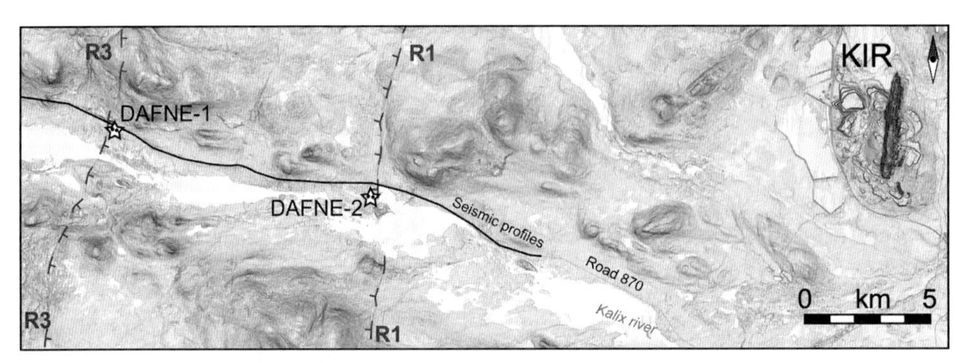

Figure 9.5 Terrain slope map showing locations of the drill sites and their proximity to the Kiruna mine (KIR). The location of the map is indicated in Figure 9.1. The scarps and the fault dip directions of the main and subsidiary faults (R3 and R1, respectively) are indicated as jagged and dashed lines. The two sites (DAFNE-1 and DAFNE-2) are shown by stars, with the proposed locations of the boreholes at each site indicated by black circles (Holes 1A and 2A are located closest to their respective fault scarp). The black line shows the location of the seismic profile presented in Figure 9.4. The terrain slope map is downloaded from the national height model of the National Land Survey of Sweden (Lantmäteriet).

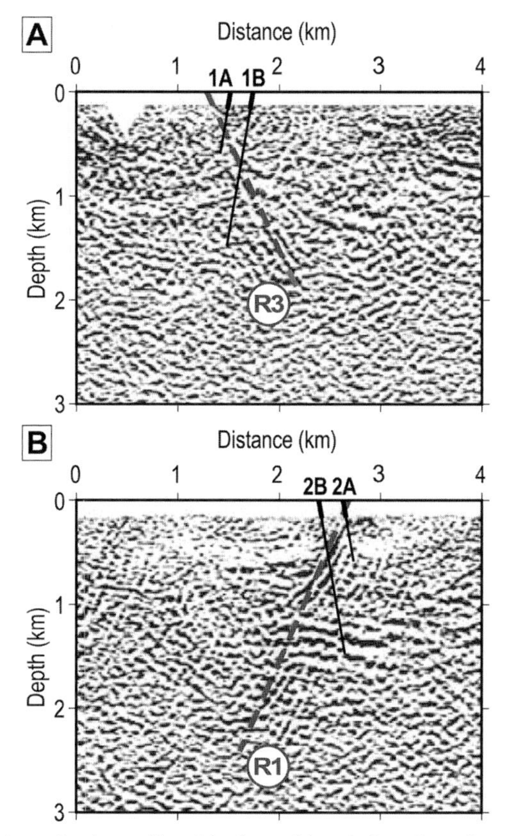

Figure 9.6 Reflection seismic profile with planned borehole trajectories. A) Site DAFNE-1. The image suggests that the main fault (reflector R3, grey dotted line) dips ~65° to the east. B) Site DAFNE-2. The image suggests that the subsidiary fault (reflector R1, grey dotted line) dips ~60° to the west. The locations of the four boreholes (1A, 1B, 2A, 2B) are shown in black, with the percussion section (200 m long) indicated by thicker line for each borehole. The borehole deviation is 10° for all four boreholes. The seismic data were originally collected by Juhlin et al. (2010), and subsequently re-migrated by Ahmadi et al. (2015) with the velocity model provided by Lindblom et al. (2015). The two site locations are indicated in Figure 9.5.

(Figure 9.4). Two drill sites (DAFNE-1 and DAFNE-2), with two boreholes each (one shallow and one deep) and two types of borehole observatories, are planned. This will allow four-dimensional sampling, testing and long-term monitoring of the fault system.

9.4.1 Drilling and Coring Operations

The field season in the arctic is short, from May to September. As a result, activities related to the proposed drilling operations will extend over three consecutive years (Table 9.1). The first year will see establishment of the two drill sites. The four boreholes will be

Table 9.1. *Planned on-site activities of the DAFNE project during years 1–3 (Y1–Y3).*

On-Site Activity	Site 1 (Holes 1A, 1B)	Site 2 (Holes 2A, 2B)
Prepare drill sites	Y1: Site 1	Y1: Site 2
Percussion drill to 200 m, case the hole (>P-size)	Y1: Holes 1A, 1B	Y1: Holes 2A, 2B
Drill flushing water wells	Y1: Site 1	Y1: Site 2
H-size coring: Holes A from 200–600 mbl* and Holes B from 200–1,500 mbl	Y2: Holes 1A, 1B	Y3: Holes 2A, 2B
Flowing fluid electric testing**	Y2: Holes 1A, 1B	Y3: Holes 2A, 2B
Logging with slimhole tools from 200 mbl to TD	Y2: Holes 1A, 1B	Y3-Holes 2A, 2B
Slimwave geophone testing**	Y2: Holes 1A, 1B	Y3: Holes 2A, 2B
In situ stress measurements from 200 mbl to TD	Y2: Holes 1A, 1B	Y3: Holes 2A, 2B
Install BHO fluids above, over and below the fault	Y2: Hole 1A	Y3: Hole 2A
Install BHO seismics from 1,000 to 1,500 mbl	Y2: Hole 1B	Y3: Hole 2B

* mbl = metres borehole length

** Activity included in time plan, but not in budget yet.

percussion-drilled to 200 m of depth and then cased. After that, one site containing two boreholes will be completed each year.

The drilling will be conducted with the Swedish national research infrastructure for scientific drilling, *Riksriggen*, which is owned and operated by Lund University. Riksriggen is an Atlas Copco Christensen CT20C wireline coring rig that can handle three common hole/core sizes P, H and N (123/85 mm, 96/63 mm and 76/48 mm hole/core diameter, respectively) and has a depth capacity of around 1,050 m, 1,600 m and 2,500 m, respectively (assuming a vertical water-filled hole). The rig is equipped with a data acquisition system for monitoring operational parameters that is tightly integrated with the mobile ICDP Drilling Information System (mDIS). Tracer fluorescein will be added with the drilling fluid and special protocols for microbial sampling of core and fluid will be followed to control contamination (e.g. Claesson & Nilsson, 2005; Kallmeyer et al., 2017). Semi-continuous data of some physical properties will be acquired (pressure, temperature, electrical conductivity, fluorescein content) of the drilling fluid (flushing and return water). We also plan to use the online mud gas monitoring (OLGA) system of ICDP to characterize the gases in formation fluids. Other equipment belonging to the infrastructure are, amongst others, a mud cleaning and mixing system, cementing equipment, fishing tools, well heads and a blow-out preventer.

Mirroring operations are planned for the two DAFNE sites (Table 9.1; Figure 9.6). Both drill sites are located on the hanging wall. The boreholes will be drilled at hole azimuths of ~90° to the strike of the main and subsidiary faults, implying an eastward trend of the DAFNE-1 boreholes, and trends towards west for the DAFNE-2 boreholes. The deviation needs to allow maximum depth penetration while preventing the borehole from deviating from the planned trajectory across the fault zone. At the same time, the maximum deviation (10°) from the vertical for seismometers should not be exceeded. The shallow boreholes (Holes 1A and 2A) are projected to have a borehole length of 600 m (metre borehole length, mbl), with target depth (TD) of fault penetration at ~400 mbl (Figure 9.6).

The corresponding depths for the deeper boreholes (Holes 1B and 2B) are 1,500 mbl and ~750 mbl, respectively.

All boreholes will be continuously cored below 200 m. The coring will employ both double- and triple-tube core drilling techniques. In double-core tubing drilling, the drill core is retrieved directly from within the core barrel. The triple-tube assembly has an additional core liner in the core barrel that improves core recovery, which will be needed in fractured rock. Double-tube drilling will be used if rock quality is good enough to allow faster and cheaper drilling. Microbial sampling will be conducted with special core liners in the triple-tube core barrel. H-size drilling (96-mm hole/61-mm core diameter) will be possible down to TD. If borehole stability requires, a casing (i.e. the H-size drill string) will be installed in the borehole, and N-size drilling will be conducted to TD (76-mm hole/48-mm core diameter).

Cored sequences will be used to determine composition, age, rock strength and frictional properties, and fracture geometries and frequencies of host and fault rock. These values are important for determining the mechanics of the postglacial rupture and inferring how far from failure the fault zone is at present. Cores will also be used to constrain palaeoseismic history of fault rocks, i.e. by mapping and dating of brittle deformation structures to obtain information of the palaeostress field(s) that generated past ruptures (surface structures subjected to erosion).

9.4.2 Logging and Borehole Testing

A suite of wireline downhole logging tool runs and in situ borehole tests is included in the time schedule and budget. In addition to sampling for microbial studies from core samples and drill hole fluids, formation water will be pumped from fractures in the post-drilling monitoring phase (see below). Together with fluid geochemistry, the microbial content may provide information on fluid circulation patterns: microbial communities with predominantly oxygen-dependent metabolism would signal near-surface origin, and communities with oxygen-independent metabolism would signal a deeper origin.

A comprehensive logging campaign will be conducted subsequent to the drilling operations. This campaign will include the following slimhole logs (see further biturl.top/IBrInq): gamma spectrum (U, Th, K), dual laterolog resistivity (Rshallow, Rdeep), borehole sonic (full waveforms, Vp, maybe Vs), magnetic susceptibility, dipmeter (oriented 4-arm caliper, borehole orientation, dip, magnetic field magnitude), mud parameter (T, p, and resistivity of the borehole fluid), acoustic borehole televiewer (borehole wall images and high-resolution geometry, oriented caliper) and self (spontaneous) potential (SP). In addition, the boreholes will also be logged using an electric imager during stress measurements. Logging data will be used for identifying lithologies, mapping of fracture and fault geometries including identification of hydraulically active fractures, detecting stress-induced failure of the borehole (borehole breakouts, drilling-induced fractures), measuring the in situ physical properties and downhole temperature, as well as assessing the borehole quality (e.g. width of the damage zone arising from drilling). Logging data is important because it is less affected by drilling; either microfracturing or core recovery may be reduced in some formations.

The state of stress will be measured with the Swedish research infrastructure for stress measurements, *Spänningstrailern* (the stress trailer). The system was developed and built at Luleå University of Technology (LTU) and University of Strasbourg with support by two industry experts (Geosigma AB and Fracsinus Rock Stress Measurements AB). The infrastructure is owned and operated by LTU. It is a wire-line activated straddle packer system for slim boreholes that has an electric imager for measuring the new/existing fracture orientation permanently mounted below the packers. Three borehole dimensions can be tested (N-, H- and P-sized boreholes) to 3 km depth. Three methods may be applied: hydraulic fracturing (HF), hydraulic testing of pre-existing fractures (HTPF) and sleeve fracturing (SF). While the HF and HTPF methods are well known and included in suggested methods of the International Society of Rock Mechanics (Haimson & Cornet, 2003), the SF method (Stephansson, 1983) is less well-known. SF involves generation of axial fractures with the packer elements and allows determination of the tensile strength at a representative scale. It may be used to induce an axial fracture if the HF method results in activation of foliation planes rather than an axial fracture. SF-generated fractures are per definition very shallow, as they are generated by the packer pressure without fluid injection. The test section is logged after pressure testing, giving an immediate answer to testing outcome, as well 100 per cent overlap between tested sections of the straddle packers and the electrical imager. Logging under packers is important as it allows identification if test quality is influenced by packer-initiated fractures. The full stress tensor and its spatial variation with depth within a continuous rock mass can be determined in a scientifically unambiguous way by integrating results from the three test methods. The test plan is to conduct sufficient measurements above and below the fault zone. The exact number of tests depends on borehole conditions and fracture distribution, but a rough estimate is about one test per 25-m borehole. The aim is to determine the full stress tensor and its gradient with depth on both sides of the fault zone. Further stress integration is subsequently planned to include other stress indicators (e.g. focal mechanisms and logging data of stress-induced borehole failure). The effective stress state, which considers total stress and fluid pressure, influences the stability of the rock mass and is important for determining how far from failure the fault zone is.

Time has been allocated, but no costs are yet included in the budget for flowing fluid electric conductivity drill string testing (FFEC; e.g. Doughty et al., 2017) and to employ the GFZ slimwave geophone chain (SGC) and fluid sampler (with a 0.6-liter displacement type mercury-free sampler). The FFEC is used to identify inflow zones along the borehole and to estimate their hydraulic transmissivity and salinity. The SGC provides data on the borehole velocity structure, which helps to identify the location of the fault scarps. The fluid sampler collects water from the borehole for geochemical and microbial sampling.

9.4.3 Borehole Observatories

Borehole observatories are very powerful tools for addressing active processes because time-dependent properties are monitored. A growing body of evidence of seasonally modulated seismicity, in a variety of both intra- and interplate tectonic environments, identifies increased pore pressure caused by groundwater recharge as a trigger of seismicity (e.g. Saar & Manga, 2003; Westaway, 2006; Costain, 2016; Johnson et al., 2017; Ueda & Kato, 2019). Monitoring on a multi-year temporal timescale may capture variations in fluid, stress and seismic properties related to local seismicity that may be used for various types of thermal, hydraulic, mechanical and chemical modeling as well as to model fault system response to deglaciation.

Two types of borehole observatories will be installed. In Holes DAFNE-1A and 2A, the type referred to as BHO fluids will be installed for capturing transient hydrogeologic, geoelectric, temperature and pore fluid pressure properties at three levels in each borehole. Three packer elements will seal off sections of the borehole from surface water and above and below the fault zone. Sensors for pressure, temperature and electrical conductivity will monitor borehole conditions in each section, and water can be pumped to surface through check valves. BHO fluids are prepared for installation of optical fibres. BHO fluids observatory is a modernized version of the observatories in the Greenland Analogue Project (GAP) (Harper et al., 2016; Claesson Liljedahl et al., 2016), which aimed to increase knowledge of rock mass conditions during an ice age.

In Holes DAFNE-1B and -2B, a 6-level seismic array will be installed (BHO seismics). At each borehole, a 6-level seismic cable with 3-component 15-Hz omni-geophones will be installed at a spacing of 100 m. The total cable length of 1,500 m will position the geophone array at the bottom 500 m of each of Holes 1B and 2B. Coupling between the array and the formation is achieved with bow spring, which means that the array will not be cemented. This leaves the hole open for later studies and/or repair of the seismometers. The array will provide enhanced detection and location capabilities, making it possible to use high-accuracy relative location to pinpoint micro-earthquake locations and relate this to the faults and fault system. The downhole systems will also make it possible to study seismic source mechanisms and their relations to fault properties and hydraulics in more detail and offer the ability to derive focal mechanisms. The array is designed to record micro-earthquakes at least down to $M_L = -0.5$.

The two BHOs are anticipated to produce data that will allow improved understanding of the dynamics of the PFS seismogenic zone. The data records will allow comparison with other past, ongoing or planned ICDP projects (e.g. Zoback et al., 2011; Lin et al., 2012; Bohnhoff et al., 2017; Gupta et al., 2017; Townend et al., 2017), as well as International Ocean Discovery Program borehole observatories (e.g. Peng & Gomberg, 2010; Sugihara et al., 2014; Araki et al., 2017), some of which are available on-line (iodp.tamu.edu/database/borehole.html).

9.4.4 Target Drilling Depths and Fault Zone Identification

A number of constraints control the final decision on what the target drilling depths will be, and thus the design and layout of the planned boreholes. ICDP's depth-and-cost criterion implies that the project should be cost effective by minimizing the depth and difficulty of the drilling targets selected. Riksriggen provides a particularly good value service for scientific drilling because it is run as a national infrastructure. It also has a very small environmental footprint, which is an important factor to consider for the sensitive subarctic flora and fauna.

The main- and subsidiary faults will be penetrated at two depths, making it possible to obtain good core recovery of fresh unweathered fault rocks and to document and measure variation of downhole fault properties for the two fault branches. Two boreholes per site are also required to allow separate installation of the BHOs.

The plans for in situ stress measurements and seismic monitoring significantly control what target depths and borehole design will be. First, it is well known that the stress field becomes more uniform with depth (e.g. Stephansson, 1989). This implies that the shallow boreholes likely penetrate heterogeneous stress conditions in the hanging wall and uniform stress conditions in the footwall. The deeper boreholes will reveal the transition from heterogeneous to uniform stress conditions in the hanging wall. Second, it is important to collect stress data at a distance above and below the fault zones. Therefore, the fault zones are targeted to be penetrated at about 400 mbl and 750 mbl, respectively, for the shallow and deep boreholes. The TD of the boreholes will allow stress measurements in the footwall away from the fault zone. Third, the objective for seismic monitoring is to record micro-seismicity with good accuracy. The planned depth of installation ($\sim 1,000 - 1,500$ mbl) is well into the uniform stress field and considered to be sufficient to resolve earthquake activity to at least 10 km of depth. In addition, a local seismic network will be installed in the area surrounding the two drill sites prior to the start of drilling in order to further improve seismic location accuracy and quality of focal mechanism determination. Seismic monitoring will provide precise localization of events (on which fault strand the events fall and/or if they occur along the intersection of the faults), track changes to the seismicity rate during the drilling operations, aid in the analysis of shallow events that may be future drilling targets and detect larger events occurring in the nearby underground mine. BHO seismics and local network data will also be combined with data from the Swedish National Seismic Network. One topic that may be explored is whether slow earthquakes, widely observed in plate boundary regions (e.g. Peng & Gomberg, 2010), can be recorded in an intraplate setting.

One common challenge for fault zone drilling is identifying the location and extent of the fault zones, especially if the fault lacks lithological control. However, recognition of PFS faults in a core should be readily achievable, even if the there is no lithological contrast across the fault, thanks to the following: Firstly, faults reported in previous mining exploration boreholes are typically oblique to local lithological boundaries (Gerdin, 1979; Carlson & Lundqvist, 1984), so PGFs may therefore be distinguishable from older features by their orientation (Figure 9.2). PGFs are also expected to contain fractures and/or fault gouges lacking greenschist to amphibolite facies precipitates and/or overprints, whereas such metamorphic features are expected and observed in outcrops of older structures. We also expect to encounter clay zones in the shallow parts of the PGFs, resulting from older weathering processes as observed in other PGFs in northern Fennoscandia (Kukkonen et al., 2010; Olesen et al., 2013).

Near the DAFNE-1 drill site in the Laukujärvi prospect in 1950–1980 (Carlson & Lundqvist, 1984), it was recorded that lithological layering and schistosity dip about

50–70° SW, and the contact between volcanic and igneous rocks trends SE-NW. The main fault, however, is oblique to the inferred plutonic-volcanic contact. In the area of the DAFNE-2 site, surface geology and drilling data from the 1970s (Gerdin, 1979) show that the main contact between granite and felsic volcanics is running approximately E-W at a high angle to the subsidiary fault scarp (R1). Faults intersected by boreholes dip 70–90° NW (Gerdin, 1979). North of the DAFNE-2 site, a complex, dominantly volcanogenic, lithological assemblage terminates against granite at a sharp N-S-striking contact (Figure 9.2). This boundary coincides with the surface scarp of the W-dipping reflector R1 in our data and with a sharp change in aeromagnetic anomaly. These data overall suggest PGFs should be recognizable as cross-cutting volcanic and igneous rocks and pre-existing structures. Operations should expect the cored rock formations to be low- to medium-grade metamorphic rock with moderate to steeply dipping schistosity.

9.5 Concluding Remarks

In this chapter, we presented the background, motivation and strategy for addressing the objectives for the DAFNE project. The fundamental objective has been to make in situ observations of a postglacial fault and follow it up with a programme of laboratory experiments, numerical models and borehole monitoring. Through these data, which can only be obtained by drilling, we expect to reach a better understanding of large postglacial earthquakes and subsequent, ongoing, persistent microseismicity. Furthermore, borehole measurements will document thermal and hydrogeological regimes within a PGF and in the surrounding bedrock. The local data from the drill sites will be used to calibrate and test regional models of glacial isostatic adjustment and distal effects of the passive margin development along the western Fennoscandian Shield.

The scientific objectives have societal relevance in allowing assessment of seismic risk to mines, tailing dams and hydropower dams near PGFs, prevalent in Fennoscandia and other northern regions globally. This risk assessment is particularly important in the current warming climate, making it possible to predict the future behaviour of bedrock during forthcoming glaciations/deglaciations at high latitudes worldwide, in particular for safe geological disposal of toxic and nuclear waste. The data are, however, not solely for risk assessment: increased understanding of thermal, hydrological and mechanical bedrock properties also supports the development of sustainable energy production from enhanced geothermal systems in crystalline basement.

The proposal has been accepted for ICDP funding after detailed evaluation by the ICDP Science Advisory Group, Executive Committee and Assembly of Governors, and their positive review provides a stamp of approval with respect to quality of science and societal objectives, qualification of proponents and science team, cost effectiveness and need-to-drill with an appropriate target. The next step, to attract co-funding, is an ongoing activity for the proponents and the science team. The data and monitoring infrastructure will be instrumental in furthering our understanding of PGFs and intraplate seismicity.

Acknowledgements

The DAFNE project lies within the International Continental Drilling Program. We thank personnel from the Operations Support Group for their support. We extend special thanks to John Townend, Victoria University of Wellington, Matt Ikari, Bremen University, and an anonymous reviewer for their constructive comments. Sadly, the DAFNE science team lost two of its more senior members in 2020, Ove Stephansson, GFZ, and Francois Cornet, University of Strasbourg. Both are missed and here greatly acknowledged for their contribution to the development of the DAFNE project; Francois Cornet is as well for playing a central role in the development of the LTU stress trailer.

References

Ahmadi, O., Juhlin, C., Ask, M. V. S. and Lund B. (2015). Revealing the deeper structure of the end-glacial Pärvie fault system in northern Sweden by seismic reflection profiling. *Solid Earth*, **6**, 621–632, doi.org/10.5194/se-6-621-2015.

Andersson, J. B. H. (2019). *Structural Evolution of Two Ore-Bearing Palaeoproterozoic Metasupracrustal Belts in the Kiruna Area, Northwestern Fennoscandian Shield.* Licentiate thesis, Luleå University of Technology, Sweden, 91 pp.

Araki, E., Saffer, D. M., Kopf, A. et al. (2017). Recurring and triggered slow-slip events near the trench at the Nankai Trough subduction megathrust. *Science*, **356**, 1157–1160, doi.org/10.1126/science.aan3120.

Arvidsson, R. (1996). Fennoscandian earthquakes: whole crustal rupturing related to postglacial rebound. *Science*, **274**, 744–746, doi.org/10.1126/science.274.5288.744.

Bäckström, A., Giulio, V., Rantakokko, N., Jonsson, E. and Ask, M. (2013). *Preliminary Results from Fault-Slip Analysis of the Pärvie Neotectonic Postglacial Fault Zone, Northern Sweden.* EGU General Assembly 2013, 7–12 April 2013 in Vienna, Austria, id. EGU2013–1751.

Bauer, T., Andersson, J., Sarlus, Z., Lund, C. and Kearney, T. (2018). Structural controls on the setting, shape and hydrothermal alteration of the Malmberget IOA deposit, northern Sweden. *Economic Geology*, **113**(2), 377–395, doi.org/10.5382/econgeo.2018.4554.

Bergman, S., Kübler, L. and Martinsson, O. (2001). Regional geological and geophysical maps of northern Norrbotten County: bedrock map (east of the Caledonian orogen). *Sveriges Geologiska Undersökning*, Ba 56.

Bohnhoff, M., Dresen, G., Ceken, U. et al. (2017). GONAF – the borehole Geophysical Observatory at the North Anatolian Fault in the eastern Sea of Marmara. *Scientific Drilling*, **22**, 19–28, doi.org/10.5194/sd-22-19-2017.

Calais, E., Camelbeeck, T., Stein, S., Liu, M. and Craig, T. J. (2016). A new paradigm for large earthquakes in stable continental plate interiors. *Geophysical Research Letters*, **43**, 10621–10637, doi.org/10.1002/2016GL070815.

Campell, D. L. (1978). Investigation of the stress-concentration mechanism for intraplate earthquakes. *Geophysical Research Letters*, **5**, 477–479.

Carlson, L. and Lundqvist, A. (1984). *Laukujärvi kopparfyndighet. Prospekteringsförslag 1984 och geologisk sammanfattning* [Laukujärvi copper deposit. Exploration proposal 1984 and geological summary]. Sveriges Geologiska AB, PRAP 84067, 57 pp. (in Swedish).

Claesson, L.-Å. and Nilsson, G. (2005). *Forsmark Site investigation. Drilling of the Borehole KFM01B at Drilling Site DS1*. SKB Report P-04-302, Swedish Nuclear Fuel and Waste Management Co., Stockholm, 32 pp.

Claesson Liljedahl, L., Kontula, A., Harper, J. et al. (2016). *The Greenland Analogue Project: Final Report*, SKB Technical Report TR-14-13, Swedish Nuclear Fuel and Waste Management Co., Stockholm, 142 pp.

Costain, J. K. (2017). Groundwater recharge as the trigger of naturally occurring intraplate earthquakes. In A. Landgraf, S. Kuebler, E. Hintersberger and S. Stein, eds., *Seismicity, Fault Rupture and Earthquake Hazards in Slowly Deforming Regions*. Geological Society, London, Special Publication, Vol. 432, pp. 91–118, doi.org/10.1144/SP432.9.

Doughty, C., Tsang, C.-F., Rosberg, J.-E. et al. (2017). Flowing fluid electrical conductivity logging of a deep borehole during and following drilling: estimation of transmissivity, water salinity and hydraulic head of conductive zones. *Hydrogeology Journal*, **25**(2): 501–517, doi.org/10.1007/s10040–016-1497-5.

England, R. W. and Ebbing, J. (2012). Crustal structure of central Norway and Sweden from integrated modelling of teleseismic receiver functions and the gravity anomaly. *Geophysical Journal International*, **191**(1), 1–11, doi.org/10.1111/j.1365-246X.2012.05607.x.

England, P. and Jackson, J. (2011). Uncharted seismic risk. *Nature Geoscience*, **4**, 348–349, doi.org/10.1038/ngeo1168.

Gaál, G. and Gorbatschev, R. (1987). Precambrian geology and of the Central Baltic Shield. *Precambrian Research*, **35**, 382 pp.

Gerdin, P. (1979). *Vieto resultat av utförda prospekteringsarbeten* [Vieto Results of Exploration Work Performed]. Rapport för NSG. Berggrundsbyrån, Geological Survey of Sweden.

Grad, M., Tiira, T. and the ESC Working Group (2009). The Moho depth map of the European Plate. *Geophysical Journal International*, **176**, 279–292, doi.org/10.1111/j.1365-246X.2008.03919.x.

Gupta, H. K., Arora, K., Rao, N. P. et al. (2017). Investigations of continued reservoir triggered seismicity at Koyna, India. In S. Mukherjee, A. A. Misra, G. Calvès and M. Nemčok, eds., *Tectonics of the Deccan Large Igneous Province*. Geological Society, London, Special Publication, Vol. 445, pp. 151–188, doi.org/10.1144/SP445.11.

Haimson, B. C. and Cornet, F. H. (2003). ISRM SM for rock stress estimation – part 3: hydraulic fracturing (HF) and/or hydraulic testing of pre-existing fractures (HTPF). *International Journal of Rock Mechanics and Mining Sciences*, **40**, 1011–1020, doi.org/10.1016/j.ijrmms.2003.08.002.

Harper, J., Hubbard, A., Ruskeeniemi, T. et al. (2016). *The Greenland Analogue Project*. SKB Report R-14-13, 387 pp.

Hirose, T., Kawagucci, S. and Suzuki, K. (2011). Mechanoradical H_2 generation during simulated faulting: implications for an earthquake-driven subsurface biosphere. *Geophysical Research Letters*, **38**, L17303, doi.org/10.1029/2011GL048850.

Johnson, C. W., Fu, Y. and Bürgmann, R. (2017). Seasonal water storage, stress modulation, and California seismicity. *Science*, **356**(6343), 1161–1164, doi.org/10.1126/science.aak9547.

Juhlin, C., Dehghannejad, M., Lund, B., Malehmir, A. and Pratt, G. (2010). Reflection seismic imaging of the end-glacial Pärvie Fault system, Sweden. *Journal of Applied Geophysics*, **70**, 307–316, doi.org/10.1016/j.jappgeo.2009.06.004.

Kallmeyer, J. (2017). Contamination control for scientific drilling operations. *Advances in Applied Microbiology*, **98**, 61–91, doi.org/10.1016/bs.aambs.2016.09.003.

Keiding, M., Kreemer, C., Lindholm, C. D. et al. (2015). A comparison of strain rates and seismicity for Fennoscandia: depth dependency of deformation from glacial isostatic adjustment. *Geophysical Journal International*, **202**, 1021–1028, doi.org/10.1093/gji/ggv207.

Kierulf, H. P., Steffen, H., Simpson, M. J. R. et al. (2014). A GPS velocity field for Fennoscandia and a consistent comparison to glacial isostatic adjustment models. *Journal of Geophysical Research*, **119**(8), 6613–6629, doi.org/10.1002/2013JB010889.

Korja, T. (2007). How is the European lithosphere imaged by magnetotellurics? *Surveys in Geophysics*, **28**(2–3), 239–272, doi.org/10.1007/s10712-007-9024-9.

Kreemer, C., Blewitt, G. and Klein, E. C. (2014). A geodetic plate motion and Global Strain Rate Model. *Geochemistry, Geophysics, Geosystems*, **15**, 3849–3889, doi.org/10.1002/2014GC005407.

Kuivamäki, A., Vuorela, P. and Paananen, M. (1998). *Indications of Postglacial and Recent Bedrock Movements in Finland and Russian Karelia*. Geological Survey of Finland Nuclear Waste Disposal Research Report YST-99, Espoo, Finland, 92 pp.

Kujansuu, R. (1972). The deglaciation of Finnish Lapland. In L. K. Kauranne, ed., *Glacial Stratigraphy, Engineering Geology and Earth Construction*. Geological Survey of Finland Special Paper 15, pp. 21–31.

Kukkonen, I. T., Olesen, O., Ask, M. V. S. and the PFDP Working Group (2010). Postglacial faults in Fennoscandia: targets for scientific drilling. *GFF*, **132**(1), 71–81, doi.org/10.1080/11035891003692934.

Lagerbäck, R. (1978). Neotectonic structures in northern Sweden. *Geologiska Föreningens i Stockholm Förhandlingar*, **100**(3), 263–269, doi.org/10.1080/11035897809452533.

Lagerbäck, R. (1992). Dating of Late Quaternary faulting in northern Sweden. *Journal of the Geological Society, London*, **149**, 285–291, doi.org/10.1144/gsjgs.149.2.0285.

Lagerbäck, R. and Sundh, M. (2008). Early Holocene Faulting and Paleoseismicity in Northern Sweden. Geological Survey of Sweden Research Paper, **C836**, 84 pp.

Lagerbäck, R. and Witschard, F. (1983). *Neotectonics in Northern Sweden – Geological Investigations*. SKBF/KBS Technical Report 83–58, Svensk Kärnbränslehantering AB, Stockholm, 58 pp.

Lahtinen, R., Korja, A. and Nironen, M. (2005). Palaeoproterozoic tectonic evolution. In M. Lehtinen, P. Nurmi and T. Rämö, eds., *Precambrian Geology of Finland – Key to the Evolution of the Fennoscandian Shield*. Elsevier Science Publishers, Amsterdam, pp. 481–531, doi.org/10.1016/S0166-2635(05)809012-X.

Leonard, M. (2010). Earthquake fault scaling: self-consistent relating of rupture length, width, average displacement, and moment release. *Bulletin of the Seismological Society of America*, **100**, 1971–1988, doi.org/10.1785/0120090189.

Li, Q., Liu, M. and Stein, S. (2009). Spatiotemporal complexity of continental intraplate seismicity: insights from geodynamic modeling and implications for seismic hazard estimation. *Bulletin of the Seismological Society of America*, **99**, 52–99, doi.org/10.1785/0120080005.

Lin, Y.-Y., Ma, K.-F. and Oye, V. (2012). Observation and scaling of microearthquakes from the Taiwan Chelungpu-fault borehole seismometers. *Geophysical Journal International*, **190**, 665–676, doi.org/10.1111/j.1365-246X.2012.05513.x.

Lindblom, E., Lund, B., Tryggvason, A. et al. (2015). Microearthquakes illuminate the deep structure of the endglacial Pärvie fault, northern Sweden, *Geophysical Journal International*, **201**, 1704–1716, doi.org/10.1093/gji/ggv112.

Lund, B. (2015). Palaeoseismology of glaciated terrain. In M. Beer, I. A. Kougioumtzoglou, E. Patelli and I. K. Au, eds., *Encyclopedia of Earthquake Engineering*. Springer, Berlin/Heidelberg, doi.org/10.1007/978-3-642-36197-5_25-1.

Cover Image

Figure 1.1

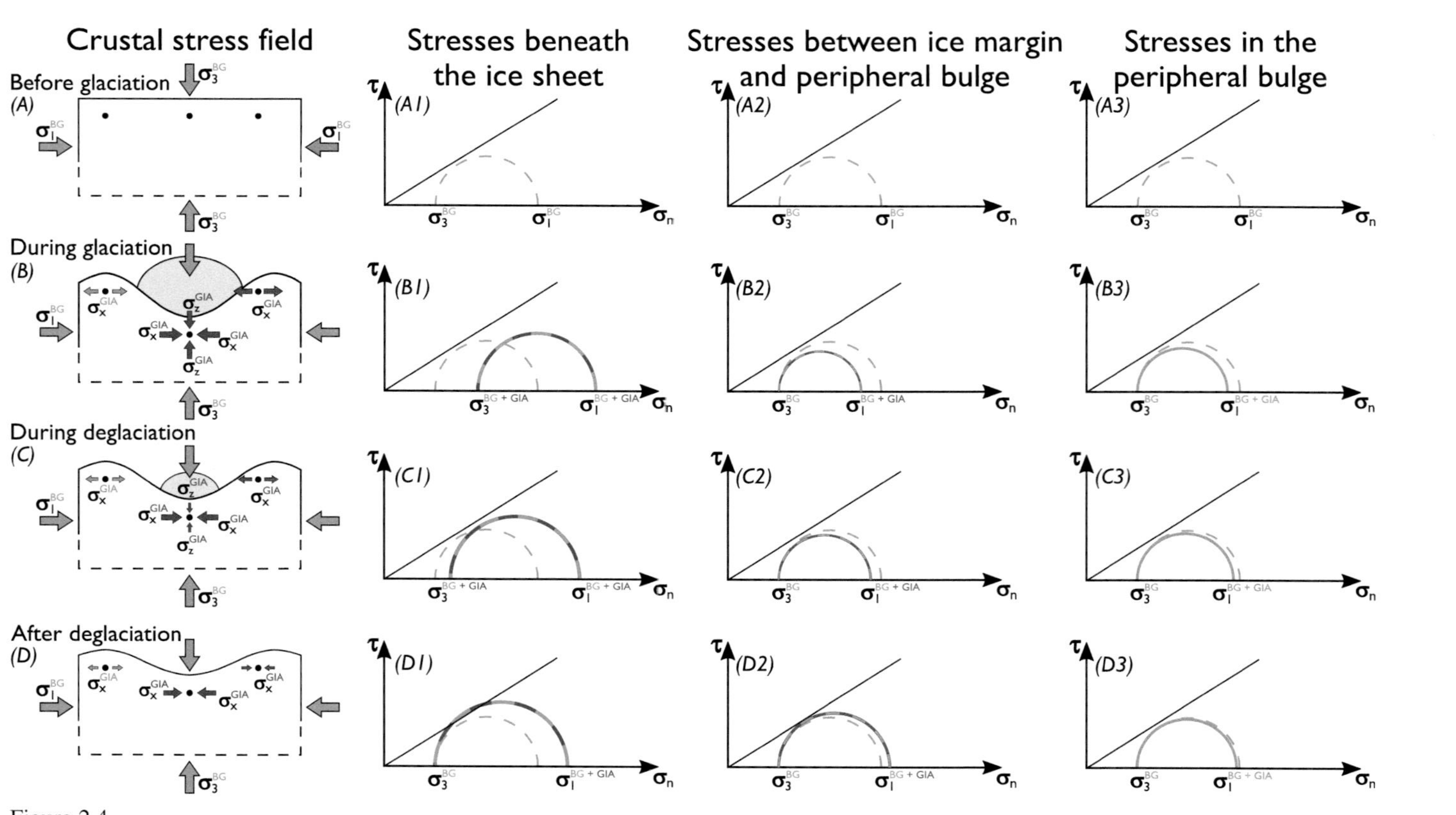
Crustal stress field
Stresses beneath the ice sheet
Stresses between ice margin and peripheral bulge
Stresses in the peripheral bulge
Before glaciation (A)
During glaciation (B)
During deglaciation (C)
After deglaciation (D)
(A1)
(A2)
(A3)
(B1)
(B2)
(B3)
(C1)
(C2)
(C3)
(D1)
(D2)
(D3)
σ_1^{BG}
σ_3^{BG}
σ_x^{GIA}
σ_z^{GIA}
σ_3^{BG+GIA}
σ_1^{BG+GIA}
τ
σ_n

Figure 2.4

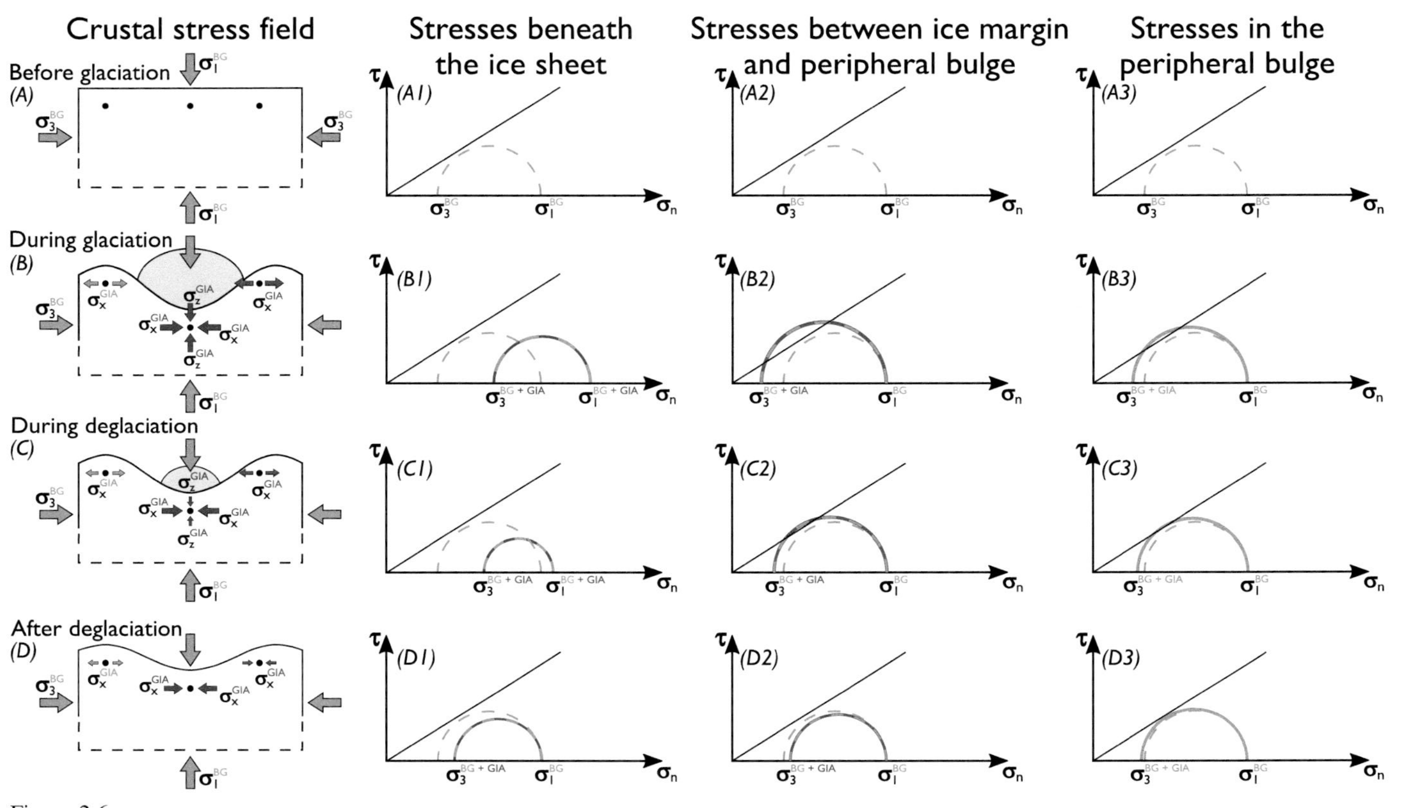

Crustal stress field
Stresses beneath the ice sheet
Stresses between ice margin and peripheral bulge
Stresses in the peripheral bulge
Before glaciation
(A)
During glaciation
(B)
During deglaciation
(C)
After deglaciation
(D)
(A1)
(A2)
(A3)
(B1)
(B2)
(B3)
(C1)
(C2)
(C3)
(D1)
(D2)
(D3)

Figure 2.6

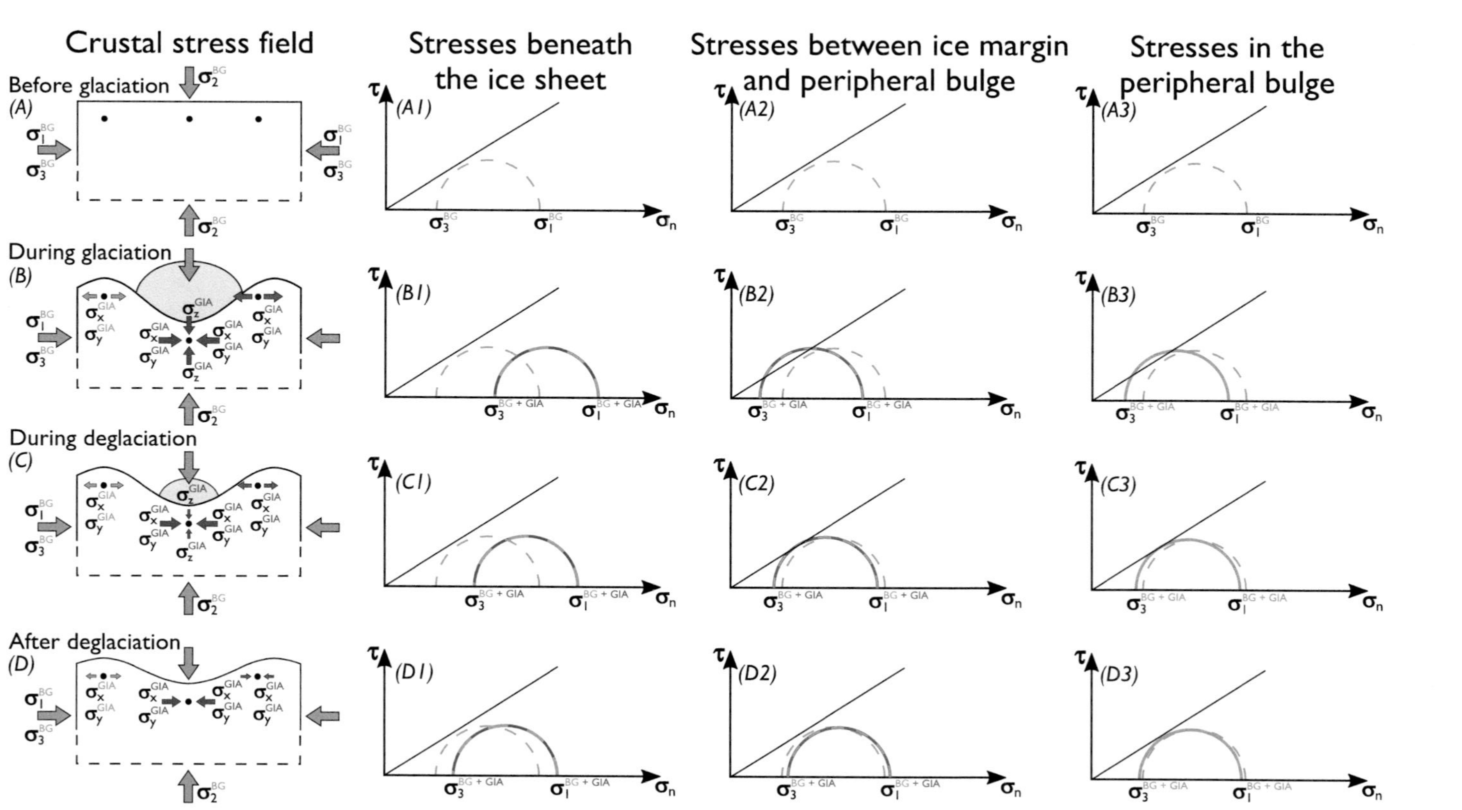

Crustal stress field
Stresses beneath the ice sheet
Stresses between ice margin and peripheral bulge
Stresses in the peripheral bulge
Before glaciation
(A)
During glaciation
(B)
During deglaciation
(C)
After deglaciation
(D)
(A1)
(A2)
(A3)
(B1)
(B2)
(B3)
(C1)
(C2)
(C3)
(D1)
(D2)
(D3)

Figure 2.7

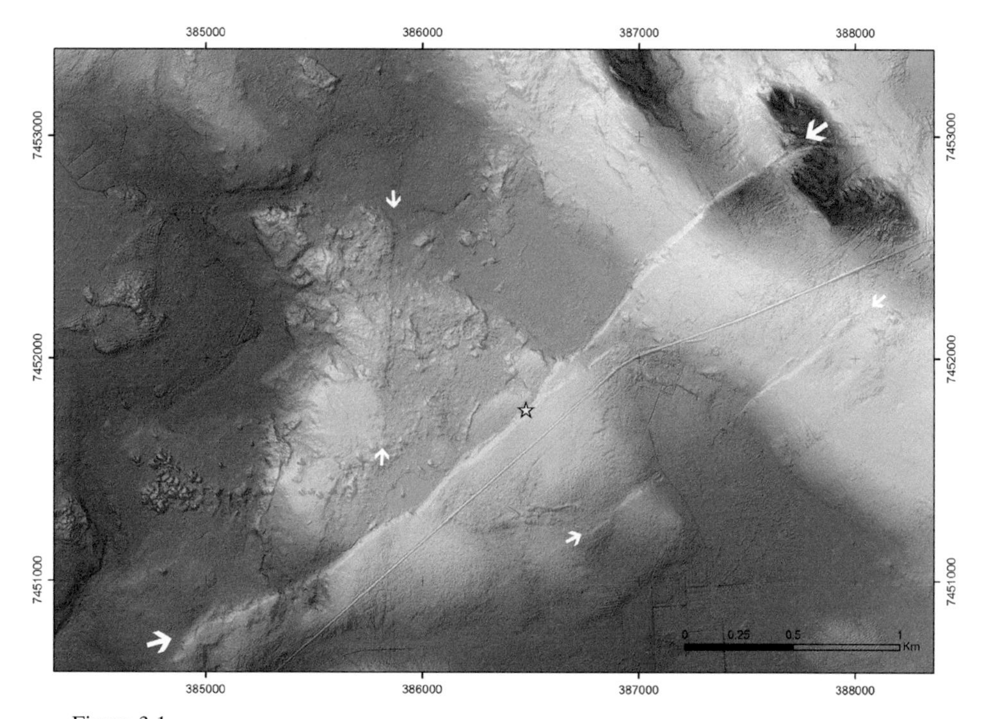

Figure 3.1

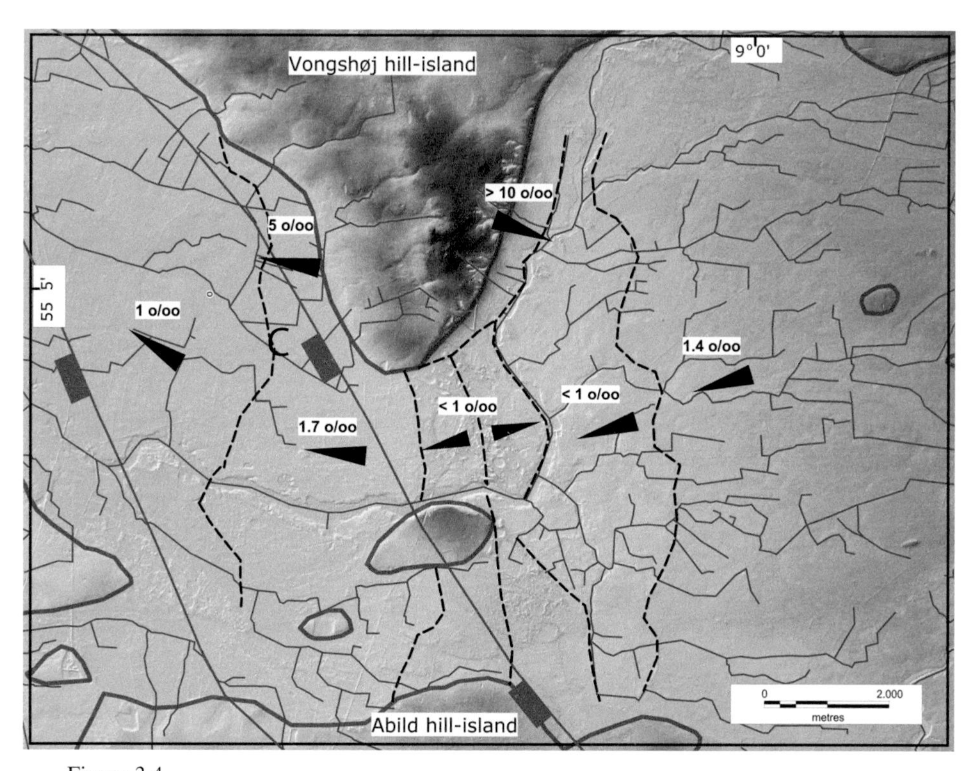

Figure 3.4

Glaciotectonic complex

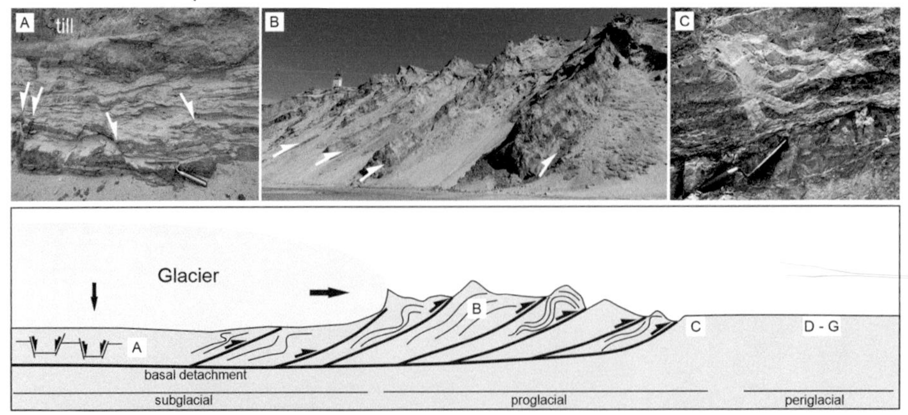

Periglacial soft-sediment deformation structures

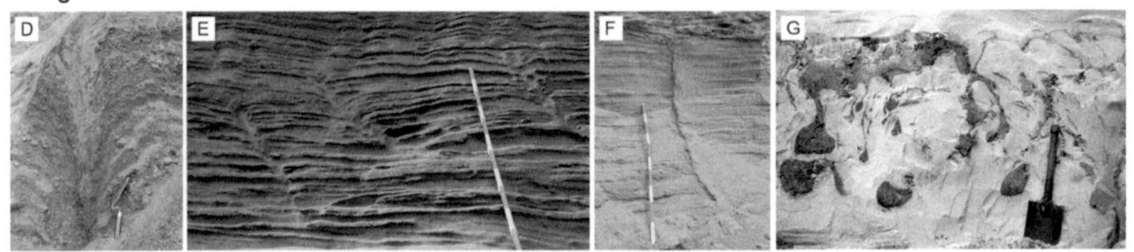

Earthquake-induced soft-sediment deformation structures

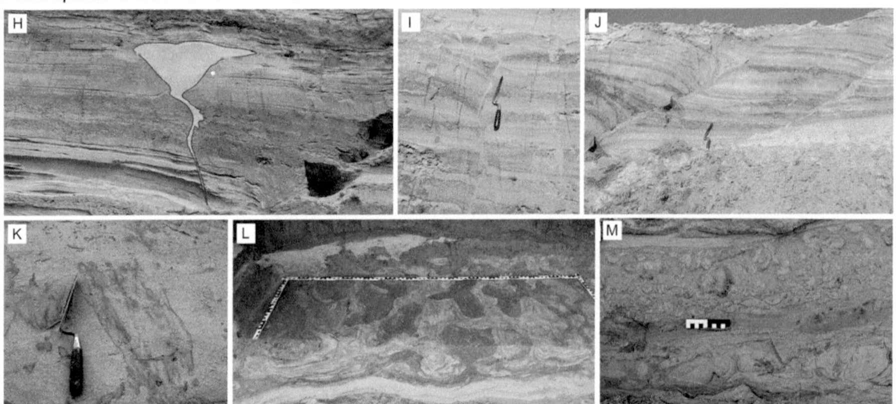

Deformation bands

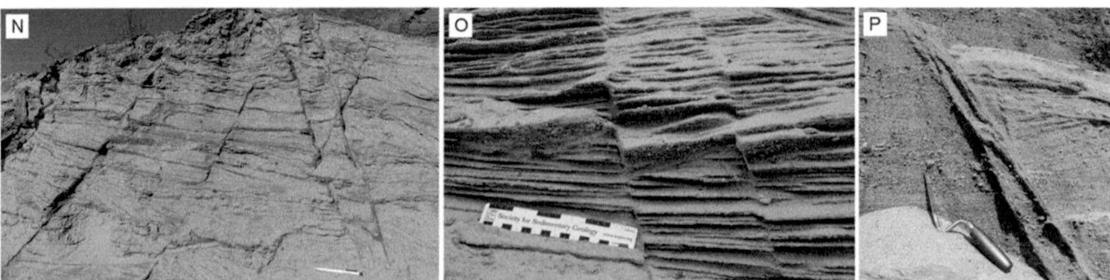

Figure 4.3

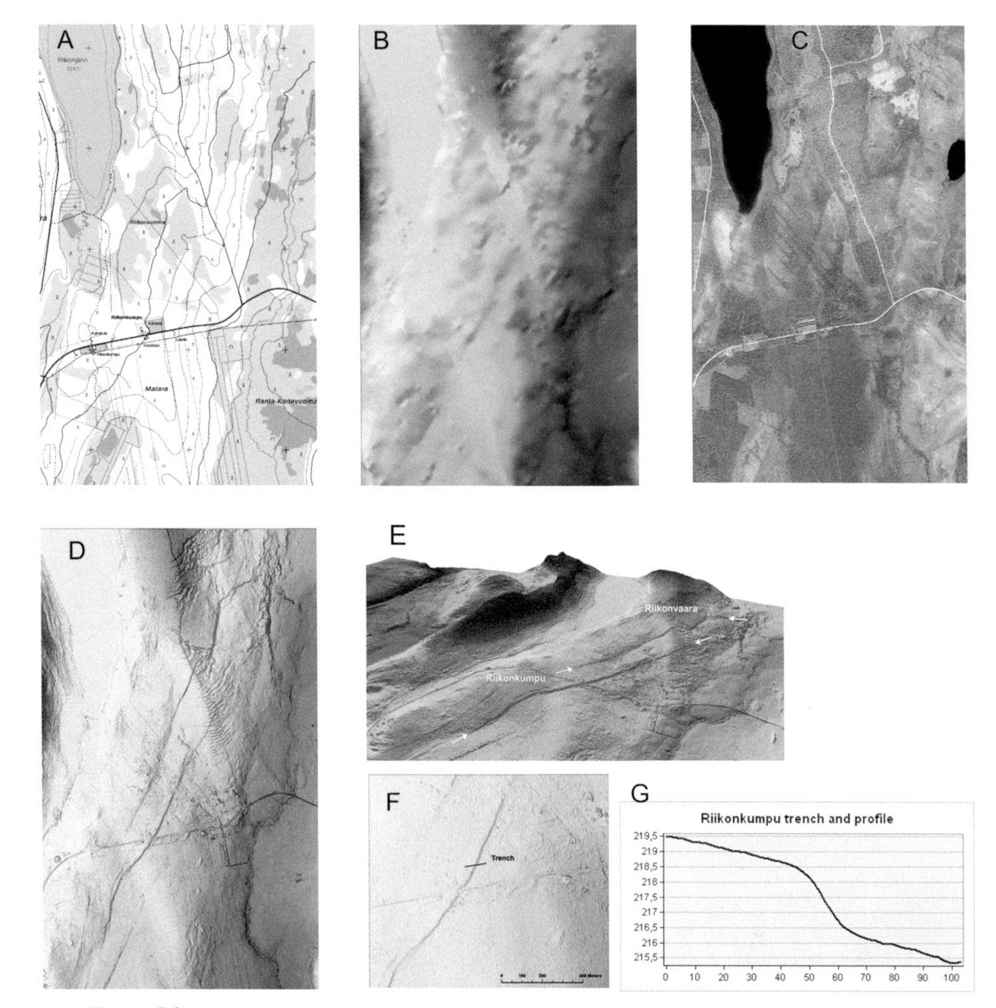

A
Matara
Ranta-Kortevuoma
B
C
D
E
Riikonvaara
Riikonkumpu
F
Trench
G
Riikonkumpu trench and profile
219,5
219
218,5
218
217,5
217
216,5
216
215,5
0
10
20
30
40
50
60
70
80
90
100

Figure 5.2

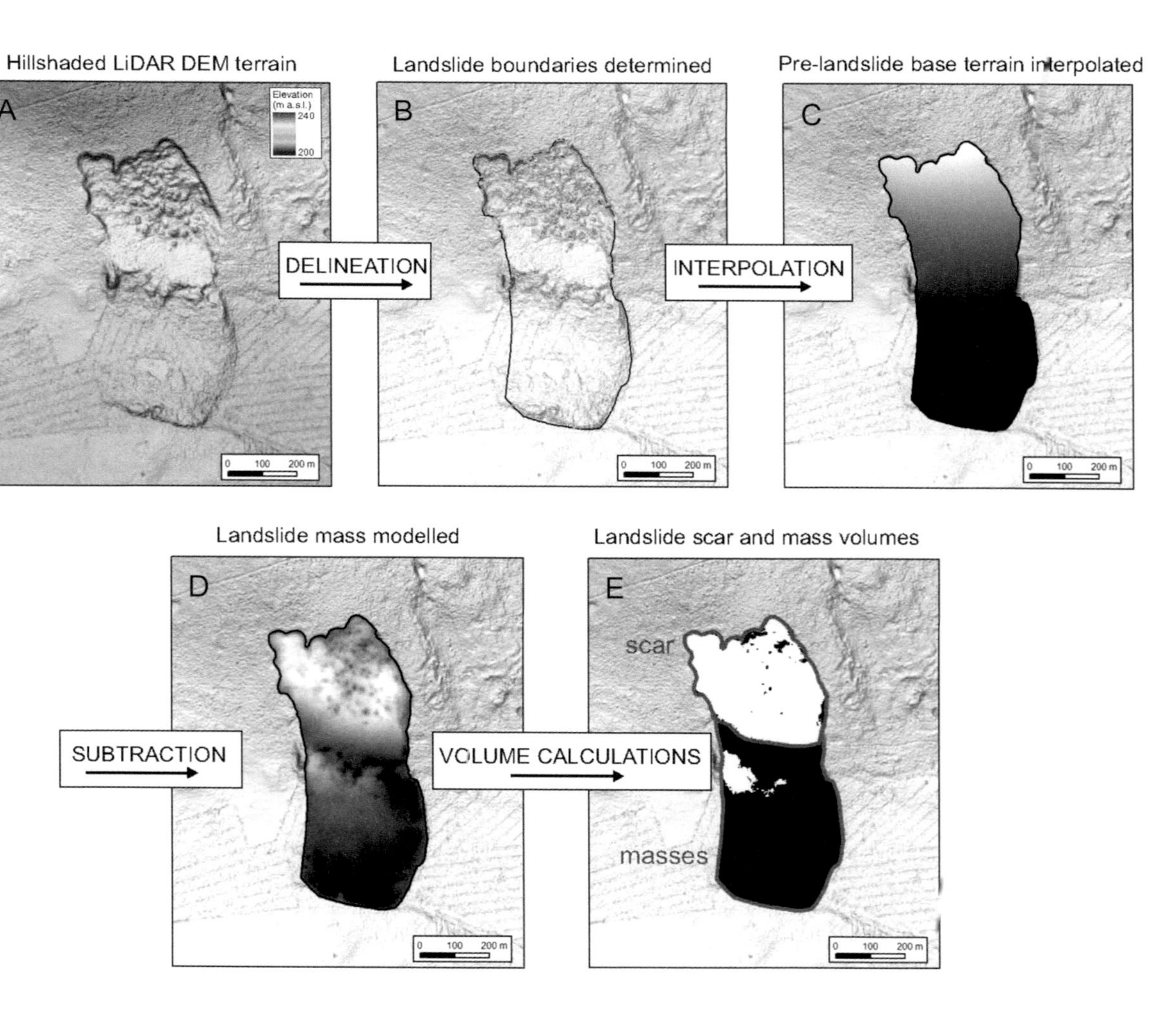
Hillshaded LiDAR DEM terrain
A
Elevation (m a.s.l.)
240
200
0 100 200 m
DELINEATION
Landslide boundaries determined
B
0 100 200 m
INTERPOLATION
Pre-landslide base terrain interpolated
C
0 100 200 m
SUBTRACTION
Landslide mass modelled
D
0 100 200 m
VOLUME CALCULATIONS
Landslide scar and mass volumes
E
scar
masses
0 100 200 m

Figure 5.5

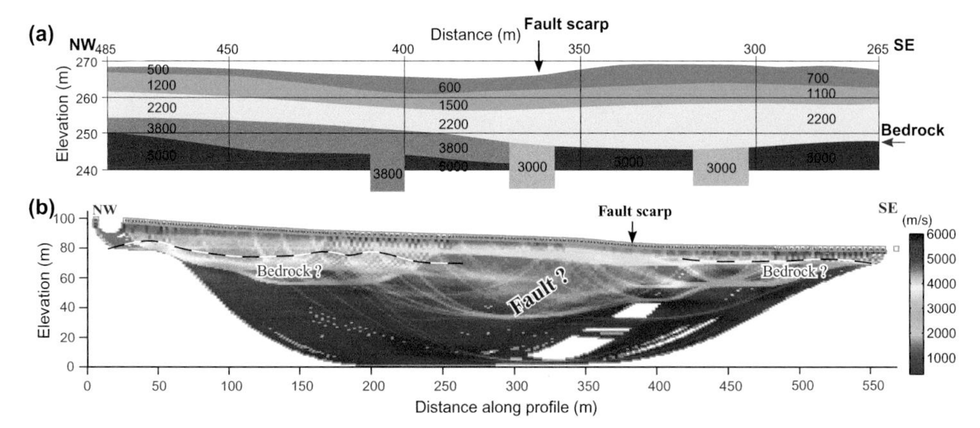

Figure 7.3

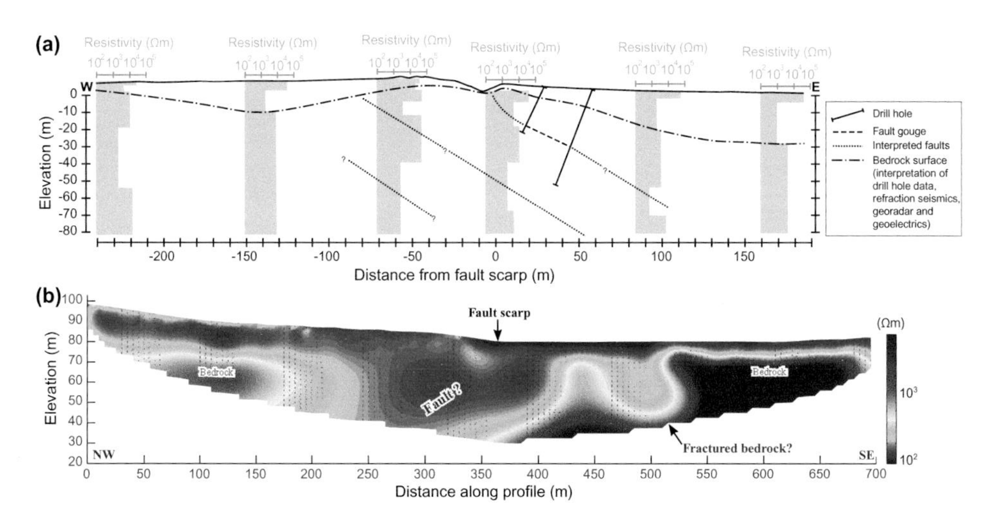

Figure 7.5

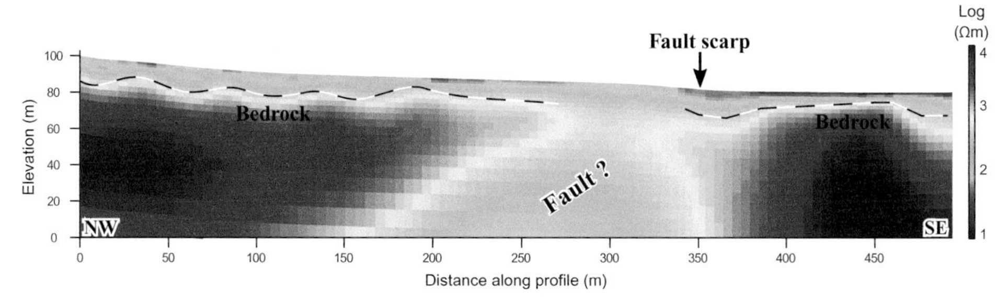

Figure 7.6

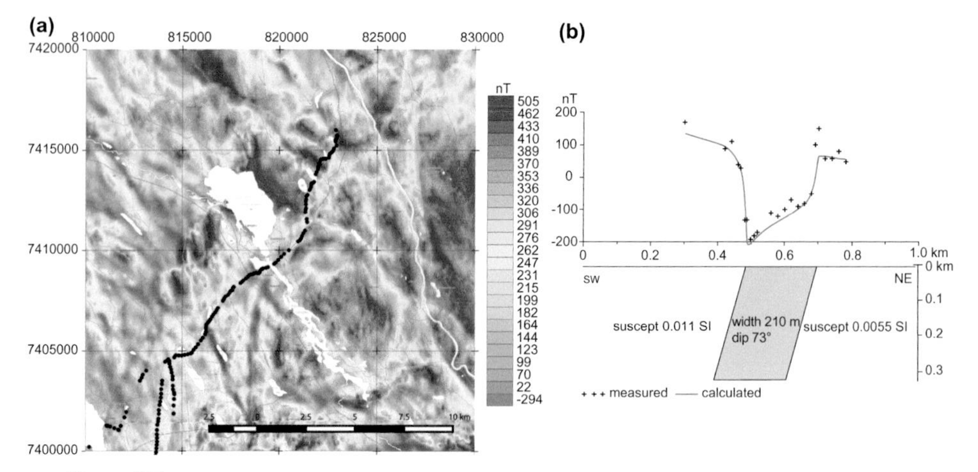

Figure 7.7

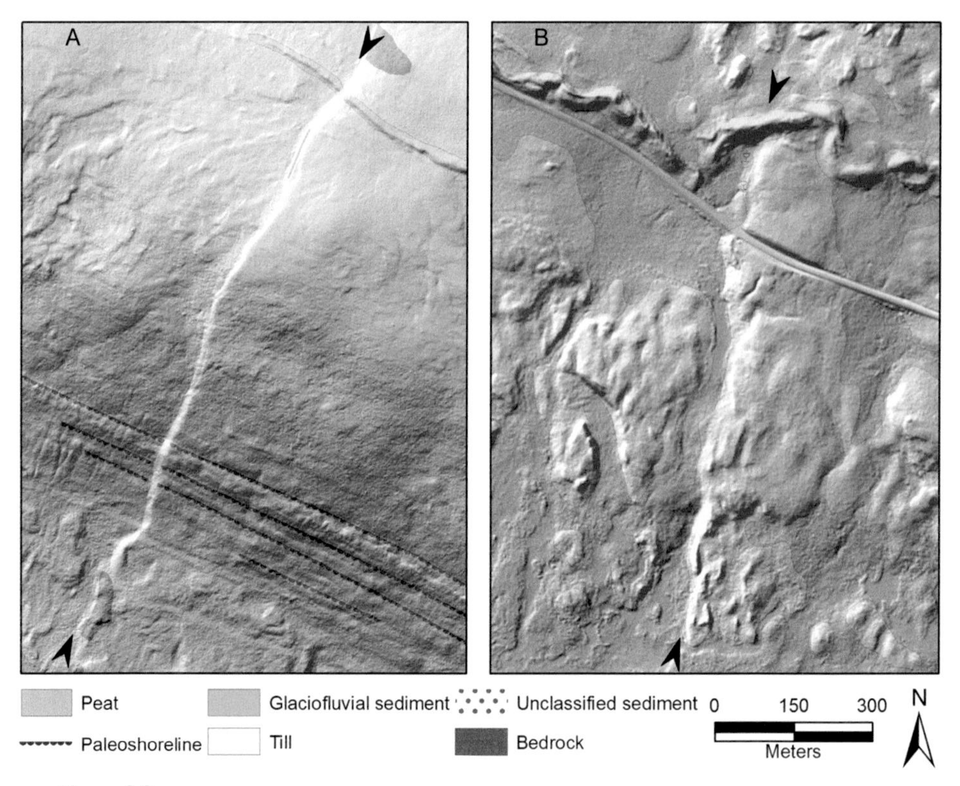

Figure 8.2

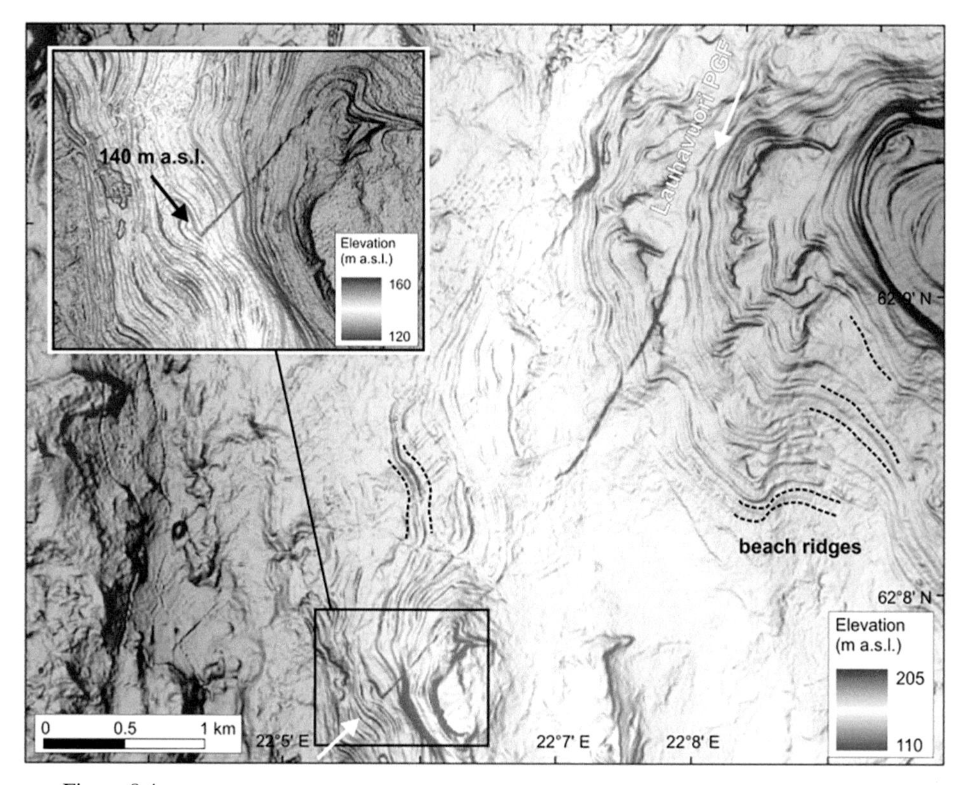

140 m a.s.l.
Elevation
(m a.s.l.)
160
120
Lauhavuori PGF
62°9' N
beach ridges
62°8' N
Elevation
(m a.s.l.)
205
110
0
0.5
1 km
22°5' E
22°7' E
22°8' E

Figure 8.4

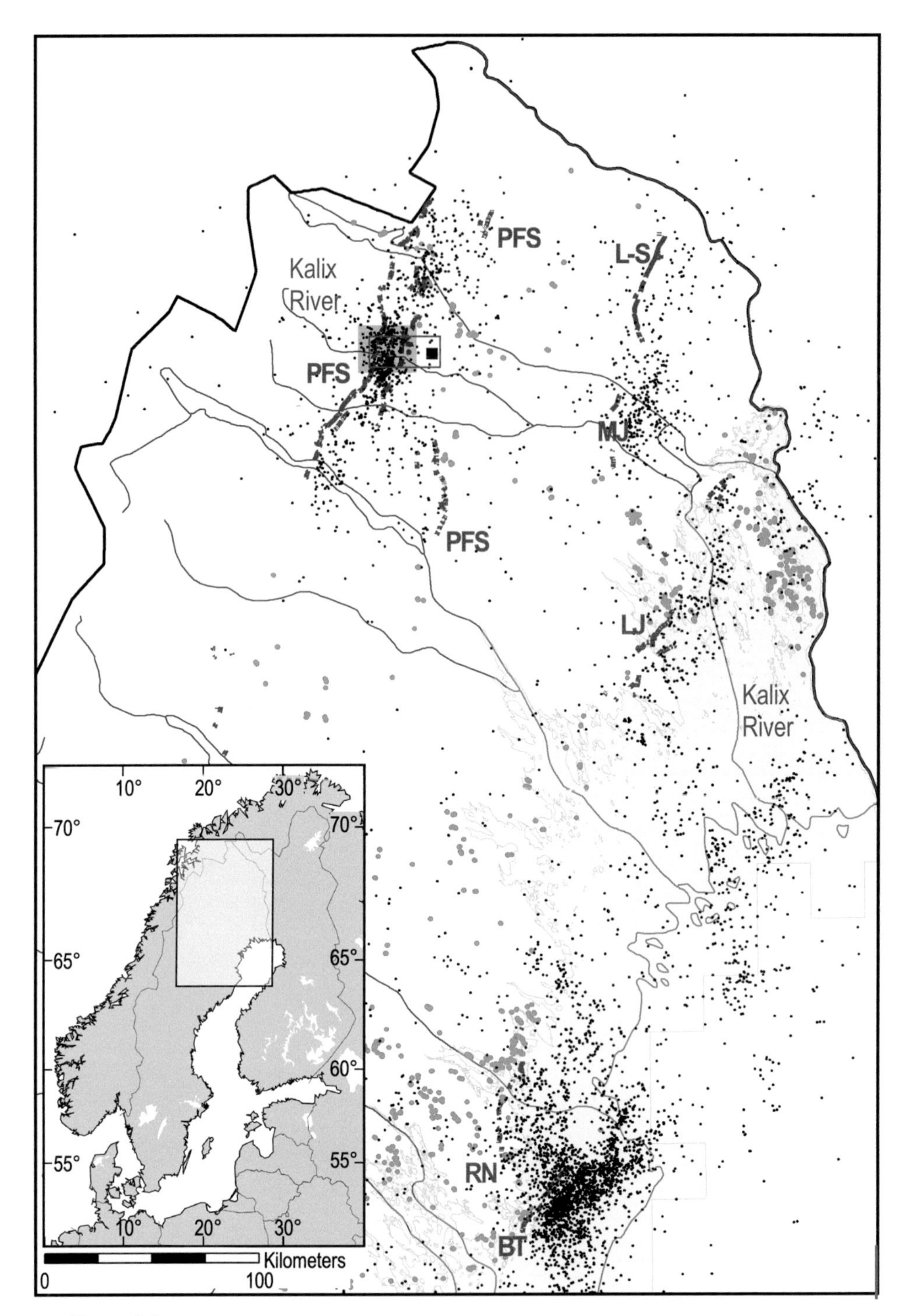
PFS
L-S
Kalix
River
PFS
MJ
PFS
LJ
Kalix
River
10°
20°
30°
70°
70°
65°
65°
60°
55°
55°
10°
20°
30°
RN
BT
Kilometers
0
100

Figure 9.1

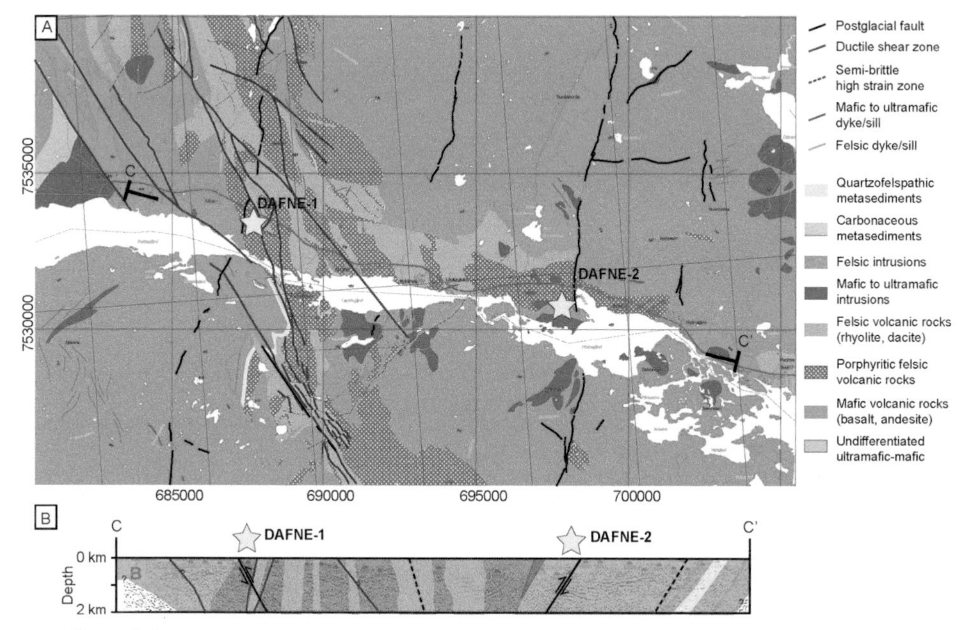

Figure 9.2

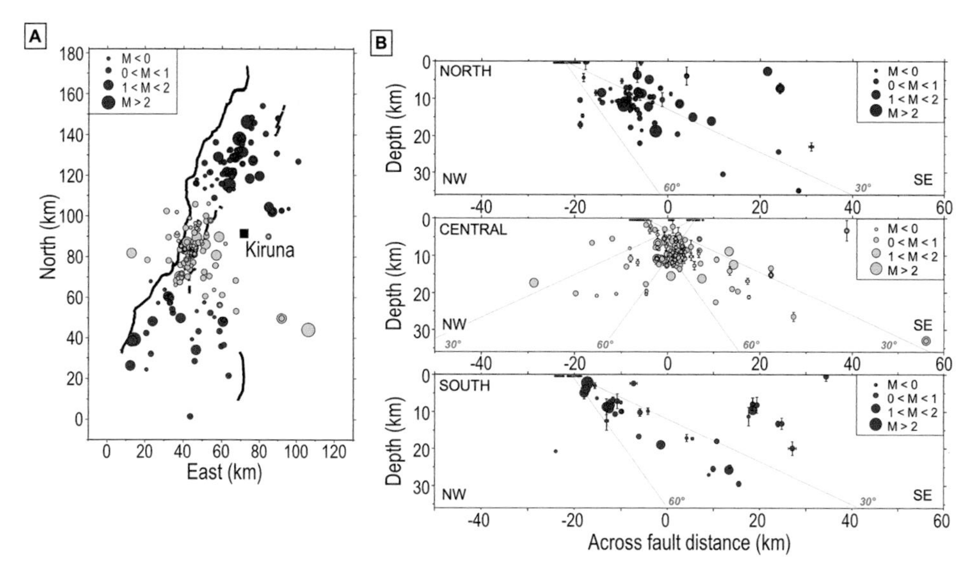

Figure 9.3

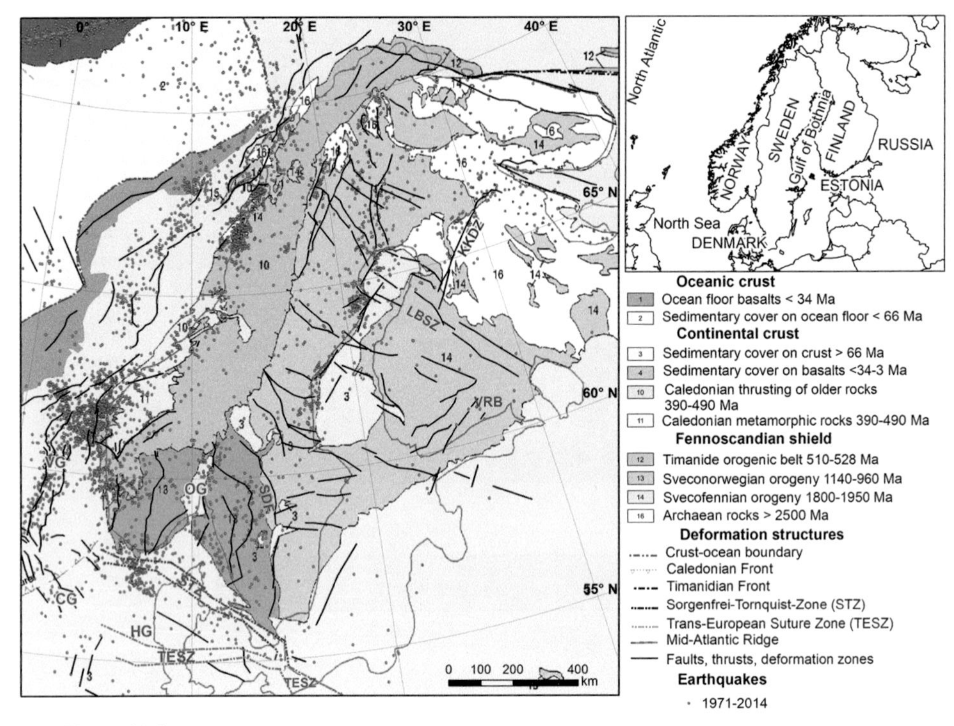

0°
10° E
20° E
30° E
40° E
65° N
60° N
55° N
KKDZ
LBSZ
VRB
VG
OG
SDF
CG
HG
STZ
TESZ
0 100 200 400 km
North Atlantic
NORWAY
SWEDEN
Gulf of Bothnia
FINLAND
RUSSIA
ESTONIA
North Sea
DENMARK
Oceanic crust
Ocean floor basalts < 34 Ma
Sedimentary cover on ocean floor < 66 Ma
Continental crust
Sedimentary cover on crust > 66 Ma
Sedimentary cover on basalts <34-3 Ma
Caledonian thrusting of older rocks 390-490 Ma
Caledonian metamorphic rocks 390-490 Ma
Fennoscandian shield
Timanide orogenic belt 510-528 Ma
Sveconorwegian orogeny 1140-960 Ma
Svecofennian orogeny 1800-1950 Ma
Archaean rocks > 2500 Ma
Deformation structures
Crust-ocean boundary
Caledonian Front
Timanidian Front
Sorgenfrei-Tornquist-Zone (STZ)
Trans-European Suture Zone (TESZ)
Mid-Atlantic Ridge
Faults, thrusts, deformation zones
Earthquakes
1971-2014

Figure 10.1

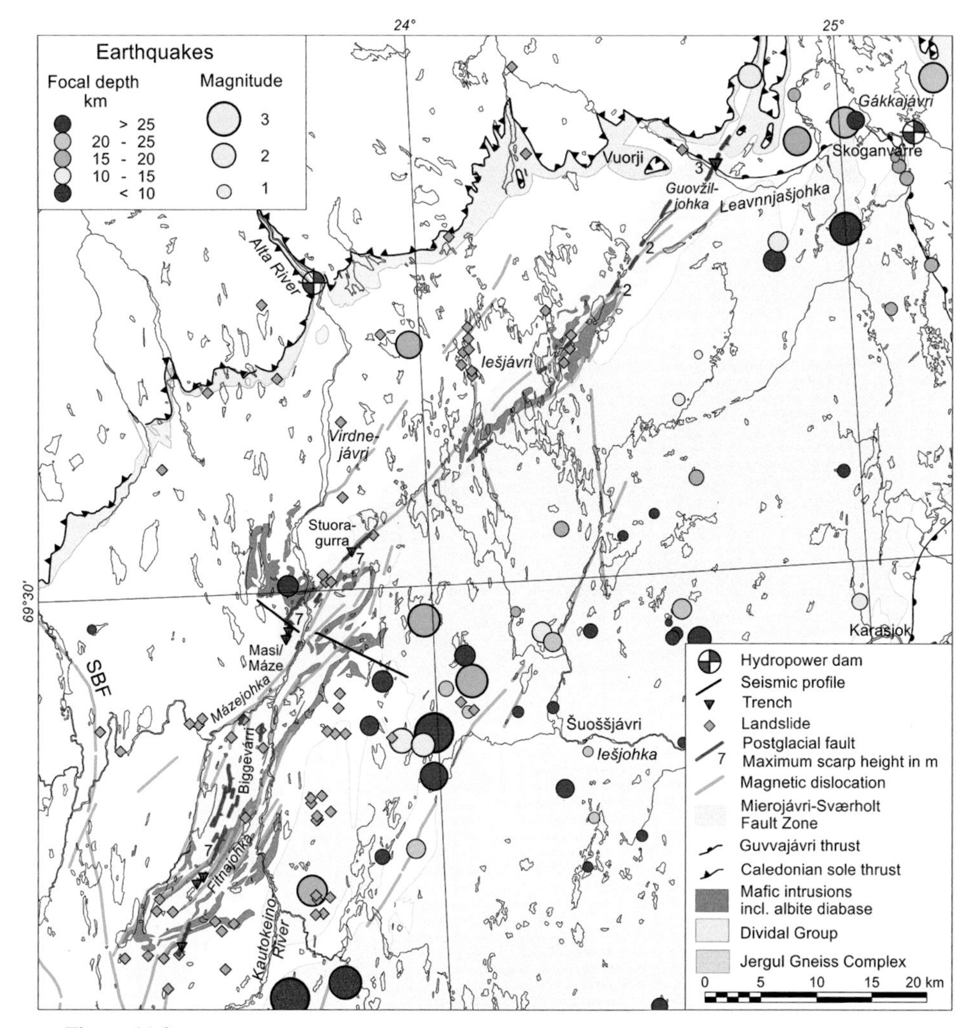
Earthquakes
Focal depth km
> 25
20 - 25
15 - 20
10 - 15
< 10
Magnitude
3
2
1
24°
25°
69°30'
Gákkajávri
Skoganvarre
Vuorji
Guovžil-johka
Leavnnjašjohka
Alta River
Iešjávri
Virdne-jávri
Stuora-gurra
Masi/Máze
SBF
Mázejohka
Biggevárri
Fitnajohka
Kautokeino River
Šuoššjávri
Iešjohka
Karasjok
Hydropower dam
Seismic profile
Trench
Landslide
Postglacial fault
Maximum scarp height in m
Magnetic dislocation
Mierojávri-Sværholt Fault Zone
Guvvajávri thrust
Caledonian sole thrust
Mafic intrusions incl. albite diabase
Dividal Group
Jergul Gneiss Complex
0 5 10 15 20 km

Figure 11.2

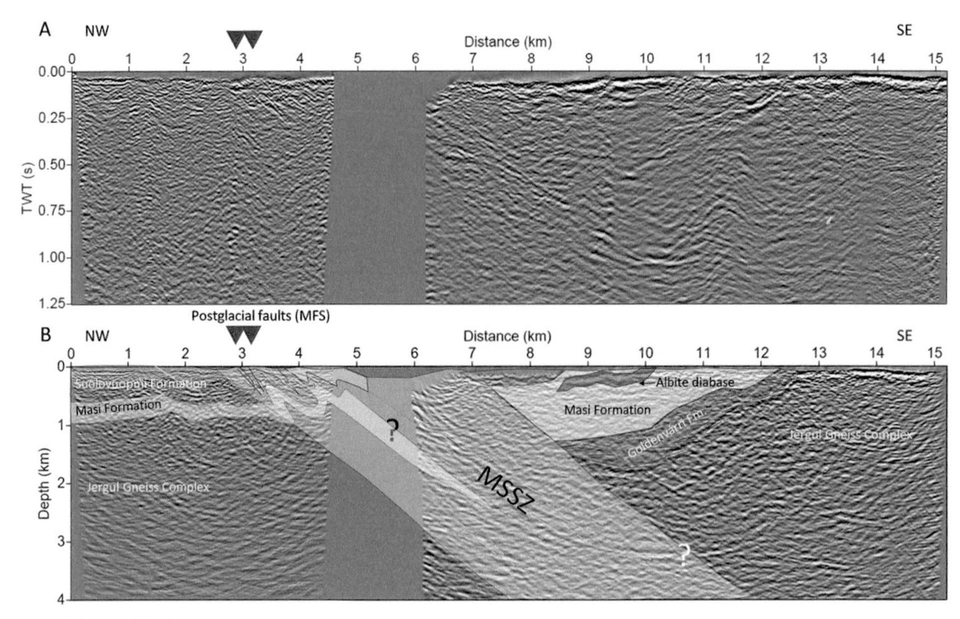

Figure 11.3

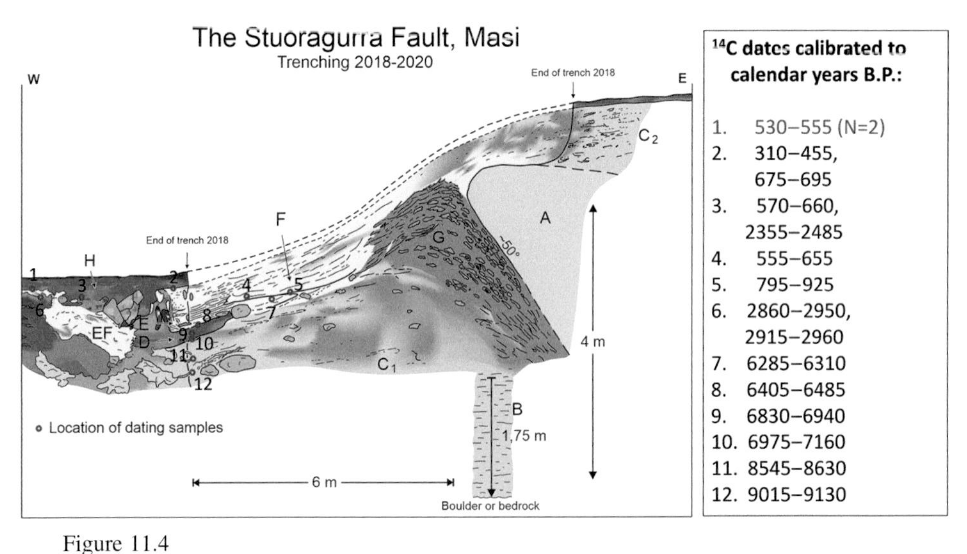

Figure 11.4

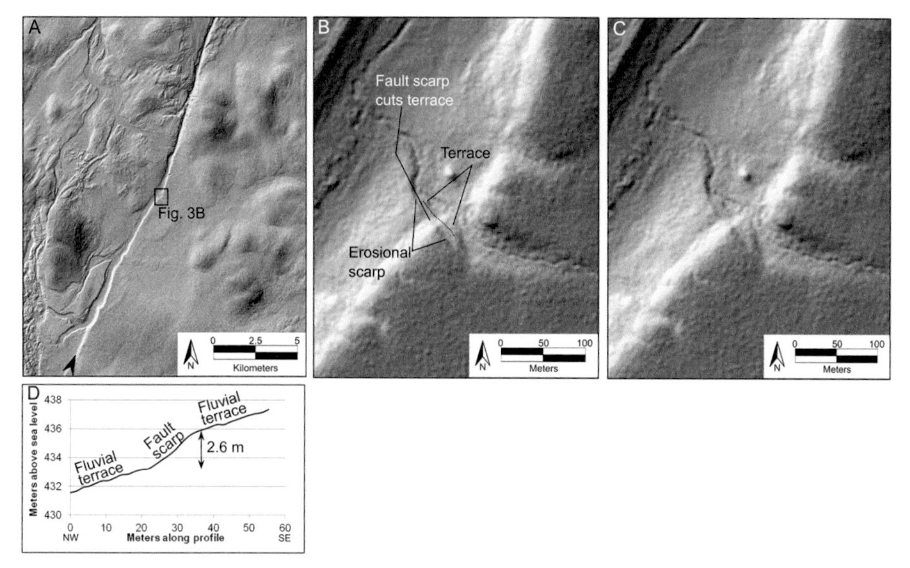

Figure 12.3

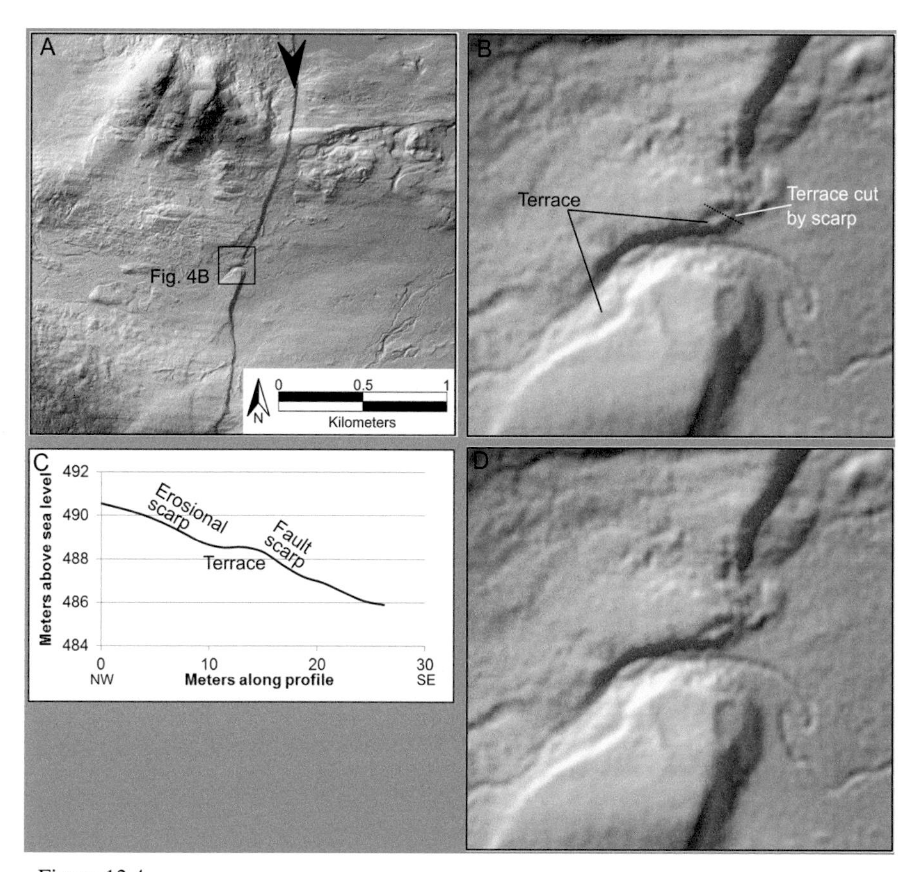

Figure 12.4

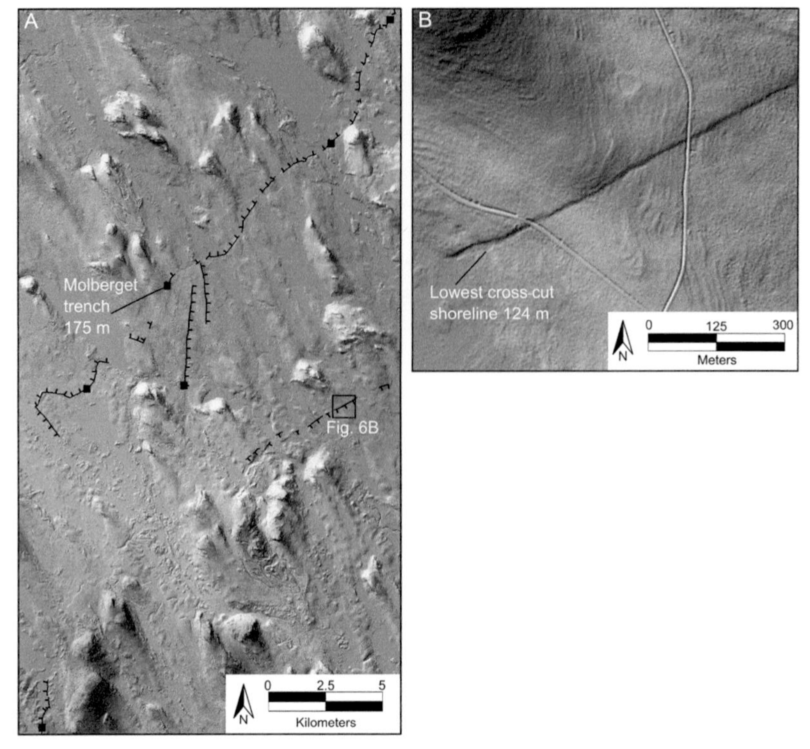

Figure 12.6

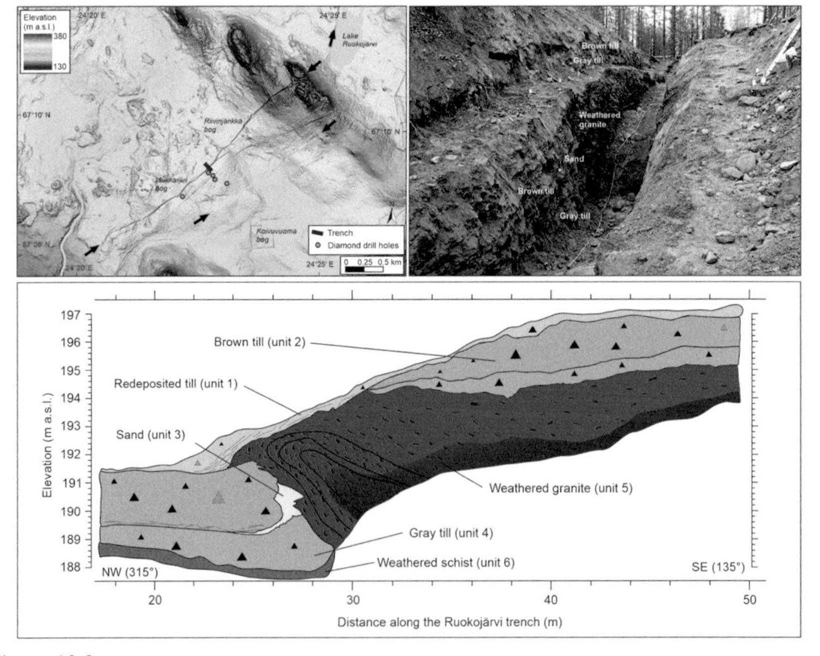

Figure 13.2

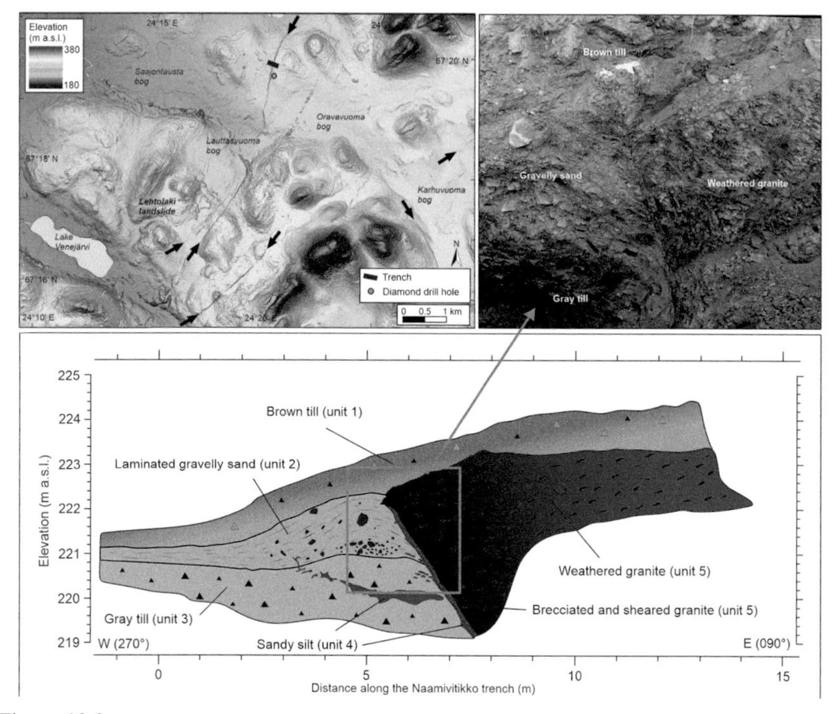

Figure 13.3

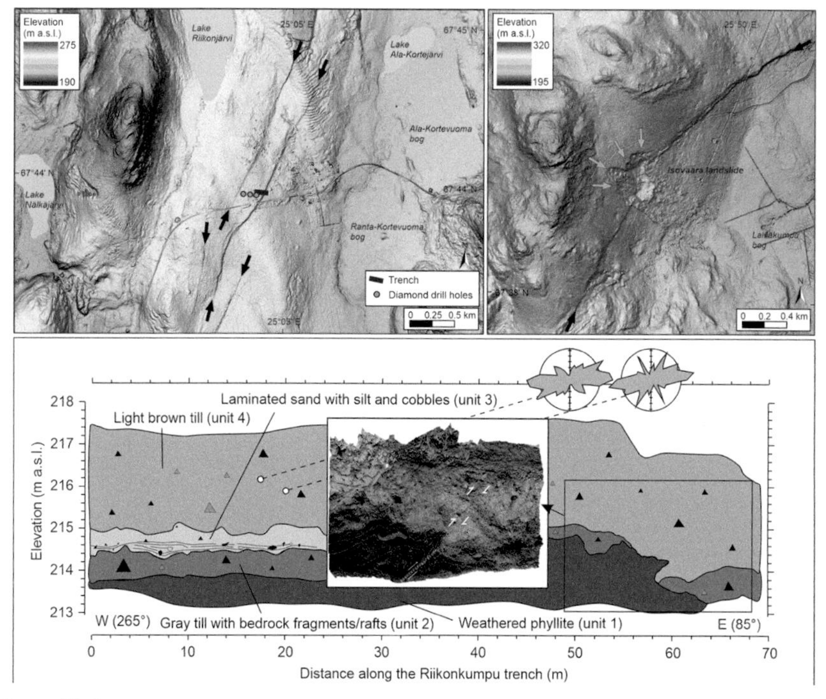

Figure 13.4

Figure 14.4

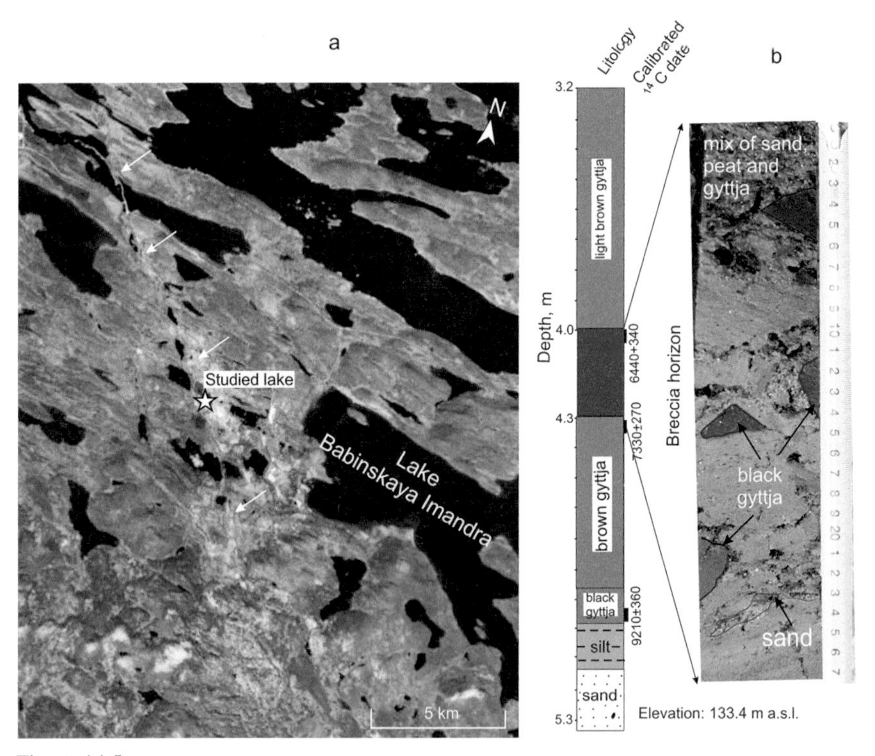

Figure 14.5

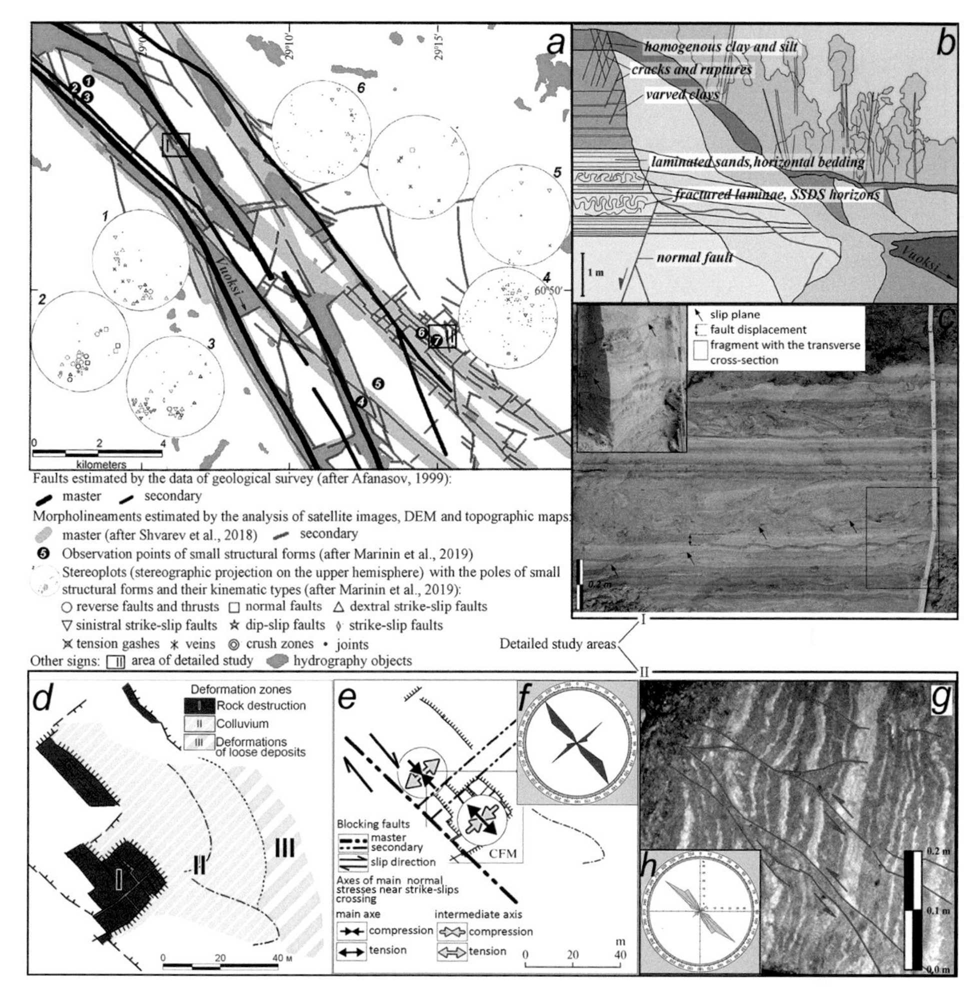
a
b
homogenous clay and silt
cracks and ruptures
varved clays
laminated sands, horizontal bedding
fractured laminae, SSDS horizons
normal fault
1 m
Vuoksi
slip plane
fault displacement
fragment with the transverse cross-section
0.2 m
kilometers
Faults estimated by the data of geological survey (after Afanasov, 1999):
master
secondary
Morpholineaments estimated by the analysis of satellite images, DEM and topographic maps:
master (after Shvarev et al., 2018)
secondary
Observation points of small structural forms (after Marinin et al., 2019)
Stereoplots (stereographic projection on the upper hemisphere) with the poles of small structural forms and their kinematic types (after Marinin et al., 2019):
reverse faults and thrusts
normal faults
dextral strike-slip faults
sinistral strike-slip faults
dip-slip faults
strike-slip faults
tension gashes
veins
crush zones
joints
Other signs:
area of detailed study
hydrography objects
Detailed study areas
d
Deformation zones
Rock destruction
Colluvium
Deformations of loose deposits
e
f
g
h
Blocking faults
master
secondary
slip direction
Axes of main normal stresses near strike-slips crossing
main axe
intermediate axis
compression
tension
CFM
0.2 m
0.1 m
0.0 m

Figure 14.6

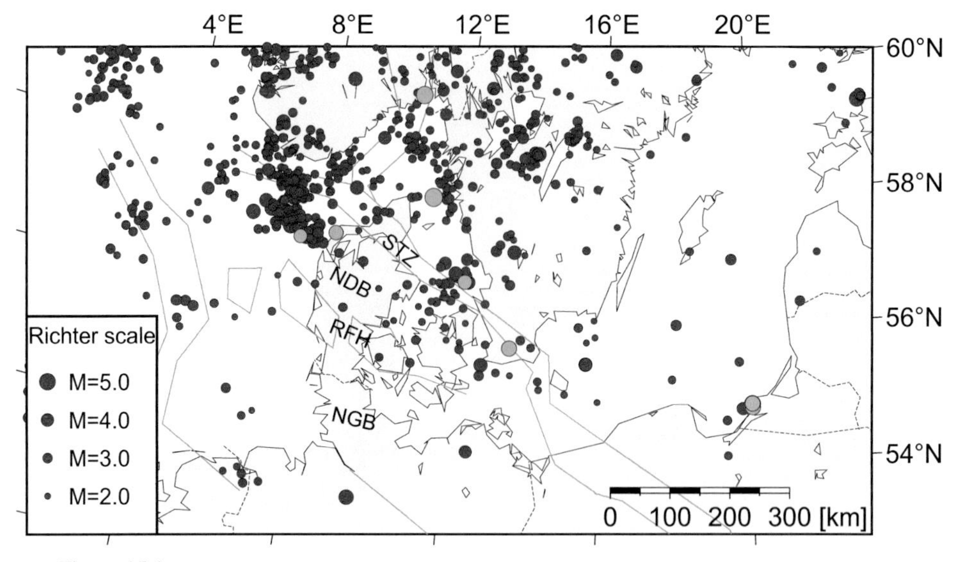

Figure 15.1

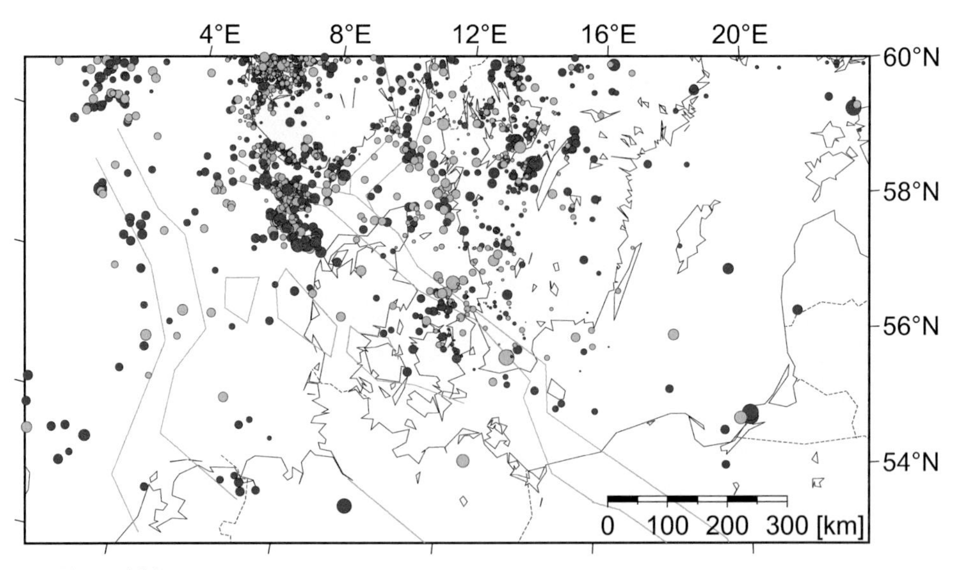

Figure 15.2

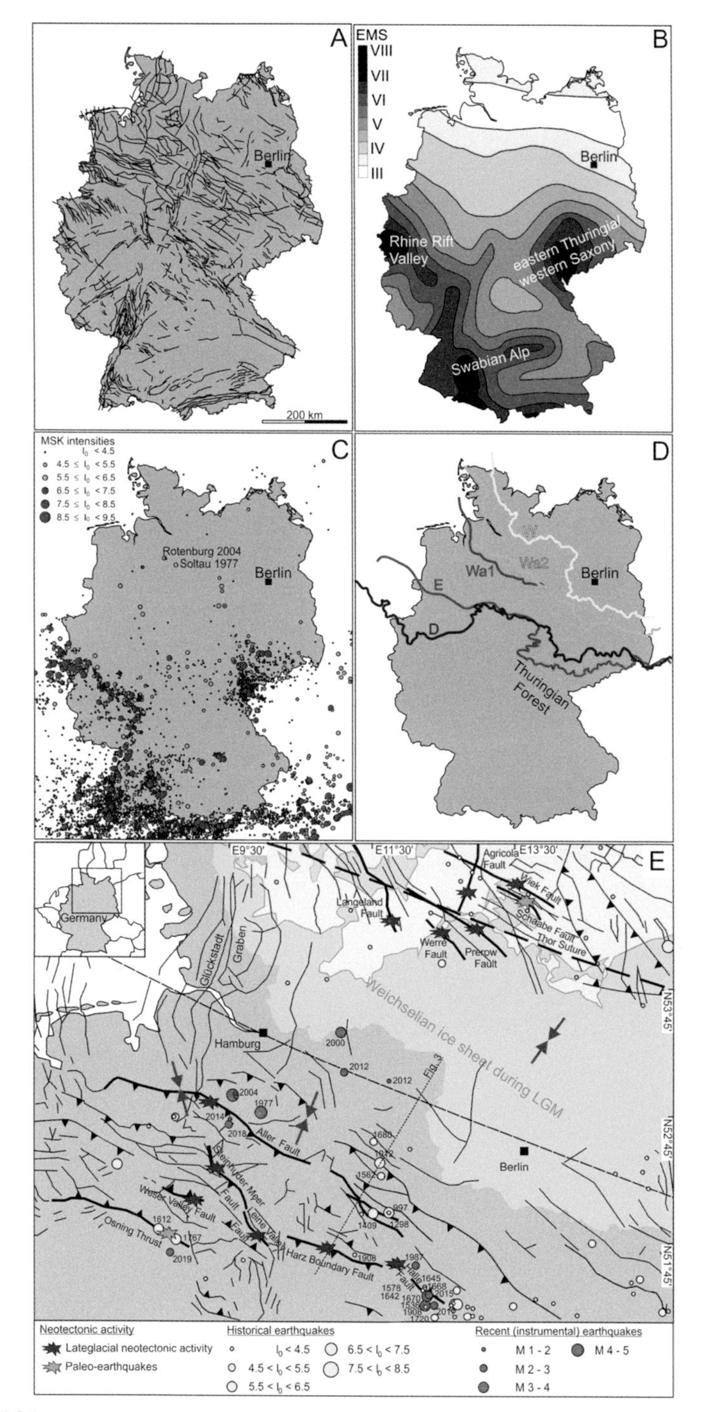
A
Berlin
200 km
B
EMS
VIII
VII
VI
V
IV
III
Berlin
Rhine Rift Valley
eastern Thuringia/ western Saxony
Swabian Alp
C
MSK intensities
Rotenburg 2004
Soltau 1977
Berlin
D
W
Wa2
Wa1
E
D
Berlin
Thuringian Forest
E
E9°30'
E11°30'
E13°30'
Agricola Fault
Wiek Fault
Langeland Fault
Schaabe Fault
Thor Suture
Werre Fault
Prerow Fault
Germany
Glückstadt Graben
Weichselian ice sheet during LGM
Hamburg
Berlin
Aller Fault
Weser Valley Fault
Osning Thrust
Harz Boundary Fault
N53°45'
N52°45'
N51°45'
Neotectonic activity
Lateglacial neotectonic activity
Paleo-earthquakes
Historical earthquakes
Recent (instrumental) earthquakes
M 1 - 2
M 2 - 3
M 3 - 4
M 4 - 5

Figure 16.1

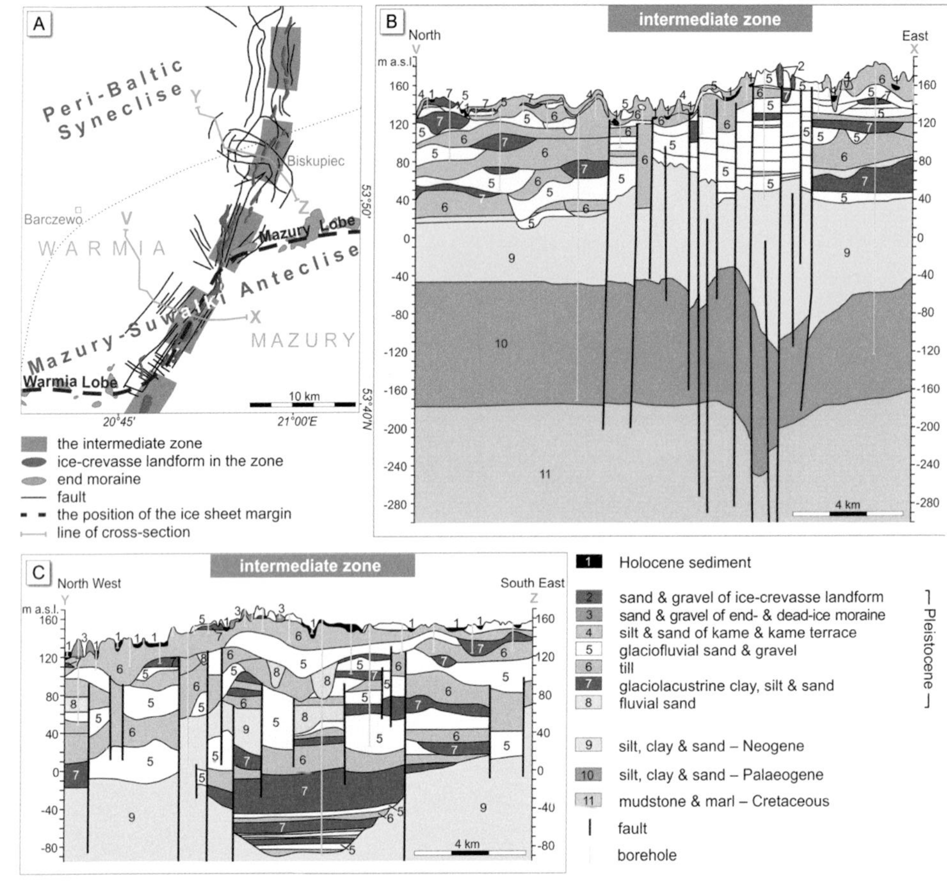

A
Peri-Baltic Syneclise
Biskupiec
Barczewo
WARMIA
Mazury Lobe
Mazury-Suwałki Anteclise
MAZURY
Warmia Lobe
10 km
20°45'
21°00'E
53°50'
53°40'N
the intermediate zone
ice-crevasse landform in the zone
end moraine
fault
the position of the ice sheet margin
line of cross-section
B
intermediate zone
North
East
m a.s.l.
4 km
C
intermediate zone
North West
South East
m a.s.l.
4 km
Holocene sediment
sand & gravel of ice-crevasse landform
sand & gravel of end- & dead-ice moraine
silt & sand of kame & kame terrace
glaciofluvial sand & gravel
till
glaciolacustrine clay, silt & sand
fluvial sand
Pleistocene
silt, clay & sand – Neogene
silt, clay & sand – Palaeogene
mudstone & marl – Cretaceous
fault
borehole

Figure 17.4

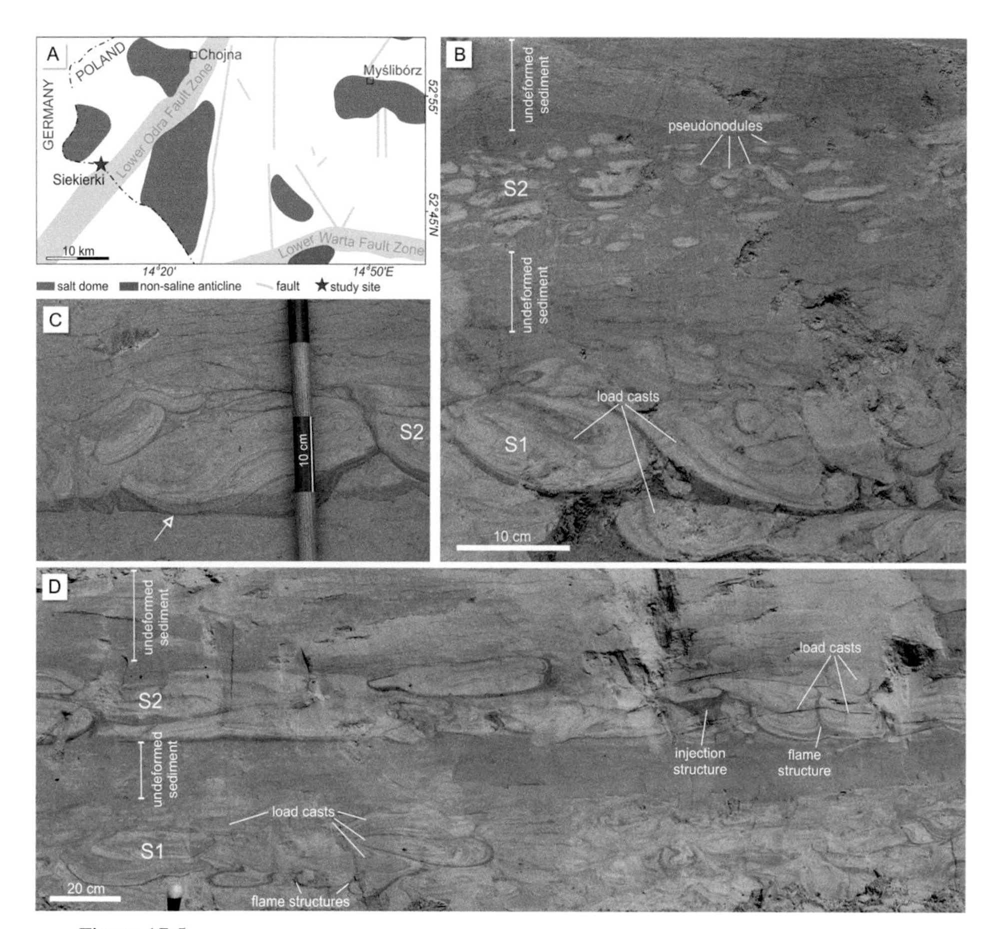
A
POLAND
GERMANY
Chojna
Myślibórz
Lower Odra Fault Zone
Lower Warta Fault Zone
Siekierki
10 km
14°20'
14°50'E
52°55'
52°45'N
salt dome
non-saline anticline
fault
study site
B
undeformed sediment
pseudonodules
S2
undeformed sediment
load casts
S1
10 cm
C
10 cm
S2
D
undeformed sediment
S2
undeformed sediment
load casts
S1
20 cm
flame structures
injection structure
load casts
flame structure

Figure 17.5

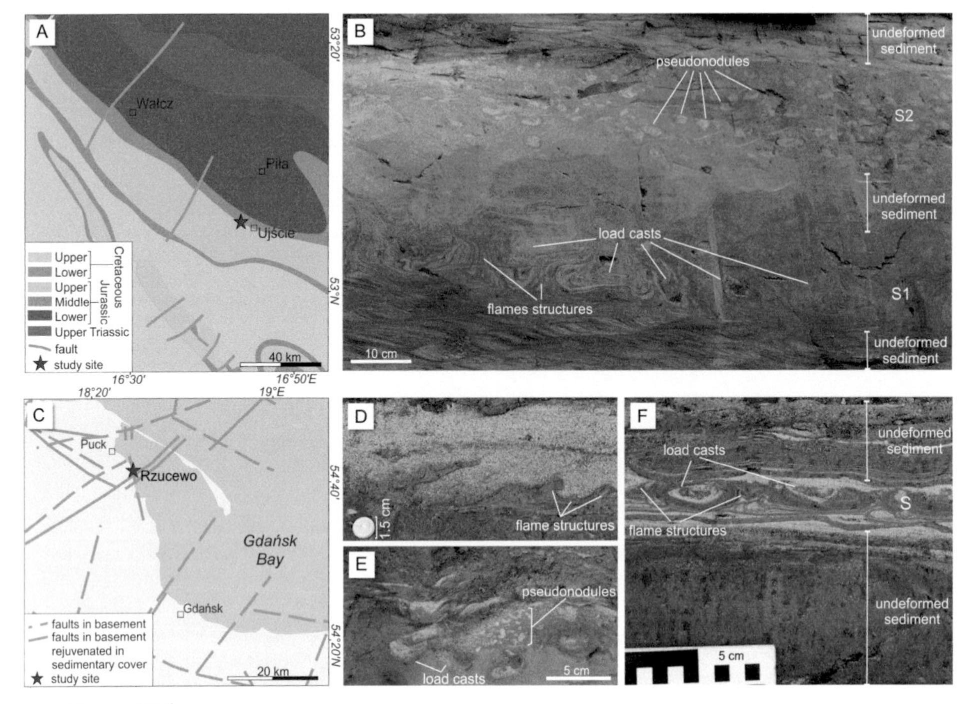
A
Wałcz
Piła
Ujście
Upper
Lower
Cretaceous
Upper
Middle
Lower
Jurassic
Upper Triassic
fault
study site
40 km
53°20'
53°N
16°30'
16°50'E
B
undeformed sediment
pseudonodules
S2
undeformed sediment
load casts
flames structures
S1
undeformed sediment
10 cm
C
18°20'
19°E
Puck
Rzucewo
Gdańsk Bay
Gdańsk
54°40'
54°20'N
faults in basement
faults in basement rejuvenated in sedimentary cover
study site
20 km
D
1.5 cm
flame structures
E
pseudonodules
load casts
5 cm
F
undeformed sediment
load casts
S
flame structures
undeformed sediment
5 cm

Figure 17.6

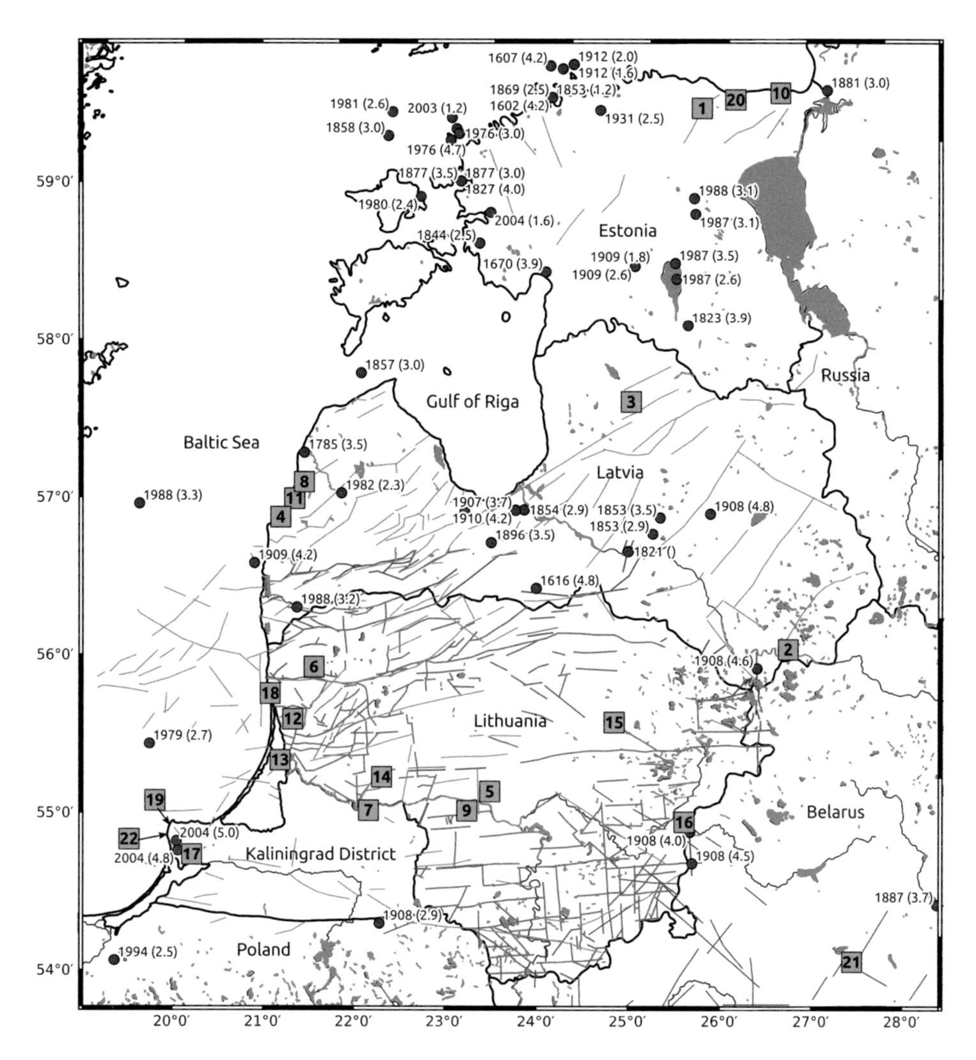

Baltic Sea
Gulf of Riga
Estonia
Latvia
Lithuania
Russia
Belarus
Poland
Kaliningrad District
59°0′
58°0′
57°0′
56°0′
55°0′
54°0′
20°0′
21°0′
22°0′
23°0′
24°0′
25°0′
26°0′
27°0′
28°0′

Figure 18.1

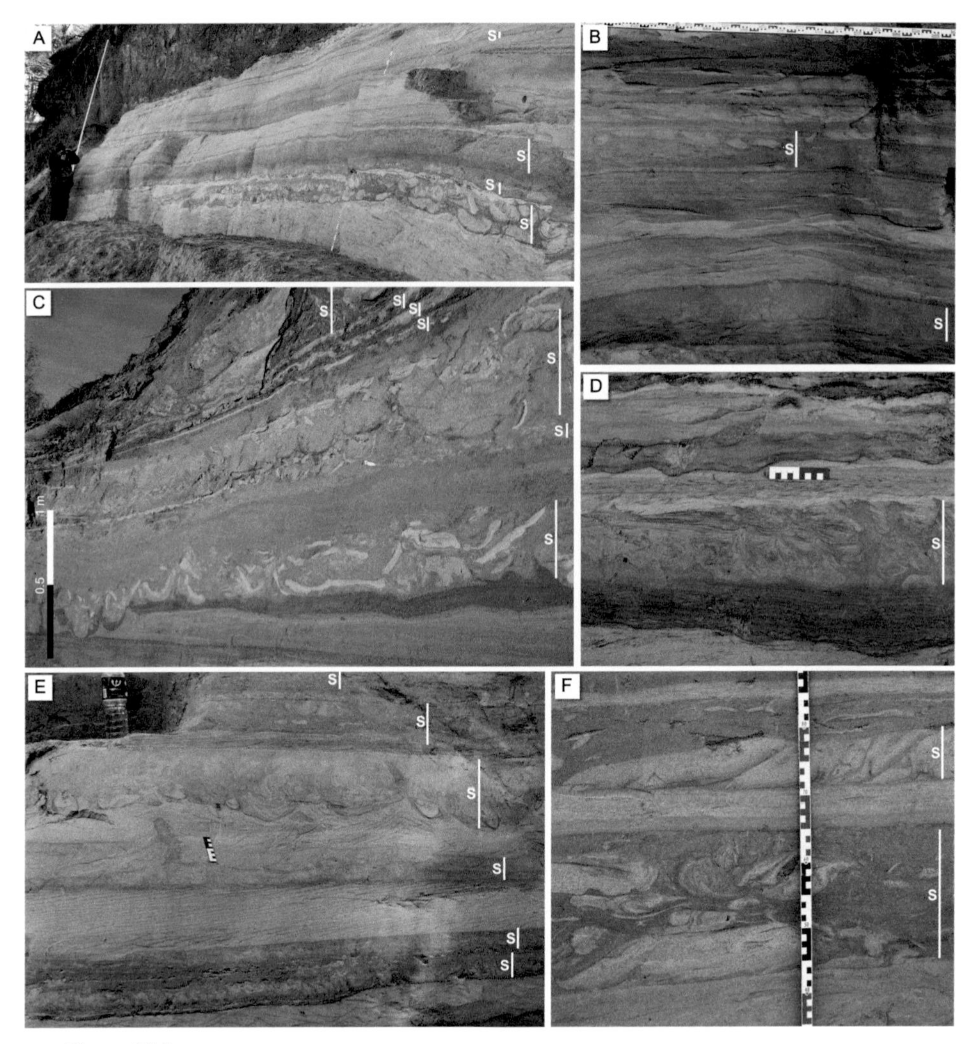

Figure 18.2

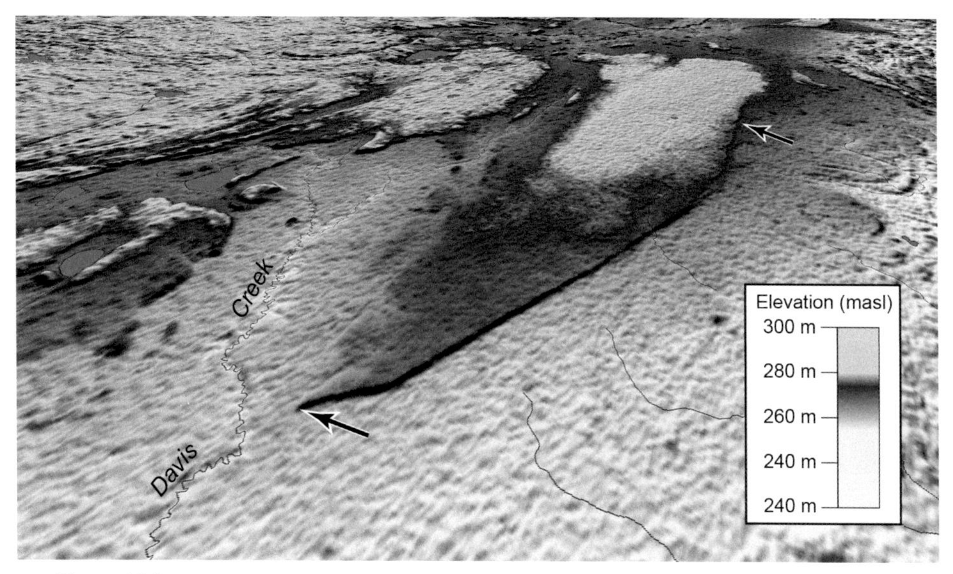

Figure 19.2

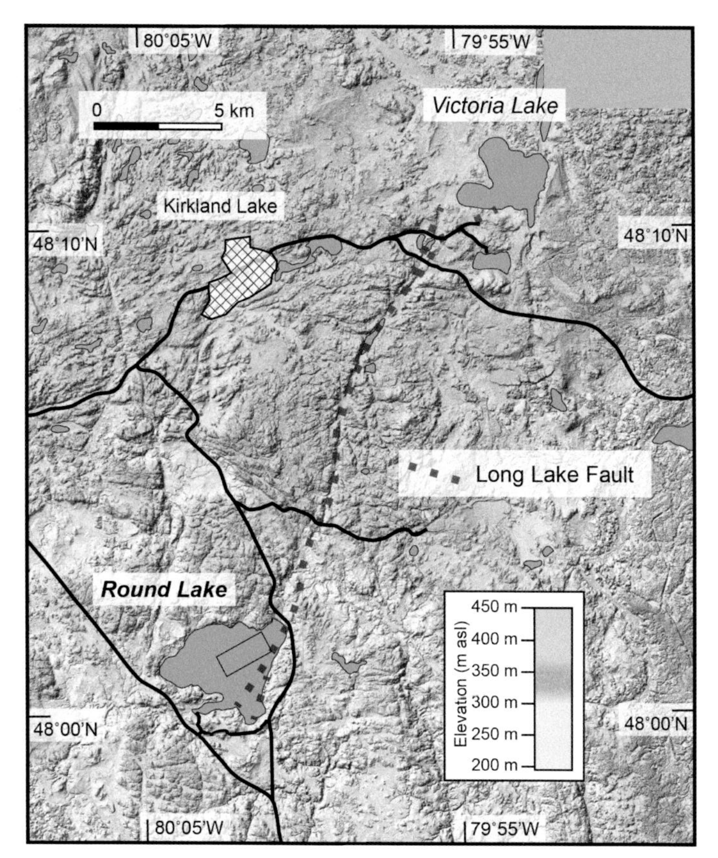

Figure 19.3

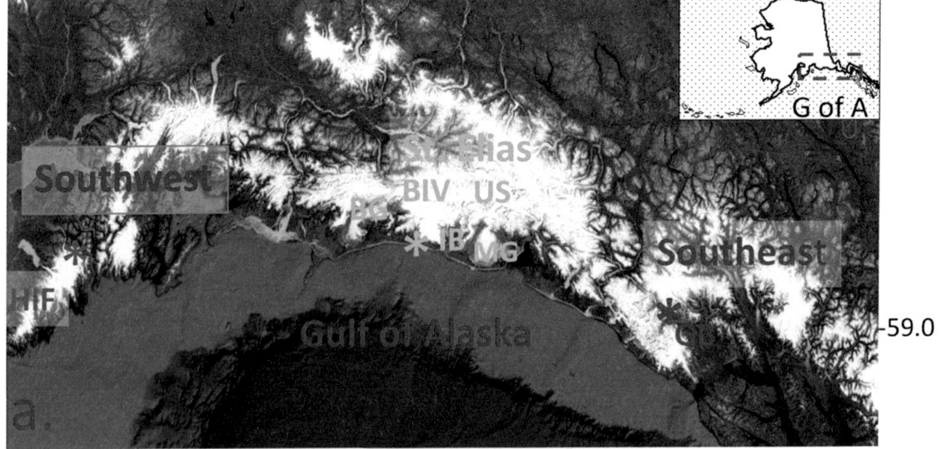
G of A
Southwest
BIV
US
IB
MG
Southeast
HIF
-59.0
a.

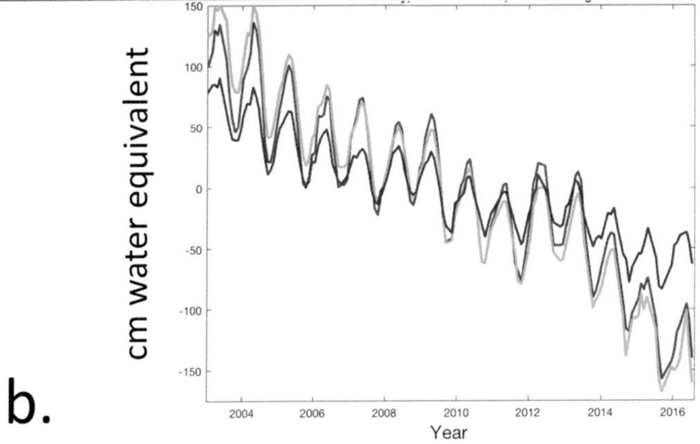
cm water equivalent
150
100
50
0
-50
-100
-150
2004
2006
2008
2010
2012
2014
2016
Year
b.

Figure 20.1

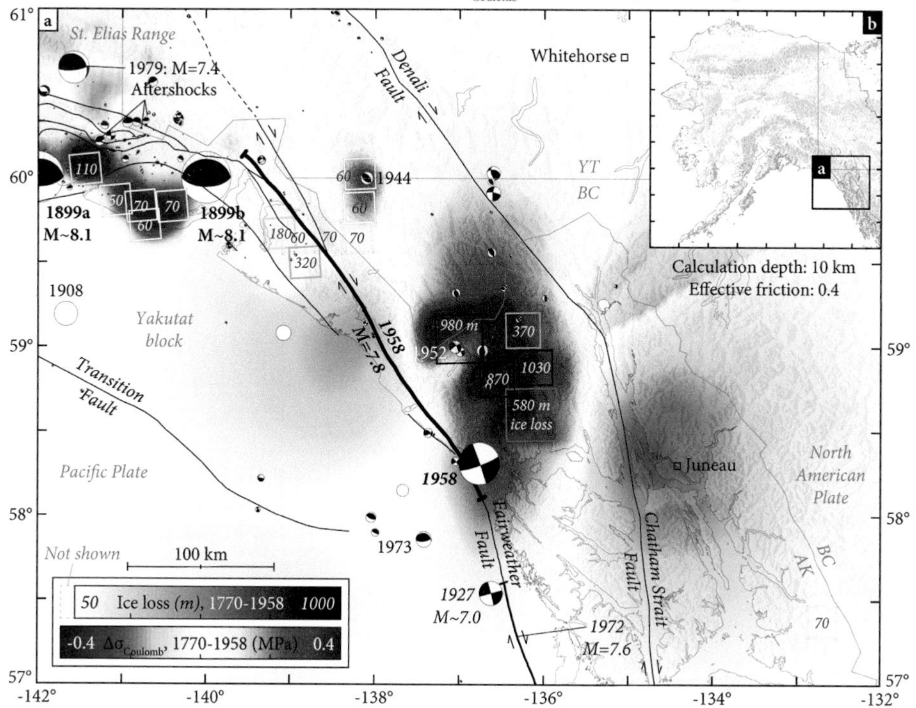

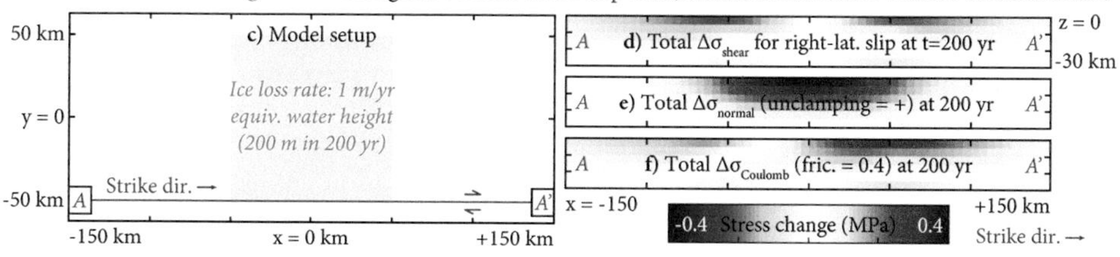

Figure 20.2

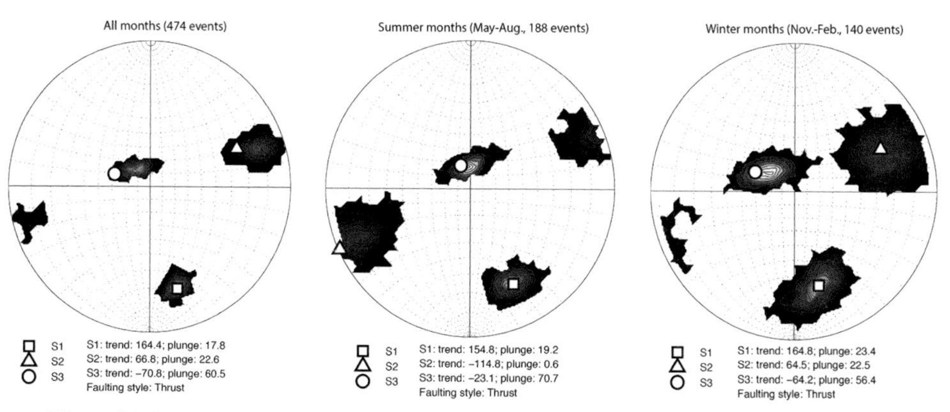

Figure 20.5

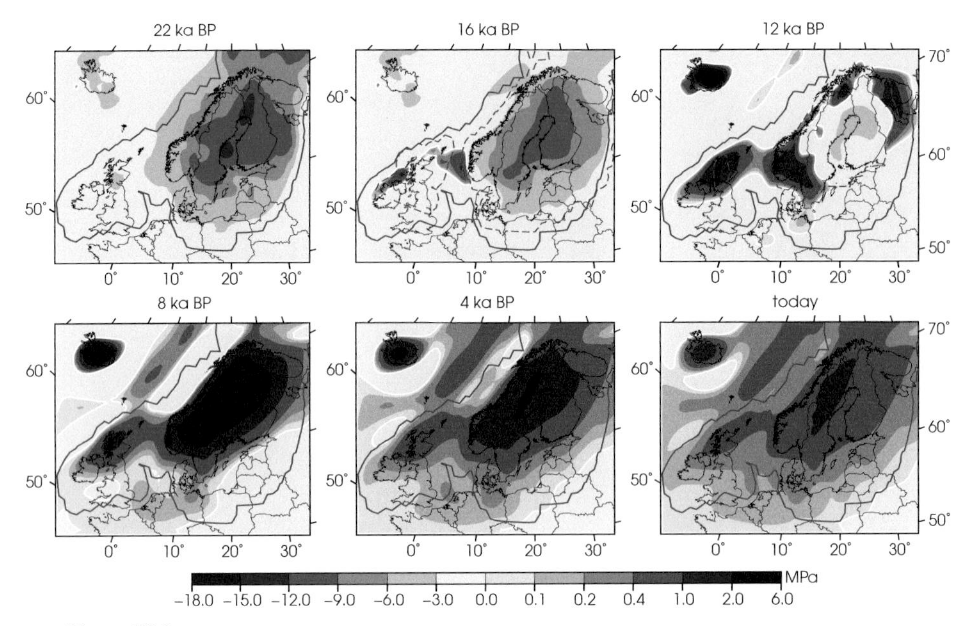

Figure 22.2

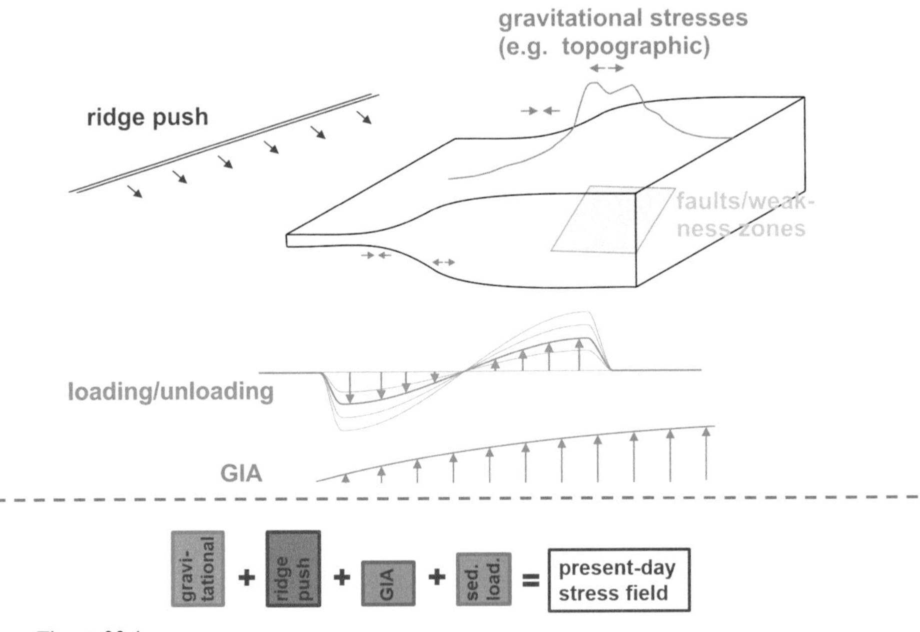

Figure 23.1

Lundqvist, J. and Lagerbäck, R. (1976). The Pärve Fault: a late-glacial fault in the Precambrian of Swedish Lapland. *Geologiska Föreningens i Stockholm Förhandlingar*, **98**, 45–51, doi.org/10.1080/11035897609454337.

Mather, W. W. (1843). *Geology of New-york. Part I. Comprising the Geology of the First Geological District*. Carroll & Cook, Albany.

Matthew, G. F. (1894). Movements of the Earth's crust at St. John, N. B., in post-glacial times. *Bulletin of the Natural History Society of New Brunswick*, **12**, 34–42.

Mikko, H., Smith, C. A., Lund, B., Ask, M. V. S. and Munier, R. (2015). LiDAR-derived inventory of post-glacial fault scarps in Sweden. *GFF*, **137**, 344–352, doi.org/10.1080/11035897.2015.1036360.

Muir Wood, R. (1989). Extraordinary deglaciation reverse faulting in northern Fennoscandia. In S. Gregersen and P. W. Basham, eds., *Earthquakes at North-Atlantic Passive Margins: Neotectonics and Postglacial Rebound*. Kluwer Academic Publishers, Dordrecht, pp. 141–173, doi.org/10.1007/978-94-009-2311-9_10.

Ojala, A. E. K., Mattila, J., Ruskeeniemi, T. et al. (2017). Postglacial seismic activity along the Isovaara–Riikonkumpu fault complex. *Global and Planetary Change*, **157**, 59–72, doi.org/10.1016/j.gloplacha.2017.08.015.

Ojala, A. E. K., Markovaa-Koivisto, M., Middleton, M. et al. (2018). Dating of paleolandslides in western Finnish Lapland. *Earth Surface Processes and Landforms*, **43**, 2449–2462, doi.org/10.1002/esp.4408.

Ojala, A. E. K., Mattila, J., Markovaara-Koivisto, M. et al. (2019). Distribution and morphology of landslides in northern Finland: an analysis of postglacial seismic activity. *Geomorphology*, **326**, 190–201, doi.org/10.1016/j.geomorph.2017.08.045.

Olesen, O., Henkel, H., Lile, O. B., Mauring, E. and Rønning, J. S. (1992). Geophysical investigations of the Stuoragurra postglacial fault, Finnmark, northern Norway. *Journal of Applied Geophysics*, **29**, 95–118, doi.org/10.1016/0926-9851(92)90001-2.

Olesen, O., Bungum, H., Lindholm, C. et al. (2013). Neotectonics, seismicity and contemporary stress field in Norway – mechanisms and implications. In L. Olsen, O. Fredin and O. Olesen, eds., *Quaternary Geology of Norway*. Geological Survey of Norway Special Publication 13, pp. 145–174.

Page, M. T. and Hough, S. E. (2014). The New Madrid seismic zone: not dead yet. *Science*, **343**(6172), 762–764, doi.org/10.1126/science.1248215.

Peng, Z. and Gomberg, J. (2010). An integrated perspective of the continuum between earthquakes and slow-slip phenomena. *Nature Geoscience*, **3**(9), 599–607, doi.org/10.1038/ngeo940.

Plomerova, J. and Babuska, V. (2010). Long memory of mantle lithosphere fabric – European LAB constrained from seismic anisotropy. *Lithos*, **120**, 131–143, doi.org/10.1016/j.lithos.2010.01.008.

Riad, L. (1990). *The Pärvie Fault, Northern Sweden*. Research Report 63, Minerology, Department of Mineralogy and Petrology, Uppsala University, 48 pp.

Saar, M. O. and Manga, M. (2003). Seismicity induced by seasonal groundwater recharge at Mt. Hood, Oregon. *Earth and Planetary Science Letters*, **214**(3–4), 605–618, doi.org/10.1016/S0012-821X(03)00418-7.

Sarlus, Z., Andersson, U. B., Bauer, T. E. et al. (2018). Timing of plutonism in the Gällivare area: implications for Proterozoic crustal development in the northern Norrbotten ore district, Sweden. *Geological Magazine*, **155**(6), 1351–1376, doi.org/10.1017/S0016756817000280.

Smith, C., Sundh, M. and Mikko, H. (2014). Surficial geology indicates early Holocene faulting and seismicity, central Sweden. *International Journal of Earth Sciences*, **103**, 1711–1724, doi.org/10.1007/s00531–014-1025-6.

Stein, S. and Liu, M. (2009). Long aftershock sequences within continents and implications for earthquake hazard assessment. *Nature*, **462**, 87–89, doi.org/10.1038/nature08502.
Stein, S., Liu, M., Calais, E. and Li, Q. (2009). Mid-continent earthquakes as a complex system. *Seismological Research Letters*, **80**(4), 551–553, doi.org/10.1785/gssrl.80.4.551.
Stephansson, O. (1983). *Rock Stress Measurements by Sleeve Fracturing*. Proceedings of the 5th Congress of the International Society for Rock Mechanics and Rock Engineering, 10–15 April 1983, Melbourne, Australia, F129–F137.
Stephansson, O. (1989). Stress measurements and modeling of crustal rock mechanics in Fennoscandia. In S. Gregersen and P. W. Basham, eds., *Earthquake at North-Atlantic Passive Margins: Neotectonics and Postglacial Rebound*. NATO Advanced Studies Institute Series, Series C, Vol. 266, pp. 213–229.
Stroeven, A. P., Hättestrand, C., Kleman, J. et al. (2016). Deglaciation of Fennoscandia. *Quaternary Science Reviews*, **147**, 91–121, doi.org/10.1016/j.quascirev.2015.09.016.
Sugihara, T., Kinoshita, M., Araki, E. et al. (2014). Re-evaluation of temperature at the updip limit of locked portion of Nankai megasplay inferred from IODP Site C0002 temperature observatory. *Earth, Planets and Space*, **66**, 107, doi.org/10.1186/1880-5981-66-107.
Sutinen, R. (2005). Timing of early Holocene landslides in Kittilä, Finnish Lapland. In A. E. K. Ojala, ed., *Quaternary Studies in the Northern and Arctic Regions of Finland*, Proceedings of the workshop organized within the Finnish National Committee for Quaternary Research (INQUA), Kilpisjärvi Biological Station, Finland, January 13–14th 2005. Geological Survey of Finland Special Paper 40, Espoo, Finland, pp. 53–58.
Sutinen, R., Hyvönen, E., Middleton, M. and Ruskeeniemi, T. (2014). Airborne LiDAR detection of postglacial faults and Pulju moraine in Palojärvi, Finnish Lapland. *Global and Planetary Change*, **115**, 24–32, doi.org/10.1016/j.gloplacha.2014.01.007
Townend, J., Sutherland, R., Toy, V. G. et al. (2017). Petrophysical, geochemical, and hydrological evidence for extensive fracture-mediated fluid and heat transport in the Alpine Fault's hanging-wall damage zone. *Geochemistry, Geophysics, Geosystems*, **18**(12), 4709–4732, doi.org/10.1002/2017GC007202.
Ueda, T. and Kato, A. (2019). Seasonal variations in crustal seismicity in San-in district, southwest Japan. *Geophysical Research Letters*, **46**, 3172–3179, doi.org/ 10.1029/2018GL081789.
Vestøl O. (2006). Determination of postglacial land uplift in Fennoscandia from leveling, tide-gauges and continuous GPS stations using least squares collocation. *Journal of Geodesy*, **80**(5), 248–258, doi.org/10.1007/s00190-006-0063-7.
Wanke, A. and Melezhik, V. (2005). Sedimentary and volcanic facies recording the Neoarchaean continent breakup and decline of the positive δ^{13} C_{carb} excursion. *Precambrian Research*, **140**(1–2), 1–35, doi.org/10.1016/j.precamres.2005.05.003.
Westaway, R. (2006). Investigation of coupling between surface processes and induced flow in the lower continental crust as a cause of intraplate seismicity. *Earth Surface Processes and Landforms*, **31**, 1480–1509, doi.org/10.1002/esp.1366.
Witschard, F. (1984). The geological and tectonic evolution of the Precambrian of northern Sweden – a case for basement reactivation? *Precambrian Research*, **23**(3–4), 273–315, doi.org/10.1016/0301-9268(84)90047-0.

Wu, P. and Hasegawa, H. S. (1996a). Induced stresses and fault potential in eastern Canada due to a disc load: a preliminary analysis. *Geophysical Journal International*, **125**, 415–430, doi.org/10.1111/j.1365-246X.1996.tb00008.x.

Wu, P. and Hasegawa, H. S. (1996b). Induced stresses and fault potential in eastern Canada due to a realistic load: a preliminary analysis. *Geophysical Journal International*, **127**, 215–229, doi.org/10.1111/j.1365-246X.1996.tb01546.x.

Zoback, M., Hickman, S., Ellsworth, W. and the SAFOD Science Team (2011). Scientific drilling into the San Andreas Fault Zone – an overview of SAFOD's first five years. *Scientific Drilling*, **11**, 14–28, doi.org/10.2204/iodp.sd.11.02.2011.

Part III
Glacially Triggered Faulting in the Fennoscandian Shield

The following chapters will give an overview of the stress situation in Northern Europe and of the most prominent fault scarps found in northern Fennoscandia in the northernmost parts of Norway, Sweden and Finland. In addition, signs of glacially triggered faulting have been identified in adjacent Russia. The reader will find an overview of these faults from the time of their identification up to the very recent results that include, among other things, new reactivation dating and revised fault geometries at the surface, obtained from laser scanning.

10

Seismicity and Sources of Stress in Fennoscandia

SØREN GREGERSEN, CONRAD LINDHOLM, ANNAKAISA KORJA, BJÖRN LUND,
MARJA USKI, KATI OINONEN, PETER H. VOSS AND MARIE KEIDING

ABSTRACT

The stress field in the Earth's crust and lithosphere is caused by several geological and geophysical factors. This chapter investigates the Fennoscandian area of uplift since the latest Ice Age and addresses the question of whether glacial isostatic adjustment may influence current seismicity. The region is far from plate tectonic boundaries, so investigation occurs in an intraplate area, with stresses caused by the lithospheric relative plate motions, as have been investigated by many authors over the years. Discussions on whether uplift and plate tectonics are the only causes of stress have been going on for many years in the scientific community. We present the earthquake distribution, the uplift pattern, the coast lines and the large postglacial faults, in a geographical overview. This is compared with the geological zones and zone boundaries. This review takes into account the improved sensitivity of the seismograph networks, and at the same time attempts to omit manmade explosions and mining events in the pattern, in order to present the best possible earthquake pattern.

From the earthquake data, focal mechanisms are derived that give indications of the present-day stress orientations. Supplemented by other stress measurements, it is possible to evaluate the stress orientations and their connection to the uplift pattern and known tectonics. Besides plate motion and uplift, one finds that some regions are affected stresswise by differences in geographical sediment loading as well as by topography variations. The stress release in the present-day earthquakes shows a pattern that deviates from that of the time right after the Ice Age. This chapter treats the stress pattern generalized for Fennoscandia and guides the interested reader to more details in the following chapters of this book.

10.1 Introduction and Geological Setting

Fennoscandia is, in a global perspective, a low seismicity region in the NW part of the Eurasian lithospheric plate. It has nevertheless exhibited some of the largest earthquakes in continental Western Europe north of the Alps (latitudes above 38°N) over the past few hundred years (FENCAT, 2020). The current tectonic regimes comprise continental intra-plate regions in the east and the Caledonian mountain range and the passive continental

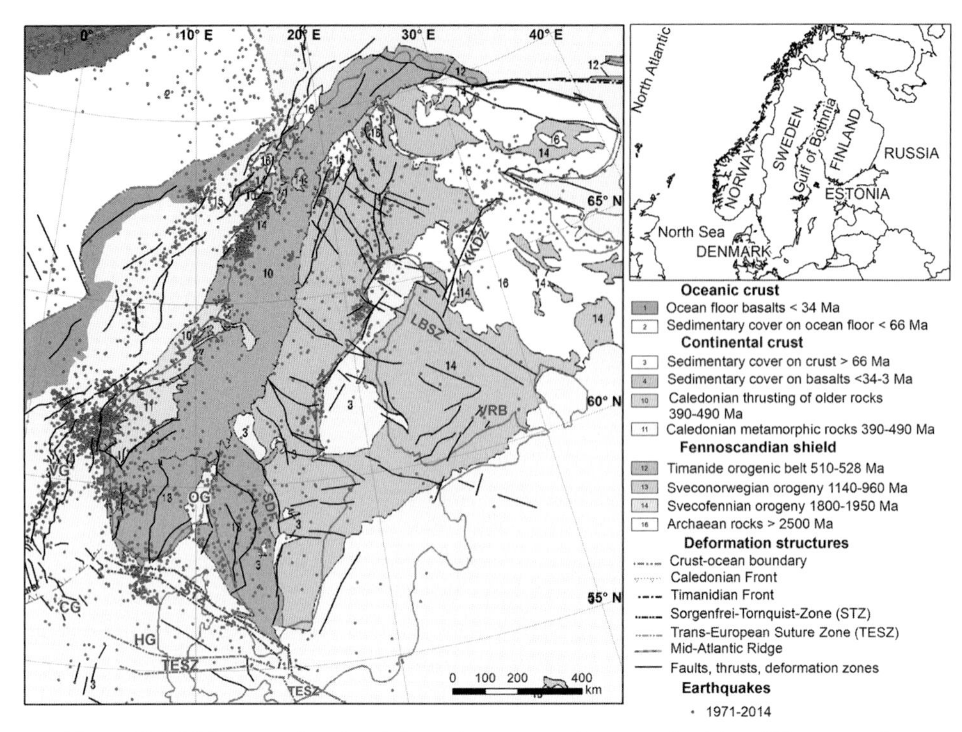

Figure 10.1 Major tectonic units, deformation zones and recent seismicity of Fennoscandia. Tectonic units and simplified deformation zones after Sigmond (2002). Instrumental seismicity, M $\geq$ 2 earthquakes in 1971–2014 from FENCAT (2020). Deformation zones and geological units: KKDZ – Kuusamo-Kandalaksha, LBSZ – Ladoga-Bothnian Bay, SDF – Sveconorwegian Deformation Front, STZ – Sorgenfrei-Tornquist Zone, TESZ – Trans-European Suture Zone and VRB – Vyborg rapakivi granite batholith. Grabens: CG – Central, HG – Horn, OG – Oslo, VG – Viking. (A black and white version of this figure will appear in some formats. For the colour version, please refer to the plate section.)

margin in the west. The continental part consists of the East European Craton (EEC) flanked by the Caledonides in the west to south-west. The craton is partly overlain by Mesoproterozoic to Phanerozoic platform sedimentary cover (Figure 10.1). In the south and south-west, the craton is separated from the Phanerozoic Europe by several deformation zones through Denmark defining the craton boundary via thickness changes in (1) lithosphere, (2) crust and (3) sedimentary cover (see Sandersen et al., Chapter 15). The latest major tectonic event that affected Fennoscandia was the Cenozoic opening and the spreading of the North Atlantic Ocean initiating 60 Ma ago (e.g. Ramberg et al., 2013) with ridge push force affecting all of Northern Europe and with the creation of major faults on- and off-shore western Norway in addition to the major oceanic transform faults.

The complex compressional and extensional evolution left the region with a multitude of deformation zones, faults and fractures that are and have been potential locations of geological reactivation. During the Pleistocene the region was subjected to repeated glaciations with varying duration and ice thickness. The latest/Weichselian glacial retreat started some 19,000 years before present (19 ka BP), and abrupt warming some 11.5 ka BP accelerated the ice retreat (Ramberg et al., 2013). In Fennoscandia, the glacial rebound is still ongoing, with a maximum uplift rate of about 1 cm/year on the Swedish north-east coast.

This chapter provides a regional overview of the seismicity of Fennoscandia and the stresses that drive that seismicity.

10.2 Present-Day Seismicity

Seismicity and sources of seismicity in Fennoscandia have been studied by numerous authors (e.g. Kolderup, 1905; Kjellén, 1912; Renqvist, 1930, and later Stephansson et al., 1986; Slunga, 1989; Bungum et al., 1991; Gregersen, 1992; Fejerskov & Lindholm, 2000; Byrkjeland et al., 2000; Muir Wood, 2000; Fjeldskaar et al., 2000; Pascal & Cloetingh, 2009; Gregersen & Voss, 2009; and many others). The older studies are based primarily on macroseismic observations, whereas the later ones are based primarily on instrumental recordings.

Fennoscandian earthquake observations are quite heterogeneous with respect to the level of magnitude of completeness and uncertainties of source parameters. Although the first seismograph stations in Fennoscandia were already installed in the early 1900s, density of the seismic stations remained for a long time rather sparse, and spatial coverage has been heterogeneous during most of the century. Most earthquakes have taken place on blind faults at significant focal depths. Accuracy of routine earthquake location usually has not been sufficient for identifying an individual fault as an earthquake causative fault. The situation prevails today with a few notable exceptions.

A marked expansion of the national seismic networks has taken place since year 2000, significantly decreasing magnitude detection thresholds and improving event location accuracy. Today, earthquake observations are based on continuous online seismic monitoring by the national seismological networks in Denmark (GEUS), Finland (University of Helsinki and University of Oulu), Norway (University of Bergen and NORSAR) and Sweden (University of Uppsala). Parametric earthquake data are kept in national databases and are also compiled into the Fennoscandian Earthquake Catalogue (FENCAT, 2020). In the following we are using two subsets of the FENCAT data to illustrate regional seismicity patterns in Fennoscandia. The first subset covers historical earthquakes with magnitude $M \geq 4.0$ in 1400–1970 and the second subset $M \geq 2$ instrumentally recorded earthquakes in 1971–2014. The data have been filtered for human-induced events (explosions, rock bursts, collapses, etc.) as well as events of questionable seismic origin (cf. Korja et al., 2016). We recognize that some explosions may still remain in the database, but these are so few that they do not bias the overall spatial distribution. Figure 10.2 includes both datasets and illustrates the spatial variations as well as magnitude variations of the data. Although some of the locations and magnitudes of the larger historical earthquakes are disputed, Figure 10.2 provides a fair overview of the spatial distribution of the earthquake activity as well as the locations of the largest earthquakes.

In a global perspective, Northern Europe is tectonically and seismically quiet, but western Scandinavia is still the most earthquake-active region north of the Alpine mountain

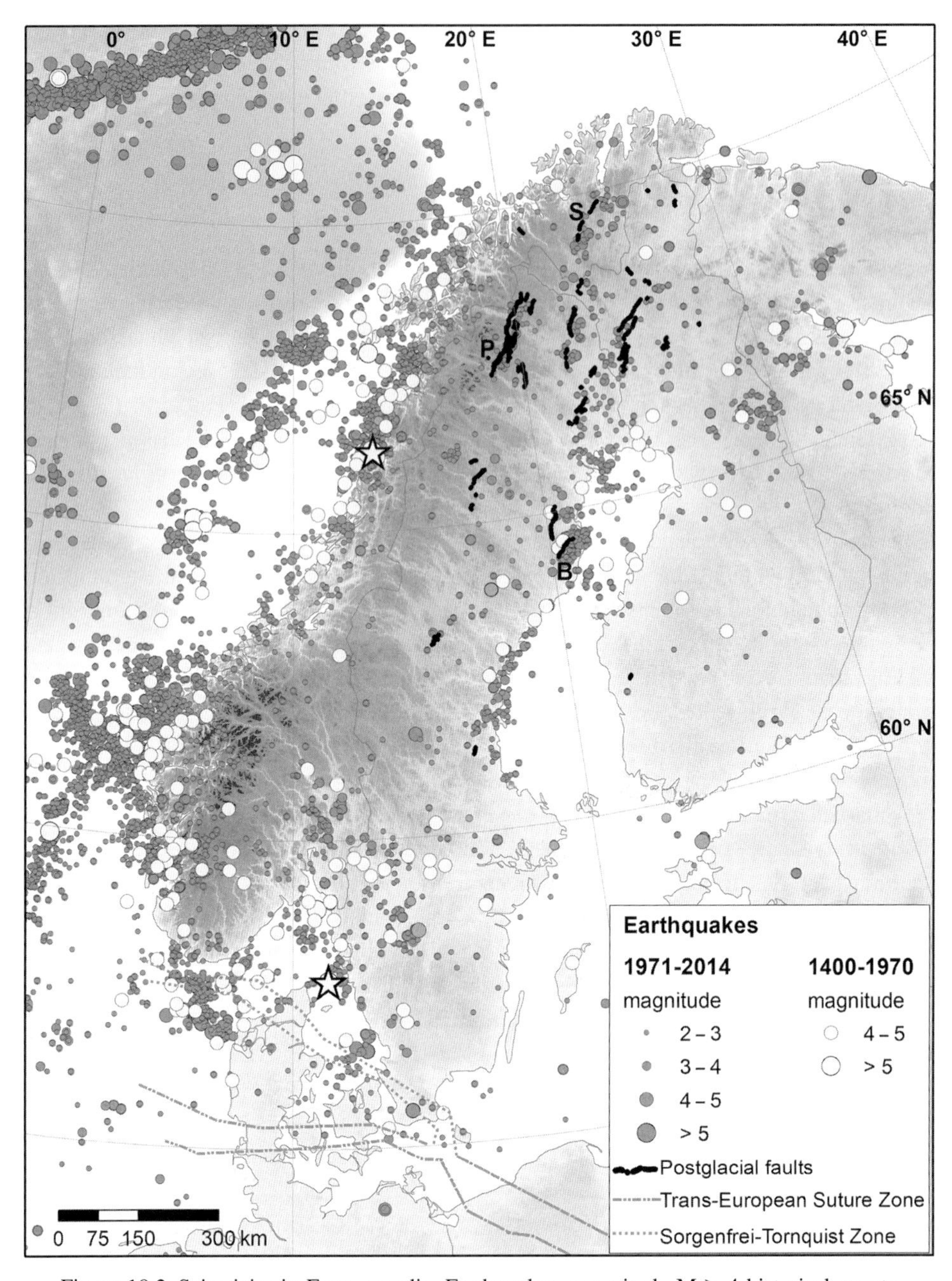

Figure 10.2 Seismicity in Fennoscandia. Earthquakes, magnitude M $\geq$ 4 historical events in 1400–1970 and M $\geq$ 2 instrumental events in 1971–2014 from FENCAT (2020). See legend for more information. The 1759 Kattegat and 1819 Lurøy earthquakes are marked with asterisks and the Burträsk, Pärvie and Stuoragurra faults with letters B, P and S. Postglacial faults from Munier et al. (2020), topographic data from GLOBE Task Team (1999) and bathymetric data from NOAA (Amante & Eakins, 2009).

chain. Figure 10.2 demonstrates the spatial relationship between earthquake activity and topography, with less activity in the eastern lowlands and the central mountain chain and higher activity in the westernmost offshore and coastal regions.

Seismicity in Fennoscandia is classified as intraplate within the Fennoscandian Shield, seismicity of various kinds at the craton boundary in Denmark, failed rift seismicity in the North Sea, Skagerrak and Oslo Graben and passive margin seismicity on- and off-shore the Norwegian coast (Figure 10.2). The largest earthquakes are localized along the failed rifts (Oslo and Viking grabens) and along the passive margin off-shore the Norwegian coast. Norway recognized in 2019 the '200-year anniversary' for the largest earthquake in continental Western Europe north of the Alps: the 1819 Lurøy earthquake (Figure 10.2). Although rare, earthquakes with magnitude $M > 5$ are found both in the historical and the instrumental datasets (e.g. Hansen et al., 1989; Pirli et al., 2010).

As can be seen in Figure 10.2, the seismicity in the Archean and Palaeoproterozoic part of Fennoscandia is to some extent diffusely distributed and, in some areas, more well-defined. In large parts of Fennoscandia, earthquakes and mapped faults are only marginally overlapping (Gregersen & Voss, 2010, 2014), whereas for other regions, e.g. Norway shelf regions, earthquake and fault density correlate well. Some of the large well-known post-glacial faults (PGFs) are still very active (e.g. Lindblom et al., 2015): the Burträsk Fault (Figure 10.2) is the most seismically active area in Sweden. Clusters of seismic activity are found along the north-east coast of Sweden and the NE-SW-trending Kuusamo–Kandalaksha deformation zone in Finland and north-west Russia (Figure 10.1). Recently, very shallow microearthquake clusters have been observed along minor faults in the Vyborg rapakivi granite batholith in south-eastern Finland and adjacent Russia (Uski et al., 2006; Smedberg et al., 2012; Assinovskaya et al., 2019). The Neoproterozoic Sveconorwegian Vänern region in south-western Sweden has also an elevated level of seismicity, associated with the Sveconorwegian Deformation Front.

The Sorgenfrei–Tornquist Zone (STZ) is the southern boundary of the Fennoscandian Shield. The STZ is a major deformation zone, across which the crustal and lithosphere thickness change abruptly (Berthelsen, 1998; Wilde-Piórko et al., 2002; Gregersen et al., 2006) and which has been suggested to be an old plate boundary (e.g. Mazur et al., 2015). Together with the Trans-European Suture Zone (TESZ), STZ outlines a wedge-shaped block in the south-western most corner of the craton. The block is highly deformed and has a thick Phanerozoic sedimentary cover (Gregersen et al., 2008). Seismicity is broadly associated with the STZ and its south-eastern extensions. South-west of the Danish activity there are, essentially, no earthquakes.

The failed Permian rifts, i.e. Oslo Graben in southern Norway and in the Viking Graben and Central Graben in the North Sea west of Norway and Denmark, seem to exhibit enhanced seismicity compared to surrounding regions. These regions have thinned crust and large normal faults, and the observation of enhanced seismicity in these regions is corroborated by global observations from stable continental regions (SCR), e.g. Landgraf et al. (2017). An outlier in this respect is the large 1759 Kattegat earthquake (e.g. Muir Wood, 1989), which is not easily associated with a graben structure, albeit by a steep crustal thinning in the region of the earthquake. In the sea between Sweden, Denmark and

Norway, very small earthquakes take place; their tectonic significance has been energetically debated (Hansen, 1986; Gregersen & Voss, 2010; Mörner, 2003).

The most intense earthquake activity in Fennoscandia (both in terms of regularity and the largest magnitudes) is found (i) offshore western Norway, (ii) in the Norwegian coastal region between steep mountains and offshore sedimentary basins and (iii) along the western shelf edge west of mid Norway. The passive margin is greatly extended, and it comprises seismically active zones along the coast and shelf edge.

Information on the focal depth distribution is not optimal due to the combination of sparse station density and large lateral variations in the crustal structure. Routine source depth estimates may contain significant uncertainties, and fixed depth estimates are frequently used by some seismic observatories.

Lindholm et al. (2000) made a detailed study for the west coast of Norway where the focal depth was somewhat shallower (10–15 km) near shore and onshore than what was found offshore (>20 km). NORSAR and NGI (1998) furthermore investigated Norwegian focal mechanisms and found that reverse faulting earthquakes had a median depth of 20 km, whereas the normal faulting events were shallower at 15 km median depth. Based on a subset of the most recent intraplate earthquakes in the Fennoscandian Shield, Korja and Kosonen (2015) and Korja et al. (2016) suggested that shallow seismicity (down to ~15 km in depth) dominates in most of Finland and northern Sweden. A trend of westward deepening of seismic sources was observed in the Gulf of Bothnia and onshore Sweden, but only a few events occurred below ~35 km in depth. They further suggested that a detachment zone controlling the depth extent of local fault zones is also controlling the depth distribution of seismicity. Deeper earthquakes, near the bottom of the crust or just below the crust, seem to be associated with major/crustal scale deformation zones, like the STZ in Denmark.

The seismic activity along the PGFs in northern Norway, Sweden and Finland is distinct from the regional background seismicity in terms of location, number of events and magnitude (Figures 10.2 and 10.3). There is a remarkable correlation between activity of low-magnitude seismicity and the mapped PGFs (Lagerbäck, 1978; Arvidsson, 1996, Lindblom et al., 2015). In the map view, the seismicity mainly clusters south-east of the main fault scarps, in accordance with the south-easterly dip of the faults. A study based on earthquake data from permanent stations and a local seismic network around the Pärvie Fault concluded that 71 per cent of the observed earthquakes north of 66° N in Sweden locate within 30 km to the south-east and 10 km to the north-west of the PGFs (Lindblom et al., 2015). Reflection seismic surveys have shown that the reverse Pärvie Fault dips steeply 50–65° to the south-east (Juhlin et al., 2010; Ahmadi et al., 2015), suggesting that the main event reactivated an old weakness zone in the crust. The seismicity near the Pärvie Fault does not correlate with a simple fault plane, but rather occurs in a zone dipping 30–60° to the south-east (Lindblom et al., 2015), that is, seismicity mainly locates in the crustal volume above the fault plane. Well-constrained earthquakes near the Pärvie Fault locate down to a depth of 35 km, indicating that the crust is seismogenic to at least 35 km depth (Lindblom et al., 2015). The focal mechanisms around the PGFs are mainly oblique reverse to strike-slip (see Figure 10.3).

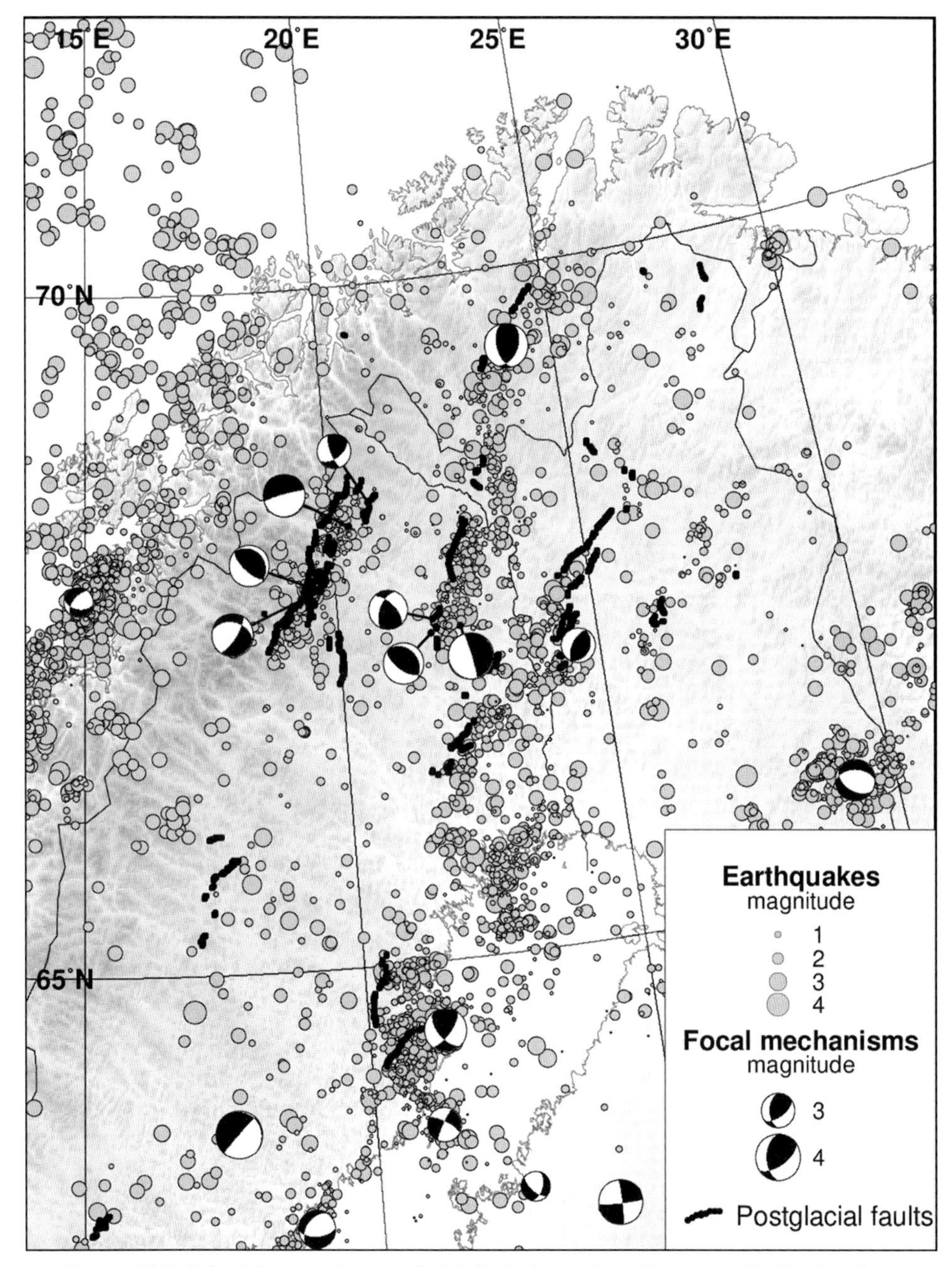

Figure 10.3 Seismicity near the postglacial faults in northern Fennoscandia. Earthquakes, all recorded events during 1971–2014 from FENCAT (2020). Earthquake focal mechanisms, M $\geq$ 2.5 events from a compilation in Keiding et al. (2015). Postglacial faults from Munier et al. (2020).

Interestingly, only the PGFs north of 64° N correlate with elevated seismicity. The proposed PGFs in southern Sweden, Denmark and Germany are not associated with current seismicity. The correlation between PGFs and seismicity is strongest in northern Sweden, corresponding to the region with the highest modelled glacial isostatic adjustment (GIA) stresses (Lund et al., 2009). The cause of the elevated seismicity is not fully understood. Lindblom (2011) showed that focal mechanisms on the Pärvie Fault imply a mainly tectonic stress regime. An alternative suggestion is that the elevated seismicity is caused by the static stress change induced by the large earthquakes that created the PGFs (Ronald Arvidsson, personal communication, 2016). Such a long-term effect would indicate a very long lithospheric relaxation time, or a very low tectonic stressing rate, as Stein and Liu (2009) argued for very long intraplate aftershock sequences. Yet another possibility is that the seismicity occurs as a result of the remaining GIA stresses or possibly a combination of all three processes.

10.3 The Regional Stress Direction in Fennoscandia

Sources of information on crustal stress are diverse and reflect the different stress conditions at different depths. Earthquake focal mechanisms reflect conditions at the hypocentre, and other methods such as overcoring and borehole breakouts reflect stresses closer to the surface. In the following, we will put more weight on the regionally significant stress derived from earthquake focal mechanisms.

An earthquake is the result of shear stresses on a specific structure. If it is possible to estimate reliably, then the earthquake source parameters provide important information on the crustal stress and the rupture process. For large earthquakes (M = 6+) a highly detailed model of the rupture and of the causative stress is possible (like what is routinely done by many international agencies today). However, for many smaller earthquakes, frequent in Northern Europe, focal mechanism qualities remain uncertain. In the study area, each country has a national database of focal mechanisms where magnitudes, quality and determination methods vary greatly.

From Norwegian regions some 200 focal mechanisms have been established, mostly from small and local earthquakes on- and offshore, with large uncertainty in source mechanisms. However, even if the individual mechanisms are very uncertain, the general pattern seems to be consistent. The focal mechanisms suggest a regionally consistent NW-SE compression with significant regional deviations. In Norway, reverse faulting dominates with pockets of normal faulting locally (Fjeldskaar et al., 2000; Hicks et al., 2000; Hicks & Ottemøller, 2001; Janutyte & Lindholm, 2017). In Sweden and Finland, there are some focal mechanisms from small earthquakes (Slunga, 1991; Lindblom, 2011; Uski et al., 2003; Uski et al., 2006). These point to a NW-SE to WNW-ESE direction for the maximum horizontal compressional stress release. Oblique strike-slip to reverse mechanisms with some local variations dominate in Finland, Estonia and north-west Russia, whereas in and around the Gulf of Bothnia, strike-slip is the dominant component of motion, although normal faulting also occurs. Recent focal mechanisms for small mid-crustal earthquakes along the eastern coast of Bothnian Bay suggest a transtensional setting where strike-slip faults have a small to significant component of extension. These events are spatially associated with the major Ladoga–Bothnian Bay shear zone. Reverse mechanisms occur more frequently in southern Finland and north of the Bothnian Bay area (cf. Lindblom et al., 2015; Korja & Kosonen, 2015; Korja et al., 2016 and references therein).

In south-central Sweden, earthquake focal mechanisms are generally of strike-slip type, with a clear NW-SE direction of compression (Slunga, 1991). In Denmark only few focal mechanisms are available. They show large uncertainty but are judged to be of the same class as those of southern Norway and southern Sweden.

In addition to plate-tectonic stress sources, the isostatic uplift (Figure 10.4) since the last glaciation, which ended about 9,000 years ago, is expected to contribute to the observed crustal stress in Fennoscandia. Discussions on whether uplift and plate tectonics are the only causes of stress and seismicity have been going on for many years in the scientific community (e.g. Gregersen & Basham, 1989; Arvidsson et al., 1991; NORSAR & NGI, 1998; Hicks et al., 2000; Gregersen, 2002; Lund, 2015; Pascal et al., 2010; Bungum et al., 2010; Redfield & Osmundsen, 2015; Korja & Kosonen, 2015; Brandes et al., 2015; Keiding et al., 2015) and are also to some extent discussed in Chapters 11–18 of this book.

Earthquake focal mechanisms reflect the stress field present at depth in the crust. Other methods, such as overcoring and borehole breakouts, reflect stresses closer to the surface. Stress magnitudes are significantly more difficult to estimate than stress directions. Therefore, most techniques only give the directions of the principal stresses, sometimes rotated into the maximum σ_H and minimum σ_h horizontal stresses. Observations on regional stress field indicators have been collected and standardized in the global World Stress Map Project (www.world-stress-map.org/). We are using data from Northern Europe (Figure 10.5; Heidbach et al., 2016) comprising earthquake focal mechanisms, borehole breakouts and in situ stress measurements (overcoring, hydraulic fracturing, borehole slotter). Heidbach et al. (2016) classified the data based on their reliability and precision into four classes A, B, C and D, with standard deviation of $\sigma_H \pm 15°$, $\pm 15-20°$, $\pm 20-25°$ and $\pm 25-40°$, respectively. Northern European observations in classes A–C are shown in Figure 10.5. The data are largely based on earthquake focal mechanisms but are supplemented by three other datasets (mainly offshore borehole breakouts). It is furthermore important to recognize that much more stress information exists in each Nordic country that has not yet been added to the WSM global database. Nevertheless, the maximum horizontal stress field indicator data indicate that NW-SE compression prevails throughout Northern Europe even at shallow depths. This NW-SE compression is most often attributed to the ongoing spreading process at the Mid-Atlantic Ridge, which exhibits a continuous ridge-push tectonic stress field with maximum horizontal compression oriented approximately NW-SE.

In evaluating the stress directions in Figure 10.5 it is, however, important to recognize that stresses are deduced from quite different methods, crustal depths, topographic terrains and rock conditions, which all have their influence on the derived stresses. In SCR like Fennoscandia the relative stress magnitudes are small, and the magnitude-wise similarity between σ_H and σ_h consequently results in a small deviatoric stress regime. This was lately

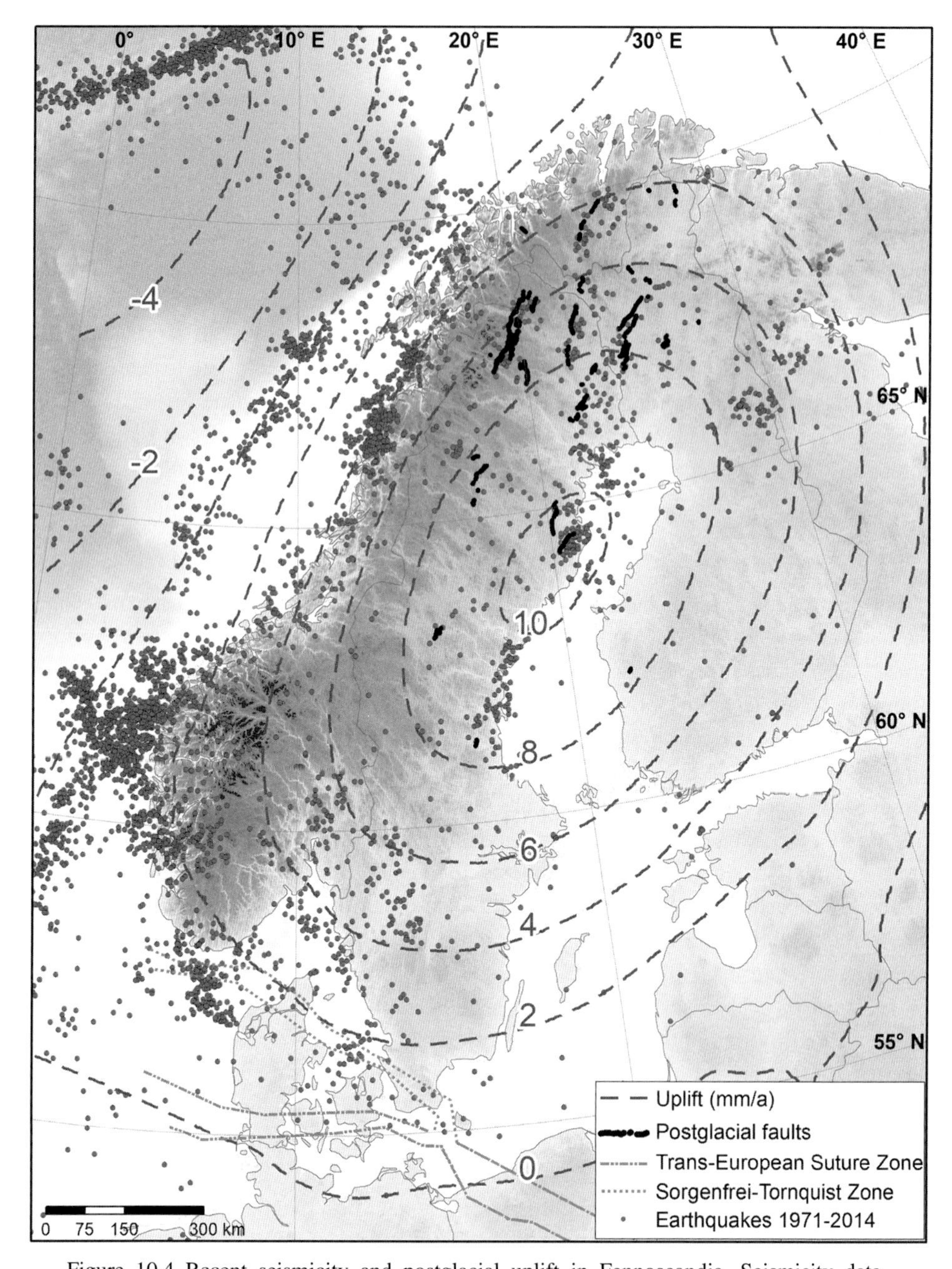

Figure 10.4 Recent seismicity and postglacial uplift in Fennoscandia. Seismicity data, M $\geq$ 2 earthquakes in 1971–2014 from FENCAT (2020). The absolute land uplift velocity (mm/a) model for Fennoscandia NKG2016LU_abs of the Nordic Geodetic Commission (Vestøl et al., 2019). Postglacial faults from Munier et al. (2020), topographic data from GLOBE Task Team (1999) and bathymetric data from NOAA (Amante & Eakins 2009).

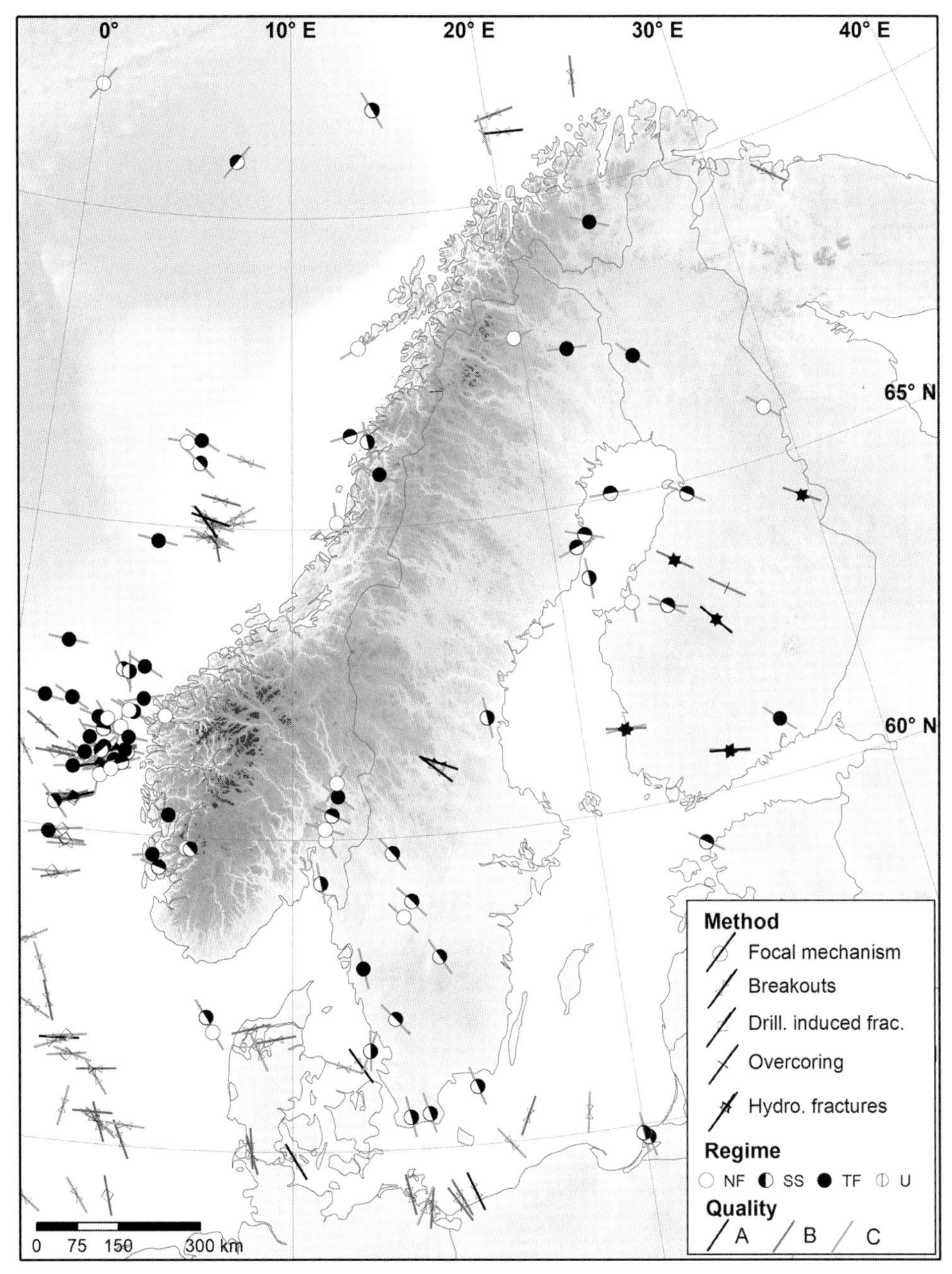

Figure 10.5 Contemporary stress field in Fennoscandia. The map displays quality A–C records from the 2016 World Stress Map database, complemented with some quality C earthquake focal mechanisms in Sweden and Finland. Shallow measurements (depth < 500 m) are excluded. Lines show the orientation of maximum horizontal stress $S_{H\max}$ and line shade represents quality ranking. Different stress regimes (NF – normal faulting, SS – strike-slip faulting, TF – thrust faulting, U – unknown) and stress indicators are marked with different symbols, see legend for more information. Topographic data from GLOBE Task Team (1999) and bathymetric data from NOAA (Amante & Eakins, 2009).

confirmed for northern Norway in the comprehensive NEONOR 2 project, where also regions of 90° stress rotation are identified (Olesen et al., 2018; Janutyte & Lindholm, 2017).

The other primary source of deformation relates to the Pleistocene glaciations. Northern Europe has been subjected to repeated glacial cycles and associated loading and unloading events resulting in reshaping of the geomorphology (Donner, 1995; Fredén, 2002). Loading and unloading of the lithosphere with an ice sheet produces an isostatic imbalance which is compensated by GIA, where the viscous mantle flows to accommodate variations in ice load and thereby produces subsidence or uplift. Today, the remaining isostatic imbalance is causing slow land uplift centred in north-eastern Sweden and the Bay of Bothnia, and according to a recent absolute land uplift model by the Nordic Geodetic Commission (NKG) NKG2016LU (Vestøl et al., 2019), the maximum rate of absolute uplift is 10 mm/year (Figure 10.4). The model demonstrates that GIA is still taking place (e.g. Janutyte & Lindholm, 2017; Keiding et al., 2015; Kierulf et al., 2014; Fjeldskaar et al., 2000; Fejerskov & Lindholm, 2000), causing an additional stress contribution in Fennoscandia (see also Steffen et al., Chapter 2).

10.4 Discussion on Earthquake Causes

Already Lagerbäck (1978) and Muir Wood (1989) suggested that landslides, faulting and associated enhanced earthquake activity in Lapland is related to the deglaciation process. This has been confirmed by Lagerbäck and Sundh (2008), Ojala et al. (2018) and Sutinen et al. (2019). In Figure 10.4 it is observed how the northern large postglacial faults are associated with present-day microseismic activity and how the pattern of seismic activity has poor correlation with the uplift curves in general. A good correlation between observed seismic activity and the uplift curves gradient is only found in the Norwegian coastal areas, where the uplift curves align with the coastlines and where the crust bending is expected to be at its maximum. Recently Lindholm (2019) observed, how the present-day earthquake activity is spatially much better correlated with the deglaciation isochrons than with the uplift model (and with present-day glaciated regions).

A key question raised early on was the source of the stress causing deformations in plate interiors. Pioneering work on global distribution of crustal earthquake-driving stresses were published from the mid-1970s by the seminal contributions of Solomon et al. (1975), Richardson et al. (1979), Harper (1989) and Zoback et al. (1989), and for Fennoscandia by Stephansson et al. (1986), Slunga (1989), Bott (1991) and Bungum et al. (1991) and many others. The relative importance of various crustal stress generating mechanisms in Fennoscandia, such as ridge-push, GIA, sedimentation and crustal density contrasts, have been discussed and modelled (e.g. Gregersen & Basham, 1989; Arvidsson et al., 1991; Hicks et al., 2000; Fejerskov & Lindholm, 2000; Muir Wood, 2000; Gregersen, 2002; Pascal & Cloetingh, 2009; Lund et al., 2009; Pascal et al., 2010; Lund, 2015; Redfield & Osmundsen, 2015; Brandes et al., 2015; Korja & Kosonen, 2015; Olesen et al., 2018). On a regional scale it has been difficult to conclude on the relative importance of these four stress sources. Recently, however, modelling has shown that GIA and ridge push may act constructively and increase crustal stresses in regions like middle Norway (Muir Wood, 2000; Fejerskov & Lindholm, 2000; Gregersen & Voss, 2009 and others).

The seismicity in Northern Europe can be classified as intraplate with preferred earthquake sources along (a) rifted passive margins, (b) palaeosutures and (c) failed rifts. The seismicity distribution is largely consistent with conclusions from global studies of so-called SCR (Johnston et al., 1994; Schulte & Mooney, 2005; Landgraf et al., 2017), maintaining that rifted passive margins and failed rifts are the two main types of host structures responsible for the largest earthquakes in such areas. A new concept that challenges the 'ridge-push concept' as the primary earthquake cause for the observed earthquake activity in western Fennoscandia has been proposed by Redfield and Osmundsen (2015). The concept takes its outset from the geological development perspective (not a stress perspective) and suggests that the first order patterns of Fennoscandian seismicity reflect the domain boundaries of the Mesozoic rifted margin. Redfield and Osmundsen (2015) identify three distinct belts of earthquakes striking subparallel to the generalized line of breakup, and its originality is that it refers to large crustal weakness zones and boundaries as a first order explanation to the observed seismicity (rather than stress). We will in this chapter not pursue these ideas further, but the reader should be aware of this alternative/complementary perspective that emphasizes pre-existing weakness zones in the crust as important for the earthquake generation.

The crustal stress generated by the GIA uplift has been modelled by many groups (e.g. Fjeldskaar, 2000, and a comprehensive review by Steffen & Wu, 2011), and they have demonstrated that major compressional stresses in the lower crust are large in the western coastal region of Norway (lately in Olesen et al., 2018). See more detailed discussion in this book.

A fundamental observation is that the exposed bedrock areas of Fennoscandia (Figures 10.2 and 10.4) exhibit more earthquakes than the surrounding sediment-covered platform areas of Denmark, Germany, Poland, Baltic States and Russia. This could be caused by crust/lithosphere differences or by the uplift/downwelling. But already in the discussion of Figures 10.1 and 10.5, we have concluded that the spatial distribution of earthquakes does not correlate well with the overall uplift pattern. So even if the postglacial uplift contributes to crustal stress, it cannot alone drive the present-day seismicity. It can, however, trigger seismicity.

Zoback et al. (1989) confirmed the earlier models of Richardson et al. (1979), where stress in the plate interiors is attributed to the relative plate motions at the plate boundaries. For Fennoscandia, the ridge-push has been interpreted as the dominating source of stress (Stephansson et al., 1986; Fejerskov & Lindholm, 2000; Korja & Kosonen, 2015; Gregersen & Voss, 2014). While the regionally observed stress directions largely support the plate boundary interpretation, the detailed analysis of focal mechanisms clearly indicates that other stress-generating mechanisms are also at work in some regions (Hicks et al., 2000; Janutyte & Lindholm, 2017; Olesen et al., 2018; Lindholm, 2019). A sometimes overlooked stress-generating factor is the gravitational effect caused by lateral changes in lithosphere thickness, as e.g. at the transition between oceanic and continental crust

(Fejerskov & Lindholm, 2000; Pascal & Cloetingh, 2009). Specifically, the relief of the western Scandinavian mountain range as well as the sharp crustal thickness variations play a vital role, as already pointed out by Fejerskov and Lindholm (2000). The Figure 10.4 uplift contours show a poor correlation with the pattern of the earthquake activity; however, when uplift isochrons are plotted with earthquake distribution some correlation is striking (Lindholm & Bungum, 2019; Lindholm, 2019). This indicates that postglacial uplift is influential in a different way: as shown by Lindholm (2019), the regions with the highest gradient in the deglaciation isochrons are clearly showing more present-day earthquakes than the regions of highest uplift gradients show. The mentioned curves on the maps are dependent on mathematical interpolation methods, so a discussion will have to be sorted out via comparison of mathematical interpolation methods.

While the existence of crustal stress is a natural prerequisite for earthquake generation, the existence of 'lubricated' faults that are favourably oriented in the stress field is another prerequisite (Copley, 2017; Landgraf et al., 2017). The existence of lubricated faults is in many cases more important than regional stress because large structures may themselves alter the surrounding stress field (e.g. Fjeldskaar et al., 2000) and observed local σ_H may deviate from the regional stress field.

An enigma in terms of stress and earthquake activity is presented by the large northern PGFs. The structures are all NE-SW striking, and they presumably ruptured during large earthquakes (less than 10,000 years ago) in postglacial times. Very recent (2018–2019) new trenching analyses from the Stuoragurra Fault, in Norway, surprisingly revealed rupture activities as late as 4,000–600 years ago (Olsen et al., 2020). These findings are so new that the full implication of these results is yet to come. Today, microseismicity is observed and clearly associated with these faults. Large earthquakes such as the ones that ruptured the PGFs are not expected in the current stress field.

We have above concentrated our short discussions of the PGFs to the main structures in the north and refrained from discussing the more recent claims of PGFs in central and southern Fennoscandia (and a few others in the north). These claims (among them, Mörner, 2003; Olesen et al., 2013; Mikko et al., 2015; Brandes et al., 2015) are important indicators that more surface deformations of postglacial origin may exist, and these are also analyzed and discussed more in this book.

The estimated tectonic strain rates for Fennoscandia are very small, and measuring these (attempted by Scherneck et al., 2001; Keiding et al., 2015) is further complicated by the fact that the surface velocity field is dominated by GIA, in both vertical and horizontal directions (e.g. Lidberg et al., 2010; Kierulf et al., 2014). Any estimate of the tectonic deformation signal in the Fennoscandian strain rate field must therefore attempt to quantify the GIA effect, which is very difficult within the uncertainty limits of the Global Positioning System (GPS) signal (e.g. Scherneck et al., 2001; Keiding et al., 2015). Previous geodetic studies concluded that glacial rebound drives earthquake activity in Fennoscandia (e.g. Gudmundsson, 1999), but mounting evidence from focal mechanisms and stress measurements as well as more detailed GIA modelling indicate that the remaining glacially induced stresses are small in magnitude (e.g. Wu et al., 1999; Lund et al., 2009). This implies that even though GIA may act as a stress contributor to the seismicity the lithospheric plate motions or other stress sources such as sediment loading or topographical loading are probably the main driving forces of Fennoscandian seismicity.

10.5 Summary

The Fennoscandian intraplate area can, in terms of seismicity, be divided between the more active mountainous region to the west (Norway with Caledonian overthrust belt) and the low-active region of the Fennoscandian Shield to the east (Sweden and Finland). The seismically most active zones appear to be optimally oriented for reverse and strike-slip faulting driven by the opening of the Atlantic.

The current strain rates in Fennoscandia are low but high enough to reactivate old structures, joints and extension fractures. As is well known, the reactivation of pre-existing deformation zones and faults depends on the stress field and the fault frictional state. What is often not sufficiently recognized is that major structures also alter the crustal stresses so that nearby structures may experience enhanced or decreased shear stress and with a direction different from the regional trends.

Finally, we may highlight the many-sided observations on the locations of modern Fennoscandian earthquakes. Many earthquakes are concentrated within and along mapped fault zones, and the larger earthquakes seem to favour aborted rift zones with thinned crust. Many earthquakes occur along the passive continental margin where the crustal thickness is subject to significant lateral variations, and along parts of the Norwegian coast. In other regions several of the scattered earthquakes occur unexpectedly on unknown or disregarded deformation zones. The Fennoscandian earthquake activity takes place in the middle to lower crust in the shelf regions and at shallower depths in the coastal regions, and in the upper to middle crust in the shield area (Finland and Sweden). The northern PGFs are associated mostly with shallow upper crust earthquakes.

The glacial isostatic adjustment contribution cannot be ignored, but the contribution is, on the regional scale, considered less than the plate tectonic contribution. In certain regions the GIA effect may, however, be important.

References

Ahmadi, O., Juhlin, C., Ask, M. V. S. and Lund, B. (2015). Revealing the deeper structure of the end-glacial Pärvie fault system in northern Sweden by seismic reflection profiling. *Solid Earth*, **6**, 621–632, doi.org/10.5194/se-6-621-2015.

Amante, C. and Eakins, B. W. (2009). *ETOPO1 1 Arc-Minute Global Relief Model: Procedures, Data Sources and Analysis*. NOAA Technical Memorandum, NESDIS NGDC-24. National Geophysical Data Center, NOAA, doi.org/10.7289/V5C8276M [26.8.2019].

Arvidsson, R. (1996). Fennoscandian earthquakes: whole crustal rupturing related to postglacial rebound. *Science*, **274** (5288), 744–746, doi.org/10.1126/science.274.5288.744.

Arvidsson, R., Gregersen, S., Kulhánek, O. and Wahlström, R. (1991). Recent Kattegat earthquakes – evidence of active intraplate tectonics in southern Scandinavia. *Physics of the Earth and Planetary Interiors*, **67**(3–4), 275–287, doi.org/10.1016/0031-9201(91)90024-C.

Assinovskaya, B. A., Gabsatarova, I. P., Panas, N. M. and Uski, M. (2019). Seismic events in 2014–2016 around the Karelian Isthmus and their nature. *Seismic Instruments*, **55**(1), 24–40, doi.org/10.3103/S074792391901002X.

Berthelsen, A. (1998). The Tornquist Zone northwest of the Carpathians: an intraplate pseudosuture. *Geologiska Föreningen i Stockholm Förhandlingar*, **120**, 223–230, doi.org/10.1080/11035899801202223.

Bott, M. H. P. (1991). Ridge push and associated plate interior stress in normal and hot spot regions. *Tectonophysics*, **200**(1–3), 17–32, doi.org/10.1016/0040-1951(91)90003-B.

Brandes, C., Steffen, H., Steffen, R. and Wu, P. (2015). Intraplate seismicity in northern Central Europe is induced by the last glaciation. *Geology*, **43**(7), doi.org/10.1130/G36710.1.

Bungum, H., Alsaker, A., Kvamme, L. B. and Hansen, R. A. (1991). Seismicity and seismotectonics of Norway and surrounding continental shelf areas. *Journal of Geophysical Research*, **96**, 2249–2265, doi.org/10.1029/90JB02010.

Bungum, H. and Lindholm, C. (1996). Seismo- and neotectonics in Finnmark, Kola and the southern Barents Sea, part 2: seismological analysis and seismotectonics. *Tectonophysics*, **270**, 15–28, doi.org/10.1016/S0040-1951(96)00139-4.

Bungum, H., Lindholm, C. and Faleide, J. I. (2005). Postglacial seismicity offshore mid-Norway with emphasis on spatio-temporal-magnitudal variations. *Marine and Petroleum Geology*, **22**, 137–148, doi.org/10.1016/j.marpetgeo.2004.10.007.

Bungum, H., Pascal, C., Olesen et al. (2010). To what extent is the present seismicity of Norway driven by postglacial rebound? *Journal of the Geological Society of London*, **167**, 373–384, doi.org/10.1144/0016-76492009-009.

Byrkjeland, U., Bungum, H. and Eldholm, O. (2000). Seismotectonics of the Norwegian continental margin. *Journal of Geophysical Research*, **105**(B3), 6221–6236, doi.org/10.1029/1999JB900275.

Copley, A. (2017). The strength of earthquake-generating faults. *Journal of the Geological Society*, **175**, 1–12, doi.org/10.1144/jgs2017-037.

Donner, J. (1995). *The Quaternary History of Scandinavia*. Cambridge University Press, Cambridge.

Fejerskov, M. and Lindholm, C. (2000). Crustal stress in and around Norway: an evaluation of stress-generating mechanisms. In Nøttvedt et al., eds., *Dynamics of the Norwegian Margin*. Geological Society, London, Special Publication, Vol. 167, pp. 451–467, doi.org/10.1144/GSL.SP.2000.167.01.19.

FENCAT (2020). Fennoscandian earthquake catalogue for 1375-2014, www.seismo.helsinki.fi/bulletin/list/catalog/FENCAT.html.

Fjeldskaar, W. (2000). How important are elastic deflections in the Fennoscandian postglacial uplift? *Norsk Geologisk Tidsskrift*, **80**, 57–62, doi.org/10.1080/002919600750042681.

Fjeldskaar, W., Lindholm, C., Dehls, J. F. and Fjeldskaar, I. (2000). Postglacial uplift, neotectonics and seismicity in Fennoscandia. *Quaternary Science Reviews*, **19**, 1413–1422, doi.org/10.1016/S0277-3791(00)00070-6.

Fredén, C. (2002). *Sveriges nationalatlas* [National Atlas of Sweden]. Geological Survey of Sweden, 208 pp. (in Swedish).

GLOBE Task Team (Hastings, D. A., Dunbar, P. K., Elphingstone, G. M. et al.). (1999). *The Global Land One-Kilometer Base Elevation (GLOBE) Digital Elevation Model, Version 1.0*. National Oceanic and Atmospheric Administration, National Geophysical Data Center, Boulder, Colorado.

Gregersen, S. (1992). Crustal stress regime in Fennoscandia from focal mechanisms. *Journal of Geophysical Research*, **97**(B8), 11,821–11,827, doi.org/10.1029/91JB02011.

Gregersen, S. (2002). Earthquakes and change of stress since the Ice Age in Scandinavia. *Bulletin of the Geological Society Denmark*, **49**, 73–78.

Gregersen, S. and Basham, P. V. (1989). *Earthquakes at North Atlantic Margins: Neotectonics and Postglacial Rebound*. Nato ASI Series, Vol. 266. Kluwer Academic Publishers, Dordrecht, doi.org/10.1007/978-94-009-2311-9.

Gregersen, S. and Voss, P. (2009). Stress change over short geological time: case of Scandinavia over 9,000 years since the Ice Age. In K. Reicherter, A. Michetti and P. G. Silva, eds., *Paleoseismology. Historical and Prehistorical Records of Earthquake Ground Effects for Seismic Hazard Assessment*. Geological Society, London, Special Publication, Vol. 316, pp. 173–178, doi.org/10.1144/SP316.10.

Gregersen, S. and Voss, P. (2010). Irregularities in Scandinavian postglacial uplift/subsidence in time scales tens, hundreds, thousands of years. *Journal of Geodynamics*, **50**(1), 27–31, doi.org/10.1016/j.jog.2009.11.004.

Gregersen, S. and Voss, P. H. (2014). Review of some significant claimed irregularities in Scandinavian postglacial uplift on timescales of tens to thousands of years – earthquakes in Denmark. *Solid Earth*, **5**, 109–118, doi.org/10.5194/se-5-109-2014.

Gregersen, S., Voss, P., Shomali, H. et al. (2006). Physical differences in the deep lithosphere of Northern and Central Europe. In D. G. Gee and R. A. Stephenson, eds., *European Lithosphere Dynamics*. Geological Society, London, Memoirs, Vol. 32, pp. 313–322, doi.org/10.1144/GSL.MEM.2006.032.01.18.

Gregersen, S., Nielsen, L. V. and Voss, P. (2008). Evidence of stretching of the lithosphere under Denmark. *Geological Survey of Denmark and Greenland Bulletin*, **15**, 53–56.

Gudmundsson, A. (1999). Postglacial crustal doming, stresses and fracture formation with application to Norway. *Tectonophysics*, **307**, 407–419, doi.org/10.1016/S0040–1951(99)00107-9.

Hansen, J. M. (1986). Læsø: a result of fault displacements, earthquakes and level changes. *Danish Geological Society, D*, **6**, 47–72 (in Danish).

Hansen R., Bungum, H. and Alasker, A. (1989). Three recent larger earthquakes offshore Norway. *Terra Nova*, **1**(3), 284–295, doi.org/10.1111/j.1365-3121.1989.tb00371.x.

Harper, J. F. (1989). Forces driving plate tectonics: the use of simple dynamic models. *Reviews in Aquatic Science*, **1**, 319–336.

Heidbach, O., Rajabi, M., Reiter, K. and Ziegler, M. (2016). World Stress Map 2016. GFZ Data Services, doi.org/10.5880/WSM.2016.002.

Hicks, E. and Ottemöller, L. (2001). The ML 4.5 Stord/Bømlo, southwestern Norway, earthquake of 12 August 2000. *Norsk Geologisk Tidsskrift*, **81**, 293–304.

Hicks, E., Bungum, H. and Lindholm, C. (2000). Stress inversions of earthquake focal mechanism solutions from onshore and offshore Norway. *Norsk Geologisk Tidsskrift*, **80**, 235–250.

Janutyte, I. and Lindholm, C. (2017). Earthquake source mechanisms in onshore and offshore Nordland, northern Norway. *Norwegian Journal of Geology*, **97**, 177–189, doi.org/10.17850/njg97-3-03.

Johnston, A. C., Coppersmith, K. J., Kanter, L. R. and Cornell, C. A. (1994). *The Earthquakes of Stable Continental Regions*. Technical Report EPRI TR-102261s-V1-V5. Electric Power Research Institute (EPRI), Palo Alto, California.

Juhlin, C., Dehghannejad, M., Lund, B., Malehmir, A. and Pratt, G. (2010). Reflection seismic imaging of the end-glacial Pärvie Fault system, northern Sweden. *Journal of Applied Geophysics*, **70**(4), 307–316, doi.org/10.1016/j.jappgeo.2009.06.004.

Keiding, M., Kreemer, C., Lindholm, C. D. et al. (2015). A comparison of strain rates and seismicity for Fennoscandia: depth dependency of deformation from glacial isostatic adjustment. *Geophysical Journal International*, **202**, 1021–1028, doi.org/10.1093/gji/ggv207.

Kierulf, H. P., Steffen, H., Simpson, M. J. R. et al. (2014). A GPS velocity field for Fennoscandia and a consistent comparison to glacial isostatic adjustment models. *Journal of Geophysical Research*, **119**(8), 6613–6629, doi.org/10.1002/2013JB010889.

Kjellén, R. (1912). Sveriges jordskalf, försök till en seismisk landsgeografi. Göteborg 1910 [Sweden's earthquakes, attempt for a national seismic geography]. *Geologiska Föreningen i Stockholm Förhandlingar*, **34**(6), 211 pp.

Kolderup, C. F. (1905). *Norges Jordskjelv* [Norway's Earthquakes]. Bergen Museums Årbog.

Korja, A. and Kosonen, E. (2015). *Seismotectonic Framework and Seismic Source Area Models in Fennoscandia, Northern Europe*. Institute of Seismology, University of Helsinki Report S-63, 284 pp.

Korja, A., Kihlman, S. and Oinonen, K. (2016). *Seismic Source Areas in Central Fennoscandia.* Institute of Seismology, University of Helsinki Report S-64, 315 pp.

Lagerbäck, R. (1978). Neotectonic structures in northern Sweden. *Geologiska Föreningens i Stockholm Förhandlingar*, **100**(3), 263–269, doi.org/10.1080/11035897809452533.

Lagerbäck, R. and M. Sundh (2008). *Early Holocene Faulting and Paleoseismicity in Northern Sweden.* Technical Report C836, Geological Survey of Sweden, Uppsala, Sweden.

Landgraf, A., Kübler, S., Hintersberger, E. and Stein, S. (eds.) (2017). *Seismicity, Fault Rupture and Earthquake Hazards in Slowly Deforming Regions*. Geological Society, London, Special Publication, Vol. 432, doi.org/10.1144/SP432.

Lidberg, M., Johansson, J. M., Scherneck, H.-G. and Milne, G. A. (2010). Recent results based on continuous GPS observations of the GIA process in Fennoscandia from BIFROST. *Journal of Geodynamics*, **50**(1), 8–18, doi.org/10.1016/j.jog.2009.11.010.

Lindblom, E. (2011). *Microearthquake Study of End-Glacial Faults in Northern Sweden.* Phil lic thesis in seismology, University of Uppsala, Sweden.

Lindblom, E., Lund, B., Tryggvason, A. et al. (2015). Microearthquakes illuminate the deep structure of the endglacial Pärvie fault, northern Sweden. *Geophysical Journal International*, **201**, 1704–1716, doi.org/10.1093/gji/ggv112.

Lindholm, C. (2019). *Earthquakes in Norway. Fjellsprengningskonferansen 2019, Oslo, Norway.* Fjellsprengningsteknikk Bergmekanikk/Geoteknikk 2019, 8.1–8.13.

Lindholm, C. and Bungum, H. (2019). *Seismic Zonation and Earthquake Loading for Norway and Svalbard; Load Estimates as Basis for Eurocode 8 Applications.* NORSAR Report, 19-005 (confidential), 176 pp.

Lindholm, C. D., Bungum, H., Hicks, E. and Villagran, M. (2000). Crustal stress and tectonics in Norwegian regions determined from earthquake focal mechanisms. In Nøttvedt et al., eds., *Dynamics of the Norwegian Margin.* Geological Society, London, Special Publication, Vol. 167, pp. 429–439, doi.org/10.1144/GSL.SP.2000.167.01.17.

Lund, B. (2015). Paleoseismology of glaciated terrain. In M. Beer, I. A. Kougioumtzoglou, E. Patelli and S.-K. Au, eds., *Encyclopedia of Earthquake Engineering*. Springer Verlag, Berlin/Heidelberg, pp. 1765–1779, doi.org/10.1007/978-3-642-36197-5_25-1.

Lund, B., Schmidt, P. and Hieronymus, C. (2009). *Stress Evolution and Fault Stability during the Weichselian Glacial Cycle.* SKB Technical Report TR-09-15, Swedish Nuclear Fuel and Waste Management Co., Stockholm, 106 pp.

Mazur, S., Mikolajczak, M., Krzywiec, P. et al. (2015). Is the Teisseyre Tornquist Zone an ancient plate boundary of Baltica? *Tectonics*, **34**, 2465–2477, doi.org/10.1002/2015TC003934.

Mikko, H., Smith, C. A., Lund, B., Ask, M. V. S. and Munier, R. (2015). LiDAR-derived inventory of post-glacial fault scarps in Sweden. *GFF*, **137**(4), 334–338, doi.org/10.1080/11035897.2015.1036360.

Mörner, N.-A. (2003). *Paleoseismicity of Sweden. A Novel Paradigm.* JOFO Grafiska AB, Stockholm.

Muir Wood, R. (1989). Extraordinary deglaciation reverse faulting in northern Scandinavia. In S. Gregersen and P. V. Basham, eds., *Earthquakes at North Atlantic Margins: Neotectonics and Postglacial Rebound.* Nato ASI Series, Vol. 266. Kluwer Academic Publishers, Dordrecht, pp. 141–173, doi.org/10.1007/978-94-009-2311-9_10.

Muir Wood, R. (1989). The Scandinavian earthquakes of 22 December 1759 and 31 August 1819. *Disasters*, **12**(3), 223–236, doi.org/10.1111/j.1467-7717.1988.tb00672.x.

Muir Wood, R. (2000). Deglaciation seismotectonics: a principal influence on intraplate seismogenesis at high latitudes. *Quaternary Science Reviews*, **19**, 1399–1411, doi.org/10.1016/S0277-3791(00)00069-X.

Munier, R., Adams, J., Brandes, C. et al. (2020). International database of Glacially Induced Faults. *PANGAEA*, doi.org/10.1594/PANGAEA.922705.

NORSAR and NGI (1998). *Development of Seismic Zonation for Norway.* Final Report for Norwegian Council for Building Standardization (NBR) (on behalf of a consortium of industrial partners), NORSAR, 187 pp.

Ojala, A. E. K., Markovaara-Koivisto, M., Middleton, M. et al. (2018). Dating of paleo-landslides in western Finnish Lapland. *Earth Surface Processes and Landforms*, **43**, 2449–2462, doi.org/10.1002/esp.4408.

Olesen, O., Bungum, H., Dehls, J. et al. (2013). Neotectonics, seismicity and contemporary stress field in Norway – mechanisms and implications. In L. Olsen, O. Fredin and O. Olesen, eds., *Quaternary Geology of Norway, Geological Survey of Norway Special Publication 13.* Geological Survey of Norway, Trondheim, pp. 145–174.

Olesen, O., Janutyte, I., Michálek, J. et al. (2018). *Neotectonics in Nordland – Implications for Petroleum Exploration (NEONOR2).* NGU Report, 2018.010, 329 pp.

Olsen, L., Olesen, O. and Høgaas, F. (2020). Dating of the Stuoragurra Fault at Finnmarksvidda, northern Norway. In H. A. Nakrem and A. M. Husås, eds., *34th Nordic Geological Winter Meeting January 8th–10th 2020, Oslo, Norway.* Abstracts and Proceedings of the Geological Society of Norway, No. 1, pp. 157–158.

Pascal, C. and Cloetingh, S. (2009). Gravitational potential stresses and stress field of passive continental margins: insights from the south-Norway shelf. *Earth and Planetary Science Letters*, **277**, 464–473, doi.org/10.1016/j.epsl.2008.11.014.

Pascal, C., Roberts, D. and Gabrielsen, R. H. (2010). Tectonic significance of present-day stress relief phenomena in formerly glaciated regions. *Journal of the Geological Society of London*, **167**, 363–371, doi.org/10.1144/0016-76492009-136.

Pirli, M., Schweitzer, J., Ottemöller, L. et al. (2010). Preliminary analysis of the 21 February 2008 Svalbard (Norway) Seismic Sequence. *Seismological Research Letters*, **81**(1), 63–75, doi.org/10.1785/gssrl.81.1.63.

Ramberg, I. B., Bryhni, I., Forening, N. G. and Nøttvedt, A. (2013). *Landet blir til: Norges geologi* [*The Land Arises: Norway's Geology*]. Norsk geologisk forening, Trondheim, Norway.

Redfield, T. F. and Osmundsen, P. T. (2015). Some remarks on the earthquakes of Fennoscandia: a conceptual seismological model drawn from the perspective of hyperextension. *Norwegian Journal of Geology*, **94**, 233–262.

Renqvist, H. (1930). Finlands Jordskalv [Finland's earthquakes] (in Swedish). *Fennia*, **54**, 113 pp.

Richardson, R. M., Solomon S. C. and Sleep, N. H. (1979). Tectonic stress in the plates. *Reviews in Geophysics*, **17**, 981–1019, doi.org/10.1029/RG017i005p00981.

Scherneck, H.-G., Johanson, J. M., Vermeer, M. et al. (2001). BIFROST project: 3-D crustal deformation rates derived from GPS confirm postglacial rebound in Fennoscandia. *Earth, Planets and Space*, **53**, 703–708, doi.org/10.1186/BF03352398.

Schulte, S. and Mooney, W. (2005). An updated global earthquake catalogue for stable continental regions: reassessing the correlation with ancient rifts. *Geophysical Journal International*, **161**, 707–721, doi.org/10.1111/j.1365-246X.2005.02554.

Sigmond, E. M. O. (2002). Geological map of land and sea areas of Northern Europe. Scale 1:4 million. Geological Survey of Norway, Trondheim.

Slunga, R. S. (1989). Focal mechanisms and crustal stresses in the Baltic Shield. In S. Gregersen and P. W. Basham, eds., *Earthquakes at North Atlantic Margins: Neotectonics and Postglacial Rebound.* Nato ASI Series, Vol. 266, Kluwer Academic Publishers, Dordrecht, pp. 261–276, doi.org/10.1007/978-94-009-2311-9_15.

Slunga, R. S. (1991). The Baltic Shield earthquakes. *Tectonophysics*, **189**(1–4), 323–331, doi.org/10.1016/0040-1951(91)90505-M.

Smedberg, I., Uski, M., Tiira, T., Komminaho, K. and Korja, A. (2012). Intraplate earthquake swarm in Kouvola, south-eastern Finland. *Geophysical Research Abstracts*, **14**, EGU2012–8446.

Solomon, S. C., Sleep, N. H. and Richardson, R. M. (1975). On the forces driving plate tectonics: inferences from absolute plate velocities and intraplate stress. *Geophysical Journal of the Royal Astronomical Society*, 769–801, doi.org/10.1111/j.1365-246X.1975.tb05891.x.

Steffen, H. and Wu, P. (2011). Glacial isostatic adjustment in Fennoscandia – a review of data and modelling. *Journal of Geodynamics*, **52**, 169–204, doi.org/10.1016/j.jog.2011.03.002.

Stein, S. and Liu, M. (2009). Long aftershock sequences within continents and implications for earthquake hazard assessment, *Nature*, **462**, 87–89, doi.org/10.1038/nature08502.

Stephansson, O., Särkkä, P. and Myrvang, A. (1986). *State of Stress in Fennoscandia.* Proceedings of the International Symposium on Rock Stress and Rock Stress Measurements, Centek, Luleå, Sweden, 21–32.

Sutinen, R., Andreani, L. and Middleton, M. (2019). Post-Younger Dryas fault instability and deformations on ice lineations in Finnish Lapland. *Geomorphology*, **326**, 202–212, doi.org/10.1016/j.geomorph.2018.08.034.

Uski, M., Hyvönen, T., Korja, A. and Airo, M.-L. (2003). Focal mechanisms of three earthquakes in Finland and their relation to surface faults. *Tectonophysics*, **363**(1–2), 141–157, doi.org/10.1016/S0040-1951(02)00669-8.

Uski, M., Tiira, T., Korja, A. and Elo, S. (2006). The 2003 earthquake swarm in Anjalankoski, south-eastern Finland. *Tectonophysics*, **422**(1–4), 55–69, doi.org/10.1016/j.tecto.2006.05.014.

Vestøl, O., Ågren, J., Steffen, H., Kierulf, H. and Tarasov, L. (2019). NKG2016LU: a new land uplift model for Fennoscandia and the Baltic Region. *Journal of Geodesy*, **93**, 1759–1779, doi.org/10.1007/s00190-019-01280-8.

Wilde-Piórko, M. Grad, M. and TOR Working Group (2002). Crustal structure variation from the Precambrian to Palaeozoic platforms in Europe imaged by the inversion of teleseismic receiver functions – project TOR. *Geophysical Journal International*, **150**, 261–270, doi.org/10.1046/j.1365-246X.2002.01699.x.

Wu, P., Johnston, P. and Lambeck, K. (1999). Postglacial rebound and fault instability in Fennoscandia. *Geophysical Journal International*, **139**, 657–670, doi.org/10.1046/j.1365-246x.1999.00963.x.

Zoback, M. L., Zoback, M. D., Adams, J. et al. (1989). Global patterns of tectonic stress. *Nature*, **341**, 291–298, doi.org/10.1038/341291a0.

11

Postglacial Faulting in Norway

Large Magnitude Earthquakes of the Late Holocene Age

ODLEIV OLESEN, LARS OLSEN, STEVEN J. GIBBONS, BENT OLE RUUD, FREDRIK HØGAAS, TOR ARNE JOHANSEN AND TORMOD KVÆRNA

ABSTRACT

The Stuoragurra Fault Complex (SFC) constitutes the Norwegian part of the larger Lapland province of postglacial faults. The 90-km-long SFC consists of three separate fault systems: the Fitnajohka Fault System in the south-west, the Máze Fault System in the central area and the Iešjávri Fault System to the north-east. The distance between the three fault systems is 7–12 km. The faults dip at an angle of 30–75° to the SE and can be traced on reflection seismic data to a depth of around 500 m. Postglacial faults occur within the ~4–5-km-wide Precambrian Mierojávri–Sværholt shear zone in the northern Fennoscandian Shield. Deep seismic profiling shows that the shear zone dips at an angle of ~43° to the south-east and can be traced down to a depth of ~3 km. A total of approximately 80 earthquakes were registered along the SFC between 1991 and 2019. Most occurred to the south-east of the fault scarps and less than 10 km from the extrapolated Mierojávri–Sværholt shear zone at depth. The maximum moment magnitude was 4.0. The formation of postglacial faults in northern Fennoscandia was previously associated with the deglaciation of the last inland ice. Trenching of different sections of the fault complex and radiocarbon dating of buried and deformed organic material reveal, however, a late Holocene age (between around 600 and 4000 years before present day). The reverse displacement of ~9 m and fault system lengths of 14 and 21 km of the two southernmost fault systems indicate a moment magnitude of ~7. A total of around 60 landslides occurred along the SFC. The results from this study indicate that the expected maximum magnitude of future earthquakes in Fennoscandia is ~7.

11.1 Introduction

Two postglacial faults have previously been reported from Norway: the Nordmannvikdalen Fault in northern Troms (Bakken, 1983; Tolgensbakk & Sollid, 1988; Dehls et al., 2000; Olsen et al., 2018) and the Stuoragurra Fault Complex in western Finnmark (Olesen, 1988; Olesen et al., 1992a,b, 2004, 2013; Bungum & Lindholm, 1997; Roberts et al., 1997; Dehls et al., 2000). The faults have been assumed to be of the Younger Dryas and postglacial age, respectively, and constitute the northern part of the Lapland postglacial fault province (Lagerbäck & Sundh, 2008; Olesen et al., 2013: Smith et al., 2014; Sutinen et al., 2014; Mikko et al., 2015; Palmu et al., 2015). The postglacial faults are assumed to have formed as a result of released excess horizontal stresses generated during glaciation and activated by deglaciation and glacial isostatic rebound (Wu et al., 1999; Lund et al., 2009; Steffen et al., 2014a,b). Lagerbäck and Sundh (2008) argue further that the outburst of Lateglacial faulting in northern Sweden was restricted to the last deglaciation. They argued that the modest glacial erosion in this part of Sweden would allow older fault scarps of similar magnitude to be preserved. The postglacial faults in northern Fennoscandia (Figure 11.1) have been considered to represent single rupture events at the end of, or after, the Last Weichselian glaciation. Lagerbäck and Sundh (2008), Smith et al. (2018) and Mattila et al. (2019) have, however, provided field evidence for multiple slip events on some of these postglacial faults.

The present study represents a review of the present-day knowledge of the structure and deformation along the postglacial Stuoragurra Fault Complex. Analyses of digital aerial photographs, LiDAR data, seismicity, electrical resistivity, seismic profiling, trenching and radiocarbon dating have been conducted to characterize the geological structure, deformation and age of the postglacial faults.

We have adapted the term 'complex' for the Stuoragurra Fault since it consists of more than 30 segments, which we now think were partly formed at different times (Olsen et al., 2020). We have further grouped the segments into three fault systems. The terms 'fault complex' and 'fault system' have also been introduced for other postglacial faults in northern Fennoscandia (e.g. by Ojala et al., 2017). A review of the Nordmannvikdalen postglacial fault has recently been published by Olsen et al. (2018) and is not reiterated in the present paper.

11.2 Geological and Geophysical Setting

The Stuoragurra Fault Complex (SFC) constitutes the northernmost part of the Lapland province of postglacial faults. The SFC (Figure 11.2) is located within the regional about 4–5-km-wide Mierojávri–Sværholt shear zone (MSSZ) in the Precambrian of Finnmark, northern Norway (Olesen et al., 1992a; Olesen & Sandstad, 1993; Siedlecka & Roberts, 1996; Bingen et al., 2015; Henderson et al., 2015). The central part of the SFC in the Masi area occurs at the north-western boundary of the MSSZ, while the southern and northern parts are located in the middle of the fault zone. MSSZ (Figure 11.2) constitutes the north-western margin of the Jergul Gneiss Complex (Siedlecka & Roberts, 1996). Mylonites along the MSSZ can be observed in a roadcut immediately to the north of Nieidagorži along the main road between Masi and Kautokeino. Olesen et al. (1992a) interpreted a dextral displacement along the MSSZ and related the contractional Biggevárri duplex to the Palaeoproterozoic deformation (Figure 11.2). Structural mapping by Henderson et al. (2015) later verified this interpretation. Diabases several hundred metres thick and albite diabases (Figure 11.2) intruded along the MSSZ 2,220 ± 7 Ma ago (Bingen et al., 2015). These dykes and sills are deformed due to later movements along the fault zone (Solli, 1988; Olesen et al., 1992a). The deformation is mostly related to the Svecofennian orogeny (circa 1,700–1,900 Ma).

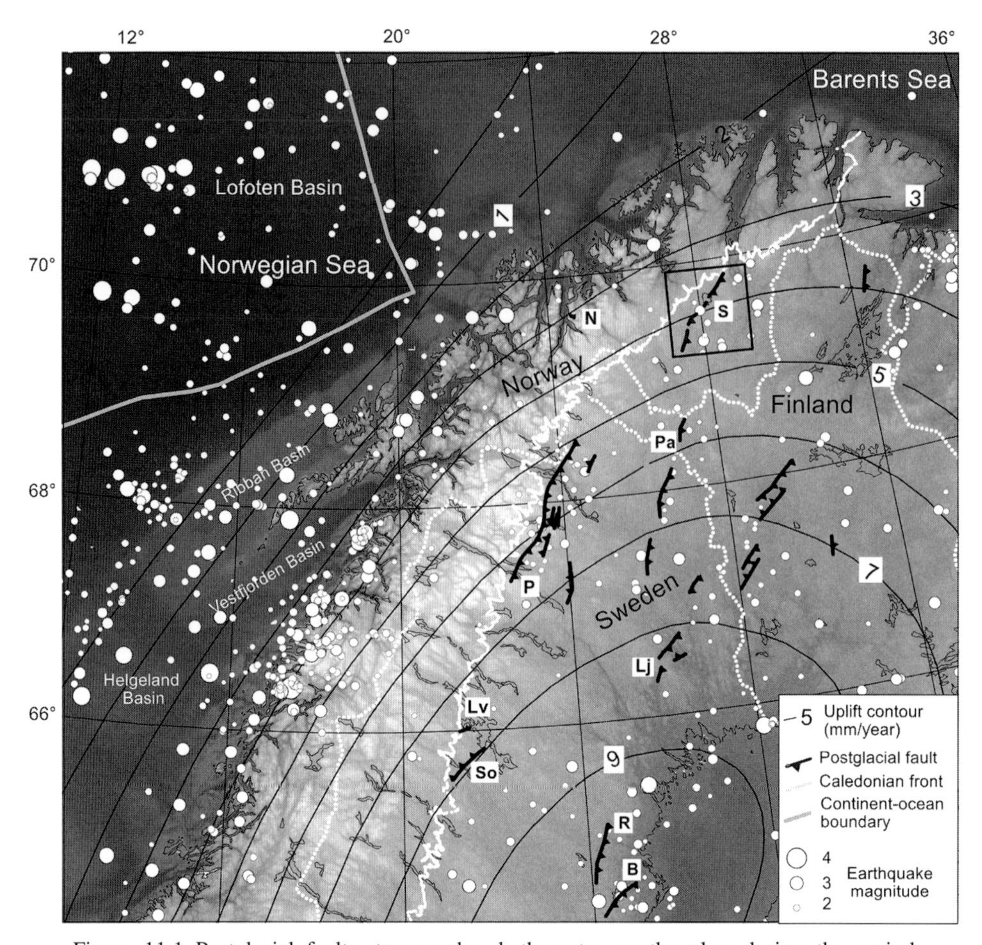

Figure 11.1 Postglacial faults, topography, bathymetry, earthquakes during the period 1750–2011 and present-day uplift (mm/year) in northern and central Fennoscandia (modified from Dehls et al., 2000; Olesen et al., 2004; Lagerbäck & Sundh, 2008; Sutinen et al., 2014; Mikko et al., 2015; Palmu et al., 2015; Vestøl et al., 2019). The Norwegian National Seismological Network at the University of Bergen is the source of the earthquake data in Norway, Svalbard and NE Atlantic. Data on the other earthquakes in Finland and Sweden are downloaded from the web pages of the Institute of Seismology at the University of Helsinki; www.seismo.helsinki.fi/english/bulletins/index.html. The topography and bathymetry are compiled from various sources by Olesen et al. (2010). Most of the faults within the Lapland province of postglacial faults cuts the present-day uplift contours at an angle of 30–70°. The postglacial faults occur in areas with increased seismicity, indicating they are active at depth. The Stuoragurra, Pärvie, Laisvall and Sorsele faults are located immediately to the east of the Caledonian front. The black frame shows the location of Figure 11.2 and the Stuoragurra Fault Complex. N – Nordmannvikdalen Fault; S – Stuoragurra Fault Complex; B – Burträsk Fault; Lv – Laisvall Fault; Lj – Lansjärv Fault; P – Pärvie Fault; Pa – Palojärvi Fault; R – Röjnoret Fault; So – Sorsele Fault.

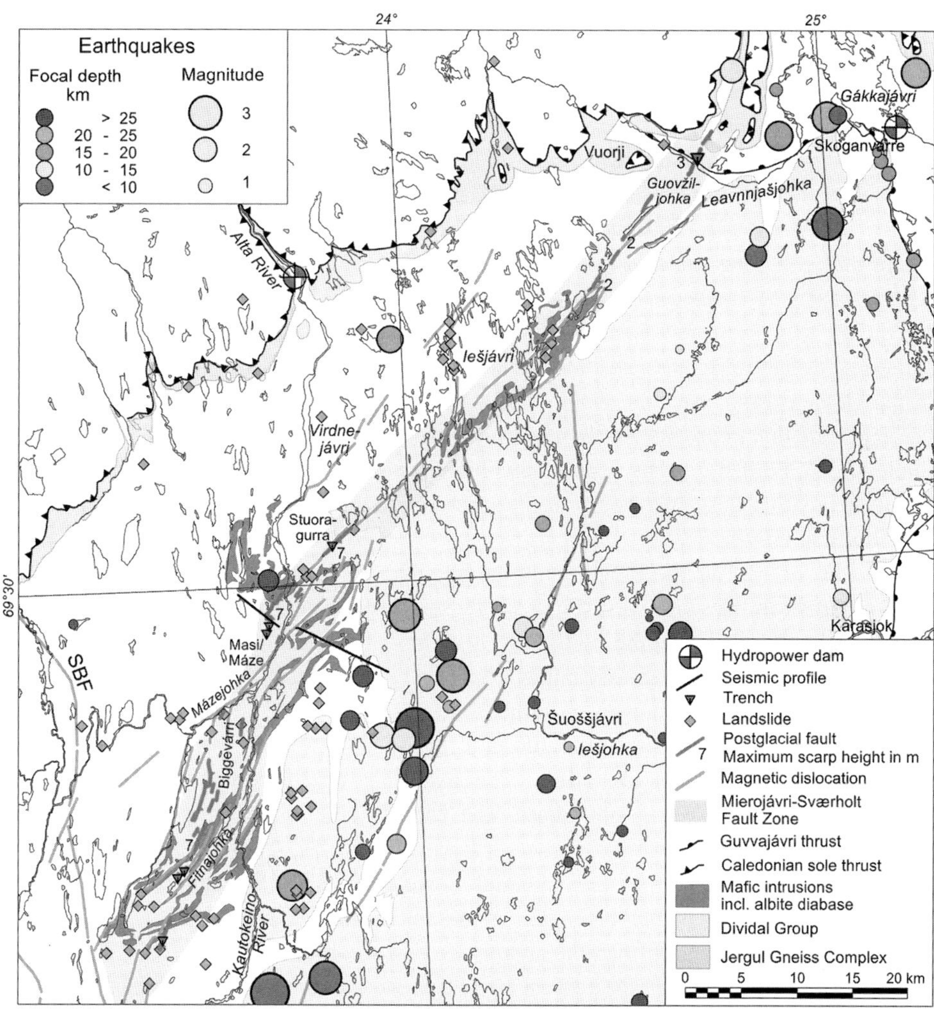

Figure 11.2 Simplified geological map of Finnmarksvidda (from Zwaan, 1985; Henriksen, 1986; Siedlecka, 1987; Solli, 1988; Olesen & Sandstad, 1993; Siedlecka & Roberts, 1996; Siedlecka et al., 2011). The postglacial faults are adapted from Olesen (1988). The 90-km-long Stuoragurra Fault Complex (SFC) consists of three separate fault systems: the Fitnajohka Fault System in the south-west, the Máze Fault System in the central area and the Iešjávri Fault System to the north-east. The SFC occurs within the 4–5 km-wide Mierojávri–Sværholt shear zone (MSSZ), located along the north-western boundary of the Jergul Gneiss Complex (Olesen et al., 1992a). The MSSZ is also characterized by magnetic anomalies produced by the highly magnetic mafic intrusions (diabase, albite diabase and gabbro). An albite diabase south of Masi has been dated by Bingen et al. (2015) to 2,220 ± 7 Ma. The mafic intrusions are partly adapted from the 1:50,000 bedrock maps (Fundal, 1967; Zwaan, 1985; Solli, 1988; Siedlecka, 1987) and partly interpreted from aeromagnetic maps (Nasuti et al., 2015). SBF – Soagŋojávri-Bajášjávri Fault (Olesen & Sandstad, 1993). Evidence of a total of 60 landslides is found within 20 km of the fault scarp. The earthquake epicentres are based on recordings at the ARCES seismic array immediately to the north of Karasjok, in addition to seismic stations in northern Norway and northern Finland (A black and white version of this figure will appear in some formats. For the colour version, please refer to the plate section.).

The MSSZ and SFC are mainly situated within the quartzites of the Masi Formation (Solli, 1988) and the Skuvvanvarri Formation (Siedlecka, 1985). The SFC terminates towards amphibolites above the Guvvajávri thrust (Henriksen, 1986; Skaar, 2014) within the Karasjok Greenstone Belt. The MSSZ terminates in the south-west against the Soagŋojávri–Bajášjávri Fault, which constitutes the eastern fault of the regional Bothnian–Kvænangen Fault Complex (Berthelsen & Marker, 1986; Henkel, 1991; Olesen & Sandstad, 1993). The MSSZ can be traced on the aeromagnetic map below the Caledonian nappes, where the fault zone truncates the Levajok Granulite Belt below the Sværholt Peninsula (Olesen et al., 1990). Olesen (1988) noted that the SFC occurs where the gravity field is low. According to Olesen and Sandstad (1993), this implies that the granitoid basement is situated at a shallow depth beneath the Masi Formation. This is to say that the amphibolites in the Goldenvárri Formation (Solli, 1983), which often occur in the position above the basement, are thin or absent. North of Masi (Figure 11.3), the postglacial fault cross-cuts the Suolovuopmi Formation. The exposed amphibolites in this area are flat-lying (Solli, 1988) and relatively thin (less than 200 m thick) according to the gravity interpretations of Olesen and Sandstad (1993).

The SFC as well as the Pärvie Fault are sub-parallel to the Caledonian front, located 125 km farther to the north-west (Figures 11.1 and 11.2). The SFC seems to die out along the Guovžiljohka valley in the north-east. It is, however, possible that the fault complex extends farther to the north-east but is concealed underneath large boulder fields. The regional MSSZ continues, however, farther to the north-east. The Cryogenian to Cambrian Dividal Group is down-faulted by ~100 m to the south-east along the MSSZ (Townsend et al., 1989; Siedlecka et al., 2011). The overlying Caledonian sole thrust (Silurian Gaissa Thrust) is, however, not offset. The MSSZ represents a long-lived fault zone (Olesen et al., 1992a).

The SFC and the Pärvie, Laisvall and Sorsele faults in Sweden coincide with a physiographic border (Figure 11.2). The mountainous area to the west of the Caledonian front has a higher elevation than the area to the east/south-east. The continental ice sheet was consequently thicker in the eastern area. This would involve more depression during the glacial period and consequently a greater contribution to the subsequent postglacial stress regime. The differential loading of ice across a pre-stressed fault line might consequently be sufficient to cause a reactivation of the faults and weakness zones (Olesen, 1988; Muir Wood, 1989). This process cannot, however, explain the formation of the other postglacial faults in northern Fennoscandia. The young ages obtained from radiocarbon dating of the SFC (Olsen et al., 2020, 2021) are also less compatible with this model.

The SFC cross-cuts till, an esker and other glaciofluvial deposits on Finnmarksvidda (Olesen 1988, 1992b). A total of 60 landslides (Figure 11.2) have occurred within 20 km of the fault scarp (modified from Sletten et al., 2000). Most of these landslides occur in the south. They may be better preserved in this area due to the younger faulting.

The Stuoragurra faulting has produced distinct deformation and fracturing of the host rock. Percussion and core drilling in the Fitnajohka area revealed a dip of ~40° to the SE at a shallow depth of ~100 m (Olesen et al., 1992a,b, 2013; Roberts et al., 1997). The

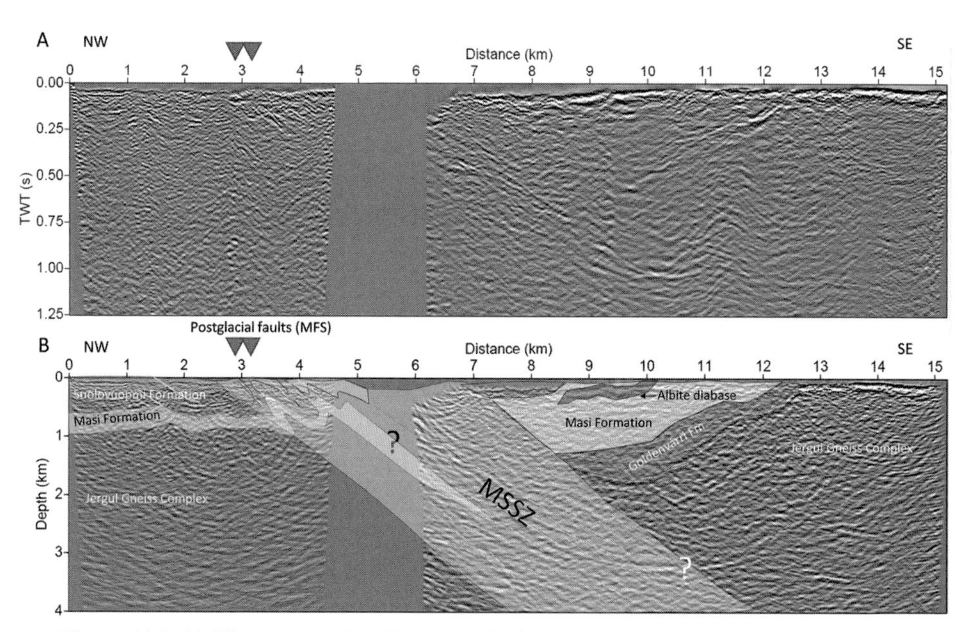

Figure 11.3 A) Dip-moveout (DMO) corrected seismic stack in TWT across the Mierojávri–Sværholt shear zone (MSSZ) and the postglacial Máze Fault System (MFS). B) Migrated and depth-converted version of seismic line shown in (A). The MSSZ and MFS are shown by the pale grey raster and two red lines, respectively. The interpretation is partly based on earlier geological and geophysical studies (Olesen, 1988; Solli, 1988; Olesen et al., 1992a,b; Henderson et al., 2015; Mrope et al., 2019). The seismic profile cuts the MFS and MSSZ at an angle of 70°. The apparent dip of the MSSZ is $\sim$ 40°to the SE. A correction for the oblique crossing of the MSSZ increases the dip angle to 43°. The MFS is located within the regional $\sim$ 4-km-wide MSSZ. The dips of the western and eastern segments of the MFS within the SFC as read from the migrated seismic section are 48° and 59°, respectively. The dip increases to 52° and 65° after correcting for oblique crossing. The two postglacial fault segments seem to merge at a depth of $\sim$ 500 m. The postglacial faults are partly parallel to the axial plane of the folds along the MSSZ (Olesen et al., 1992a,b; Roberts et al., 1997). The flat-lying reflector along the south-eastern margin of the MSSZ at a depth of 3–4 km may represent a tectonic ramp associated with a phase of thrusting or detachment faulting along the fault zone. (A black and white version of this figure will appear in some formats. For the colour version, please refer to the plate section.)

postglacial fault consists of zones of clay minerals, a few centimetres thick, within a 1.5-m-thick interval of fractured quartzite. The clay zones contain kaolinite, vermiculite and smectite (Åm, 1994) and most likely represent a weathered fault gouge. Several 2–3-m-thick zones of breccia occur within a 25-m-wide interval and reveal that the postglacial faults were formed within an old zone of weakness partly coinciding with the margins of deformed Palaeoproterozoic albite diabases. Magnetic modelling of the albite diabase in the vicinity of the drill holes shows a dip of $\sim$ 40° to the SE (Olesen et al., 1992a), consistent with the results from the drilling. Refraction seismic and resistivity profiling (Olesen et al., 1992a) show both low seismic P-wave velocity ($\sim 3,800$ m/s) and resistivity ($\sim 900\ \Omega$m) and indicate a high degree of fracturing. Trenching across the Fitnajohka Fault System was first performed in 1998, in a location that did not include buried organic materials usable for ^{14}C dating (Dehls et al., 2000). The till above the SF was folded, forming a blind thrust. Drilling and geophysical data revealed that the fault has a high groundwater yield (Olesen et al., 1992a, 2004). Dehls et al. (2000) showed that high-pressure groundwater was injected more than 15 m into the till during the main rupture event in Fitnajohka.

Bungum and Lindholm (1997) recorded a large number of earthquakes along the SFC. Focal mechanism solutions for five earthquakes recorded along or close to the SFC indicated that the maximum principal compressive stress, $S_{H\max}$, is oriented approximately NW-SE. The individual focal mechanisms were located at shallow depths south-east of the SFC surface expression. The reverse/oblique nodal planes were oriented so that one plane could be associated with the fault strike for all events; however, σ_H varied from N-S to E-W (averaging to NW-SE).

11.3 Methods

11.3.1 Electrical Resistivity Tomography (ERT)

The 2D electric resistivity tomography method (ERT) was used to study the shallow part of the Máze and Fitnajohka fault systems (Dalsegg & Olesen, 2014). Data were collected using a cable system developed by the Institute of Technology of Lund University (Dahlin, 1993). The system consists of a relay box (Electrode Selector ES10–64 C) and two or four multi-electrode cables. The ABEM Terrameter ABEM-LS (ABEM, 2012) resistivity instrument contains an integrated PC for full control of the data acquisition process and storage of data. In this survey, four cables were used with the gradient electrode configuration and 10 m electrode spacing. The electrical current was split in pulses of 1 s with alternating polarity (Dalsegg & Olesen, 2014). A maximum depth range of about 120 m can be reached with the latter configuration. The resolution decreases with depth, and resistivity data deeper than about 80 m are, from experience, less reliable. All resistivity measurements give an apparent resistivity value that represents a weighted average resistivity resulting from the resistivity of each heterogeneous volume in the surroundings of the measurement points. So that specific resistivity of each part of the heterogeneous investigated volume can be found, the data are inverted. This is done by dividing the profile into blocks, each characterized by specific resistivity values. The resistivity values of the blocks are adjusted following an iterative procedure until the theoretical model fits the measured data. The resistivity of the bedrock, which is obtained by ERT, depends on the porosity, water saturation and resistivity of groundwater in cracks and fracture zones, as well as on the clay content that is produced by weathering and other geological phenomena. Resistivity measurements were inverted using the computer program RES2DINV (Loke, 2010).

11.3.2 Reflection Seismic Data Acquisition and Processing

The seismic data were acquired across the Máze Fault System by the University of Bergen in 2012 (Figure 11.2). The data were acquired using a combination of a snow-streamer and autonomous nodes. Snow-streamer technology (Eiken et al., 1989; Johansen et al., 2011) is an efficient way to do seismic profiling in snow-covered and relatively flat terrain. The seismic source used is a detonating cord which, besides being very efficient and giving a favourable source directivity, has the advantage of leaving no permanent footprint on the ground. Detonating cord was deployed in intervals of 50 m, and each sensor consisted of eight gimballed geophones connected in strings of length 25 m. The seismic data were acquired with 2-ms sampling and 6 s recording time. The snow-streamer recorded 60 traces with offsets from 125 to 1,625 m. The purpose of the autonomous nodes was mainly to record traces with higher offsets and to acquire data in areas with rough terrain where the snow-streamer could not be used. The nodes recorded data from the same sources and with the same type of sensors as used for the snow-streamer. Up to 40 nodes were used simultaneously and two geophone strings were connected to each node.

The data processing followed the usual steps for seismic reflection profiling with trace editing, elevation statics, muting, amplitude corrections, surface consistent deconvolution and various kinds of filtering and noise reduction techniques, followed by velocity analysis, normal moveout corrections and stacking of CMP gathers (traces with a common source/receiver midpoint). In the noise reduction process, velocity filtering (both in the frequency-wavenumber (f-k) and intercept time-ray parameter (t-p) domains) were used to enhance P-wave reflections while reducing surface waves, S waves, ambient noise and incoherent energy. Based on observations of refracted waves, P-wave velocities were determined to be around 6,000 m/s, except for the upper 100–200 m, where velocities could be down to between 4,000 and 5,000 m/s. Dip-moveout (DMO) corrections and migration were also applied, but since the total seismic section is composed of several short line segments, migration artefacts were quite strong.

11.3.3 Recording and Processing of Seismicity

The earthquake locations are based on recordings on the NORSAR ARCES seismic array immediately north of Karasjok, in addition to many other seismic stations in northern Fennoscandia. All earthquake location estimates displayed have been recalculated following a re-analysis of the seismic signals and using the probabilistic, multiple-event, Bayesloc algorithm of Myers et al. (2007). Bayesloc has been demonstrated to improve location estimates for clustered seismic events significantly by calculating corrections to travel-time predictions, implicitly compensating for deficiencies in the velocity model applied (e.g. Gibbons & Kværna, 2017).

11.4 Results

11.4.1 Trenching and Radiocarbon Dating

The 1998 trenching across the Fitnajohka Fault System south of Masi did not reveal any buried organic materials usable for ^{14}C dating. New trenching was carried out in 2018, 2019 and 2020 at six locations (Guovžiljohka, Stuoragurra, Masi and Fitnajohka in

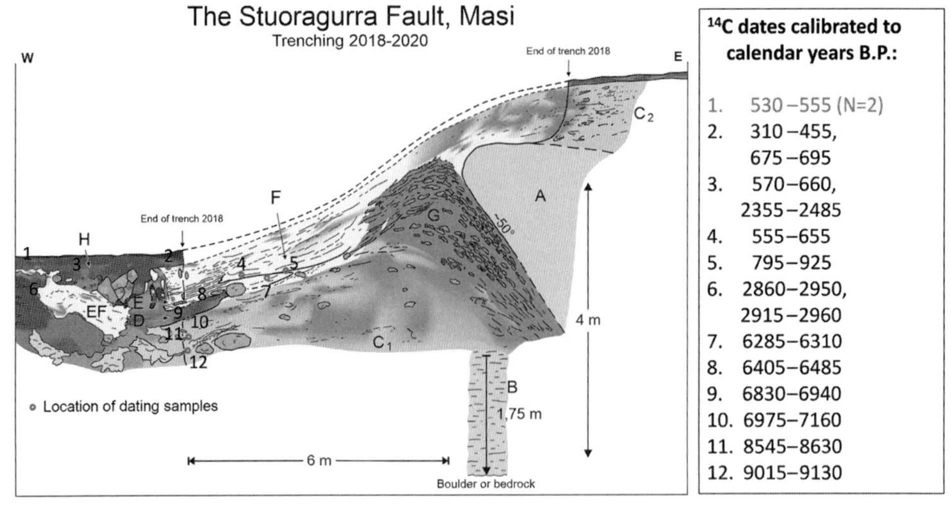

^{14}C dates calibrated to calendar years B.P.:
1. 530–555 (N=2)
2. 310–455, 675–695
3. 570–660, 2355–2485
4. 555–655
5. 795–925
6. 2860–2950, 2915–2960
7. 6285–6310
8. 6405–6485
9. 6830–6940
10. 6975–7160
11. 8545–8630
12. 9015–9130

Figure 11.4 Outline of the trench across the Máze Fault System (within the postglacial SFC) in the southern part of the Juŋkorajeaggi swamp located 1.5 km to the north of Masi (Figure 11.2). The orientation of the section is normal to the fault. The nose of the up-thrown block of bedrock (A) is buried by deformed basal till (C1 and C2) with deformed lacustrine sand (with thin gyttja layers) (E and F) and modern, partly deformed peat (H) on top. B represents glaciolacustrine or glaciofluvial sediments (silt and fine-grained sand). Unit D is buried and deformed peat, and unit G is crushed and fragmented amphibolite injected into the sediment overburden during the fault event. The base layer of the modern peat (H), as well as organic material, mainly plant remains, from the deformed and buried gyttja and peat (from units C1, D, E and F) are sampled for ^{14}C dating. The results are shown in the table to the right. The uncertainties (age interval) represent a 67 per cent confidence interval. (A black and white version of this figure will appear in some formats. For the colour version, please refer to the plate section.)

Figure 11.2) in the Iešjávri, Máze and Fitnajohka fault systems, where basin sediments with peat and gyttja were supposed to have pre-existed and thus predated the fault event. The trenching at Masi revealed gyttja and peat layers (Figure 11.4), which were buried and deformed during the main fault event. ^{14}C dates of macro plant remains from the buried organics and from the base of the undeformed modern surficial peat indicate that the age is less than 600 calibrated years before present (a BP) for the faulting event (Olsen et al., 2021). If true, faulting occurred around 10,000 years after deglaciation and is not a direct result of rapid, initial rebound following deglaciation. In addition, ^{14}C datings from Guovžiljohka and Fitnajohka indicate that the fault events might have happened at different times in all three faults, most recently at the Máze Fault System and slightly before 1,280 a BP at the Fitnajohka Fault System and after 4,000 a BP at Guovžiljohka on the northern part of the Iešjávri Fault System (Olsen et al., 2020). The apparently different ages of the three main fault systems in Finnmark are supported by empirical data from historical surface ruptures reported from other faults where distances more than 3–5 km between fault segments often indicate different rupture ages of each fault system (e.g. Wesnousky, 2008). Different periods of postglacial faulting are also evident from age dating of landslides in northern Finland (Ojala et al., 2018). Radiocarbon age data revealed three episodes of increased landslide formation associated with postglacial faulting, from 9 to 11 ka BP, from 5 to 6 ka BP and from 1 to 3 ka BP.

11.4.2 Geophysical Studies

The seismic profile across the Máze Fault System and the MSSZ in two-way travel-time (TWT) is shown in Figure 11.3a. The interpretation of the migrated version in Figure 11.3b is based on the seismic profile as well as on earlier geological and geophysical studies (Olesen, 1988; Solli, 1988; Olesen et al., 1992a,b; Henderson et al., 2015; Mrope et al., 2019). The SFC is located within the regional ~4-km-wide MSSZ. The dip of the western and eastern segments (marked by red lines in Figure 11.3) of the Máze Fault System is 52° and 65°, respectively. The two postglacial fault segments seem to merge at a depth of ~500 m. An ERT profile across the easternmost segment of the Máze Fault System indicates a water-bearing fracture zone with a dip of ~70° to the SE (Figure 11.5). The results from the seismic and electrical methods are quite consistent.

The flat-lying reflector (marked with a white question mark) immediately to the right of the MSSZ at a depth of ~35 km in Figure 11.3 may represent a tectonic ramp associated with a phase of thrusting or detachment faulting along the fault zone. Alternatively, the flat-lying reflector can originate from a structure located out of the plane. Magnetotelluric (MT) data were acquired along the seismic profile in the summer of 2015 and interpreted by Mrope et al. (2019). The approximately 1-km-wide and vertical low-resistivity zone along the postglacial faults extends to a depth of ~1.3 km. The vertical dip of the low-resistivity zone is most likely caused by a reduced sensitivity of the MT inversion method to the dip of

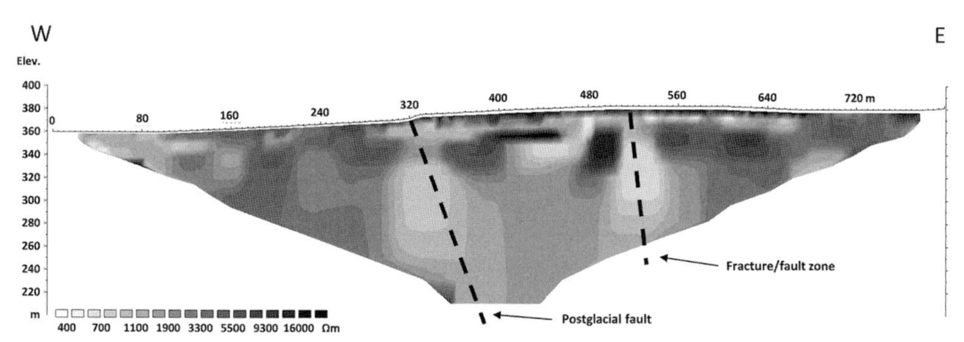

Figure 11.5 Resistivity (ERT) profile located eastwards from the southern shore of lake Nillorjávri (3 km to the north of Masi). The profile is situated close to the interception of the seismic profile and the postglacial fault in Figure 11.2. The profile is located perpendicular to the postglacial fault segments. The two dashed black lines indicate possible faults, represented by low-resistivity zones. The westernmost anomaly dipping at an angle of ~70° to the south-east coincides with the easternmost postglacial fault segment in Figure 11.2. The ERT method is not very sensitive to the dip of electrical conductors, and the estimated dip has therefore some uncertainty. The low-resistivity zone to the far west on the profile coincides with a swamp and the westernmost postglacial fault segment. The uppermost high-resistivity layer represents dry Quaternary overburden (till). Modified from Dalsegg & Olesen (2014).

electrical conductors. One vertical low-resistivity zone also occurs on either side of the SFC within the MSSZ. Mrope et al. (2019) also interpreted an older processed version of the seismic line and produced a combined interpretation with the MT resistivity magnetic and gravity modelling.

The estimated moment magnitudes of the Nordmannvikdalen Fault and the Fitnajohka, Máze and Iešjávri fault systems within the SFC are shown in Table 11.1. The moment magnitudes are calculated from fault offset and length utilizing formulas by Wells and Coppersmith (1994) and Moss and Ross (2011). The estimates vary considerably, from 6.4 to 8.0 on Richter's scale. The estimates using the Moss and Ross (2011) equation for reverse faults give the highest magnitudes. We conclude that the three earthquakes associated with the formation of the three fault systems along the SFC were on the order of 6.4 to 8.0.

Approximately 80 earthquakes were registered along the postglacial faults between 1991 and 2019 (Figures 11.2 and 11.6). The maximum moment magnitude was 4.0 and, due to the proximity of sensitive seismic stations, the catalogue is likely complete down to around magnitude 1.5. Focal depths varied between 5 and 32 km. Based on the joint probability distributions for event hypocentres obtained, most events were located with a lateral and vertical uncertainty of 4 km or less. On the assumption of a dip of $\sim 43^\circ$ of the SFC and the MSSZ, most of the earthquakes occurred at a distance of less than 10 km from these structures. Some of the earthquakes might represent aftershocks of a large-magnitude earthquake that occurred about 600 years before today. Aftershock sequences can, according to seismological observations and models (Stein & Liu, 2009), last for centuries in intraplate regions.

11.5 Discussion

The Stuoragurra Fault Complex consists of more than 30 fault segments (Figure 11.2). It is not known whether they formed during a single large earthquake or during several smaller events. Trenching of one of the southernmost segments in the Fitnajohka area in 1998 (Dehls et al., 2000) reveals one seismic event deforming the lodgement till into a large 5-m-high fold above a blind thrust below the hanging wall block. Comparing the SFC with other postglacial faults in northern Fennoscandia (Lagerbäck & Sundh, 2008; Palmu et al., 2015) shows evidence of at least three separate seismic events, i.e. the Fitnajohka Fault System in the south-west, the Máze Fault System in the central area and the Iešjávri Fault System to the north-east. Fault parameters are listed in Table 11.1. Each of the three fault systems consists of 8 to 14 segments. The three fault systems constitute the SFC and occur within the Palaeoproterozoic Mierojávri–Sværholt shear zone (Olesen et al., 1992a). Some of the drainage systems in the area, especially north-east of Iešjávri, are controlled by the same weakness zones. Landslides and erosion along streams and rivers may have modified the fault escarpments following their formation. The escarpments in the two southernmost fault systems have maximum scarp heights of 7 m and occur partly in an en échelon pattern. The distance between the Fitnajohka and Máze fault systems is 7 km, whereas the distance

Table 11.1. *Summary of properties of postglacial faults in Norway.*

Fault/fault system	Length (km)	Number of segments	Trend	Maximum scarp height (m)	Height/ length	Estimated maximum displacement	Magnitude from fault length	Magnitude from max. displacement
Nordmann-vikdalen	1.5	3	NW-SE	1.5	0.0010	2.3	6.0 [1]	6.6 [1] 6.8 [2]
Fitnajohka	21	14	NE-SW	7	0.0004	9	6.5 [1]	6.9 [1] 8.0 [2]
Máze	14	8	NE-SW	7	0.0005	9	6.4 [1]	6.9 [1] 8.0 [2]
Iešjávri	34	9	NE-SW	3	0.0001	4	6.8 [1]	6.8 [1] 7.4 [2]

The major faults in the Precambrian of Finnmarksvidda are NE–SW-trending and reverse. The Nordmannvikdalen Fault in northern Troms is a smaller normal fault trending perpendicular to most of the reverse faults in northern Fennoscandia. The scarp height-to-length ratio is generally less than 0.001. Moment magnitudes are calculated from fault offset and length utilizing formulas by [1]Wells & Coppersmith (1994) and [2] Moss & Ross (2011).

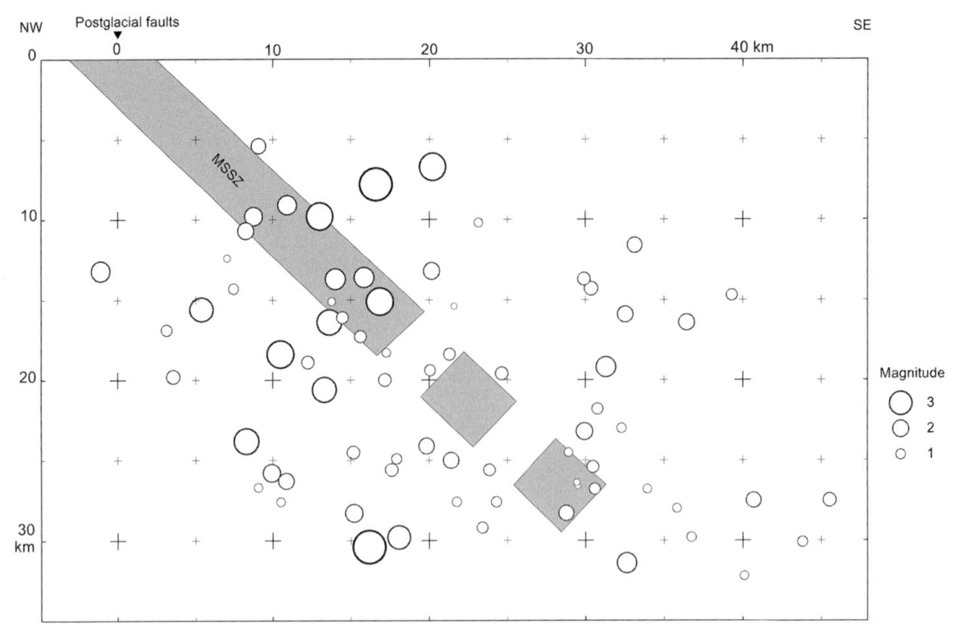

Figure 11.6 Focal depths and magnitudes of the earthquakes in Figure 11.2 projected into a profile perpendicular to the postglacial Stuoragurra Fault Complex. The fault is located at coordinate 0 km along the profile. The focal depths vary between 5 and 32 km. The earthquakes show a significant spread along the extrapolated MSSZ dipping at ~43° to the SE. Most of the earthquakes occurred, however, at a distance of less than 10 km from this structure. Some of the earthquakes might represent aftershocks of a large-magnitude earthquake that occurred ~600 years before today.

between the Máze and the Iešjávri fault systems is 12 km. We assume that the Iešjávri Fault System is more or less continuous at the floor of the Iešjávri lake, since postglacial faults occur on either side of the lake (Figure 11.2). Wesnousky (2008) suggests that earthquakes do not typically jump from one segment to another if the distance between the segments is more than 3–5 km, basing his conclusion on a total of 37 historical earthquakes with observed surface ruptures. A spacing of more than 5 km between surface ruptures therefore likely indicates that the earthquakes occurred in separate events. This would support the idea that the three fault systems along the SFC might have ruptured independently. Mattila et al. (2019) arrived at similar conclusions for the postglacial faults in northern Finland.

The nine segments of the Iešjávri Fault System occur with a spacing of 1–5 km and have a maximum scarp height of 2–3 m. The length of the individual segments is maximum 2.5 km. We cannot conclude with the present-day knowledge whether they are formed in one large event or during two or several smaller events.

Large earthquakes in mid-continents such as Australia, Eastern United States, North China and Northwestern Europe show complex patterns in time and space that do not fit existing seismotectonic models (Clark et al., 2012; Calais et al., 2016; Liu & Stein, 2016). Individual faults tend to fall into thousands of years quiescence following a cluster of ruptures, whereas large earthquakes seem to migrate between widespread faults. This model can also be applied to the postglacial faults within the Lapland fault province. The end-glacial faults in the Bay of Bothnia area, e.g. the Lansjärv, Röjnoret and Burträsk faults (Lagerbäck & Sundh, 2008), could have transferred stress to neighbouring faults and disturbed conditions on distant faults in Finnmark and Finnish Lapland. This model can explain the much younger age of the SFC and the northernmost postglacial faults in Finland (Ojala et al., 2019) compared with the Gulf of Bothnia faults in Sweden. It is also compatible with the GIA modelling by Steffen et al. (2014a,b) who predicted faulting along a 45° dipping weakness zone in a pre-stressed region beneath the centre of the ice sheet. In such a system, widespread mid-continental faults can accommodate slow tectonic loading from the far field (including ridge push). Calais et al. (2016) and Craig et al. (2016) argue that the lithosphere can store elastic strain over long timescales, the release of which may be triggered by rapid, local and transient stress changes caused by erosion, fluid migration or ice loading, resulting in the intermittent occurrence of intraplate seismicity. Liu and Stein (2016) and Calais et al. (2016) conclude that this paradigm shift would make some commonly used concepts such as recurrence intervals and characteristic earthquakes inadequate in many mid-continental areas. This has significant bearings on the seismic hazard estimates for Fennoscandia and implies that large earthquakes on the order of magnitude 7 can occur today.

11.6 Conclusions

1. The 90-km-long Stuoragurra Fault Complex (SFC) dips at an angle of 30–75° to the SE and can be traced on reflection seismic data to a depth of $\sim$500 m.
2. The fault segments of the SFC can be grouped into three major systems with approximately 7- and 12-km-wide gaps without any apparent faulting. They were most likely formed during three or more separate earthquakes.
3. The fault systems extend most likely to a depth of several kilometres because of the large rupture lengths ($\geq$14 km).
4. The postglacial faults occur within the $\sim$4–5-km-wide Precambrian Mierojávri-Sværholt shear zone (MSSZ).
5. The deep seismic profiling shows that the regional MSSZ dips at an angle of $\sim$43° to the SE and can be traced down to a depth of $\sim$3 km on the reflection seismic profile.
6. The dip of the reverse postglacial faults as observed in trenches is 50–55°, implying a maximum reverse displacement of approximately 9 m, which together with the 14–21-km length of the fault systems indicates associated earthquakes with a moment magnitude of 6.4–8.0.
7. Approximately 80 earthquakes were registered along the fault between 1991 and 2019. The maximum moment magnitude was 4.0. Focal depths varied between 5 and 32 km. Most of the earthquakes occurred at a distance of less than 10 km from the extrapolated MSSZ at depth.

8. Trenching of different sections of the fault and radiocarbon dating of buried and deformed organic material reveal a late Holocene age (between about 600 and 4,000 years before present at the three separate fault systems). If true, faulting occurred ~10 ka after deglaciation and is not a direct result of rapid, initial rebound following deglaciation.
9. A total of 60 landslides have been observed in relatively flat terrain on aerial photographs and LiDAR data. They occurred within 20 km from the fault scarps. Most of them occurred in the south and may be well preserved here because of the young faulting.
10. The present study indicates that the expected maximum magnitude of future earthquakes in Fennoscandia is about 7.
11. There is a need for seismic, palaeoseismic and geodetic studies extending beyond areas where postglacial faulting occurred and for hazard assessments to reduce the possible effect of large earthquakes dominating the short-term earthquake records.

Acknowledgements

The acquisition of the snow-streamer seismic profile in 2012 was funded by the Mineral Resources in North Norway (MINN) program initiated by the Norwegian government. Helge Johnsen at the University of Bergen was the leader for the field work during the acquisition of the snow-streamer seismic profile. Alf Kristian Nilsen, Espen Lehn-Nilsen and Malin Waage, University of Bergen, Jomar Gellein and Geir Viken, NGU, Karstein Rød, Bergen Oilfield Services, Roar Helge Iversen, Dolphin Geophysical, Nils Per Nilsen Sara, Isak Anders Nilsen Eira, Klemet Anders Bals, Mikkel Per Bongo, Eskil Hermansen Hætta and Lars Johan Larsen Sara participated in the field campaign. Klemet Johansen Hætta, Nils Erik Eriksen, Nils Peder Eira, Joar Førster, Svein Are Eira and Joar Henning Hætta excavated the 2018, 2019 and 2020 trenches in Masi, Fitnajohna, Guovžiljohka and Stuoragurra. Poznan Radiocarbon Laboratory, Poland and Beta Analytic, USA, carried out AMS ^{14}C dating of macro plant remains. Chistopher Juhlin and Antti E. K. Ojala made a thorough review of the manuscript and made suggestions towards its improvement. We express our gratitude to these persons, companies and institutions. Earthquake relocations were calculated using the Bayesloc program, available from www-gs.llnl.gov/nuclear-threat-reduction/nuclear-explosion-monitoring/bayesloc.

References

ABEM (2012). *ABEM Terrameter LS. Instruction Manual.* ABEM 20120109, based on release 1.10. ABEM, Sweden.

Åm, M. (1994). *Mineralogisk og petrologisk karakterisering av vitrings/sleppemateriale fra Stuoragurraforkastningen Finnmark* [Mineralogical and Petrological Characterization of Weathering/Drag-Along Material from the Stuoragurra Fault, Finnmark]. MSc thesis, Norwegian University of Science and Technology, 102 pp.

Bakken, A. J. H. (1983). *Nordmannvikdalen kvartærgeologi og geomorfologi* [Quaternary Geology and Geomorphology of Nordmanvikdalen]. MSc thesis, University of Oslo, 126 pp.

Berthelsen, A. and Marker, M. (1986). 1.9–1.8 Ga old strike-slip megashears in the Baltic Shield, and their plate tectonic implications. In D. A. Galson and S. Mueller, eds., *The European Geotraverse, Part 2. Tectonophysics*, **128**(3–4), pp. 163–181, doi.org/10.1016/0040-1951(86)90292-1.

Bingen, B., Solli, A., Viola, G. et al. (2015). Geochronology of the Palaeoproterozoic Kautokeino Greenstone Belt, Finnmark, Norway: tectonic implications in a Fennoscandia context. *Norwegian Journal of Geology*, **95**, 365–396, doi.org/10.17850/njg95-3-09.

Bungum, H. and Lindholm, C. (1997). Seismo- and neotectonics in Finnmark, Kola Peninsula and the southern Barents Sea. Part 2: seismological analysis and seismotectonics. *Tectonophysics*, **270**, 15–28, doi.org/10.1016/S0040-1951(96)00139-4.

Calais, E., Camelbeeck, T., Stein, S., Liu, M. and Craig, T. J. (2016). A new paradigm for large earthquakes in stable continental plate interiors. *Geophysical Research Letters*, **43**, 10,621–10,637, doi.org/10.1002/2016GL070815.

Clark, D., McPherson, A. and Van Dissen, R. (2012). Long-term behaviour of Australian stable continental region (SCR) faults. *Tectonophysics*, **566**, 1–30, doi.org/10.1016/j.tecto.2012.07.004.

Craig, T. J., Calais, E., Fleitout, L., Bollinger, L. and Scotti, O. (2016). Evidence for the release of long-term tectonic strain stored in continental interiors through intraplate earthquakes, *Geophysical Research Letters*, **43**, 6826–6836, doi.org/10.1002/2016GL069359.

Dahlin, T. (1993). *On the Automation of 2D Resistivity Surveying for Engineering and Environmental Applications.* PhD thesis, Department of Engineering Geology, Lund Institute of Technology, Lund University.

Dalsegg, E. and Olesen, O. (2014). *Resistivitetsmålinger ved Masi, Fitnajohka and Riednajávre og implikasjoner for malmleting* [Resistivity Measurements at Masi, Fitnajohka and Riednajávre and Implications for Ore Exploration]. NGU Report 2014.021, 28 pp.

Dehls, J., Olesen, O., Olsen, L. and Blikra, L. H. (2000). Neotectonic faulting in northern Norway; the Stuoragurra and Nordmannvikdalen postglacial faults. *Quaternary Science Reviews*, **19**, 1445–1460, doi.org/10.1016/S0277-3791(00)00073-1.

Eiken, O., Degutsch, M., Riste, P. and Rød, K. (1989). Snowstreamer: an efficient tool in seismic acquisition. *First Break*, **7**(9), 374–378, doi.org/10.3997/1365-2397.1989021.

Fundal, E. (1967). *En undersøkelse i det prekambriske Biggevarre område i Finnmark, Nord-Norge med særlig henblikk på de såkalde albitdiabasers geologi og petrografi* [An Investigation in the Precambrian Biggevarre Area in Finnmark, Northern Norway with Special Attention to the Geology and Petrography of the So-Called Albitdiabas]. NGU Report 680, 81 pp.

Gibbons, S. J. and Kværna, T. (2017). Illuminating the seismicity pattern of the October 8, 2005, M = 7.6 Kashmir earthquake aftershocks. *Physics of the Earth and Planetary Interiors*, **270**, 1–8, doi.org/10.1016/j.pepi.2017.06.008.

Henderson, I. H. C., Viola, G. and Nasuti, A. (2015). A new tectonic model for the Kautokeino Greenstone Belt, northern Norway, based on high resolution airborne magnetic data and field structural analysis and implications for mineral potential. *Norwegian Journal of Geology*, **95**, 339–363, doi.org/10.17850/njg95-3-05.

Henkel, H. (1991). Magnetic crustal structures in northern Fennoscandia. In P. Wasilewski and P. Hood, eds., *Magnetic Anomalies – Land and Sea. Tectonophysics*, **192**, pp. 57–79, doi.org/10.1016/0040-1951(91)90246-O.

Henriksen, H. (1986). Bedrock map Iddjajavri 2034 II M 1:50 000, preliminary edition. Geological Survey of Norway, Trondheim.

Johansen, T. A., Ruud, B. O., Bakke, N. E. et al. (2011). Seismic profiling on Arctic glaciers. *First Break*, **29**(2), 29–35.

Lagerbäck, R. and Sundh, M. (2008). *Early Holocene Faulting and Paleoseismicity in Northern Sweden*. Geological Survey of Sweden Research Paper, Series C, Vol. 836, 80 pp.

Liu, M. and Stein, S. (2016). Mid-continental earthquakes: spatiotemporal occurrences, causes, and hazards. *Earth-Science Reviews*, **162**, 364–386, doi.org/10.1016/j .earscirev.2016.09.016.

Loke, M. H. (2010). *RES2INV ver. 3.59. Geoelectrical Imaging 2D & 3D. Instruction Manual*. 151 pp. www.geoelectrical.com.

Lund, B., Schmidt, P. and Hieronymus, C. (2009). *Stress Evolution and Fault Instability during the Weichselian Glacial Cycle*. SKB Technical Report TR-09-15, Swedish Nuclear Fuel and Waste Management Co., Stockholm, Sweden, 106 pp.

Mattila, J., Ojala, A. E. K., Ruskeeniemi, T. et al. (2019). Evidence of multiple slip events on postglacial faults in northern Fennoscandia. *Quaternary Science Reviews*, **215**, 242–252, doi.org/10.1016/j.quascirev.2019.05.022

Mikko, H., Smith, C. A., Lund, B., Ask, M. V. S. and Munier, R. (2015). LiDAR-derived inventory of post-glacial fault scarps in Sweden. *GFF*, **137**, 334–338, doi.org/10 .1080/11035897.2015.1036360.

Moss, E. S. and Ross, Z. E. (2011). Probabilistic fault displacement hazard analysis for reverse faults. *Bulletin of the Seismological Society of America*, **101**, 1542–1553, doi .org/10.1785/0120100248.

Mrope, F. M., Becken, M., Ruud, B. O. et al. (2019). *Magnetotelluric 2D Inversion and Joint Interpretation of MT, Seismic, Magnetic and Gravity Data from Masi, Kautokeino Municipality, Finnmark*. NGU Report, 2019.009, 64 pp.

Muir Wood, R. (1989). Extraordinary deglaciation reverse faulting in northern Fennoscandia. In S. Gregersen and P. W. Basham, eds., *Earthquakes at North-Atlantic Passive Margins: Neotectonics and Postglacial Rebound*. Kluwer Academic Publishers, Dordrecht, pp. 141–173.

Myers, S. C., Johannesson, G. and Hanley, W. (2007). A Bayesian hierarchical method for multiple-event seismic location. *Geophysical Journal International*, **171**, 1049–1063, doi.org/10.1111/j.1365-246X.2007.03555.x.

Nasuti, A., Roberts, D., Dumais, M.-A. et al. (2015). New high-resolution aeromagnetic and radiometric surveys in Finnmark and North Troms: linking anomaly patterns to bedrock geology and structure. *Norwegian Journal of Geology*, **95**, 217–243, doi.org/ 10.17850/njg95-3-10.

Ojala, A. E. K., Mattila, J., Ruskeeniemi, T. et al. (2017). Postglacial seismic activity along the Isovaarae–Riikonkumpu fault complex. *Global and Planetary Change*, **157**, 59–72, doi.org/10.1016/j.gloplacha.2017.08.015.

Ojala, A. E. K., Markovaara-Koivisto, M., Middleton, M. et al. (2018). Dating of paleo-landslides in western Finnish Lapland. *Earth Surface Processes and Landforms*, **43**, 2449–2462, doi.org/10.1002/esp.4408.

Ojala, A. E. K., Mattila, J., Hämäläinen, J. and Sutinen, R. (2019). Lake sediment evidence of paleoseismicity: timing and spatial occurrence of late- and postglacial earthquakes in Finland. *Tectonophysics*, **771**(228227), doi.org/10.1016/j.tecto.2019.228227.

Olesen, O. (1988). The Stuoragurra Fault, evidence of neotectonics in the Precambrian of Finnmark, northern Norway. *Norsk Geologisk Tidsskrift*, **68**, 107–118.
Olesen, O. and Sandstad, J. (1993). Interpretation of the Proterozoic Kautokeino Greenstone Belt, Finnmark, Norway from combined geophysical and geological data. *Geological Survey of Norway Bulletin*, **425**, 43–64.
Olesen, O., Roberts, D., Henkel, H., Lile, O. B. and Torsvik, T. H. (1990). Aeromagnetic and gravimetric interpretation of regional structural features in the Caledonides of West Finnmark and Northern Troms, north Norway. *Geological Survey of Norway Bulletin*, **419**, 1–24.
Olesen, O., Henkel, H., Lile, O.B., Mauring, E. and Rønning, J. S. (1992a). Geophysical investigations of the Stuoragurra postglacial fault, Finnmark, northern Norway. *Journal of Applied Geophysics*, **29**, 95–118, doi.org/10.1016/0926-9851(92)90001-2.
Olesen, O., Henkel, H., Lile, O.B. et al. (1992b). Neotectonics in the Precambrian of Finnmark, northern Norway. *Norsk Geologisk Tidsskrift*, **72**, 301–306.
Olesen, O., Blikra, L. H., Braathen, A. et al. (2004). Neotectonic deformation in Norway and its implications: a review. *Norwegian Journal of Geology*, **84**, 3–34.
Olesen, O., Brönner, M., Ebbing, J. et al. (2010). New aeromagnetic and gravity compilations from Norway and adjacent areas – methods and applications. In B. A. Vining and S. C. Pickering, eds., *Petroleum Geology: From Mature Basins to New Frontiers. Proceedings of the 7th Petroleum Geology Conference*. Petroleum Geology Conference Series 7, Geological Society of London, pp. 559–586, doi.org/10.1144/0070559.
Olesen, O., Bungum, H., Lindholm, C. et al. (2013). Neotectonics, seismicity and contemporary stress field in Norway – mechanisms and implications. In L. Olsen, O. Fredin and O. Olesen, eds., *Quaternary Geology of Norway*. Geological Survey of Norway Special Publication, 13, pp. 145–174.
Olsen, L., Olesen, O., Dehls, J. and Tassis, G. (2018). Late-/postglacial age and tectonic origin of the Nordmannvikdalen Fault, northern Norway. *Norwegian Journal of Geology*, **98**, 483–500, doi.org/10.17850/njg98-3-09.
Olsen, L., Olesen, O. and Høgaas, F. (2020). Dating of the Stuoragurra Fault at Finnmarksvidda, northern Norway. In H. A. Nakrem and A. M. Husås, eds., *34th Nordic Geological Winter Meeting January 8th–10th 2020, Oslo, Norway*. Abstracts and Proceedings of the Geological Society of Norway, **1**, pp. 157–158.
Olsen, L., Olesen, O., Høgaas, F. and Tassis, G. (2021). A part of the Stuoragurra postglacial fault complex, at Máze in N-Norway, is less than 600 yrs old. In H. A. Nakrem and A. M. Husås (eds.), Vinterkonferansen 2021, Digital, January 6–8, 2020. Abstracts and Proceedings of the Geological Society of Norway, **1**, p. 55.
Palmu, J.-P., Ojala, A. E. K., Ruskeeniemi, T., Sutinen, R. and Mattila, J. (2015). LiDAR DEM detection and classification of postglacial faults and seismically-induced landforms in Finland: a paleoseismic database. *GFF*, **137**, 344–352, doi.org/10.1080/11035897.2015.1068370.
Roberts, D., Olesen, O. and Karpuz, M. R. (1997). Seismo- and neotectonics in Finnmark, Kola Peninsula and the southern Barents Sea; part 1, geological and neotectonic framework. *Tectonophysics*, **270**, 1–13, doi.org/10.1016/S0040-1951(96)00173-4.
Siedlecka, A. (1985). Geology of the Iešjávri-Skoganvarre area, northern Finnmarksvidda, North Norway. *Geological Survey of Norway Bulletin*, **403**, 103–112.
Siedlecka, A. (1987). Berggrunnskart Iešjávri; 1934 II, foreløpig utgave, M 1:50 000 [Iešjávri bedrock map; 1934 II, preliminary edition, M 1:50,000]. Norges geologiske undersøkelse, Trondheim.

Siedlecka, A., Davidsen, B., Rice, A. H. N. and Townsend, C. (2011). Berggrunnskart; Skoganvarri 2034 IV, M 1:50 000, revidert foreløpig utgave [Bedrock map; Skoganvarri 2034 IV, M 1:50,000, revised preliminary edition]. Norges geologiske undersøkelse, Trondheim.

Siedlecka, A. and Roberts, D. (1996). Finnmark Fylke. Berggrunnsgeologi Finnmark Fylke M 1:500 000 [Finnmark Fylke; bedrock map M 1:500 000]. Norges geologiske undersøkelse, Trondheim.

Skaar, J. A. Å. (2014). *3D Geophysical and Geological Modelling of the Karasjok Greenstone Belt*. PhD thesis, Norwegian University of Science and Technology, 170 pp.

Sletten, K., Olsen, L. and Blikra, L.H. (2000). Slides in low-gradient areas of Finnmarksvidda. In J. Dehls and O. Olesen, eds., *Neotectonics in Norway, Annual Technical Report 1999*. NGU Report 2000.001, 41–42.

Smith, C. A., Grigull, S. and Mikko, H. (2018). Geomorphic evidence for multiple surface ruptures of the Merasjärvi "postglacial fault," northern Sweden. *GFF*, **140**, doi.org/10.1080/11035897.2018.1492963

Smith, C. A., Sundh, M. and Mikko, H. (2014). Surficial geologic evidence for early Holocene faulting and seismicity. *International Journal of Earth Sciences*, **103**, 1711–1724, doi.org/10.1007/s00531-014-1025-6.

Solli, A. (1983). Precambrian stratigraphy in the Masi area, Southwestern Finnmark, Norway. *Geological Survey of Norway Bulletin*, **380**, 97–105.

Solli, A. (1988). Masi, 1933 IV – berggrunnsgeologisk kart – M 1:50,000 [Masi, 1933 IV – Map of bedrock geology – M 1:50,000]. Norges geologiske undersøkelse, Trondheim.

Steffen, R., Steffen, H., Wu, P. and Eaton, D. W. (2014a). Stress and fault parameters affecting fault slip magnitude and activation time during a glacial cycle. *Tectonics*, **33**, doi.org/10.1002/2013TC003450.

Steffen, R., Wu, P., Steffen, H. and Eaton, D. W. (2014b). On the implementation of faults in finite-element glacial isostatic adjustment models. *Computers & Geosciences*, **62**, 150–159, doi.org/10.1016/j.cageo.2013.06.012.

Stein, S. and Liu, M. (2009). Long aftershock sequences within continents and implications for earthquake hazard assessment. *Nature*, **462**, 87–89, doi.org/10.1038/nature08502.

Sutinen, R., Hyvönen, E., Middleton, M. and Ruskeeniemi, T. (2014). Airborne LiDAR detection of postglacial faults and Pulju moraine in Palojärvi, Finnish Lapland. *Global and Planetary Change*, **115**, 24–32, doi.org/10.1016/j.gloplacha.2014.01.007.

Tolgensbakk, J. and Sollid, J. L. (1988). Kåfjord, kvartærgeologi og geomorfologi 1:50,000, 1634 II [Kåfjord, Quaternary geology and geomorphology, 1:50,000, 1634 II]. Geografisk Institutt, University of Oslo.

Townsend, C., Rice, A. H. N. and Mackay, A. (1989). The structure and stratigraphy of the southwestern portion of the Gaissa Thrust Belt and the adjacent Kalak Nappe Complex, N Norway. In R. A. Gayer, ed., *The Caledonide Geology of Scandinavia.* Graham & Trotman, London, pp. 111–126.

Vestøl, O., Ågren, J., Steffen, H., Kierulf, H. and Tarasov, L. (2019). NKG2016LU: a new land uplift model for Fennoscandia and the Baltic Region. *Journal of Geodesy*, **93**(9), 1759–1779, doi.org/10.1007/s00190-019-01280-8.

Wells, D. L. and Coppersmith, K. J. (1994). New empirical relationships among magnitude, rupture length, rupture area, and surface displacement. *Bulletin of the Seismological Society of America*, **84**, 974–1002.

Wesnousky, S. G. (2008). Displacement and geometrical characteristics of earthquake surface ruptures: issues and implications for seismic-hazard analysis and the process of earthquake rupture. *Bulletin of the Seismological Society of America*, **98**, 1609–1632, doi.org/10.1785/0120070111.

Wu, P., Johnston, P. and Lambeck, K. (1999). Postglacial rebound and fault instability in Fennoscandia. *Geophysical Journal International*, **139**(3), 657–670, doi.org/10.1046/j.1365-246x.1999.00963.x.

Zwaan, K. B. (1985). Berggrunnskart Suolovuopmi 1934 III, M 1:50 000, foreløpig utgave [Map of bedrock geology Suolovuompi 1934 III, M 1:50,000, preliminary edition]. Norges geologiske undersøkelse, Trondheim.

12

Glacially Induced Faults in Sweden

The Rise and Reassessment of the Single-Rupture Hypothesis

COLBY A. SMITH, HENRIK MIKKO AND SUSANNE GRIGULL

ABSTRACT

Despite early studies indicating fault rupture both before and after deglaciation, it has long been hypothesized that glacially induced faults in Fennoscandia ruptured only once. The now widespread availability of high-resolution digital elevation models allows for testing this hypothesis by examining cross-cutting relationships between the scarps and both glacial and postglacial landforms. Although not widespread, such cross-cutting relationships indicate that segments of the Merasjärvi, Laino and Pärvie faults ruptured at least twice. The timing of the Merasjärvi ruptures is unknown; the Laino ruptures occurred both before and after deglaciation, and at least one of the Pärvie ruptures is postglacial.

Additionally, it can be demonstrated that parallel segments of the Pärvie and Lansjärv faults ruptured at different times, despite being only a few kilometres from each other. Given these results, the single rupture hypothesis must be reassessed for the high-relief scarps in northern Sweden, but it may still hold true for some of the low-relief scarps.

12.1 Introduction

Glacially induced faults have been recognized in Sweden for over 40 years (Lundqvist & Lagerbäck, 1976). Despite geologic evidence to the contrary, a hypothesis developed that each glacially induced fault scarp formed as the result of a single rupture (Muir Wood, 1989). If true, such ruptures produced scarps two to three times higher than scarps that formed during historical earthquakes. Nevertheless, this hypothesis gained in popularity as more supporting data became available.

This chapter reviews the discovery of glacially induced faults in Sweden and the development of the single rupture hypothesis, before discussing evidence, both new and old, that refutes the hypothesis.

12.2 Background

Lundqvist and Lagerbäck (1976) used cross-cutting relationships observed in the field to demonstrate that some segments of the Pärvie Fault ruptured after the Late Weichselian deglaciation and other segments ruptured prior to deglaciation. Subsequently, an aerial photographic survey located the Lainio, Merasjärvi, Lansjärv, Röjnoret and Burträsk faults (Figure 12.1; Table 12.1) (Lagerbäck, 1978). The cross-cutting relationships between the glacial geomorphology and the fault scarps were visible in aerial photographs and demonstrated that some of the scarps were younger than the glacial features. Such relationships

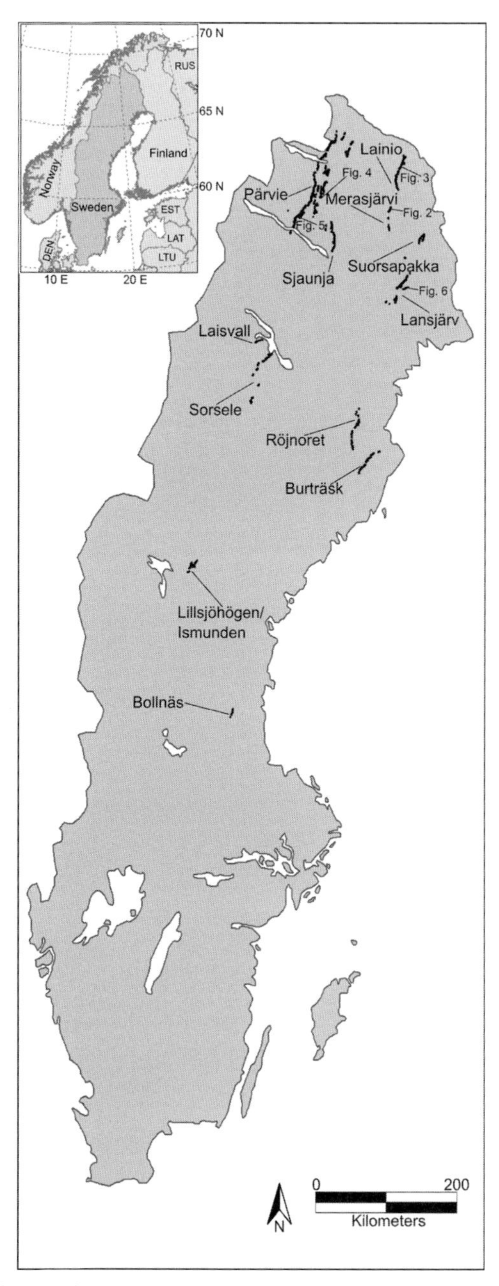

Figure 12.1 Locations and names of glacially induced fault scarps in Sweden.

Table 12.1. *Summary of major glacially induced fault scarps in Sweden.*

Fault	Discontinuous length (km)	Strike	Primary scarp aspect	General height (m)	Maximum height (m)	References
Bollnäs	10	NS	E	2–4	5	Smith et al., 2014
Burträsk	45	NE-SW	NW	5–10	15	Lagerbäck & Sundh, 2008
Ismunden	20	NE-SW	SE	3.5	6	Berglund & Dahlström, 2015; Mikko et al., 2015
Lainio	50	NE-SW	NW	10–20	30	Lagerbäck & Sundh, 2008
Laisvall	11	NE-SW	NW	5	5	Mikko et al., 2015
Lansjärv	50	NE-SW	SE	5–10	20	Lagerbäck, 1990, 1992
Lillsjöhögen	6	N-S	E	1–4	8	Berglund & Dahlström, 2015; Mikko et al., 2015
Merasjärvi	30	NE-SW	NW	10–15	20	Lagerbäck & Sundh, 2008; Mikko et al., 2015; Smith et al., 2018
Pärvie	150	NE-SW	NW	5–10	35	Lundqvist & Lagerbäck, 1976; Lagerbäck & Witschard, 1983
Röjnoret	60	N-S	W	5–10	10	Lagerbäck & Sundh, 2008
Sjaunja	40	N-S	E	1–2	3	Mikko et al., 2015; this study
Sorsele	40	NE-SW	SE	1–2	2	Ransed & Wahlroos, 2007; Mikko et al., 2015; this study
Suorsapakka	17	NE-SW	NW	2–4	4	Mikko et al., 2015; this study

led to use of the term 'postglacial faults', even though some of the scarps clearly predated deglaciation.

Refining the age of fault ruptures beyond merely naming them as postglacial required stratigraphic information (Lagerbäck, 1990, 1992). The focus of the study was on the Lansjärv Fault because it lies below the highest postglacial shoreline. Here, the widespread presence of stratified postglacial sediments proved useful for two purposes. First, faulting is easier to confirm in stratified sediments as opposed to diamictons, and second, the presence of postglacial sediments provides a chronometer for determining when the rupture occurred.

Numerous excavations across multiple segments of the fault produced the same result (Lagerbäck, 1990, 1992). The glacial sediments are faulted, but the overlying postglacial sediments are undisturbed. Thus, the faulting occurred after deglaciation, about 10,500 years ago, but before the area rose above the water in the Baltic Basin, about 10,200 years ago (Lindén et al., 2006). Since the excavations were near the highest postglacial shoreline, fault rupture was constrained to within a few hundred years after deglaciation (Lagerbäck, 1990, 1992). The stratigraphy was interpreted to indicate a single rupture. Additionally, Arvidsson (1996) argued that the modelled stress release during single ruptures would have been within normal values. Taken together, these results contributed to the hypothesis that glacially induced fault scarps reached their full lengths and heights as the result of single ruptures.

During the same period that glacially induced faults were being mapped and excavated in northern Sweden, a paradigm shift in the interpretation of the region's glacial geology was underway. Following extensive mapping, coring and excavating it was demonstrated that much of the glacial geomorphology believed to date to the Late Weichselian deglaciation was in fact older (Lagerbäck, 1988a,b; Lagerbäck & Robertsson, 1988). Widespread geologic evidence indicates that Early and Middle Weichselian glacial landforms have been preserved beneath at least one cold-based ice sheet with little or no erosive effect (Sigfúsdóttir, 2013). Landforms created during the Late Weichselian deglaciation often overlie the older landforms. The implications of these findings for glacially induced faults are twofold. First, a scarp that cross-cuts a glacial landform is not necessarily younger than the Late Weichselian deglaciation (~10 ka). Rather, it may only be younger than the early Weichselian (~80 ka). Second, if glacial landforms composed of unconsolidated sediment could be preserved beneath ice sheets, then fault scarps could be preserved beneath ice sheets.

The discovery of preserved glacial landscapes in northern Sweden, however, did not change the interpretation of glacially induced faults. Lagerbäck and Sundh (2008) point to a lack of evidence indicating multiple ruptures and the stratigraphic evidence from Lansjärv to suggest single ruptures occurred on all glacially induced faults during or shortly after the Late Weichselian deglaciation.

More recently, the availability of high-resolution digital elevation models (DEM) derived from light detection and ranging (LiDAR) data has revolutionized the way geomorphology is studied in Fennoscandia (Johnson et al., 2015). High-resolution shaded relief imagery was used in a nationwide mapping effort that refined the catalogue of scarps and scarp segments by both adding and removing features (Figure 12.1; Table 12.1) (Smith et al., 2014; Mikko et al., 2015).

DEMs also allow for detailed examination of geomorphology and cross-cutting relationships that can refine the interpretations of both the number of fault ruptures and the relative timing of the ruptures. Although based on field observations, Smith et al. (2018) use LiDAR-derived imagery to illustrate multiple ruptures of the Merasjärvi Fault. The current study expands on this work to explore evidence of multiple ruptures, differential timing of rupture on different segments and a more prolonged period of postglacial seismicity than previously understood.

12.3 Multiple Ruptures of the Same Fault Segments

12.3.1 Merasjärvi Fault

The west-facing Merasjärvi Fault scarp lies in northern Sweden. It strikes NE-SW for about 8 km, reaches a maximum height of 15 m and dams Merasjärvi Lake (Figure 12.1, Table 12.1) (Lagerbäck & Sundh, 2008).

Along the northernmost segment of the Merasjärvi Fault, multiple streams flow from the upthrown eastern block to the downthrown western block (Figure 12.2). Two of these incised courses contain terraces that lie 5 m from the top of the fault scarp. The terraces are subsequently truncated by the lowermost 8 m of the fault scarp (Smith et al., 2018). The authors interpret these relationships as indicating two fault ruptures. The first rupture occurred prior to the fluvial incision that allowed the streams to flow along the terrace treads. The second rupture truncated these fluvial surfaces and led to the incision that created the terraces. The timing of these ruptures is not well constrained because the fault lies within an area of preserved landforms (Lagerbäck, 1988b; Lagerbäck & Robertsson, 1988) that likely date to the Middle Weichselian (Sigfúsdóttir, 2013). Thus, both fault ruptures are younger than the Middle Weichselian.

12.3.2 Lainio Fault

The west-facing Lainio Fault scarp lies about 25 km north of the Merasjärvi Fault scarp. It is nearly continuous for 50 km and is generally between 10 and 20 m high (Figure 12.1, Table 12.1) (Lagerbäck & Sundh, 2008). The northern part of the Lainio Fault is known to have formed prior to deglaciation. Lagerbäck and Witschard (1983) noted that a Late Weichselian meltwater channel followed the Lainio Fault for several kilometres, implying that at least part of the scarp existed prior to deglaciation (Figure 12.3a).

The present study, however, finds evidence of postglacial faulting as well. A truncated terrace, grading to a higher ice-dammed water level, exists along the same segment of the Lainio Fault where the meltwater channel follows the strike of the scarp. The terrace lies about 2 m below the top of the scarp and is cross-cut by a 2.6-m-high fault scarp (Figure 12.3B–D). In addition to the terrace tread being cross-cut by the fault, the fluvial erosional

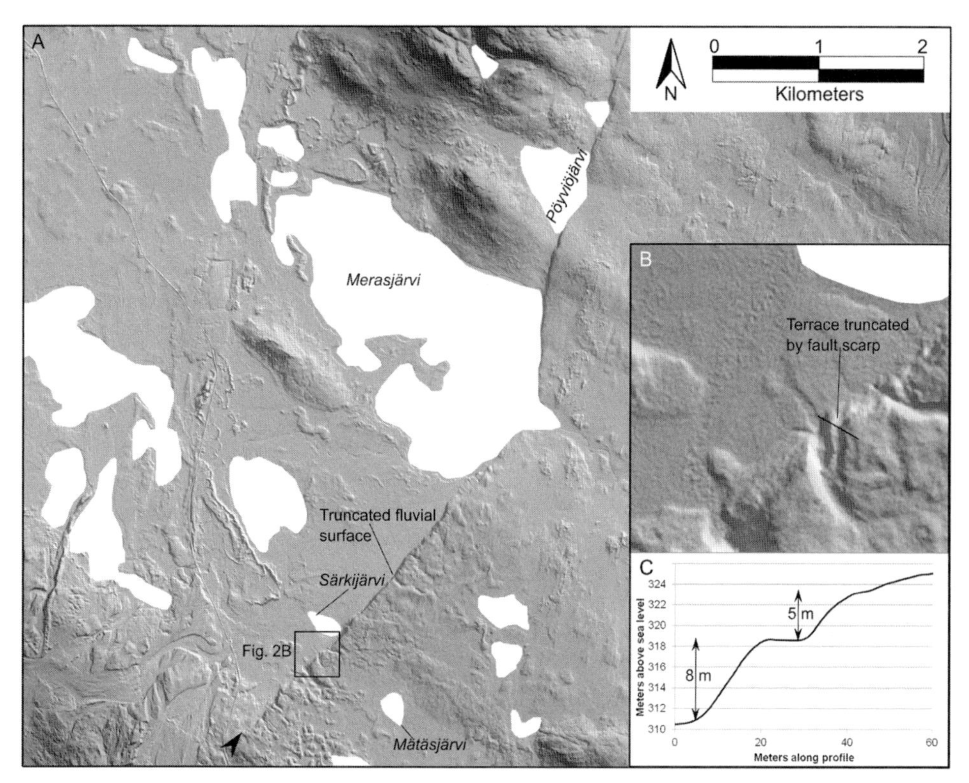

Figure 12.2 A) LiDAR-derived shaded relief imagery (Lantmäteriet, 2020) provides an overview of the northern segment of the Merasjärvi Fault scarp. The scarp segment is marked with an arrow at its southern end. The centre of the image lies at 67.5482°N and 21.9728°E. B) A fluvial terrace lies below the top of the fault scarp and is subsequently truncated by later movement of the fault. The line shows the location of the elevation profile. C) An elevation profile across the scarp and fluvial terrace.

scarp above the terrace is also cross-cut. Given the active ice flow in this area, at least during deglaciation, we suggest that both the terrace and the scarp that cross-cuts it are postglacial.

The geomorphology of the Lainio example is nearly identical to that of the Merasjärvi example. A key difference at the Lainio location is that the ice-dammed lake has drained, and the down-dropped portion of the terrace is visible in the imagery. At the Merasjärvi location, the down-dropped portion of the terrace lies below lake level. The geomorphic relationships visible at Lainio, however, were not created by the base-level lowering associated with the drainage of the ice-dammed lake. If terraces had formed during this event, they would lie between the upper terrace fragment and the current stream channel, not immediately downstream from the terrace fragment. An abrupt scarp, coinciding with an active fault, that cross-cuts both the tread of a fluvial terrace and the erosional scarp above it indicates that the fault ruptured subsequent to the formation of the terrace tread and erosional scarp.

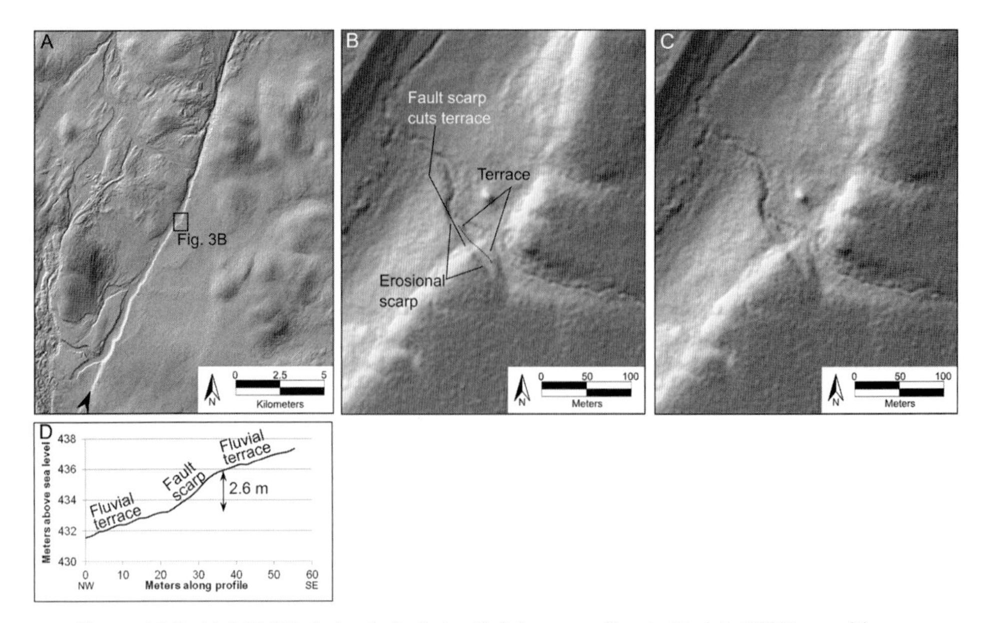

Figure 12.3 A) LiDAR-derived shaded relief imagery (Lantmäteriet, 2020) provides an overview of the Lainio Fault scarp. The centre of the image lies at 68.0931°N and 25.5719°E. B) A fluvial terrace lies below the top of the scarp and has been cross-cut by later movement on the fault. The dashed line shows the location of the profile. C) Uninterpreted shades relief imagery as in (B). D) An elevation profile along the fluvial terrace and over the fault scarp. (A black and white version of this figure will appear in some formats. For the colour version, please refer to the plate section.)

12.3.3 Pärvie Fault

The Pärvie Fault System lies in north-western Sweden and strikes generally NE-SW for 155 km with scarp heights generally between 3 and 10 m (Figure 12.1, Table 12.1) (Lagerbäck & Sundh, 2008). When mapped using LiDAR imagery, the Pärvie System consists of more than 200 individual segments (Mikko et al., 2015). The more continuous westernmost fault scarp faces west, and most of the subsidiary segments to the east face east.

An east-facing segment of the Pärvie Fault includes a truncated fluvial terrace (Figure 12.4). The terrace lies about 1.5 m below the top of the scarp and is subsequently cross-cut by a 2.5-m-high fault scarp. The geomorphology of the Pärvie location is nearly identical to that of the Merasjärvi location. The only difference being that the down-dropped portion of the terraces is below peat at the Pärvie location and below lake level at the Merasjärvi location. The scarp cuts Late Weichselian streamlined landforms formed by ice flow from WSW, indicating that at least part of the scarp height is younger than deglaciation. However, preserved landforms formed by an earlier Weichselian ice flow from NW lie within 10 km of the fault segment. Thus, the erosive work done by the Late Weichselian ice sheet was rather limited on a regional scale.

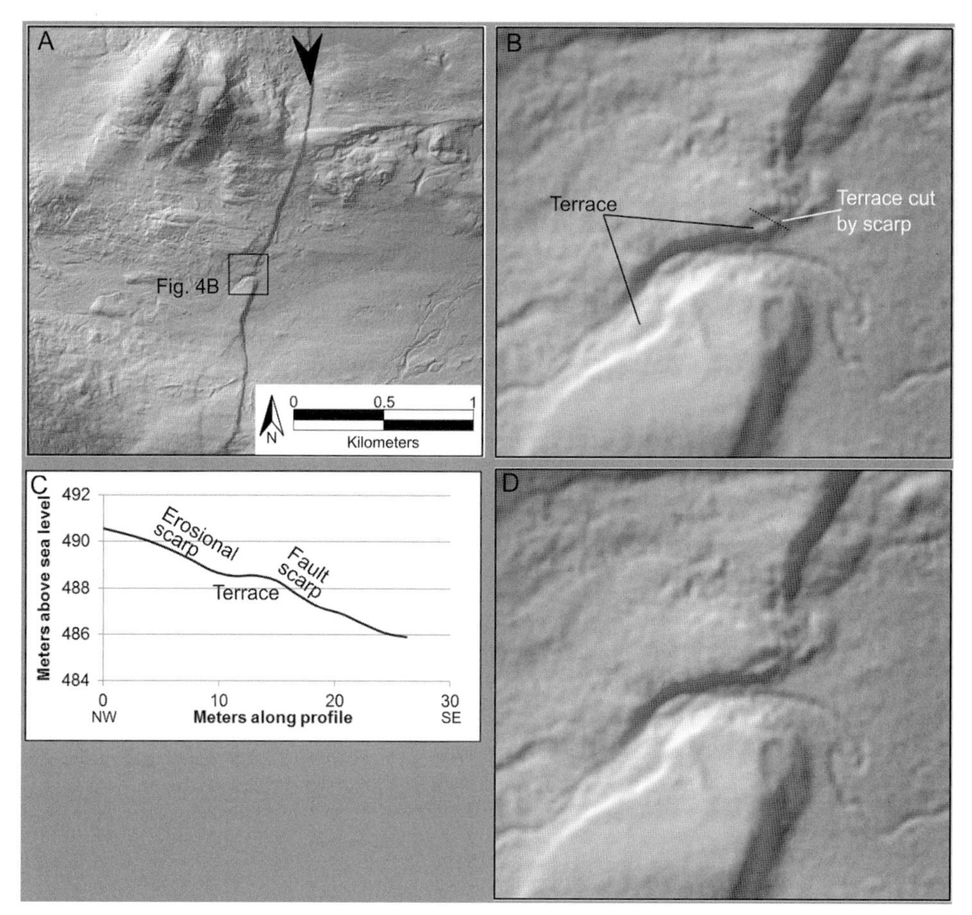

Figure 12.4 A) LiDAR-derived shaded relief imagery (Lantmäteriet, 2020) provides an overview of a segment of the Pärvie Fault scarp. Glacially streamlined landforms created by eastward flowing ice are visibly cross-cut by the fault. The centre of the image lies at 67.7930°N and 19.7029°E. B) A fluvial terrace lies below the top of the fault scarp and is subsequently truncated by later movement of the fault. The dashed line shows the location of the profile. C) An elevation profile across the scarp and the fluvial terrace. D) Uninterpreted shade relief imagery as in (B). (A black and white version of this figure will appear in some formats. For the colour version, please refer to the plate section.)

12.4 Different Ages of Ruptures on Different Segments within the Same Fault System

12.4.1 Pärvie Fault

Towards the southern end of the Pärvie Fault, two parallel scarp segments, separated by 3.5 km, strike towards the NE (Figure 12.5A). The west-facing western scarp clearly cross-cuts ice-marginal channels that date to the Late Weichselian deglaciation (Figure 12.5B). The east-facing eastern scarp does not cross-cut a N-S-trending esker (Figure 12.5C). Due to the presence of preserved landscapes and the lack of streamlined landforms in the area, it

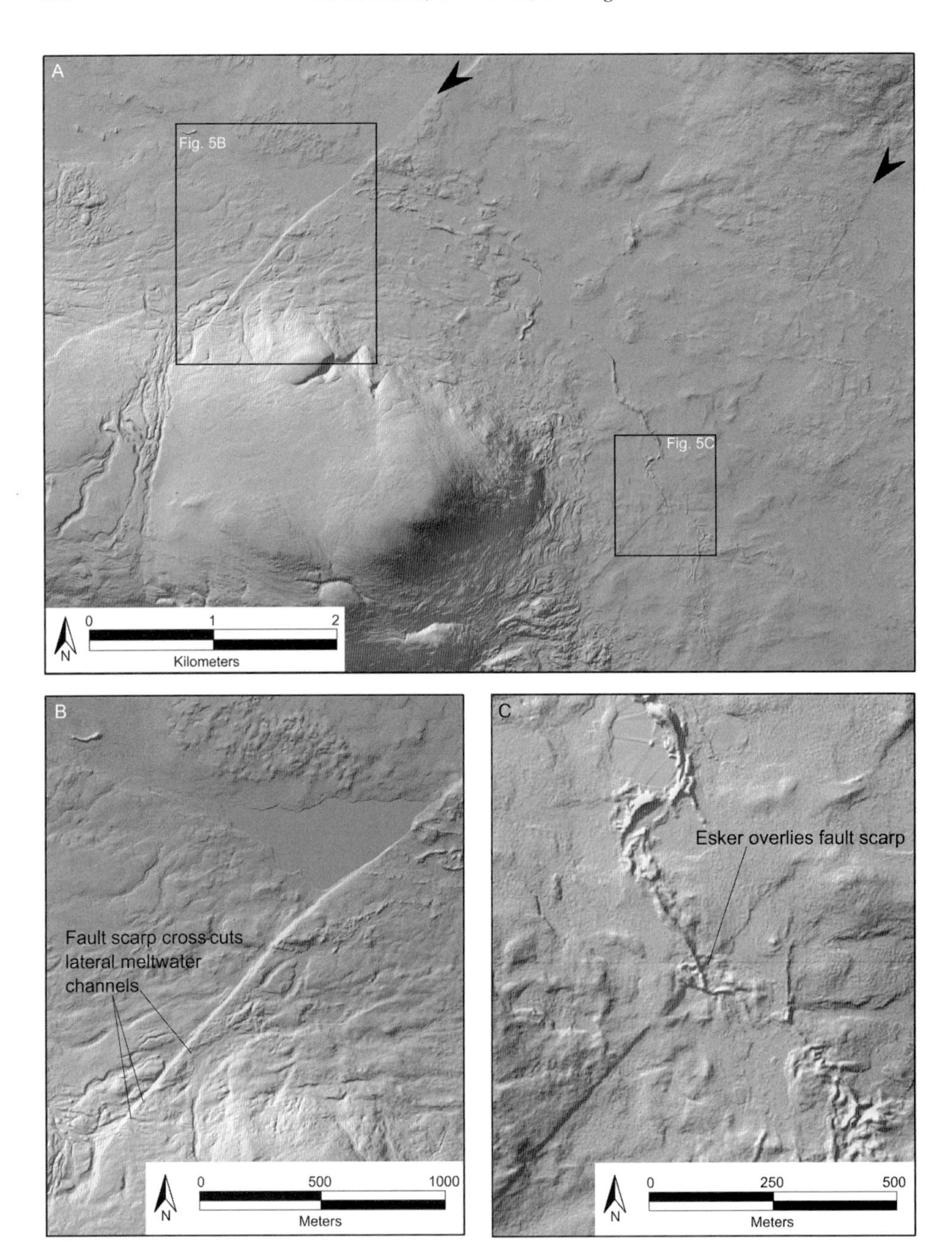

Figure 12.5 A) LiDAR-derived shaded relief imagery (Lantmäteriet, 2020) provides an overview of two parallel segments of the Pärvie Fault. The centre of the image lies at 67.5591°N and 18.9193°E. B) The western fault scarp cross-cuts ice marginal channels, indicating that faulting occurred after deglaciation. C) The esker overlying the fault scarp indicates that faulting occurred prior to deglaciation.

is impossible to determine the age of the esker. Nevertheless, the eastern scarp existed beneath at least one ice sheet and the western scarp is younger than the Late Weichselian, indicating that the faults ruptured at different times. Other examples like this exist along the Pärvie Fault and have been known for some time (Lagerbäck & Witschard, 1983).

12.4.2 Lansjärv Fault

The Lansjärv Fault System lies in north-eastern Sweden and strikes NNE-SSW for about 50 km (Figure 12.1, Table 12.1). The majority of the fault scarp segments are west-facing and up to 10 m in height. At one location, the scarp height reaches 20 m (Lagerbäck & Sundh, 2008).

The different ages of rupture along different fault segments is demonstrated best within the Lansjärv Fault System. Excavations at Molberget revealed faulted glacial sediments but undisturbed postglacial littoral sediments (Lagerbäck & Sundh, 2008), which indicates that faulting occurred after deglaciation but before emergence (Figure 12.6). According to shore displacement curves in Norrbotten, local deglaciation occurred about 10,500 years ago, and the elevation of the Molberget trench (175 m above sea level) emerged above the waters of the Baltic Basin about 10,200 years ago (Lindén et al., 2006). Thus, fault rupture occurred between these two dates.

Ten kilometres SW of Molberget lies a scarp segment not recognized in aerial photos (Lagerbäck & Sundh, 2008) but mapped by Mikko et al. (2015) using LiDAR-derived imagery (Figure 12.6). The scarp clearly cross-cuts raised shorelines and thus occurred after emergence. The lowest cross-cut shoreline lies at 124 m above sea level. Emergence occurred about 9,100 years ago (Lindén et al., 2006), and faulting occurred after that. At Lansjärv, the period of postglacial seismicity is not limited to a few hundred years after deglaciation. Rather, it lasted at least 1,000 years.

12.5 Discussion

Of the thirteen fault systems mapped by Mikko et al. (2015) and listed in Table 12.1, evidence exists for multiple ruptures on segments within four of them. Multiple ruptures on the same segment are generally found along particularly high scarp segments. Along the segments where the truncated terraces are located, the Merasjärvi, Lainio and Pärvie faults have maximum scarp heights of about 20, 25 and 10 m, respectively. Given that the maximum vertical displacement during a historical earthquake is 10 m (Plafker, 1965), it is not surprising that these scarp segments formed from multiple ruptures. While some of these ruptures occurred after the retreat of the Late Weichselian ice sheet, we hypothesize that some, perhaps most, of the displacement is not associated with the most recent deglaciation. Such results would be consistent with recent work in Finland, where stratigraphic investigations across fault scarps have indicated multiple ruptures, some of which are pre-Late Weichselian (Mattila et al., 2019; Ojala et al., 2019).

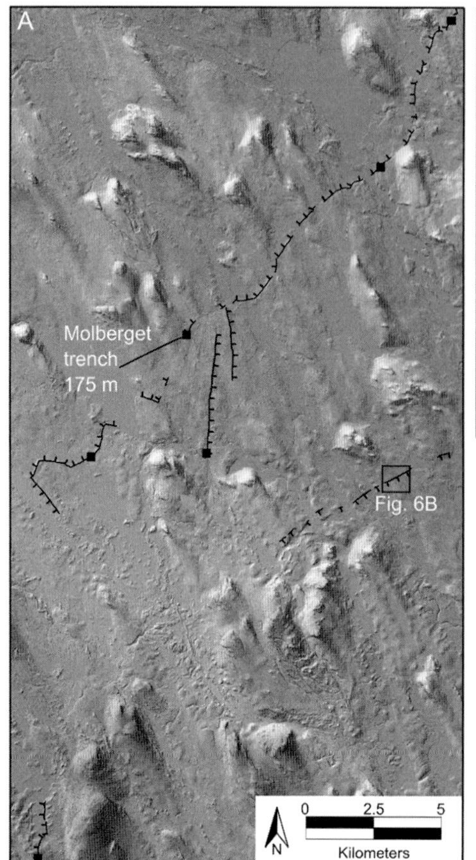

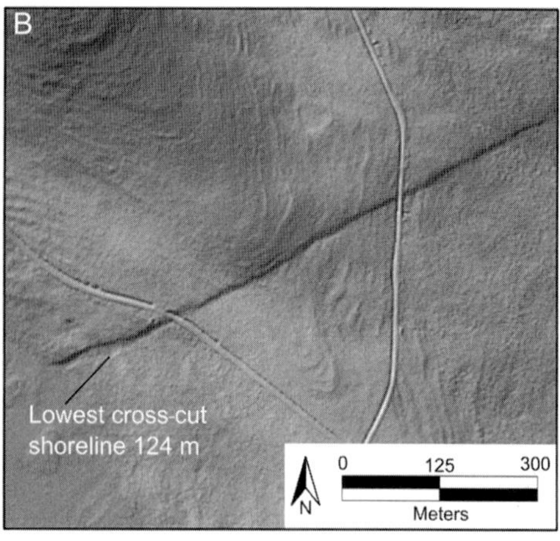

Figure 12.6 A) LiDAR-derived shaded relief imagery (Lantmäteriet, 2020) provides an overview of the Lansjärv Fault. The scarps mapped by Mikko et al. (2015) are shown. The locations of trenches across the fault, which indicated fault rupture after deglaciation but prior to emergence above the waters of the Baltic Basin, are shown as black squares (Lagerbäck & Sundh, 2008). The centre of the image lies at 66.5749°N and 22.1674°E. B) A segment of the Lansjärv Fault cross-cuts raised beaches at 124 m above sea level, indicating fault rupture after emergence above the waters of the Baltic Basin. Thus, fault rupture occurred more than 1,000 years after deglaciation. (A black and white version of this figure will appear in some formats. For the colour version, please refer to the plate section.)

This hypothesis is based on the fact that the sum of the fault heights associated with the terraces is consistently less than the maximum height of the scarp segment. For example, along the Lainio Fault, the scarp rises 2 m above the terrace and the terrace is cut by a 2.6-m scarp. The sum of these values, 4.6 m, is much less than the maximum scarp height of 25 m. If the terrace is assumed to be postglacial, then there is potentially 20 m of vertical displacement prior to deglaciation. This is consistent with the scarp existing prior to deglaciation in order to direct the meltwater channel (Lagerbäck & Sundh, 2008). Additionally, if each rupture produces a vertical displacement similar to the 2.6 m documented by the cross-cut terrace, then 7–8 ruptures are required for the scarp to reach its deglacial height of 20 m. While some of this displacement may have occurred beneath the retreating Late Weichselian ice sheet, it is reasonable to hypothesize that some of it is older. If the Lainio Fault is truly 'glacially induced', then this displacement may date to earlier interstadials or interglacials. There is, however, no existing geological evidence to support this.

Different ages of ruptures along different segments of the same fault has been known since the discovery of the Pärvie Fault and the fact that it cross-cuts some but not all glacial features (Lundqvist & Lagerbäck, 1976). The initial thinking was that the single event occurred while some areas were beneath the retreating Late Weichselian ice sheet and other areas had been only recently deglaciated. This was a reasonable explanation when it was believed that each ice sheet reset the regional geomorphology. However, given the widespread presence of preserved landscapes in northern Sweden an alternative hypothesis exists. The fault segments that do not cross-cut glacial landforms could be preserved features that predate the Late Weichselian.

Finally, the absence of evidence of multiple ruptures along more than two-thirds of the named faults does not necessarily mean they only ruptured once. Streams crossing the scarps are relatively rare, and the geomorphology adjacent to low-gradient streams is often obscured by postglacial deposits, such as peat. Additionally, multiple fault ruptures beneath the waters of the Baltic Basin would not create incised stream courses and associated terraces across the four faults that lie below the highest postglacial shoreline. Similarly, different ages of ruptures on different segments is only recorded in geomorphic settings with landforms of different ages.

12.6 Conclusions

1. In addition to the Merasjärvi Fault, segments within the Lainio Fault and Pärvie Fault System have ruptured to the surface at least twice.
2. Adjacent segments within the Pärvie and Lansjärv fault systems ruptured at different times.
3. Although it may still hold for low-relief scarps, the single rupture hypothesis needs to be reassessed for many of the high-relief fault scarps in northern Sweden.
4. It is hypothesized that some of the relief on these faults is not associated with the Late Weichselian deglaciation.

References

Arvidsson, R. (1996). Fennoscandian earthquakes: whole crustal rupturing related to postglacial rebound. *Science*, **274**(5288), 744–746, doi.org/10.1126/science.274.5288.744.

Berglund, M. and Dahlström, N. (2015). Postglacial fault scarps in Jämtland, central Sweden. *GFF*, **137**, 339–343, doi.org/10.1080/11035897.2015.1036361.

Johnson, M. D., Fredin, O., Ojala, A. E. K. and Peterson, G. (2015). Unraveling Scandinavian geomorphology: the LiDAR revolution. *GFF*, **137**(4), 245–251, doi.org/10.1080/11035897.2015.1111410.

Lagerbäck, R. (1978). Neotectonic structures in northern Sweden. *Geologiska Föreningens i Stockholm Förhandlingar*, **100**(3), 263–269, doi.org/10.1080/11035897809452533.

Lagerbäck, R. (1988a). Periglacial phenomena in the wooded areas of Northern Sweden – relicts from the Tärendö Interstadial. *Boreas*, **17**, 487–499, doi.org/10.1111/j.1502-3885.1988.tb00563.x.

Lagerbäck, R. (1988b). The Veiki moraines in northern Sweden – widespread evidence of an Early Weichselian deglaciation. *Boreas*, **17**, 469–486, doi.org/10.1111/j.1502-3885.1988.tb00562.x.

Lagerbäck, R. (1990). Late Quaternary faulting and paleoseismicity in northern Fennoscandia with particular reference to the Lansjärv area, Northern Sweden. *GFF*, **112**, 333–354, doi.org/10.1080/11035899009452733.

Lagerbäck, R. (1992). Dating of Late Quaternary faulting in northern Sweden. *Journal of the Geological Society*, **149**(2), 285–291, doi.org/10.1144/gsjgs.149.2.0285.

Lagerbäck, R. and Robertsson, A.-M. (1988). Kettle holes – stratigraphical archives for Weichselian geology and palaeoenvironment in northernmost Sweden. *Boreas*, **17**, 439–468, doi.org/10.1111/j.1502-3885.1988.tb00561.x.

Lagerbäck, R. and Sundh, M. (2008). *Early Holocene Faulting and Paleoseismicity in Northern Sweden*. Research Paper C 836. Geological Survey of Sweden.

Lagerbäck, R. and Witschard, F. (1983). *Neotectonics in Northern Sweden – Geological Investigations*. SKBF Technical Report 83-58, Stockholm, 70 pp.

Lantmäteriet (2020). Produkt beskrivning: GSD-Höjddata, grid 2+. Dokumentversion 2.7 [Product description: GSD elevation data, grid 2+. Document version 2.7]. Lantmäteriet, Gävle, 10 pp.

Linde´n, M., Möller, P., Björck, S. and Sandgren, P. (2006). Holocene shore displacement and deglaciation chronology in Norrbotten, Sweden. *Boreas*, **35**(1), 1–22, doi.org/10.1111/j.1502-3885.2006.tb01109.x.

Lundqvist, J. and Lagerbäck, R. (1976). The Pärve Fault: a late-glacial fault in the Precambrian of Swedish Lapland. *Geologiska Föreningens i Stockholm Förhandlingar*, **98**, 45–51, doi.org/10.1080/11035897609454337.

Mattila, J., Ojala, A. E. K., Ruskeeniemi, T. et al. (2019). Evidence of multiple slip events on postglacial faults in northern Fennoscandia. *Quaternary Science Reviews*, **215**, 242–252, doi.org/10.1016/j.quascirev.2019.05.022.

Mikko, H., Smith, C. A., Lund, B., Ask, M. V. S. and Munier, R. (2015). LiDAR-derived inventory of post-glacial fault scarps in Sweden. *GFF*, **137**, 334–338, doi.org/10.1080/11035897.2015.1036360.

Muir Wood, R. (1989). Extraordinary deglaciation reverse faulting in northern Fennoscandia. In S. Gregersen and P. W. Basham, eds., *Earthquakes at North-Atlantic Passive Margins: Neotectonics and Postglacial Rebound*. Kluwer, Dordrecht, pp. 141–173.

Ojala, A. E. K., Mattila, J., Ruskeeniemi, T. et al. (2019). Postglacial reactivation of the Suasselkä PGF complex in SW Finnish Lapland. *International Journal of Earth Sciences*, **108**(3), 1049–1065, doi.org/10.1007/s00531-019-01695-w.

Plafker, G. (1965). Tectonic deformation associated with the 1964 Alaska earthquake. *Science*, **148**(3678), 1675–1687, doi.org/10.1126/science.148.3678.1675.

Ransed, G. and Wahlroos, J.-E. (2007). Map of Quaternary deposits 24H Sorsele, scale 1:100 000. Geological Survey of Sweden, K42.

Sigfúsdóttir, T. (2013). *A Sedimentological and Stratigraphical Study of Veiki Moraine in Northernmost Sweden*. Dissertations in Geology at Lund University.

Smith, C. A., Grigull, S. and Mikko, H. (2018). Geomorphic evidence of multiple surface ruptures of the Merasjärvi "postglacial fault," northern Sweden. *GFF*, **140**(4), 318–322, doi.org/10.1080/11035897.2018.1492963.

Smith, C. A., Sundh, M. and Mikko, H. (2014). Surficial geology indicates early Holocene faulting and seismicity, central Sweden. *International Journal of Earth Sciences*, **103**(6), 1711–1724, doi.org/10.1007/s00531-014-1025-6.

13

Glacially Induced Faults in Finland

RAIMO SUTINEN, EIJA HYVÖNEN, MIRA MARKOVAARA-KOIVISTO, MAARIT MIDDLETON, ANTTI E. K. OJALA, JUKKA-PEKKA PALMU, TIMO RUSKEENIEMI AND JUSSI MATTILA

ABSTRACT

The zones of glacially induced faults (GIFs) in Finland are portrayed by a number of discrete < 10-km-long fault scarps, often forming multiple parallel segments and establishing longer GIF systems. A set of GIF systems further form GIF complexes, which may extend tens of kilometres cross-cutting glacial sediments. Systematic mapping has revealed 18 GIF systems forming 9 GIF complexes. Moment magnitude estimates for the earthquakes in Finnish Lapland are in the range of $M_w \approx 4.9–7.5$. The detailed trenching across fault scarps provides evidence of non-stationary seismicity and the occurrence of multiple slip events even before the Late Weichselian maximum.

13.1 Introduction

The discovery of glacially induced faults (GIFs) in Finland dates back to 1960s (Kujansuu, 1964), yet already in 1930 Tanner described a tentative GIF in Kalastajasaarento (after WW2 Rybachy Peninsula, Russia) on the coast of Barents Sea (Tanner, 1930). In western Finnish Lapland and with aerial photo-based general-scale mapping of Quaternary deposits, Kujansuu (1964) found linear features (Rautuskylä, Venejärvi, Pasmajärvi, Vaalajärvi; Figure 13.1) that were not associated with ordinary glacial bedforms but rather were cutting and eroding the glacial features. These lines were later included into the maps of the Nordkalott Project (1986) and verified by diamond drillings within the Nuclear Waste Disposal Research Project (Kuivamäki et al., 1998). Some of the minor GIFs proposed by Kujansuu (1964) in Vuotso and Kelottijärvi have not been confirmed. Landslides are considered indirect evidence of palaeoseismic events (Kujansuu, 1972). The first ^{14}C datings of basal peat on the landslide scars (Kujansuu, 1972) and landslide buried wood remains (Sutinen, 2005) found in the vicinity of GIFs suggested seismic activity between 7,495 and 9,790 calibrated years before present (BP) in western Finnish Lapland (see age compilation in Ojala et al., 2018a).

During the past decade, a systematic surveillance of GIFs has been carried out in Finland by the Geological Survey in cooperation with Posiva (Ojala et al., 2019a). These projects utilized remote sensing tools such as light detection and ranging (LiDAR) digital elevation models (DEMs) (Sutinen et al., 2014a,b; Palmu et al., 2015; Ojala et al., 2017, 2020) and

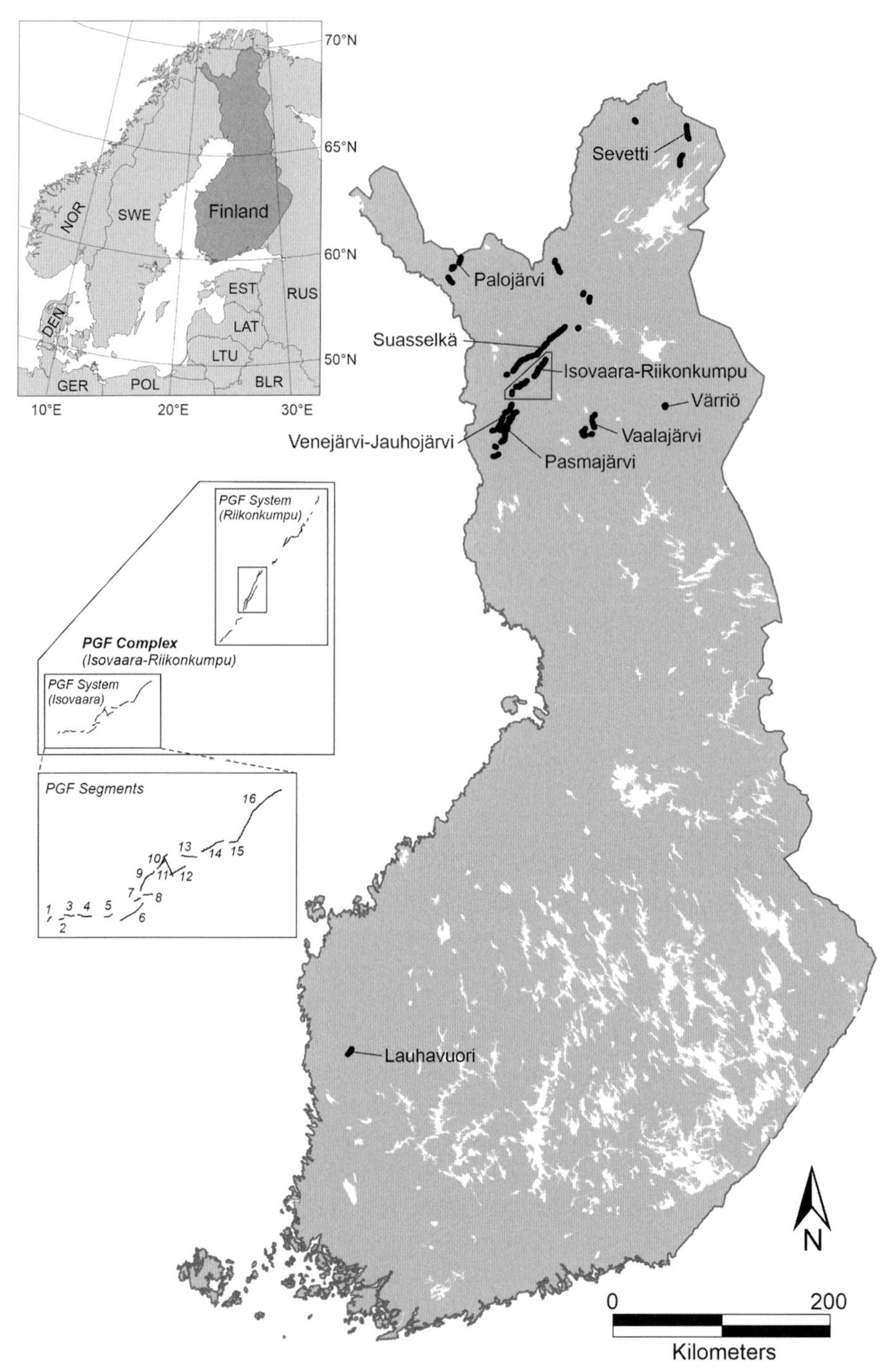

Figure 13.1 Glacially induced fault segments, systems and complexes in Finland (Sutinen et al., 2009a, 2014a,b, 2019a; Palmu et al., 2015; Ojala et al., 2017, 2019a, 2020; Mattila et al., 2019, Markovaara-Koivisto et al., 2020).

airborne geophysics (e.g. Sutinen et al., 2018; Mattila et al., 2019) for the reconnaissance of potential GIFs. Ground-penetrating radar (GPR) was systematically applied to reveal fault scarps beneath peatbogs and as a guidance for excavator trenching. Azimuthal electromagnetic surveys were applied to distinguish earthquake-induced debris flows from the glacial dispersal patterns (Sutinen et al., 2009b, 2014a, 2019b). Machine learning and object-based image analysis (OBIA) for the LiDAR point cloud data were used to identify seismically induced morphologies, such as the Pulju moraines, liquefaction spreads as well as mass-flow deposits (Nevalainen et al., 2016; Middleton et al., 2020a, 2020b). Finally, trenching and diamond drillings were applied to verify the GIFs (Sutinen et al., 2014a,b; Ojala et al., 2017, 2019b, 2020; Mattila et al., 2019; Markovaara-Koivisto et al., 2020) and to study their internal structure and stratigraphy. To tackle the timing and recurrence intervals of the palaeoseismic events, a systematic borehole campaign for landslide deposits and lake sediment beds (sonar surveyed) was carried out. The discovered buried organic sediments (peat and gyttja) and seismically induced soft-sediment disturbances were dated with ^{14}C (Ojala et al., 2018a,b, 2019d). Within the trenching operations, optically stimulated luminescence (OSL) dates were provided for the Quaternary time-stratigraphy (Markovaara-Koivisto et al., 2020). At the moment, three palaeoseismic episodes, 9–11 ka BP, 5–6 ka BP and 1–3 ka BP, have been established in Finland (Ojala et al., 2018a). Still some of the potential GIFs are either logistically (Sevetti in north-easternmost Finnish Lapland, see Sutinen 2009a, 2019a) or juristically (Lauhavuori National Park, see Palmu et al., 2015, Figure 13.1) challenging to investigate by drilling and trenching and therefore lack firm verification.

The confirmed GIF zones are portrayed by a number of discrete 0.2–9.1-km-long fault scarps forming often multiple parallel segments (e.g. Figure 13.1) and establish longer GIF systems. One set of GIF systems can further form a GIF complex, which may extend for tens of kilometres (Mattila et al., 2019; Ojala et al., 2019a; Figure 13.1). The systematic mapping has revealed 18 GIF systems forming 9 GIF complexes in Finland (Table 13.1). Following the typical GIF pattern in northern Fennoscandia (Lagerbäck & Sundh, 2008; Olesen et al., 2004; Mikko et al., 2015), the surveyed GIF systems and complexes frequently strike SW-NE direction (Figure 13.1) and represent reactivation of pre-existing bedrock fault zones. In the following are described comprehensively investigated GIFs: Pasmajärvi (Markovaara-Koivisto et al., 2020), Venejärvi-Jauhojärvi (Mattila et al., 2019), Isovaara-Riikonkumpu (Sutinen et al., 2014a; Ojala et al., 2017), Suasselkä (Ojala et al., 2019c) and Vaalajärvi-Ristonmännikkö (Kirsch et al., 2019; Ojala et al., 2020).

13.2 Pasmajärvi GIF Complex

The Ruokojärvi Fault scarp (Figure 13.2) belongs to the Pasmajärvi GIF Complex (Table 13.1) comprising seven SW-NE-striking faults that run through Late Weichselian sedimentary sequences (Markovaara-Koivisto et al., 2020). The combined length of these seven fault segments is 8.8 km, and the average vertical displacement of the fault scarp is 2.9 m with maximum offset of 6.8 m. The fault geometry in the Ruokojärvi trenching site

Table 13.1. *Glacially induced fault complexes, systems and segments in Finland (Ojala et al., 2019a).*

FAULT			Fault geometry			
Complex	*System*	Segment	Fault type	Upthrown block (side)	Total length (km)	Number of segments
Pasmajärvi				**SE/NW/SSE**	**52**	**30**
	Pasmajärvi			*SE/NW*	*34*	*25*
		Ruokojärvi	Reverse	SE/NW	3	1
	Näläntö			*SSE*	*7*	*5*
Venejärvi-Jauhojärvi				**SE/NW/ESE**	**45**	**17**
	Venejärvi			*NW/SE*	*18*	*10*
		Naamivitikko	Reverse	ESE	4.1	1
	Jauhojärvi			*NW*	*22*	*7*
Isovaara-Riikonkumpu				**NW**	**38**	**36**
	Isovaara			*NW*	*12*	*16*
	Riikonkumpu			*NW*	*19*	*20*
		Riikonkumpu	Reverse	NW	2.8	1
		Riikonvaara	Reverse	NW	3.5	1
Lauhavuori				**NW**	**6**	**6**
	Lauhavuori			*NW*	*6*	*6*
		Lauhavuori		NW	2,1	1
Suasselkä				**SE**	**72**	**38**
	Suaselkä			*SE*	*19*	*17*
		Suaspalo	Reverse	SE	2.1	1
		Retu	Reverse	SE	4.4	1
	Suasoja			*SE*	*5*	*3*
	Nilimaa			*SE*	*6*	*4*
	Sirkka			*SE*	*32*	*14*

Vaalajärvi				**WSW/SE**	**18**	**7**
	Vaalajärvi			*WSW*	*7*	*1*
		Vaalajärvi	Reverse	WSW	6.9	1
	Ristonmännikkö			*SE*	*11.4*	*6*
		Ristonmännikkö	Reverse	SE	1.8	1
Palojärvi				**ESE/NE**	**30**	**4**
	Palojärvi			*ESE*	*15*	*3*
	Kultima			*NE*	*7*	*1*
Sevettijärvi				***E***	**40**	**2**
	Laanaselkä			*E*	*9.6*	*1*
	Rovivaara			*E*	*12.4*	*1*
Värriö				**SE**	**2**	**2**
	Korteaavankumpu			*SE*	*2*	*2*

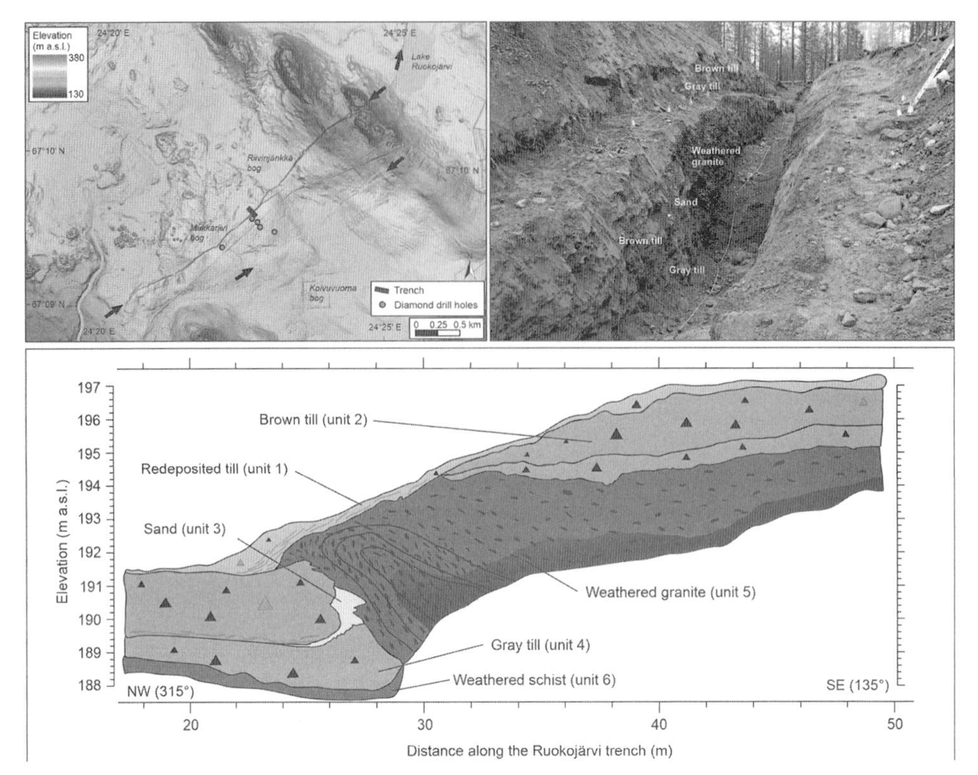

Figure 13.2 Pasmajärvi GIF System with trenched Ruokojärvi Fault ramp (Markovaara-Koivisto et al., 2020). (A black and white version of this figure will appear in some formats. For the colour version, please refer to the plate section.)

indicates that the fault is a thrust type and dips approximately 60° to SE. Moment magnitudes, estimated from surface rupture length and displacement as well as dip angle, vary between $M_w \approx 5.2$ and 7.9 (Markovaara-Koivisto et al., 2020).

Significantly reworked Weichselian till is identified in the subsurface stratigraphy at the trench wall. The collapsed esker system at the fault zone and water-escape and flame-like structures (seismites) observed in glaciofluvial sand deposits indicate that the fault zone was activated during and after the latest deglaciation (Markovaara-Koivisto et al., 2020). The ^{14}C dates acquired from basal peat at the edge of the fault scarp yielded a minimum age of 9.5 ka BP. The nearby Lehtolaki landslide buried organics suggests seismic activity around 10.2 ka BP (Ojala et al., 2018a). The organic materials and the liquefaction soft-sediment deformations as well as esker-collapse system suggest that the faulting of different segments in the Pasmajärvi GIF Complex results from more than one seismic event, which eventually occurred both subglacially and postglacially (Markovaara-Koivisto et al., 2020).

13.3 Venejärvi–Jauhojärvi GIF Complex

Analysis of high-resolution LiDAR-based DEMs, trenching and structural analysis of the bedrock as well as diamond drilling were conducted to characterize the geological structure

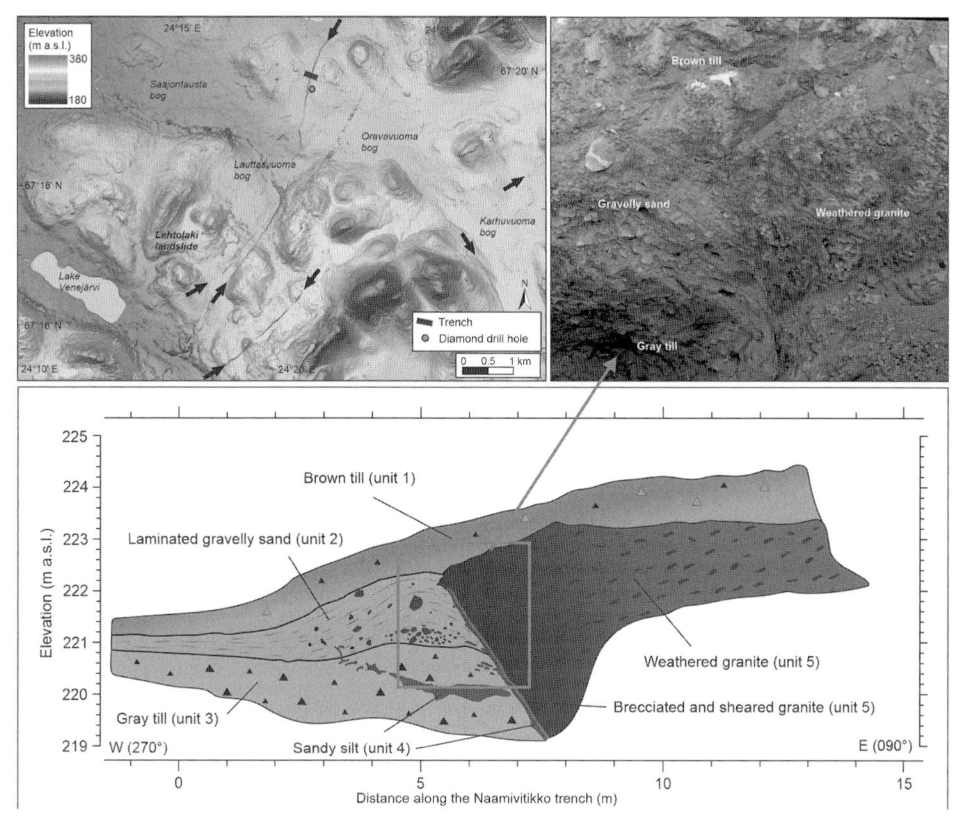

Figure 13.3 Venejärvi–Jauhojärvi GIF System with trenched Naamivittikko Fault ramp (Mattila et al., 2019). (A black and white version of this figure will appear in some formats. For the colour version, please refer to the plate section.)

and full slip profiles of the Venejärvi–Jauhojärvi Glacially Induced Fault Complex (Figure 13.3; Table 13.1). The Venejärvi–Jauhojärvi Fault Complex consists of 17 surface ruptures striking in a SW-NE direction cutting through Late Weichselian till, traceable for a distance of 29.9 km. Based on stratigraphical and structural analysis the Naamivitikko Fault Segment (Table 13.1), a part of the Venejärvi–Jauhojärvi Complex, hosted up to three distinct slip events (Mattila et al., 2019). It is therefore argued that the prevalent single-rupture hypothesis of GIFs should be treated with caution. Based on the measured surface rupture dimensions, it is conservatively estimated that the maximum moment magnitude of the palaeoseismic event that the complex hosted had a range of $M_w \approx$ 5.8–8.4. If uncertainties in the scaling laws equating surface rupture dimensions to moment magnitudes are accounted for, a more realistic maximum moment magnitude a range of $M_w = 5.8-7.6$ is given (Mattila et al., 2019). Based on geomorphological observation of a potential restraining bend at the SW end of the complex, it is further argued that strike-slip faulting and transpressional stress regime may have played a role in the formation of the Venejärvi–Jauhojärvi faults. Mattila et al. (2019) suggested that strike-slip faulting acted as a mechanism in the formation of Venejärvi–Jauhojärvi GIF.

13.4 Isovaara–Riikonkumpu GIF Complex

Analysis of airborne LiDAR-based DEMs, trenching and diamond drilling through faulted bedrock was conducted to characterize the geological structure and full slip profiles of the Isovaara–Riikonkumpu GIF Complex (Figure 13.4; Table 13.1, Ojala et al., 2017). The GIF systems are recognized as a complex of surface ruptures striking SW-NE, cutting through Weichselian till and associated with several palaeolandslides within 10 km of the surface ruptures. Evidence from the terrain rupture characteristics, the deformed and folded structure of till and the ^{14}C age of 11.3 ka BP from buried organic matter underneath the Sotka landslide indicate a postglacial origin of the Riikonkumpu GIF (see age compilation in Ojala et al., 2018a). The fracture frequency and lithology of drill cores and fault geometry in the trench log indicate that the Riikonkumpu GIF dips to WNW with a dip angle of 40–45° at the Riikonkumpu site and close to 60° at the Riikonvaara site.

A fault length of 19 km and the mean and maximum cumulative vertical displacement of 1.3 m and 4.1 m, respectively, of the Riikonkumpu GIF System indicate that the fault

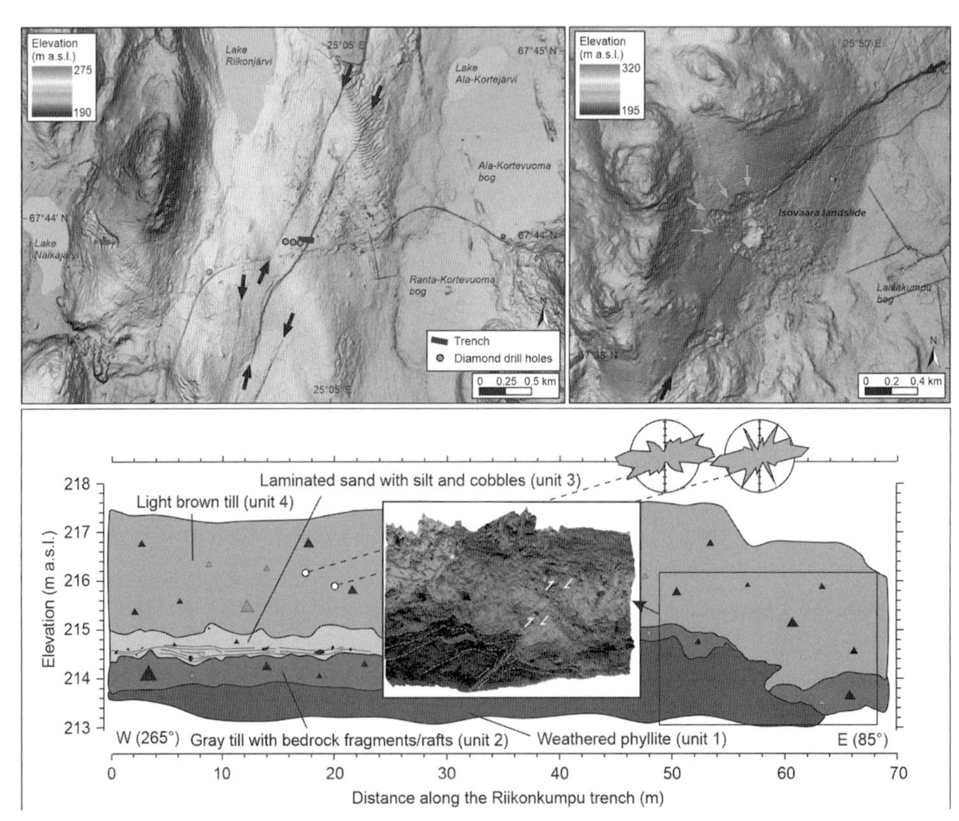

Figure 13.4 Isovaara–Riikonkumpu GIF System showing parallel scarps at Riikonkumpu site. Both Isovaara and Riikonkumpu GIFs, 15 km apart, were also verified by diamond drillings (Sutinen et al., 2014a; Ojala et al., 2017). The nearest palaeoslide, Sotka, yielding an age of 11,292 BP is located only 4 km SE of the Riikonkumpu trenching site (Ojala et al., 2018a; shown in Figure 8.7 of Smith et al., see Chapter 8). (A black and white version of this figure will appear in some formats. For the colour version, please refer to the plate section.)

potentially hosted an earthquake with a moment magnitude of $M_w = 6.7-7.3$, assuming that slip was accumulated in one event (Ojala et al., 2017). The Riikonkumpu GIF System is linked to the Isovaara GIF System, which forms a larger Isovaara–Riikonkumpu Fault Complex. Relationships between the 38-km-long rupture of the Isovaara–Riikonkumpu Complex and the fault offset parameters, with cumulative displacement of 1.5 and 8.3 m, respectively, indicate that the earthquake(s) contributing to the GIF complex potentially had a moment magnitude of $M_w \approx 6.9-7.5$. In order to adequately sample the uncertainty space, the moment magnitude was also estimated within the Isovaara–Riikonkumpu GIF Complex. The magnitude estimates for each individual segment vary roughly between $M_w \approx 5$ and 8 (Ojala et al., 2017).

13.5 Suasselkä GIF Complex

The Suasselkä GIF Complex (Rautuskylä in Kujansuu, 1964; location in Figure 13.1) is recognized from LiDAR DEMs as a complex of 37 surface ruptures striking SW-NE and cutting through Weichselian glacial sediments, confirming the postglacial nature of the rupturing event (Ojala et al., 2019b). The analysis of high-resolution digital elevation models, GPR profiles and trenching at two sites along the Suasselkä GIF Complex has provided evidence of multiple earthquakes that post- and pre-date the Late Weichselian deglaciation. At the Suaspalo trenching site, at least two different rupture events are observed: the first pre-dates the deposition of the lowermost till unit and potentially dates back to the Early or even pre-Weichselian (Figure 13.5). The second rupture event, on the other hand, deforms the overlying tills and thus dates back to post-Weichselian. The observations by Ojala et al. (2019b) and Mattila et al. (2019) confirm that multiple rupturing events can take place within GIFs and even within individual fault segments. Furthermore, the results from Suasselkä provide evidence of Early or even pre-Weichselian faulting in Finland, hence advocating that the same GIFs were eventually active after the glacial phases that pre-dated the last glacial maximum.

The Suasselkä GIF Complex is composed of four GIF systems and 37 isolated 150–7,500-m-long segments (Figure 13.1). The potential moment magnitudes are calculated based on the rupture length and mean/maximum vertical displacement for each individual segment and system, taken that they potentially ruptured independently, and finally for the entire Suasselkä GIF Complex. For the entire complex rupturing in a single event, the observed displacement and length of the complex yields a moment magnitude estimate of $M_w \approx 6.7–8.1$ (Ojala et al., 2019b). However, based on geomorphological and sedimentological evidence, it is believed that this is a conservative estimate for the palaeoearthquake. If isolated systems are considered to have ruptured individually, the moment magnitudes estimated based on surface rupture lengths range from $M_w \approx 5.5$ to $M_w \approx 6.7$, and based on mean and maximum cumulative displacement values, the estimated values range from $M_w \approx 6.2$ to $M_w \approx 7.3$ and $M_w \approx 6.7$ to $M_w \approx 8.0$, respectively. Since available data do not allow distinguishing the exact number of events, the age of events or which segments potentially ruptured at the same time, it is concluded that the

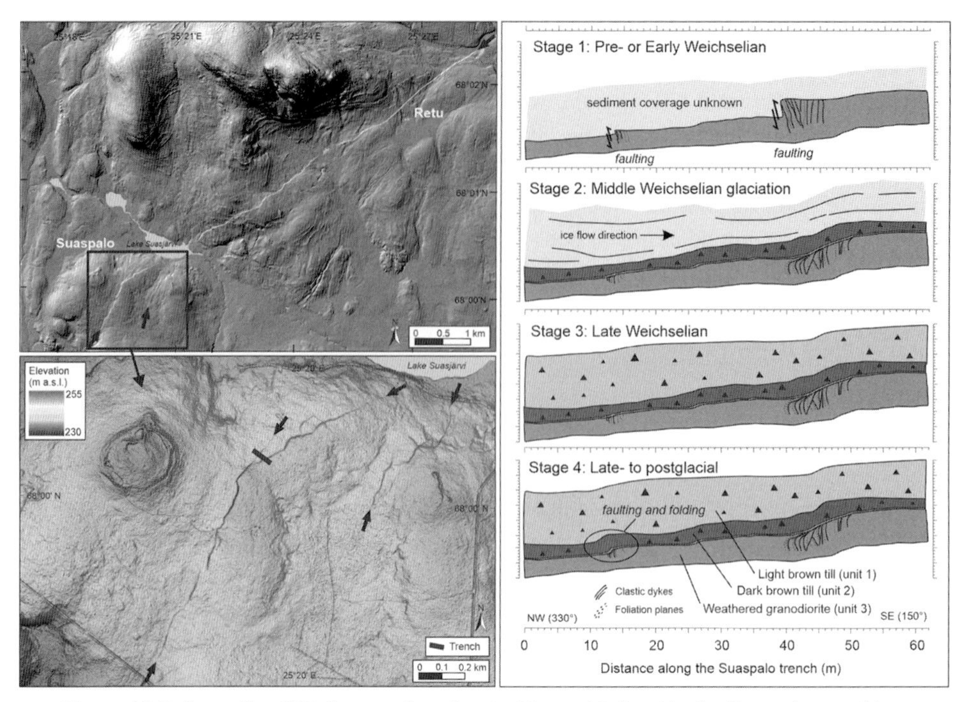

Figure 13.5 Suasselkä GIF System (location in Figure 13.1) with the Suaspalo trenching site (lower left on the LiDAR image) provides evidence of multiple slip events (sketch on the right, Ojala et al., 2019b).

events that took place in the Suasselkä GIF Complex ranged in moment magnitude from $M_w \approx 5.5$ to $M_w \approx 8.1$ (Ojala et al., 2019b).

13.6 Vaalajärvi–Ristonmännikkö GIF Complex

The rupturing history of the Vaalajärvi Fault Complex in Sodankylä, Finland (location in Figure 13.1) was based on LiDAR hyperspectral imaging, on-site geophysics (GPR) and sedimentology in excavations trenched across the faulted terrain (Figure 13.6; Kirsch et al., 2019; Ojala et al., 2020). In addition to the previously known NNW-SSE-trending Vaalajärvi Segment (Kujansuu, 1964; Kuivamäki et al., 1998), Ojala et al. (2020) discovered six new SW-NE-trending fault segments that probably belong to the same Vaalajärvi GIF Complex. The data indicate that the Vaalajärvi Fault Segment was triggered by stress change caused by ruptures on the surrounding SW-NE-trending faults and that at least two to three slip events took place in different segments of the Vaalajärvi Complex since the Early Weichselian, the most recent event(s) being postglacial. Under different scenarios of which segments or systems ruptured in a single or in separate events and with the uncertainties in the scaling laws of fault rupture length and offset, a realistic magnitude estimate for palaeoearthquake hosted by the Vaalajärvi Complex probably ranged from

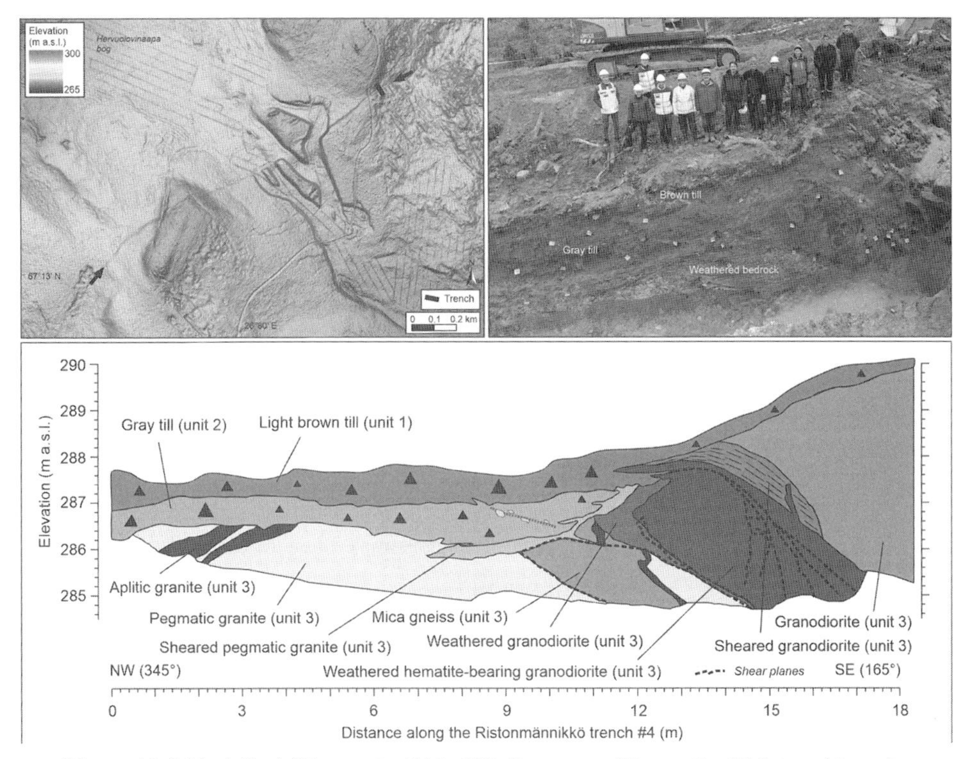

Figure 13.6 Vaalajärvi–Ristonmännikkö GIF System at Ristonmännikkö trenching site (sketch on the bottom; Kirch et al., 2019; Ojala et al., 2020) provides evidence of multiple slip events.

$M_W \approx 6.7$ to $M_W \approx 7.0$ (Ojala et al., 2020). This range is of the same order as landslide-inferred magnitudes in the Vaalajärvi area (6.7–7.4; Ojala et al., 2019c).

13.7 Discussion

Excluding Lauhavuori and Värriö, the lengths of the GIF complexes in Finland range from 22 to 52 km (Table 13.1). By orientation it may be reasoned that the Suasselkä GIF could be associated with the Landsjärv–Suorsapakka Complex in Norrbotten (see Lagerbäck & Sundh, 2008; Mikko et al., 2015) and that the Palojärvi GIF (Sutinen et al., 2014b) could be connected to the Merasjärvi–Lainio–Suijavaara Complex in Sweden (Lagerbäck & Sundh, 2008) and perhaps to the Stuorragurra Fault Complex in Finnmark, Norway (Olesen et al., 2004).

The trenched sites Naamivittikko (Mattila et al., 2019), Suaspalo (Ojala et al., 2019b) and Ristonmännikkö (Ojala et al., 2020) indicate that multiple rupturing events are common within the GIF complexes in northern Finland and that GIF segments have activated at different times during the Weichselian or even pre-Weichselian. Esker collapse geomorphology and seismites in Ruokojärvi advocate that earthquakes took place both subglacially and postglacially (Markovaara-Koivisto et al., 2020). In a similar way the Pulju moraines next to the Palojärvi GIF Complex advocate subglacial seismic activity (Sutinen et al., 2014b; see also Middleton et al., 2020a). In addition, Sutinen et al. (2021) argued that transversal moraine ridges next to the Vaalajärvi–Ristonmännikkö GIF Complex are a result of seismically induced crevasse fill-ups (see Figure 3.8 of Sandersen & Sutinen, Chapter 3).

Stratigraphic sections of the trenched GIF segments, the dating of organic matter beneath landslides and the observed soft-sediment deformations (SSDS) in lacustrine and marine sediments indicate that palaeoseismicity in the Fennoscandian Shield area is not limited to a short period during and after the Late Weichselian deglaciation (Ojala et al., 2018a,b, 2019d). Instead, the non-stationary seismicity has probably appeared (episodically) throughout the Holocene and that the same GIF complexes were potentially reactivated after glacial phases that pre-dated the Late Weichselian glacial maximum. The spatial arrangement of the data of landslide ages from northern Finland also indicates that certain ages do not cluster around specific postglacial faults, but rather they occur sporadically. Furthermore, landslides associated with specific GIF system and complexes may have different ages, such that palaeoseismic episodes are clustered around 9–11 ka BP, 5–6 ka BP and 1–3 ka BP (Sutinen, 2005; Ojala et al., 2018a). Based on the age data and geometry of the GIF systems, Ojala et al. (2019d) also suggested that the general fault propagation direction of the faults in Finland is from SW to NE, the oldest and highest magnitude events were focused on the SW sector, and the potential magnitudes decrease for younger events. However, it should be noted, according to seismological monitoring, that the GIFs are still active.

The GIF complexes are often composed of several GIF systems and isolated segments (Figure 13.1; Table 13.1); hence potential magnitudes can be calculated (i) on the basis of rupture length and mean/maximum vertical displacement of each individual GIF segment and system, (ii) assuming these potentially ruptured independently and (iii) for the entire GIF complexes, using scaling laws for faults from stable continental regions. Assuming a single rupture event for each segment, surface length values indicate moment magnitudes of $M_w \approx 2.9-6.1$ and displacement data $M_w \approx 2.9-8.0$ (Ojala et al., 2017, 2019b; Mattila et al., 2019; Markovaara-Koivisto et al., 2020). At the GIF-complex scale, the rational-length-derived moment magnitude estimates range between $M_w \approx 4.9$ and 7.4 and the displacement-derived moment magnitudes between $M_w \approx 4.6$ and 8.7. Accounting for GIF-system scale assessment of moment magnitudes, the surface rupture length data yield magnitudes in range $M_w \approx 4.9-7.0$ and displacement data values in range $M_w \approx 4.6-8.7$. Excluding the most extreme magnitude estimates, we consider that a realistic moment magnitude range for the earthquakes that took place in Finnish Lapland are in the range $M_w \approx 4.6-7.5$. Based on current understanding, the system-based estimates yield the most realistic values, consisting as they do of closely associated segments most likely linked to the same rupturing event. The defined complexes, on the other hand, may present larger linked fault zones, but due to the large non-ruptured gaps between the complexes we consider that these represent isolated rupture events. The estimates are in accordance with maximum moment magnitude calculations of $M_w \approx 6.9-7.5$, based on landslide volume-area data (Ojala et al., 2019c).

13.8 Conclusion

Glacially induced faults (GIFs) in Finland are portrayed by a number of discrete 0.2–9.1-km-long fault scarps, often forming multiple parallel segments and establishing longer GIF systems. GIF systems further form GIF complexes, which are from 22 to 52 km in length. Systematic mapping in Finland has revealed 18 GIF systems forming 9 GIF complexes. Comprehensive investigations are carried out for Pasmajärvi (Markovaara-Koivisto et al., 2020), Venejärvi–Jauhojärvi (Mattila et al., 2019), Isovaara–Riikonkumpu (Sutinen et al., 2014a; Ojala et al., 2017), Suasselkä (Ojala et al., 2019c) and Vaalajärvi–Ristonmännikkö (Kirsch et al., 2019; Ojala et al., 2020) GIFs. The moment magnitude estimates for the earthquakes in Finnish Lapland are in range $M_w \approx 4.9-7.5$ and palaeoseismic episodes are clustered around 9–11 ka BP, 5–6 ka BP and 1–3 ka BP. Detailed trenching across fault scarps provides evidence of non-stationary seismicity and occurrence of multiple slip events even before the Late Weichselian maximum.

References

Kirsch, M., Lorenz, S., Zimmermann, R. et al. (2019). Hyperspectral outcrop models for palaeoseismic studies. *The Photogrammetric Record*, **34**(168), 358–407, doi.org/10.1111/phor.12300.

Kuivamäki, A., Vuorela, P. and Paananen, M. (1998). *Indication of Postglacial and Recent Bedrock Movements in Finland and Russian Karelia*. Geological Survey of Finland Nuclear Waste Disposal Research Report YST-99, Espoo, Finland, 97 pp.

Kujansuu, R. (1964). Nuorista siirroksista Lapissa [English summary: Recent faults in Lapland]. *Geologi*, **6**, 30–36 (in Finnish).

Kujansuu, R. (1972). On landslides in Finnish Lapland. *Geological Survey of Finland Bulletin*, **256**, 22 pp., tupa.gtk.fi/julkaisu/bulletin/bt_256.pdf.

Lagerbäck, R. and Sundh, M. (2008). *Early Holocene Faulting and Paleoseismicity in Northern Sweden*. Geological Survey of Sweden Research Paper Series C, Vol. 836, 80 pp.

Markovaara-Koivisto, M., Ojala, A. E. K., Mattila, J. et al. R. (2020). Geomorphological evidence of paleoseismicity: surficial and underground structures of Pasmajärvi postglacial fault. *Earth Surface Processes and Landforms*, **45**, 3011–3024, doi.org/10.1002/esp.4948.

Mattila, J., Ojala, A. E. K., Ruskeeniemi, T. et al. (2019). Evidence of multiple slip events on postglacial faults in northern Fennoscandia. *Quaternary Science Reviews*, **215**, 242–252, doi.org/10.1016/j.quascirev.2019.05.022.

Middleton, M., Heikkonen, J., Nevalainen, P., Hyvönen, E. and Sutinen, R. (2020a). Machine learning-based mapping of micro-topographic earthquake-induced paleo Pulju moraines and liquefaction spreads. *Geomorphology*, **358**, 107099, doi.org/10.1016/j.geomorph.2020.107099.

Middleton, M., Nevalainen, P., Hyvönen, E., Heikkonen, J. and Sutinen, R. (2020b). Pattern recognition of LiDAR data and sediment anisotropy advocate polygenetic subglacial mass-flow origin of the Kemijärvi hummocky moraine field in northern Finland. *Geomorphology*, **362**, 107212, doi.org/10.1016/j.geomorph.2020.107212.

Mikko, H., Smith, C. A., Lund, B., Ask, M. V. S. and Munier, R. (2015). LiDAR-derived inventory of post-glacial fault scarps in Sweden. *GFF*, **137**(4), 334–338, doi.org/10.1080/11035897.2015.1036360.

Nevalainen, P., Middleton, M., Sutinen, R., Heikkonen, J. and Pahikkala, T. (2016). Detecting terrain stoniness from airborne laser scanning data. *Remote Sensing*, **8**, 720, doi.org/10.3390/rs8090720.

Nordkalott Project (1986). Geological map, Northern Fennoscandia, 1:1 mill. Geological Surveys of Finland, Norway and Sweden.

Ojala, A. E. K., Mattila, J., Ruskeeniemi, T. et al. (2017). Postglacial seismic activity along the Isovaara–Riikonkumpu fault complex. *Global and Planetary Change*, **157**, 59–72, doi.org/10.1016/j.gloplacha.2017.08.015.

Ojala, A. E. K., Markovaara-Koivisto, M., Middleton, M. et al. (2018a). Dating of seismically-induced paleolandslides in western Finnish Lapland. *Earth Surface Processes and Landforms*, **43**(11), 2449–2462, doi.org/10.1002/esp.4408.

Ojala, A. E. K., Mattila, J., Virtasalo, J., Kuva, J. and Luoto, T. P. (2018b). Seismic deformation of varved sediments in southern Fennoscandia at 7400 cal BP. *Tectonophysics*, **744**, 58–71, doi.org/10.1016/j.tecto.2018.06.015.

Ojala, A. E. K., Mattila, J., Ruskeeniemi, T. et al. (2019a). *Postglacial Faults in Finland – A Review of PGSdyn-Project Results*. Posiva Report 2019-1, 118 pp., Posiva Oy, Eurajoki.

Ojala, A. E. K., Mattila, J., Ruskeeniemi, T. et al. (2019b). Postglacial reactivation of the Suasselkä GIF complex in Finnish Lapland. *International Journal of Earth Sciences*, **108**, 1049–1065, doi.org/10.1007/s00531-019-01695-w.

Ojala, A. E. K., Mattila, J., Markovaara-Koivisto, M. et al. (2019c). Distribution and morphology of landslides in northern Finland: an analysis of postglacial seismic activity. *Geomorphology*, **326**, 190–201, doi.org/10.1016/j.geomorph.2017.08.045.

Ojala, A. E. K., Mattila, J., Hämäläinen, J. and Sutinen, R. (2019d). Lake sediment evidence of paleoseismicity: timing and spatial occurrence of Late- and postglacial earthquakes in Finland. *Tectonophysics*, **771**(228227), doi.org/10.1016/j.tecto.2019.228227.

Ojala, A. E. K., Mattila, J., Middlcton, M. et al. (2020). Earthquake-induced deformation structures in glacial sediments – evidence on fault reactivation and instability at the Vaalajärvi fault in northern Fennoscandia. *Journal of Seismology*, **24**, 549–571, doi .org/10.1007/s10950-020-09915-6.

Olesen, O., Blikra, L. H., Braathen, A. et al. (2004). Neotectonic deformation in Norway and its implications: a review. *Norwegian Journal of Geology*, **84**, 3–34.

Palmu, J-P., Ojala, A. E. K., Ruskeeniemi, T., Sutinen, R. and Mattila, J. (2015). LiDAR DEM detection and classification of postglacial faults and seismically-induced landforms in Finland: a paleoseismic database. *GFF*, **137**(4), 344–352, doi.org/10.1080/11035897.2015.1068370.

Sutinen, R. (2005). Timing of early Holocene landslides in Kittilä, Finnish Lapland. *Geological Survey of Finland Special Paper*, **40**, 53–58.

Sutinen, R., Piekkari, M. and Middleton, M. (2009a). Glacial geomorphology in Utsjoki, Finnish Lapland proposes Younger Dryas fault-instability. *Global and Planetary Change*, **69**, 16–28, doi.org/10.1016/j.gloplacha.2009.07.002.

Sutinen, R., Middleton, M., Liwata, M., Piekkari, M. and Hyvönen, E. (2009b). Sediment anisotropy coincides with moraine ridge trend in south-central Finnish Lapland. *Boreas*, **38**, 638–646, doi.org/10.1111/j.1502-3885.2009.00089.x.

Sutinen, R., Hyvönen, E. and Kukkonen, I. (2014a). LiDAR detection of paleolandslides in the vicinity of the Suasselkä postglacial fault, Finnish Lapland. *International Journal of Applied Earth Observation and Geoinformation*, **27**, 91–98, doi.org/10.1016/j.jag .2013.05.004.

Sutinen, R., Hyvönen, E., Middleton, M. and Ruskeeniemi, T. (2014b). Airborne LiDAR detection of postglacial faults and Pulju moraine in Palojärvi, Finnish

Lapland. *Global and Planetary Change*, **115**, 24–32, doi.org/10.1016/j.gloplacha.2014.01.007.

Sutinen, R., Hyvönen, E., Middleton, M. and Airo M-L. (2018). Earthquake-induced deformations on ice-stream landforms in Kuusamo, eastern Finnish Lapland. *Global and Planetary Change*, **160**, 46–60, doi.org/10.1016/j.gloplacha.2017.11.011.

Sutinen, R., Andreani, L. and Middleton, M. (2019a). Post-Younger Dryas fault instability and deformations on ice-lineations in Finnish Lapland. *Geomorphology*, **326**, 202–212, doi.org/10.1016/j.geomorph.2018.08.034.

Sutinen, R., Hänninen, P., Hyvönen, E. et al. (2019b). Electrical-sedimentary anisotropy of landforms adjacent to postglacial faults in Lapland. *Geomorphology*, **326**, 213–224, doi.org/10.1016/j.geomorph.2018.01.008.

Sutinen, R., Sutinen, A. and Middleton, M. (2021). Subglacial squeeze-up moraines adjacent to the Vaalajärvi-Ristonmännikkö glacially-induced fault system, Finnish Lapland. *Geomorphology*, **384**, 107716, doi.org/10.1016/j.geomorph.2021.107716.

Tanner, V. (1930). Studier över kvartärsystemet I Fennoscandias nordliga delar IV [Studies of the Quaternary system in northern Fennoscandia – IV]. *Bulletin de la Commission Géologique de Finlande*, **88**, 594 pp.

14

Lateglacial and Postglacial Faulting in the Russian Part of the Fennoscandian Shield

SVETLANA B. NIKOLAEVA, ANDREY A. NIKONOV AND SERGEY V. SHVAREV

ABSTRACT

This chapter reviews the results of studies of Late- and postglacial faults in the Russian part of the Fennoscandian Shield (Kola Peninsula, Karelia, Sankt-Petersburg region). It provides a brief overview and description from north to south of the main seismic lineaments (Murmansk and Kandalaksha), as well as results from a study of some secondary lineaments, individual Late- and postglacial faults and seismic dislocations. The obtained data allowed identifying a decrease in seismic activity from the Late Glaciation to present times, which was due to the fading glacial isostatic uplift of the shield and the change from vertically directed forces of glacial isostasy to horizontal compressive strains playing the leading role. Glacial isostasy as a factor that gave rise to stresses had nearly exhausted itself by the present time, while the tectonic factor continues to be felt.

14.1 Introduction

The eastern, Russian, part of the Fennoscandian crystalline shield (which includes the Kola Peninsula, Karelia and part of the Sankt-Petersburg region) differs from the other parts of the shield in at least two important ways. First, it shows lower rates of postglacial uplift, hence lower uplift amplitudes. Second, it contains several large old grabens, which were active during Quaternary time. These are the Kandalaksha, Onega and Ladoga grabens. Their depressions are filled with extensive water bodies that hide both major active faults and lesser lacustrine basins that were frequently created by young tectonic movements. Consequently, many major faults are hidden beneath the waters of internal basins, and often only the short terrestrial segments have been surveyed. It was possible to acquire drilling data from various organizations for only a few faults.

Sporadic observations of postglacial faults began being reported from the middle of the twentieth century. Researchers reported displacements in morainic ridges, disturbances in the heights of Lateglacial and postglacial terraces and the warping of lacustrine basins (Murzaev, 1935; Biske, 1959; Karpov, 1960; Nikonov, 1964 et al.). Faults were not dated at that time. The work of N. I. Nikolaev (1967) was one of the first reviews that put forth hypotheses about the presence of past large ruptures and earthquakes in the Kola Peninsula and in Karelia.

As new active research in the palaeoseismology of the Russian part of the Fennoscandian Shield has been carried out during recent decades, information on postglacial faults has been considerably expanded. Numerous young linear disturbances and traces of strong ancient earthquakes have been detected in Karelia, Sankt-Petersburg and the Murmansk regions (Avenarius, 1989; Lukashov, 1995; Nikolaeva, 2001, 2008; Biske et al., 2009; Nikonov et al., 2014; Shvarev & Rodkin, 2018). However, no long, continuous, postglacial fault scarps (extending for a few tens or hundreds of kilometres) have been detected on land.

More frequently, crystalline rocks were found to contain individual ruptures and cracks (or their systems) that inherited older weakened zones (deep faults). Much more extensive data have been acquired on seismic dislocations and their secondary effects.

Generalization of data on Late- and postglacial faults, as well as ancient earthquakes obtained from palaeoseismological, archaeoseismic and historical sources, made it possible to identify several seismic lineaments of different ranks (Nikonov & Shvarev, 2015). Most of these lineaments strike NW-SE, while others strike NE-SW (Figure 14.1). The NW-striking faults within the region under study run along the shorelines of sea coasts (White and Barents seas) and of the larger lakes (Ladoga and Onega).

This paper provides a brief overview and description from north to south of the main seismic lineaments (Murmansk and Kandalaksha), as well as results from a study of some secondary lineaments, individual Late- and postglacial faults and seismic dislocations in the Russian part of the Fennoscandian Shield.

14.2 The Murmansk Seismic Lineament

The fault zone that separates the Barents Sea (Svalbard) plate and the Fennoscandian Shield runs for nearly 500 km along the Murmansk coast of the Kola Peninsula. In this zone the crystalline basement stepwise plunges towards the sea. The fault system consists of a set of scarps some tens of metres in height that, starting from the Varanger Peninsula in the west, extends further east along the entire coast of the Kola Peninsula (Figure 14.1). According to movements along the fault, it is interpreted as a dextral strike-slip normal fault (Trifonov, 1999).

There is a dense fault network along the entire Murmansk coast of the Barents Sea that has recorded postglacial tectonic movements of crystalline basement blocks. Seismic exploration, sonar surveys and drilling discovered vertical offsets of Late Pleistocene and Holocene mud and pebble deposits on the sea bottom of the Teriberskaya Bay (Krapivner, 2018) (Figure 14.2). Geological evidence of displacements are the high-amplitude surface drops of crystalline rocks at a short distance (tens of metres), which are accompanied by changes in the hypsometry of the Late Pleistocene and Holocene layers, a sharp decrease (up to 0.2 m) in the thickness of gravel-sand sediments in shallow (20–30 m) areas. Deformations of mud sediments are also observed. Side-scan sonar surveys have detected several faults. The largest fault, which has been drilled at several locations, strikes 335° NNW and is 600 m long and 15 m wide. More details are given in Krapivner (2018).

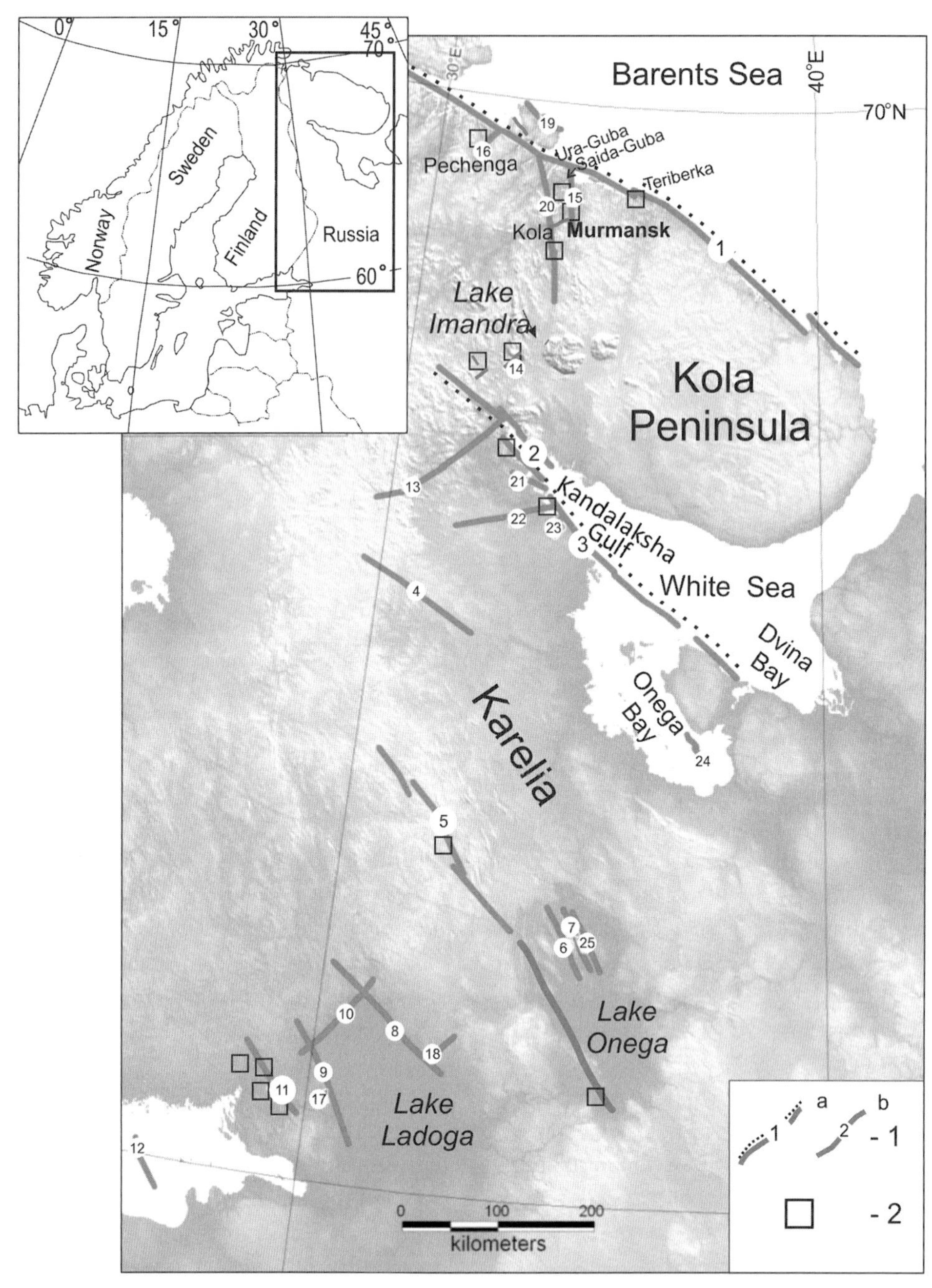

Figure 14.1 Location of seismic lineaments and Late- and postglacial faults in the eastern part of the Fennoscandian crystalline shield. (Digital elevation model based on GTOPO-30 data). Key: 1 – (a) Main seismic lineaments, (b) secondary lineaments and faults and their numbers. 2 – Sites of studied palaeoseismic dislocations. Names of lineaments: 1 – Murmansk seismic lineament; 2, 3 – Kandalaksha seismic lineament: 2 – North Kandalaksha, 3 – South Kandalaksha; 4 – Luashtangi; 5 – West Onega; 6 – Unitsa; 7 – Svyatukha; 8, 9, 10 – Ladoga; 11 – Vuoksi; 12 – Hohland; 13 – Paanajarvi; 14 – Chuna; 15 – Volshepakhk; 16 – Pechenga; 17 – Konevets; 18 – Pogrankondushi; 19 – Rybachy; 20 – Kolsky; 21 – Rugozero; 22 – Chupa-Keret; 23 – Sharapov; 24 – Kiy-ostrov; 25 – Putkozero.

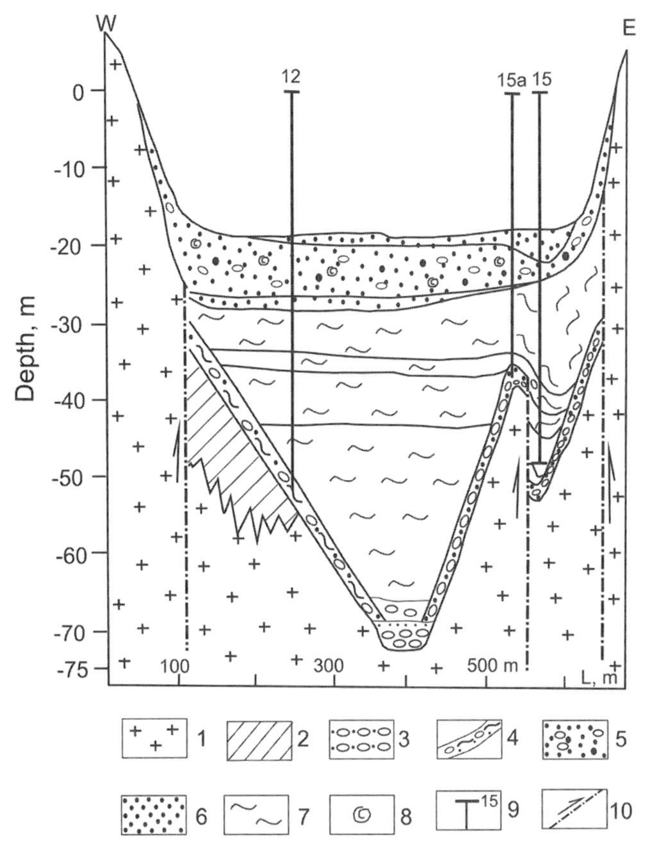

Figure 14.2 A section showing displacements of the Late Pleistocene and Holocene loose deposits on the sea bottom of the Teriberskaya Bay (after Krapivner, 2018). 1 – Precambrian rocks, 2 – estimated weathering crust, 3 – pebble-bed, 4 – pebble and gravel with sand-silt-clay aggregate, 5 – gravel with pebbles, 6 – sand, 7 – mud, 8 – mollusk fauna, 9 – core drill-hole, 10 – fault.

Earthquake-induced gravitational and vibrational deformation in rocks (strongly fractured of bedrock, open fractures, collapses, rock falls and heavily fragmented rock massifs) are observed along the entire Murmansk coast, in an up to 5–10-km-wide coastal strip (Nikolaeva, 2008, 2009, 2013; Verzilin et al., 2013; Nikonov et al., 2015). The associated earthquake intensities are estimated to be in a range from $\geq$ VI–VIII to IX–X, according to the MSK-64 and ESI-2007 scales (Nikonov & Shvarev, 2015).

Soft-sediment deformation structures (SSDS) in the Late Pleistocene and early Holocene deposits were discovered in several localities along the western part of the Murmansk lineament between the Norwegian border and the Kola Bay. Four deformed horizons contain the section of marine and fluvioglacial sediments of the Pechenga River Valley (Nikolaeva, 2009). Deformations are represented by phenomena of liquefaction, load structures (load casts and flame structures), convolute lamination, ball-and-pillow structures and folds. The deformation structures of the upper horizon were found in the lower part of the marine loam-sand interval, the age of which, according to calibrated radiocarbon dates from the mollusk fauna, is 8,821–9,159 (Beta-58712) calendar years before present (a BP). Dating of these sediments suggests correlation between the deformation features and strong earthquakes before 9 ka BP. Absence of gravity sliding, independence of deformation structures of sediment stacking patterns, stratigraphic position of deformation units between undeformed sediments and cyclic repetition of structures in the section as evidence for recurrence of the causative events testify in favor of their seismogenic genesis. Similar SSDS were found in a glacial-marine clay sequence at the Ura-Guba and in a fluvioglacial delta near the Saida-Guba of the Kola Bay (Nikolaeva, 2013). According to palaeogeographic reconstructions, the deposits are of Late Pleistocene age of approximately $\geq$ 11.5–12 ka BP.

Most seismic dislocations are localized along the seismically active western sector of the lineament. According to the initial data, there were two historical earthquakes, a moderate one in 1772 and a strong one in 1500. In the twentieth century, moderate earthquakes with $M = 4.0-4.8$ occurred in ($\sim$1907), 1917, 1926, 1936, 1956 and 1981 in different parts of the Murmansk seismogenic zone (Nikonov et al., 2015; Nikonov & Shvarev, 2015).

14.3 The Kandalaksha Seismic Lineament

The Kandalaksha lineament zone extends from the top of the Kandalaksha Gulf, towards the Dvina Bay (Figure 14.3a). The chief earthquake-generating structures of the area are concentrated in the western part of the Kandalaksha Gulf (Figure 14.3b,c) and connected with the border faults of the Kandalaksha neotectonic graben, inheriting the axial zone of the Palaeoproterozoic White Sea–Lapland belt and a Riphean palaeorift system (Baluyev et al., 2012). The system of active normal and normal-strike slip faults limits the northern and southern flanks of the Kandalaksha Graben of the White Sea (Figure 14.3a,b).

The southern fault of the graben is expressed in the relief of the bed of Riphean sediments that are as deep as 8–9 km at the location, and in the bed of Quaternary deposits and also in the sea bottom topography where one finds scarps with heights of tens to several hundred metres alongside the fault. Maximum depth (greater than 300 m) and maximum differentiation of bottom topography are observed in the western part of the Kandalaksha Gulf (Nikiforov et al., 2012).

In addition to the main longitudinal faults, there are numerous feather diagonal and transverse faults at this location that penetrate far in the surrounding mainland (Shvarev & Nikonov, 2018). Postglacial activity of the faults is proved by land and seabed observations. Marine seismic studies revealed normal faulting with offsets from a few tens of centimetres to several metres in the Holocene sediments and large-scale subaqueous landslides (Figure 14.3d) (Rybalko et al., 2011; Starovoitov et al., 2018).

Geological and tectonic observations on the coasts and islands of the Kandalaksha Gulf revealed numerous seismic dislocations that had disturbed rock surfaces with the signs of glacial and postglacial (marine) influence, including surface ruptures (Figure 14.3e,f), block landslides, rockfalls, lateral displacements of blocks and areas of vibration cracking of rocks (Lukashov, 2004; Marakhanov & Romanenko, 2014). Disturbance of moraines by

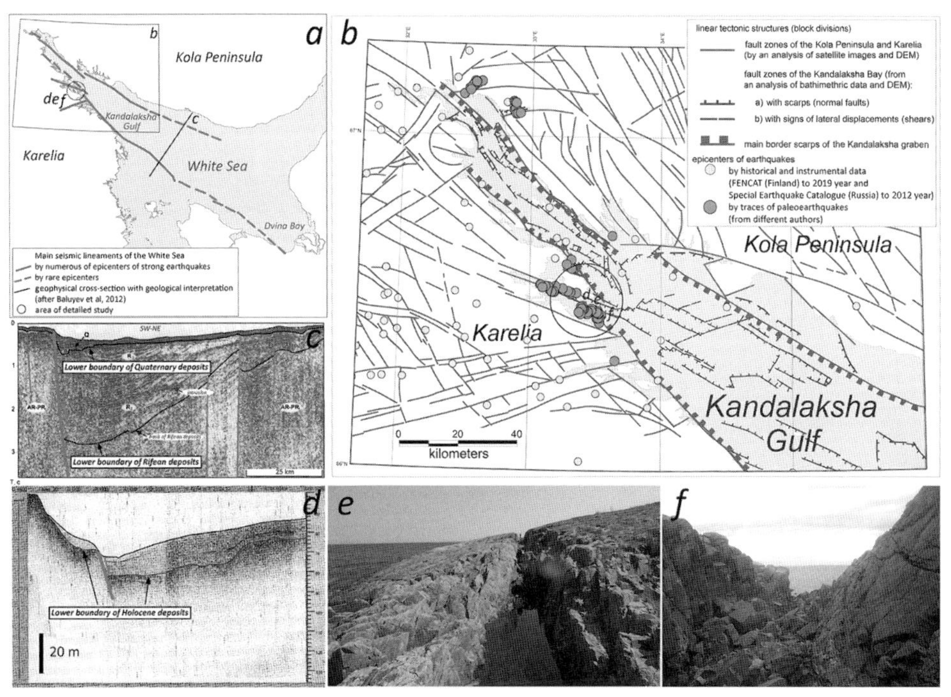

Figure 14.3 Features of structure and manifestations of the Kandalaksha seismic lineament: a) scheme of main seismic lineaments and key plots; b) scheme of fault zones manifested in the topography of the Kandalaksha Gulf bottom and adjacent land areas of the Kola Peninsula and Karelia. Geophysical cross-sections with geologic interpretation: c) across Kandalaksha Gulf with signs of displacement of Quaternary cover (after Baluyev et al., 2012); d) near the south-western shore of the Kandalaksha Gulf with the signs of displacement of Holocene cover (Rybalko et al., 2011). Photos of seismic ruptures on the islands of the Kandalaksha Gulf: e) seismic trench (man shown far back for scale); f) seismic scarp with cracking and rockfall.

active postglacial faults was recently discovered on the north-eastern coast of Kandalaksha Gulf (Kolodyazhnyi et al., 2019).

The assessment of palaeoearthquake intensity by surface signs was determined to be in the range of IX–X (Nikonov & Shvarev, 2015), using the ESI-2007 scale (Michetti et al., 2007). The age of the youngest generation of seismic dislocations is 1.8–1.9 ka BP. Data for historical earthquakes (the years 1550, 1627 and 1758 with $M_W \approx 5.0–6.5$) and from the period of instrumented observations indicate increased seismic activity in the western part of the White Sea, especially in the Kandalaksha Graben area (Nikonov, 2004; Assinovskaya, 2004).

14.4 Secondary Lineaments, Late- and Postglacial Faults and Seismic Dislocations

Kola Peninsula. Short Late- and postglacial faults and seismic dislocations have been documented not only near the main seismic lineaments that border the Kola Peninsula from

Figure 14.4 Northern segment of the Chuna Fault (view of the north). (A black and white version of this figure will appear in some formats. For the colour version, please refer to the plate section.)

the south and north, but also at the inner part of the region (Nikolaeva, 2001; Nikolaeva et al., 2018).

Detailed field work has been carried out along the Chuna Fault (Figure 14.1). This fault is 10 km long, NNE-trending (10–25°); it is located in the upper part of the western flank of the Lake Ekostrovskaya Imandra depression. Two fault segments are represented in the relief by narrow canyons with recently formed sharp walls lacking traces of glacier ice action; they are filled with large-block material (Figure 14.4). Canyons 0.5-km- and 0.4-km-long, respectively, cross an E-W-striking ridge of Late Archean granite gneisses. On both sides of the axial fault zone, at distances of 0.6–0.8 km, the massive rocks are intensely fractured and crushed.

The fault was active throughout the Lateglacial and postglacial periods, which was expressed as at least three earthquakes accompanied by different seismic deformations. The oldest event we could establish caused liquefaction of sands on the eastern flank of the canyon. The results of IR-OSL dating of sediments correspond to 14,800 (RLQG 2365-065) years ago, and the deformations themselves were formed later, at around 13.5 ka BP. The following events took place later, during Holocene, $\sim$10.3–7.1 ka BP and 2.5 ka BP, and caused the opening of the canyon, the formation of most of the fractures, rock slides, splitting-off of blocks, etc. It occurred after the final deglaciation, as evidenced by the absence of moraines from the canyon bottom (Nikolaeva et al., 2018). The youngest event has a less reliable age determination. The intensity of seismic events is estimated as $\geq$ VII–IX.

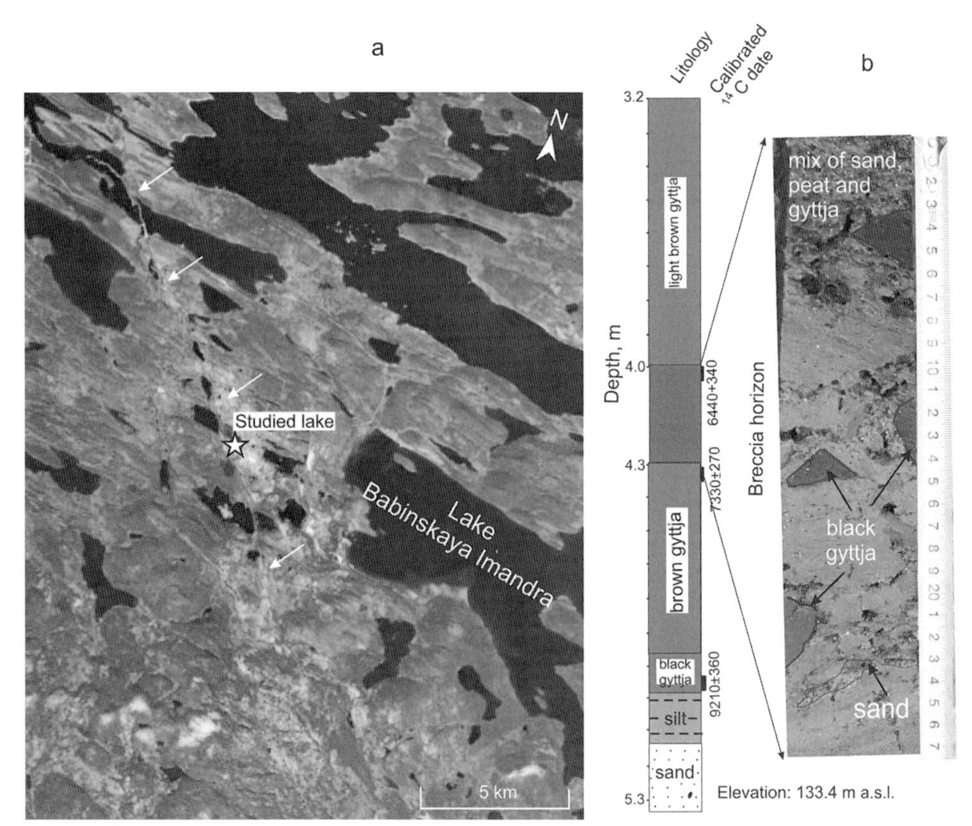

Figure 14.5 a) Location of the studied lake (shown by a star) and Lateglacial fault (shown by arrows) (Google maps) and b) stratigraphy of bottom sediments in the area west of Lake Babinskaya Imandra (Kola region). (A black and white version of this figure will appear in some formats. For the colour version, please refer to the plate section.)

Disturbances in the primary stratigraphy of bottom lake sediments localized in the NW strike lineament are also revealed (Nikolaeva et al., 2017). Sedimentary successions of a small coastal lake, located 4 km to the west of the Babinskaya Imandra Lake, reveals the presence of distinctive deposits, very different from the sediment above and below. These abnormal sediments are represented by fragments of gyttja, wood, sand and peat, which are embedded in the sapropelic matrix and look like breccia (Figure 14.5).

One of the most likely mechanisms in the formation of the breccia horizon was the catastrophic discharge of sediments from the sides of the basin as a result of shaking. In this case, the accumulated part of the near-side sediments could have moved sharply to the deeper segment of the water reservoir. According to radiocarbon analysis, a probable seismic event occurred at 6.44–7.33 ka BP (Nikolaeva et al., 2017).

In the zone of the same lineament, ground-penetrating radar survey detected vertical displacements of sand and silt, whose amplitudes vary from 1.3 to 1.7 m (Rodionov et al., 2018). This probably indicates reactivation of the lineament in the Lateglacial period.

Karelia and Sankt-Petersburg region. Local seismic dislocations were studied in Karelia on the basis of aerial photograph interpretations, airborne visual surveys and an analysis of field data in the mid-1990s (Lukashov, 1995). A later revision of these results shows frequent seismic dislocations along older fault zones that exhibit obvious signs of rejuvenation during Late Pleistocene and Holocene time. The West Onega and Vuoksi lineaments were studied in more detail by fieldwork (Nikonov et al., 2014; Shvarev & Rodkin, 2018).

The West Onega seismic lineament combines a system of activated faults of NW-SE strike (Figure 14.1). The lineament consists of individual segments (neotectonic scarps and linear hollows) stretching along the western shore of Lake Onega, extending further NW along river valleys and lake basins. The fault zone contains various seismic dislocations, which are best seen around Mount Vottovaara (Demidov et al., 1998; Lukashov, 2004; Shvarev & Rodkin, 2018). The dislocations are represented by scarps, trenches, areas of strong rock fragmentation and lateral block movements. They show several systematic signs of seismic origin: a) cracks and ruptures disturbed the glacially smoothened surface; b) the main direction of lateral block displacements is opposite to the glacial movement and coincides with the orientation and kinematics of a fault; and c) blocks demonstrate up-and-down displacements with 'jumping boulders' that are characteristic of strong earthquakes (Shvarev & Rodkin, 2018).

Geomorphological and geological data demonstrate that the fault repeatedly resumed activity during the deglaciation period and later, in early Holocene time. There is a waterlogged area in the middle of the seismogenic depression. The bottom-lake sediment core data here show evidence of deformation of sedimentary structures and the presence of overturned strata and of having formed under unstable hydrodynamic conditions. The strong earthquake led to drainage of the lake, and radiocarbon dating showed that the sedimentation process stopped 9.94–10.158 ka BP ($8,920 \pm 60$ ^{14}C years BP) (Demidov et al., 1998).

The active NW segment of the fault that dissects the Mount Vottovaara top has dextral strike-slip kinematics. According to estimations, the last strong seismic event of the early Holocene was of intensity IX–X and magnitude $M_w = 7.5$–8 (Shvarev & Rodkin, 2018).

The Vuoksi seismic lineament of the NW strike coincides with the eponymous river system, which inherits a series of fault segments (Figure 14.6a). The Vuoksi Fault Zone (VFZ) is one of the major mantle fault zones of the Ladoga–Bothnia suture (Sviridenko et al., 2017), which separates the Priozersky and Vyborg tectonic blocks. The western flank of the VFZ serves as the eastern contact of the Vyborg rapakivi granite massif with the associated Early Proterozoic enclosure. The contact was surveyed geologically with identification of displacements on NW-striking ruptures that originated during the Riphean rifting phase and resumed multiple activity later. The zone is manifested in the crystalline basement topography as a narrow straight-line trench a few tens of metres deep under the bed of the Vuoksi River filled by Quaternary deposits and, in the modern relief, as a system of parallel river channels or linear depressions.

There are numerous signs of activity of the VFZ mainly in the ruptures of NW strike and less in the secondary faults of NE and NNE directions. Seismic dislocations of different types and ages were found in bedrock and soft sediment alongside the river valley from the source of the Vuoksi River to 50–60 km downstream.

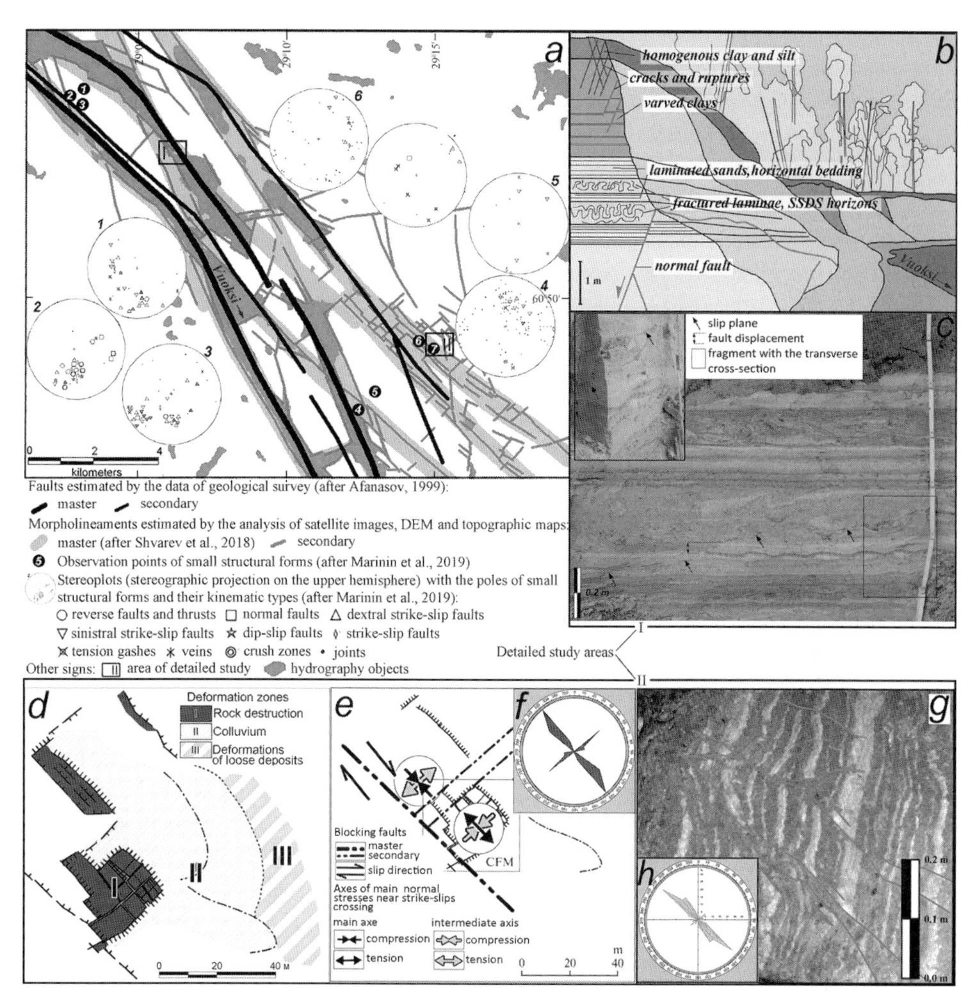

Figure 14.6 Key plot of the Vuoksi Fault Zone (VFZ) neotectonic activity study with manifestations in bedrock and sediment cover: a) scheme of VFZ near the town of Kamennogorsk with observation points of small structural forms in bedrock outcrops and areas (I and II) of detailed investigations of SSDS. Detailed study area I: b) scheme of main sediment units and types of deformations; c) fragment of sediment sequence (lower part of the cross-section on the scheme 'b') with two SSDS-horizons and normal fault; detailed study area II: d) scheme of relationship of different deformation types; e) scheme of local tectonic stress caused by palaeoseismic event; f) diagram of seismic cracks in bedrock massif; g) scheme of shear-slips in varved clays of the deformation zone III on the scheme 'd', horizontal cross-section; h) diagram of seismic cracks in sediment cover (varved clays). (A black and white version of this figure will appear in some formats. For the colour version, please refer to the plate section.)

Wide cracks were discovered in rocks on the site of the former thresholds on the Vuoksi River, where the gorge of breakthrough is located (Subetto et al., 2018). The dislocations manifest close to vertical ruptures lengthwise to the river gorge with the lateral offset of walls to several tens of centimetres and transverse cracks with displacements of adjacent blocks and fragmentation of bedrock. This set of dislocations determines the strike-slip fault ensemble caused by the postglacial seismic reactivation of an ancient fault zone. This finding allows us to presume seismic shock triggered the Vuoksi breakthrough (Subetto et al., 2018), which caused overflow of the Saimaa Lake across the Salpausselka 1 ridge ~5.7 ka BP (Saarnisto, 1970).

Fault activity led to the local earthquake with an intensity of VIII–IX, accompanied by the formation of seismic dislocations and having the character of a sinistral strike-slip that possibly accompanied a vertical component, too (Subetto et al., 2018).

The prevailing strike-slip kinematics of the reactivated ancient fault zone were confirmed recently by the study of small structural forms in bedrock downstream the Vuoksi River near the town of Kamennogorsk (former Antrea) (Figure 14.6a) (Marinin et al., 2019).

The complex of SSDS was revealed in several cross-sections of four terrace levels near the Vuoksi River. Such types of deformations as normal faults with vertical offsets to tens of centimetres, tectonic tilting, traces of catastrophic mudflow and liquefaction (clastic dykes, convolutions, fragmented laminae), micro-faulting and micro-folding with normal, reverse and strike-slip kinematics, etc. were studied. A cross-section of a terrace 6–7 m above the river level demonstrated two SSDS horizons and a normal fault oriented alongside the river channel and close (by spatial orientation) to one of the VFZ branches (Figure 14.6b,c).

Another area with closely connected spatial and kinematic parameters of deformations in rocks and loose sediments is located at the eastern branch of the VFZ (Figure 14.6d–g). Fractured and large block-fragmented granite massif is accompanied by colluvium plume and a zone of sediment deformations (Nikonov et al., 2014; Shvarev et al., 2018). It was determined that the seismogenic strike-slip displacement alongside the VFZ is the reason for the cracking of the massif (Figure 14.6e) (Nikonov et al., 2014). Such strike-slip ruptures in deformed varved clays are detected nearby (Figure 14.6g). Orientation and kinematics of rock and sediment ruptures indicate their common genesis (Figure 14.6f,h).

After analysis of OSL and ^{14}C samples, we determined dates for six strong seismic events connected with the activity of the Vuoksi Fault Zone. Three of them occurred during Early Weichselian glaciation (89.0 ± 7.3 ka (RLQG 2505-058), 85.2 ± 6.6 ka (RLQG 2504-058), 69.0 ± 5.4 ka (RLQG 2502-058) (IR-OSL age)). Three occurred in the Holocene: two with data by an analysis of plant detritus, which was deposed simultaneously with deformation (8.26 and 1.9 ka BP), and one with approximate data by the age of peat (6.6 ka BP), which was deposited shortly before the well-known event (5.7 ka BP) of the Vuoksi River bursting through the Salpausselka 1. Now this peat layer is steeply inclined by tectonic influence, so we can connect the Vuoksi breakthrough with this seismic event (Shvarev et al., 2019).

14.5 Conclusions

1. PGF lineaments and individual faults in the Russian part of the Fennoscandian Shield are represented by reactivated, previously existing fault zones in crystalline rocks. The activity of most faults was inferred from features such as shear ruptures involving

basement displacements, vertical offsets of Late Pleistocene and Holocene loose deposits, and fractures and deformations in layered sediments. Many of the discontinuities bear signs of large past seismic events (epicentral areas of earthquakes causing intensity of shaking $\geq$ VIII).

2. Young linear discontinuities in the relief and in the subsurface align themselves to form zones and lineaments of great extent. The major ones form seismic lineaments that strike NW up to a few hundred kilometres. The NNE- and NE-oriented faults are both less numerous and shorter. Young fault structures, especially larger zones, are frequently inherited from Proterozoic grabens (the White Sea, the Onega and the Ladoga grabens) and from major older fault zones.
3. The age of fault activation, largely based on the dates of strong earthquakes, lies in the time span between 13.8 ka BP and recent centuries. These dates reflect the extremely active Lateglacial (14–11 ka BP) stage and the active, without a clear decline, Holocene stage, up to recent centuries.
4. The spatial extent, the locations and orientations of Late- and postglacial faults and large earthquakes in this large region, as well as their distribution over time after the deglaciation, probably stem from the decaying glacial isostatic uplift of the shield and from the transition from vertical glacial isostatic forces to driving forces of horizontal compression. Glacial isostasy as a factor giving rise to stresses had nearly exhausted itself by the present time, while the tectonic factor continues to be felt.

References

Assinovskaya, B. A. (2004). Instrumental data on earthquakes in the Karelian region. In N. V. Sharov, ed., *Glubinnoe stroenie i seismichnost' Karel'skogo regiona i ego obramleniia* [Deep Structure and Seismicity of the Karelian Region and Its Framework]. Karelian Science Center RAS, Petrozavodsk, Russia, pp. 213–229.

Avenarius, I. G. (1989). Morphostructural analysis of the zone of neotectonic dislocations on the southern slope of the Khibiny Mountains. *Geomorfologiya*, **2**, 52–56.

Baluyev, A. S., Zhuravlyov, V. A., Terekhov, Y. N. and Przhiyalgovsky, Y. S. (2012). *Tektonika Belogo moria (Ob'iasnitel'naia zapiska k «Tektonicheskoi karte Belogo moria i prilegaiushchikh territorii masshtaba 1:1500000)* [Tectonics of the White Sea (Explanatory note to the Tectonic map of the White Sea and adjacent territories at a scale of 1:1 500 000)]. Issue 597, GEOS: Tr. GIN RAS, Moscow.

Biske, G. S. (1959). *Chetvertichnye otlozheniia i geomorfologiia Karelii* [Quaternary Sediments and Geomorphology of Karelia], Gos. izd-vo Karel'skoi ASSR, Petrozavodsk, Russia.

Biske, Yu. S., Sumareva, I. V. and Sheetov, M. V. (2009). Late Holocene seismic event in the south-eastern Ladoga area. The principles of research and deformation textures. *Vestnik of St. Petersburg University*, **7**(1), 3–25.

Demidov, I. N., Lukashov, A. D., Lavrova, N. B., Shelekhova, T. S. and Vyakhirev, S. A. (1998). Paleoecology and paleoseismology of the Mt. Vottovaara area (West Karelia) in the late glacial and postglacial time. *Paleoklimaty i evolyutsiya paleogeograficheskikh obstanovok v geologicheskoi istorii Zemli: Tezisy dokladov Mezhdunarodnogo simpoziuma* [Paleoclimates and Evolution of Paleogeographic Environments during the Geological History of the Earth: Abstracts of International Symposium Reports], Petrozavodsk, Russia, 27–31 August 1998, pp. 28–30.

Karpov, N. N. (1960). Traces of postglacial tectonic faults in the Khibiny Mountains. *Moscow University Bulletin*, **5**(4), 61.

Krapivner, R. B. (2018). Krizis lednikovoi teorii: argumenty i fakty [Crisis of the glacial theory: arguments and facts]. GEOS, Moscow.

Kolodyazhnyi, S. Y., Baluyev, A. S. and Zykov, D. S. (2019). Structure and evolution of the Belomorian–Severodvinsk shear zone in the Late Proterozoic and Phanerozoic, East European Platform. *Geotectonics*, **53**(1), 60–83, doi.org/10.1134/S0016852119010047.

Lukashov, A. D. (1995). *Paleoseismotectonics in the Northern Part of Lake Onega (Zaonezhskij Peninsula, Russian Karelia)*. Geological Survey of Finland. Nuclear Waste Disposal Research Report YST-90, Espoo, Finland, 36 pp.

Lukashov, A. D. (2004). Geodynamics of the contemporary times. In N. V. Sharov, ed., *Glubinnoe stroenie i seismichnost' Karel'skogo regiona i ego obramleniia* [Deep Structure and Seismicity of the Karelian Region and Its Framework], Karelian Science Center RAS, Petrozavodsk, Russia, pp. 150–191.

Marakhanov, A. V. and Romanenko, F. A. (2014). New data on postglacial seismic dislocations of the Northern Karelia (Karelian coast of the White Sea). Iudakhinskie chteniia. Geodinamika i ekologiia Barents-regiona v XXI v. Materialy dokladov Vserossiiskoi konferentsii s mezhdunarodnym uchastiem [Yudakhin Readings. Geodynamics and ecology of the Barents region in the XXIst century]. *Proceedings of the All-Russian Conference with International Participation, Institute of Ecological Problems of the North, Arkhangelsk, Russia, 15–18 September 2014*, pp. 137–140.

Marinin, A. V., Sim, L. A. and Bondar, I. V. (2019). Tectodynamics of the Vuoksi Fault Zone in the Karelian Isthmus. *Trudy Fersmanovskoj nauchnoj sessii GI KNC RAN*, pp. 364–368.

Michetti, A. M., Andermard, F. and Azuma, T. (2007). *Intensity Scale ESI-2007: Memory Descriptive della carta geologica d'Italia*. APAT, Rome, 74 pp.

Murzaev, P. M. (1935). On the age and formation of the gorges of the southern slope of the Khibiny massif. *Izvestiâ Leningradskogo geologo-gidrogeodezičeskogo tresta*, 6(1), 14–19.

Nikiforov, C. L., Koshel, S. M. and Froll, V. V. (2012). Digital model of the bottom of the White Sea. *Moscow University Bulletin*, **5**(4), 86–92.

Nikolaev, N. I. (1967). Neotectonics and seismicity of the East European Platform. *Izv. Academy of Sciences of the USSR*, **2**, 13–27.

Nikolaeva, S. B. (2001). Paleoseismic manifestations in the north-eastern part of the Baltic Shield and their geological and tectonic position. *Geomorfologiya*, **4**, 66–74.

Nikolaeva, S. B. (2008). Disastrous earthquakes in the vicinities of the town of Murmansk: paleoseismological and geological evidence. *Journal of Volcanology and Seismology*, **2**(3), 189–198, doi.org/10.1134/S0742046308030068.

Nikolaeva, S. B. (2009). Seismites in Late Pleistocene and Holocene deposits of the northwestern Kola region (northern Baltic Shield). *Russian Geology and Geophysics*, **50**(7), 644–650, doi.org/10.1016/j.rgg.2008.12.009.

Nikolaeva, S. B. (2013). Evidence of seismic events on the coast of Murman in the Late Glacial and Holocene. *News of the Russian Geographical Society*, **145**(4), 53–65.

Nikolaeva, S. B., Lavrova, N. B. and Denisov, D. B. (2017). A catastrophic Holocene event in the lake bottom sediments of the Kola region (northeastern Fennoscandian shield). *Doklady Earth Sciences*, **473**(1), 308–312, doi.org/10.1134/S1028334X17030072.

Nikolaeva, S. B., Nikonov, A. A., Shvarev, S. V. and Rodkin, M. V. (2018). Detailed paleoseismological research on the flank of the Lake Imandra depression (Kola region): new approaches and results. *Russian Geology and Geophysics*, **59**(6), 697–708, doi.org/10.1016/j.rgg.2018.05.008.

Nikonov, A. A. (1964), *Razvitie rel'efa i paleogeografija antropogena na zapade Kol'skogo poluostrova* [Development of the Relief and Paleogeography of the Anthropogen in the West of the Kola Peninsula], Nauka, Moscow.

Nikonov, A. A. (2004). Historical and instrumental data on seismicity of the region. Historical earthquakes. In N. V. Sharov, ed., *Glubinnoe stroenie i seismichnost' Karel'skogo regiona i ego obramleniia* [Deep Structure and Seismicity of the Karelian Region and Its Framework]. Karelian Science Center RAS, Petrozavodsk, Russia, pp. 192–213.

Nikonov, A. A. and Shvarev, S. V. (2015). Seismolineaments and destructive earthquakes in the Russian part of the Baltic shield: new solutions for the past 13 thousand years. Geologo-geofizicheskaja sreda i raznoobraznye projavlenija sejsmichnosti [Geological-geophysical environment and diverse manifestations of seismic activity]. *Proceedings of the International Conference, Neriungri, Russia, 23–25 September 2015*, pp. 243–251.

Nikonov, A. A., Shvarev, S. V., Sim, L. A. et al. (2014). Paleoseismodeformations of hard rocks in the Karelian isthmus. *Doklady Earth Sciences*, **457**, 1008–1013, doi.org/10.1134/S1028334X14080145.

Nikonov, A. A., Nikolaeva, S. B. and Shvarev, S. V. (2015). Murmansk coastal band in the Russian part of European Arctic as outstanding seismogenic zone: newest approach. In V. I. Pavlenko, ed., *Prirodnye resursy i kompleksnoe osvoenie pribrezhnyj rajonov Arkticheskoj zony. Sb. nauchnyh trudov* [Natural Resources and Integrated Development of Coastal Areas of the Arctic Zone. Collection of Scientific Papers]. Arhangel'skij nauchnyj centr Ural'skogo otdelenija RAN, Arkhangelsk, Russia, pp. 34–40.

Rodionov, A. I., Nikolaeva, S. B. and Ryazantsev, P. A. (2018). Evaluation of GPR capabilities in the study of seismogenic faulting and deformation in the bottom sediments of Lake Upoloksha (northeast of the Fennoscandian shield). *Geodynamics & Tectonophysics*, **9**(4), 1189–1203, doi.org/10.5800/GT-2018-9-4-0390.

Rybalko, A. E., Tokarev, M. Y., Fedorova, N. K. and Nikitin, M. A. (2011). New data on geology and geomorphology of the Kandalaksha Gulf from high-frequency seismoacoustic profiling and geological sampling. *Geologiya morei i okeanov: Materialy XIX Mezhdunarodnoi konferentsii (shkoly) po morskoi geologii* [Geology of Seas and Oceans: Proceedings of XIX International Conference–Workshop on Marine Geology, Moscow, Vol. 5], pp. 174–177.

Saarnisto, M. (1970). The Late Weichselian and Flandrian history of the Saimaa lake complex. *Societas Scientiarum Fennica, Commentationes Physico-Mathematicae*, **37**, 107.

Shvarev, S. V. and Nikonov, A. A. (2018). Morphotectonics of the White Sea basin in comparison with the specified characteristics of historical earthquakes. *Materialy Vserossijskoj nauchnoj konferencii: Pozdne – i postgljacial'naja istorija Belogo morja: geologija, tektonika, sedimentacionnye obstanovki, hronologija, KDU*, University Book, Moscow, pp. 174–180 (in Russian).

Shvarev, S. V. and Rodkin, M. V. (2018). Structural position and parameters of the paleoearthquakes in the area of Vottovaara Mountain (middle Karelia, eastern part of the Fennoscandian Shield). *Seismic Instruments*, **54**, 99–218, doi.org/10.3103/S0747923918020093.

Shvarev, S. V., Nikonov, A. A., Rodkin, M. V. and Poleshchuk, A. V. (2018). The active tectonics of the Vuoksi Fault Zone in the Karelian Isthmus: parameters of paleoearthquakes estimated from bedrock and soft sediment deformation features. *Bulletin of the Geological Society of Finland*, **90**, 257–273.

Shvarev, S. V., Subetto, D. A., Nikonov, A. A., Zaretskaja, N. E. and Romanov, A. O. (2019). Seismites in the pre- and postglacial sediments of the Karelian isthmus (eastern Fennoscandia). In A. Börner, H. Hüneke and S. Lorenz, eds., *Field Symposium of INQUA PeriBaltic Working Group. From Weichselian Ice-Sheet Dynamics to Holocene Land Use Development in Western Pomerania and Mecklenburg. Abstract Volume*. Scientific Technical Report STR 19/01, Potsdam GFZ German Research Centre for Geosciences, pp. 102–105, doi.org/10.2312/GFZ .b103-19012.

Starovojtov, A. V., Tokarev, M. Y., Terekhina, Y. E. and Kozupicza, N. A. (2018). The structure of the sedimentary cover of the Kandalaksha Bay of the White Sea according to seismic data. *Moscow University Bulletin*, **4**(2), 81–92.

Subetto, D. A, Shvarev, S. V., Nikonov, A. A. et al. (2018). New evidence of the Vuoksi River origin by geodynamic cataclysm. *Bulletin of the Geological Society of Finland*, **90**, 275–289.

Sviridenko, L. P., Isanina, E. V. and Sharov, N. V. (2017). Deep structure, volcano-plutonism and tectonics of Lake Ladoga region. *Trudy Karel'skogo nauchnogo centra RAN*, **2**, 73–85.

Trifonov, V. G. (1999). *Neotektonika Evrazii* [Neotectonics of Eurasia]. Scientific World, Moscow, Russia.

Verzilin, N. N., Bobkov, A. A., Kulkova, M. A. et al. (2013). On age and formation of modern dissected relief of Kola Peninsula northern part. *Vestnik of Saint-Petersburg University*, **7**(2), pp. 79–93.

Part IV

Glacially Triggered Faulting at the Edge and in the Periphery of the Fennoscandian Shield

Geological investigations in the last decade increased the number of locations with evidence or indications for glacially triggered faulting in northern Central and north-eastern Europe, i.e. in the countries of Denmark, Germany, Poland, Belarus, Lithuania, Latvia, Estonia and parts of western Russia. These locations are at the periphery, the edge or even outside of the former ice margin. They are summarized in the following chapters.

15

Lateglacial and Postglacial Faulting in Denmark

PETER B. E. SANDERSEN, SØREN GREGERSEN AND PETER H. VOSS

ABSTRACT

The Danish area is divided by the Fennoscandian Border Zone into a north-eastern part, which is generally tectonically stable with low seismicity marginally related to well-known fault zones, and the south-western non-shield part, with close to no seismicity. Stress measurements show that lithospheric plate motion is generally responsible for the modern stress pattern. However, stress changes induced by weight relief from the ice sheets during the Pleistocene appear to have created tectonic instability with maximum intensity at the time of deglaciation. The Danish area is generally covered by thick successions of unconsolidated sediments, and therefore it is often difficult to observe unambiguous examples of glacially induced faults. Nevertheless, a number of examples indicate that faults in major fault zones appear to have experienced short periods of reactivation in Lateglacial and early postglacial times. In this chapter we present an updated map of historic earthquakes in the Danish area, selected examples of Quaternary deformation of the terrain and near-surface sediments, and suggestions for intensified future monitoring, investigations and research.

15.1 Introduction

The main part of the Danish area is covered by up to some hundreds of metres of unconsolidated and complex successions of Palaeogene, Neogene and Quaternary sediments. Consequently, it is generally more difficult to observe glacially induced faults here compared to northern Scandinavia, where the basement lies close to the terrain surface. Because of a general lack of high-resolution data, evidence for Lateglacial and postglacial tectonic movements has hitherto been scarce, and some claimed examples have not been unequivocally supported by data. However, in recent years new and highly detailed data from study sites offshore as well as onshore have added several indications of fault activity following the receding of the last Scandinavian ice sheet. The data include offshore mapping of faults with high-resolution seismic data, field observations of presumably seismically deformed sediments and morphotectonic analysis of high-resolution digital elevation models. The different types of examples indicate that, although presumably not seismically active today, some of the major fault zones in Denmark appear to have experienced short periods of reactivation in Lateglacial and early postglacial times. In the following, examples from the Danish area will be presented and discussed and a summary will point towards future actions to gain additional knowledge about glacially induced deformations of the Danish subsurface.

15.2 Stress Pattern

The regional stress pattern in the Danish area has been presented in several papers (e.g. Gregersen & Voss, 2009), and the most recent information from the World Stress Map Project is displayed in Figure 10.1 (Gregersen et al., Chapter 10). The figure shows that the dominant compressional stress is horizontal and rather homogeneous. This stress pattern is caused by compression from the absolute lithospheric plate motion with direction from the north-west towards the south-east. As the rest of Northern Europe, Denmark is pushed from the Mid-Atlantic Ridge and the plate collides with other plates in southern Europe (Zoback et al., 1989; Gregersen, 1990). This is part of the worldwide pattern, and even if some individual stress measurements deviate, no consistent deviation from this pattern has been promoted. Exceptions from the dominating compression direction in Denmark scatter in several directions. Possible explanations of the local deviations include postglacial uplift of the northern-central parts of Scandinavia since the Ice Age, i.e. tilting and slight bending of the area of Denmark, and the possibility that old geological lineations may moderate the regional pattern. However, no systematic dependence on the Scandinavian uplift, which should give compression from NE to SW, is observed in the present-day stress pattern (Gregersen et al., Chapter 10). But this stress pattern appears to have changed since the last glaciation and a more seismotectonically active period is presumed to have occurred at the time of deglaciation (Gregersen & Voss, 2009).

15.3 Structure and Historic Earthquakes

The north-eastern part of Denmark is characterized by the Sorgenfrei–Tornquist Zone (STZ), being the transition between the Fennoscandian Shield with old bedrock at the surface and the stepwise deeper sedimentary basin, the 3–10-km-deep Norwegian–Danish Basin (NDB), in the middle of the country (e.g. Hansen et al., 2000). Towards the south-west the basin is bounded by the Ringkøbing–Fyn High (RFH), with basement at a depth of around 1 km. In the south-west, we find the North-German Basin (NGB), with sediment thicknesses larger than those of the NDB. The deeper structures below the sedimentary layers have been investigated in the large international cooperation projects EUGENO-S and TOR (e.g. Gregersen, 1991; Pedersen et al., 1999).

North-east of the STZ is the shield, with a thick crust (>32 km) and large lithospheric thickness (>200 km). The STZ and RFH (see Figure 15.1) comprise stepwise boundaries of the NDB, which has a rather thin crust and intermediate lithospheric thickness (~100 km). South-west of the RFH we find the NGB, with a thin crust and thin lithosphere (~50 km) (Gregersen et al., 2009).

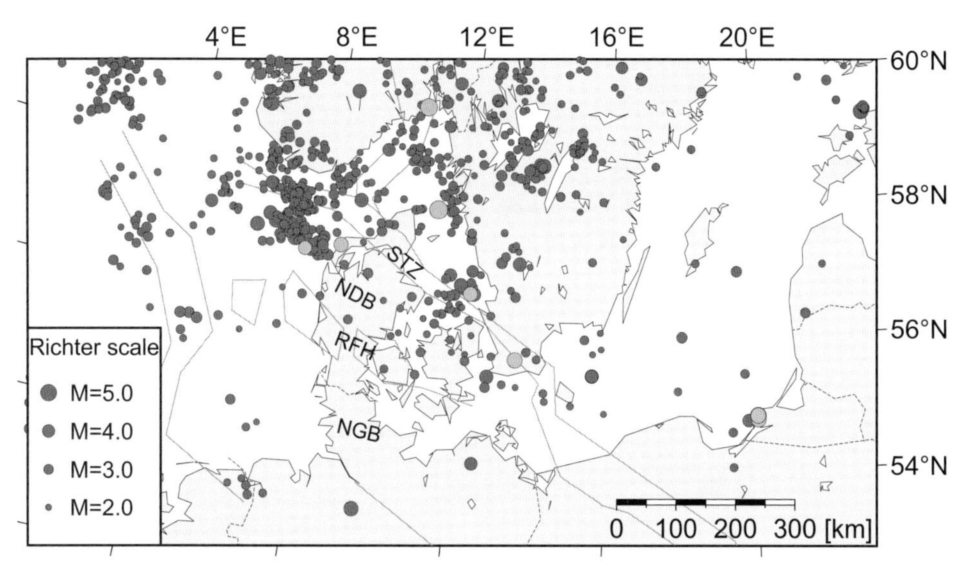

Figure 15.1 Earthquake map of Denmark 1929–2017. $M \geq 4$ earthquakes in 1929–1970 and $M \geq 2$ earthquakes in 1971–2014, from FENCAT (2020). Major tectonic structures are marked with grey lines: STZ – Sorgenfrei-Tornquist Zone; NDB – Norwegian-Danish Basin; RFH – Ringkøbing-Fyn High; NGB – North German Basin. The catalogue is marked with red circles, while a few noticeable earthquakes are marked in green: two in the NW coastal zone, one at Oslo in 1904, two in Kattegat in 1759 and 1985, one in southern Sweden in 2008 and two in Kaliningrad in 2004. (A black and white version of this figure will appear in some formats. For the colour version, please refer to the plate section.)

The Danish earthquake pattern is shown in relation to this border zone of the Fennoscandian Shield in Figures 15.1 and 15.2. The general description is one of earthquake activity scattered in the north-eastern part of Denmark, with noticeable clusters. In the sea of Kattegat, in the middle of the map in Figure 15.1, there is a cluster of earthquakes including a significant earthquake from 1985. These earthquakes are located in the STZ. The shield north-east of this zone has scattered earthquakes, and in a similar way the area to the south in the central part of Denmark has scattered, low-magnitude earthquakes. In the Skagerrak Sea up towards Norway and outside the STZ, there is a large cluster of earthquakes. To the south-west there are very few earthquakes. The border zone is not a concentrated earthquake zone; rather, it appears as a broad, gradual transition. Generally, the earthquakes in Denmark do not occur in or close to the largest known faults, and most earthquakes occur offshore towards the north-west in the deep part of the NDB/Danish–Polish Basin.

The pronounced 1985 earthquake in Kattegat is treated in special papers on earthquake patterns and known faulting along the STZ (Arvidsson et al., 1991; Gregersen et al., 1996). It is remarkable that the rest of the zone in Denmark has very few recent earthquakes. It is worth noting that only several of the rare earthquakes of magnitude $M > 4$ correlate to faults. Standard errors on horizontal locations developed over time from ~30 to ~10 km. Concerning depths, standard errors developed from 50 to ~30 km. Large standard errors are attributable to the distance between the stations (several hundreds of kilometres earlier

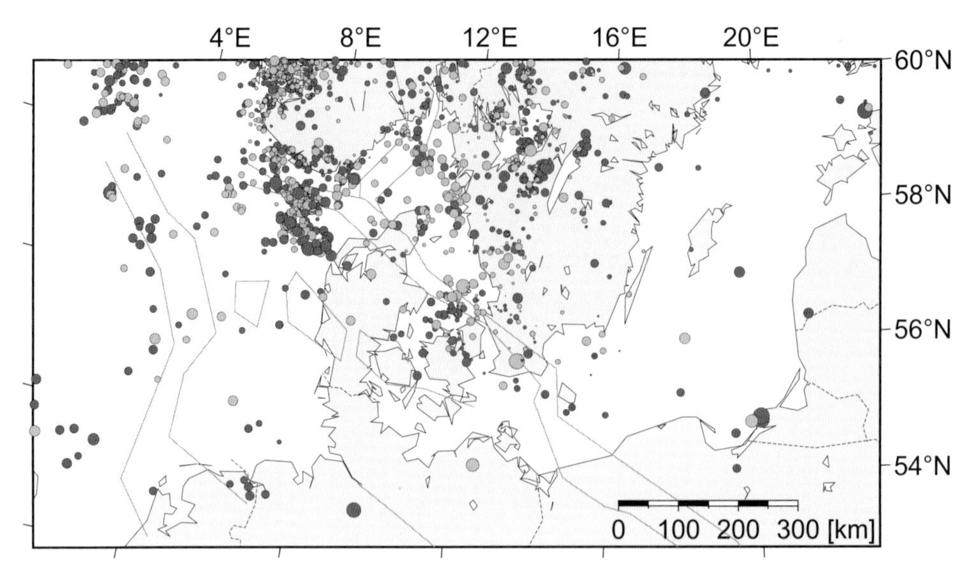

Figure 15.2 Depth of earthquakes in and around Denmark. Blue: 0–12 km, green: 12–24 km, red: below 24 km. M $\geq$ 2 earthquakes, 1971–2014, from FENCAT (2020). Earthquake magnitude key is given in Figure 15.1. (A black and white version of this figure will appear in some formats. For the colour version, please refer to the plate section.)

on to 100–200 km nowadays) and to location computations in laterally homogeneous and geologically representative models of the area (e.g. FENCAT, 2020; GEUS, 2020).

Attempts to identify trends in the depths of the earthquakes in and near the edge of the Fennoscandian Shield resulted in those seen in Figure 15.2, but no obvious pattern is seen. The standard error for the depth is very large, so intervals smaller than 12 km cannot be applied. Improved depth determination in the future will probably reveal a more detailed picture.

In conclusion, the present-day seismicity is rather small and only marginally related to the geologically well-known older fault zones in Denmark.

15.4 Observations of Late- and Postglacial Faulting in the Danish Area

Situated at the margin of the Scandinavian uplift cone, Denmark has not experienced the same amount of rebound as central and northern Fennoscandia. Still, glacially induced faulting is likely to have occurred because changes in the stress field created by loading and unloading of the ice sheet will have affected both the former ice-covered areas and large areas both close to and at considerable distances from the ice sheets (Stewart et al., 2000).

A large part of the Danish subsurface consists of unconsolidated and complex successions of Palaeogene, Neogene and Quaternary sediments (e.g. Houmark-Nielsen, 1987; Heilmann-Clausen, 1995; Rasmussen et al., 2010; Sandersen & Jørgensen, 2016). This implies that a given glacially induced fault displacement at depth will have to propagate upwards through – in many places – some hundred metres of soft sediments to have a chance of being recognized at or near the terrain surface. This makes it generally more challenging to locate glacially induced faults in the Danish area.

Before the introduction of high-resolution LiDAR data, studies of the topography demonstrating evidence of Lateglacial and postglacial tectonic movements were scarce, and some claimed examples have been considered questionable by e.g. Gregersen and Voss (2010, 2012). Despite the general lack of high-resolution data and unambiguous evidence of young deformation of near-surface sediments, several studies have pointed to observations that were difficult to explain without an influence from deeply rooted structures – the influence being either active or passive. The earliest example of this was the 'fracture valleys' described by Milthers (1916): in an area south of Hvorslev, Milthers found a remarkable system of N-S-trending, narrow and rectilinear valleys in a smooth glacial terrain (Figures 15.3 and 15.4). The orientation of the 'fractures' was not comparable to other landscape features in the area, and Milthers interpreted the valleys as being fault-generated features.

However, Milthers did not have data besides old topographic maps on which to base his interpretations (Figure 15.3; left). Jakobsen and Pedersen (2009) investigated the Hvorslev valleys further by adding highly detailed LiDAR data (Figure 15.3; right) and a seismic profile. The seismic profile indicated the presence of a series of faults displacing the Base Chalk boundary around 50 m at a depth of 1,400 m underneath the valleys. The authors concluded that the deformations were created shortly after the recession of the main Weichselian ice advance ($\sim$21,000 years ago) and that the tectonic event was probably not linked to postglacial isostatic movements but rather to the general graben subsidence of the Danish Basin.

Up through the twentieth century, several authors described other examples of large-scale topographic lineaments that they found pointed to a deep tectonic influence (e.g. Nordmann, 1958; Frost, 1977; Kronborg et al., 1978; Stockmarr, 1978; Lykke-Andersen, 1979, 1981; Hansen, 1980; Lykke-Andersen et al., 1996). Based on an analysis of large-scale lineaments in the Danish area, Frost (1977) concluded that the younger sedimentary cover seemed to be influenced by basement-controlled structures. In southern Denmark, Bruun-Petersen (1987) saw a close connection between lineaments in the surface of the pre-Quaternary sediments and the present-day landscape including the patterns of streams and brooks. This led to the conclusion that deeper structures probably have had an impact on sediment distribution and stream patterns since the Miocene.

Onshore studies based on borehole data and landscape morphology have tentatively pointed to tectonic events during the Weichselian glaciation as possible controls on deposition and erosion (e.g. Pedersen et al., 2015; Pedersen & Gravesen, 2016). In a national study, Sandersen and Jørgensen (2016) found a close correlation between the orientations of buried tunnel valleys, deep-seated faults and erosional valleys in the present-day terrain in large parts of Denmark, pointing to a relationship between the tectonic framework and erosion patterns during the Quaternary. Topographic analyses in the northern part of Denmark showed for instance that the erosional valleys had a preferred trend around WNW-ESE to NW-SE regardless of the near-surface sediments being of glacigenic or marine origin (Lateglacial or Holocene) (Sandersen & Jørgensen, 2002,

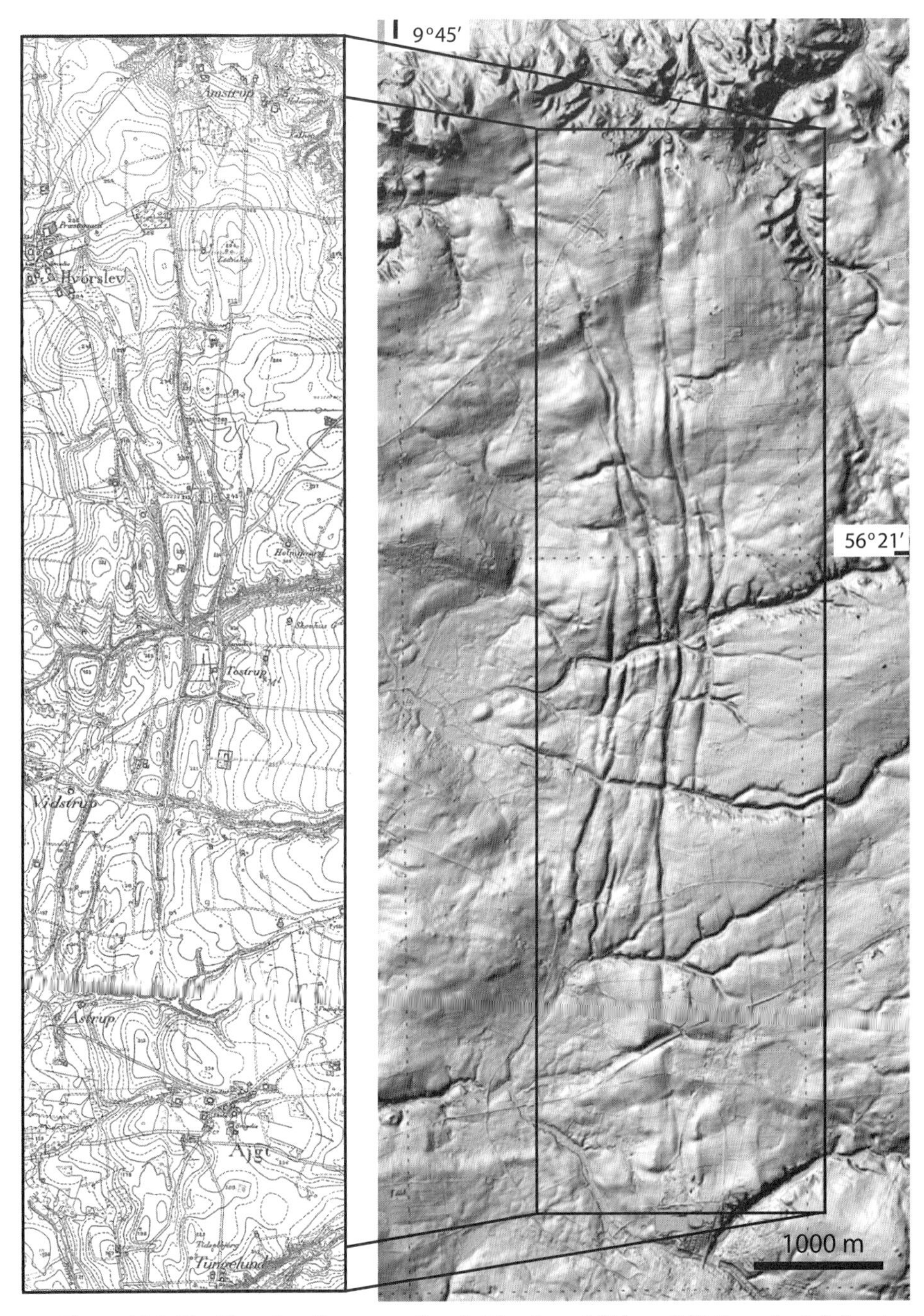

Figure 15.3 The Hvorslev 'fracture valleys'. Map from Milthers (1916) on the left (equidistance 2½ feet); LiDAR-derived hillshade map on the right. The elevations on the map range from 35 to 105 m above sea level.

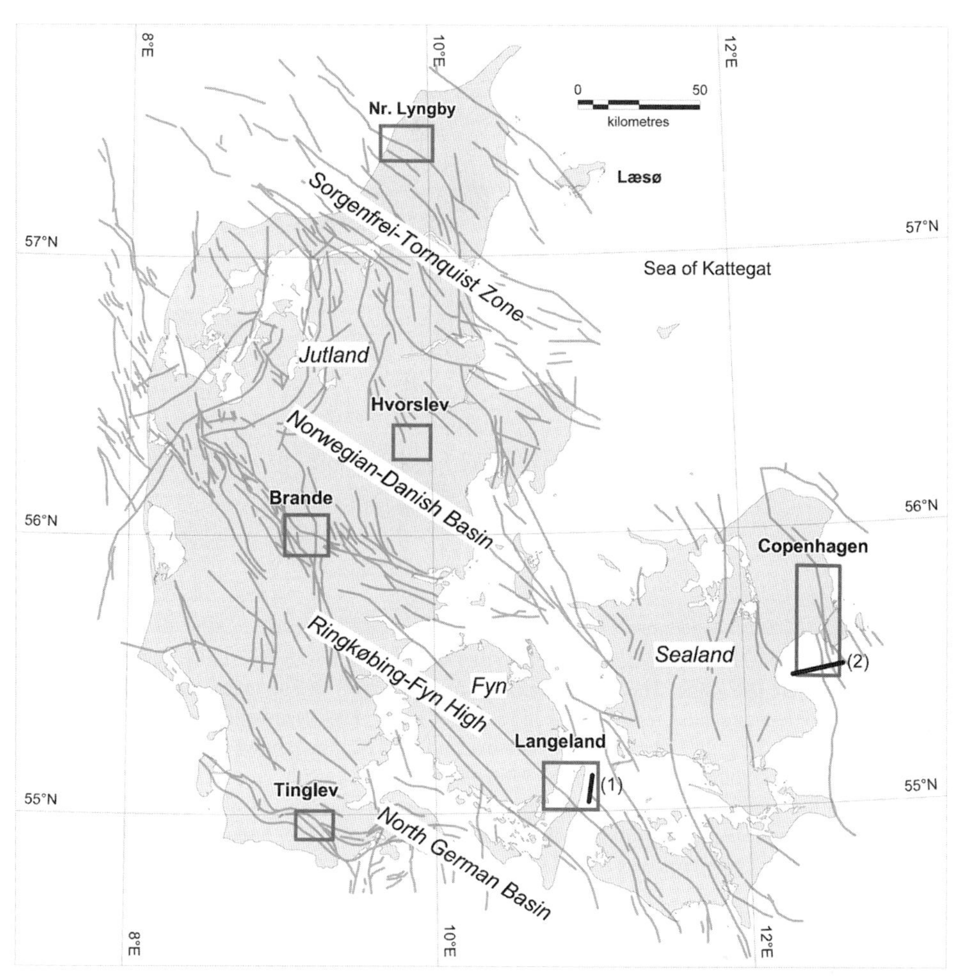

Figure 15.4 Locations of selected examples of Late- and postglacial deformations (see text). Deep-seated pre-Quaternary faults are shown as grey lines; compiled from Ter-Borch (1991); Japsen & Langtofte (1991a,b); Britze & Japsen (1991) and Vejbæk & Britze (1994). Short black lines show location of seismic lines: 1) Figure 15.7 and 2) Figure 15.8.

2016). The preferred orientations matched the known deep faults in the area, and the authors concluded that tectonic movements related to deglaciation apparently had created deformed zones in the sediments all the way to the terrain. These more easily erodible deformed zones were likely to have had a distinct control on the drainage patterns in Lateglacial and postglacial times.

The small town of Nørre Lyngby is situated in northern Denmark at the north-western segment of the Sorgenfrei–Tornquist Zone (STZ; Figure 15.4). Here, deformations in Lateglacial marine and lacustrine sediments in a narrow basin have been described in a cliff outcrop (e.g. Jessen & Nordmann, 1915; Lykke-Andersen, 1979, 1992; Fischer et al., 2013). Based on descriptions of sediment structures and analyses of seismic data,

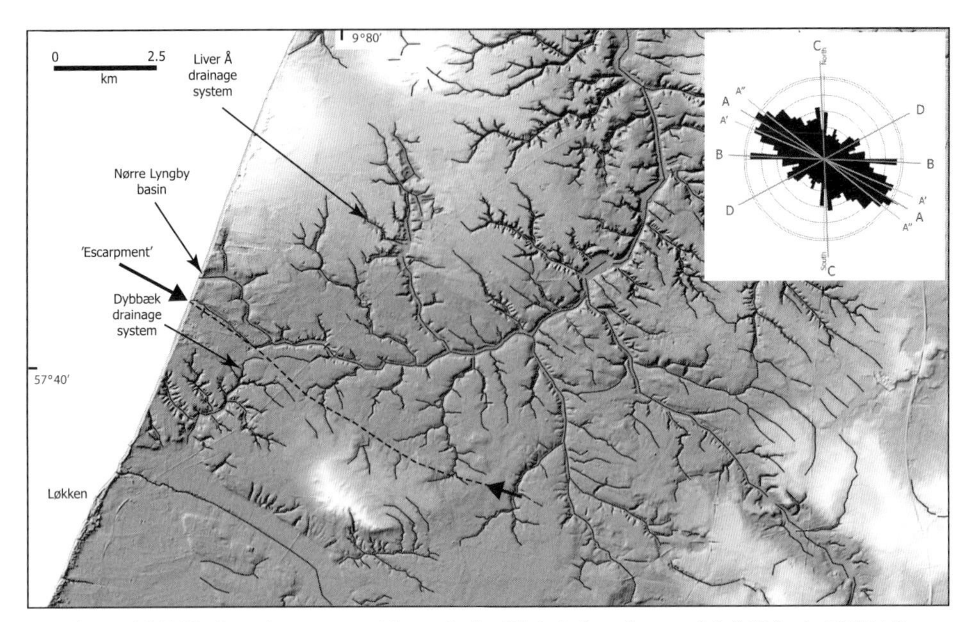

Figure 15.5 Nr. Lyngby topographic analysis. Digital elevation model (hillshaded LiDAR data) shown in greyscale. The elevations range from sea level and up to around 80 m above sea level. The white areas represent the highest elevations. Erosional valleys are digitized as black vectors along the valley bottom. The inserted rose diagram shows cumulated lengths of the vectors plotted in 5° intervals. Orientations marked with grey lines on the rose diagram represent inferred dominant orientations. A WNW-ESE 'escarpment' of 1–2-m height is shown as a hatched line highlighted with arrows. The 'escarpment' acts as a drainage divide between Liver Å and Dybbæk. Modified from Brandes et al. (2018).

Lykke-Andersen (1979, 1992) concluded that a listric fault was responsible for the formation of the basin and the deformation of the marine and lacustrine sediments therein. An earthquake was mentioned as the possible trigger of the faulting. According to Lykke-Andersen (1992) the faulting took place between 12.5 ka BP and 11.2 ka BP. Recently, Brandes et al. (2018) described a series of normal faults and soft-sediment deformation structures (SSDS) at different levels in the exposed sedimentary sequence at Nr. Lyngby (see Figure 15.5 at arrow 'Nr. Lyngby basin'). Parts of the deformed sediments are of Allerød age and Younger Dryas age and the tectonic events responsible for the deformations were placed in an interval from around 14.5 ka BP to 12 ka BP. Analyses of high-resolution LiDAR data in a large area around Nr. Lyngby showed that the drainage pattern eroded into the Lateglacial sediments was dominated by orientations around WNW-ESE. This dominant orientation was – surprisingly – perpendicular to the overall drainage direction and instead comparable to the orientation of the faults of the STZ (Figure 15.5, insert). In addition to this, an up to 2-m-high, WNW-ESE-oriented escarpment formed a drainage divide in the surface of the Lateglacial marine sediments (Figure 15.5, arrow marked 'Escarpment'). The authors interpreted the deformations in the area as caused by glacially induced faulting along the Børglum Fault, which at Nr. Lyngby represents the northern boundary fault of the STZ. As indicated by the SSDS, the movements were most likely accompanied by earthquakes with magnitudes of at least M = 4.2 or even up to M = ~7, as indicated by a fault with an exposed 3-m displacement in the cliff (Brandes et al., 2018). The topographic analysis and the SSDS observations at Nr. Lyngby were supported by numerical simulations of deglaciation-related lithospheric stress build-up.

Deformations of Lateglacial and Holocene sediments that appeared to have happened shortly after sedimentation have also been described at a few localities in Kattegat along the STZ (e.g. Gregersen et al., 1996; Jensen et al., 2002; Bendixen et al., 2013). These authors pointed to the influence of earthquakes and the possibility of reactivations of existing faults in connection with deglaciation. Based on the lithosphere modelling and on comparisons with the observations of deformed sediments in Kattegat, Brandes et al. (2018) interpreted the entire STZ as a structure where glacially induced faulting very likely occurred in Lateglacial times.

Detailed studies of shore-terraces on the island of Læsø located close to the northern boundary of the STZ in the Kattegat Sea (Figure 15.4) led Hansen (1980) to conclude that sudden NE-ward tilting of the island occurred several times during postglacial times. The author assigned the events to neotectonic movements in the Fennoscandian Border Zone (STZ). However, unequivocal postglacial faults were not found and e.g. Gregersen and Voss (2010, 2014) later questioned the conclusion. In Hansen et al. (2016), the postglacial tectonics on Læsø are described without associated earthquakes.

In the central part of Jutland (Figure 15.4), joint studies of seismic data, topography and precise geodetic levellings led Lykke-Andersen et al. (1996) to conclude that tectonic events during the Weichselian or later produced deformations of the sedimentary succession all the way to the terrain surface. The topography in the area is dominated by hills of older glacial landscape (pre-Weichselian 'hill-islands' *sensu*, Ussing, 1903) surrounded by a gently westward-sloping outwash plain formed during the Late Weichselian. However, the topography studies showed a remarkable feature: north-east of Brande a triangular part of the outwash plain measuring 6–8 km across was observed lying approximately 5 m lower than the part of the outwash plain just to the north. The seismic sections showed presence of faults below many of the stream segments in the area and positive flower structures underneath some of the surrounding hill-islands. Precise geodetic levelling even pointed to distinct jumps in measured values across the area suggesting present-day movements along certain faults. The authors concluded that tectonic activity has had an influence on the development of the landscape in the central part of Jutland during the Weichselian or maybe later. This especially concerns the varying altitude of the top of the outwash plain and the erosion pattern as revealed by the drainage geometry of brooks and streams. In addition to this, the apparent positive correlation between the flower structures and the present-day terrain implied that parts of the hill-islands might have increased their altitude because of the tectonic activity. The authors proposed a kinematic model involving reactivation of old NW-SE dextral strike-slip faults along the northern flank of the Ringkøbing–Fyn High (see Figure 15.4). Madirazza (2002) further elaborated on the ideas put forward by Lykke-Andersen et al. (1996).

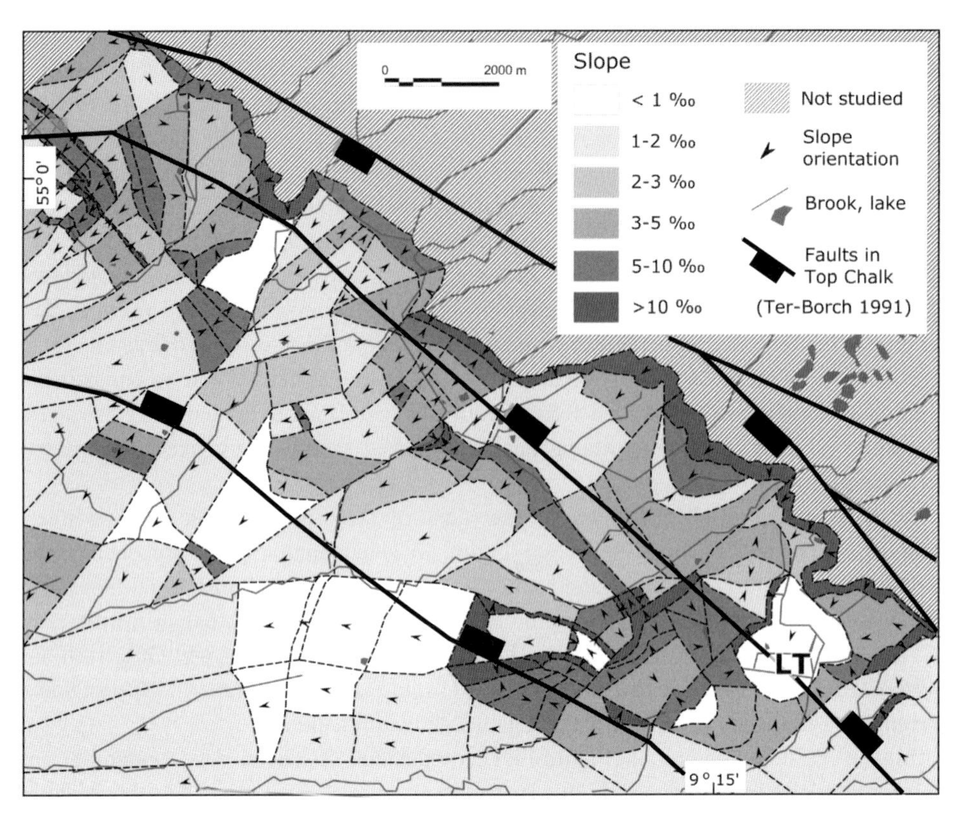

Figure 15.6 Tinglev outwash plain. Slope magnitude (grey shades) and slope orientation (arrows). The streams in the area follow the mosaic pattern and show signs of deflection relative to the general slope. 'LT' marks the location of Lake Tinglev. The faults shown represent the main faults of the Tønder Graben. Modified from Sandersen & Jørgensen (2015).

In a similar setting, Sandersen and Jørgensen (2015) analyzed high-resolution LiDAR data from the gently westward-sloping Late Weichselian Tinglev outwash plain in southern Jutland (Figures 15.4 and 15.6). The authors found that certain parts of the topography showed unusual irregularities of the outwash plain surface such as an intricate mosaic of small areas with pronounced changes in slope magnitude and orientation compared to neighbouring areas (Figure 15.6). The authors also found that the outwash plain featured notable changes in elevation across kilometre-long lineaments and areas lying both lower and higher than expected. The large depressions on the outwash plain were previously interpreted as the result of melting dead-ice blocks left by the Late Weichselian ice sheet. However, borehole samples from Lake Tinglev ('LT' on Figure 15.6) showed that lake sedimentation did not begin until the early Holocene (9,000 years ago) – a mere 9,000 years after the ice sheet had retreated from the area. The authors found that the topographic irregularities were located above or at the flank of the Tønder Graben structure below (Figure 15.6) and the deformations of the outwash plain were interpreted as the result of strike-slip movements along the graben faults, resulting in subsidence as well as uplift. The deformations of the outwash plain were interpreted as being the effect of tectonic events related to deglaciation acting within a presumably short time interval at the beginning of the Holocene.

In a study of the south-western Baltic Sea, Al Hseinat and Hübscher (2017) found several deep-seated faults propagating upwards through the unconsolidated Pleistocene sediments all the way to the sea floor, e.g. a graben structure mapped at sea just north-east of the island of Langeland (Figures 15.4 and 15.7). Here, two conjugate faults cut the succession from Triassic into Pleistocene deposits, resulting in a graben structure around 3 km wide. The faulting and deformation of the Pleistocene and Holocene sediments are not documented by other types of data, but the high-resolution data provides a picture of a sedimentary succession very recently deformed. The authors concluded that ice-sheet loading and unloading, large-scale glacial isostatic adjustment and the present-day stress

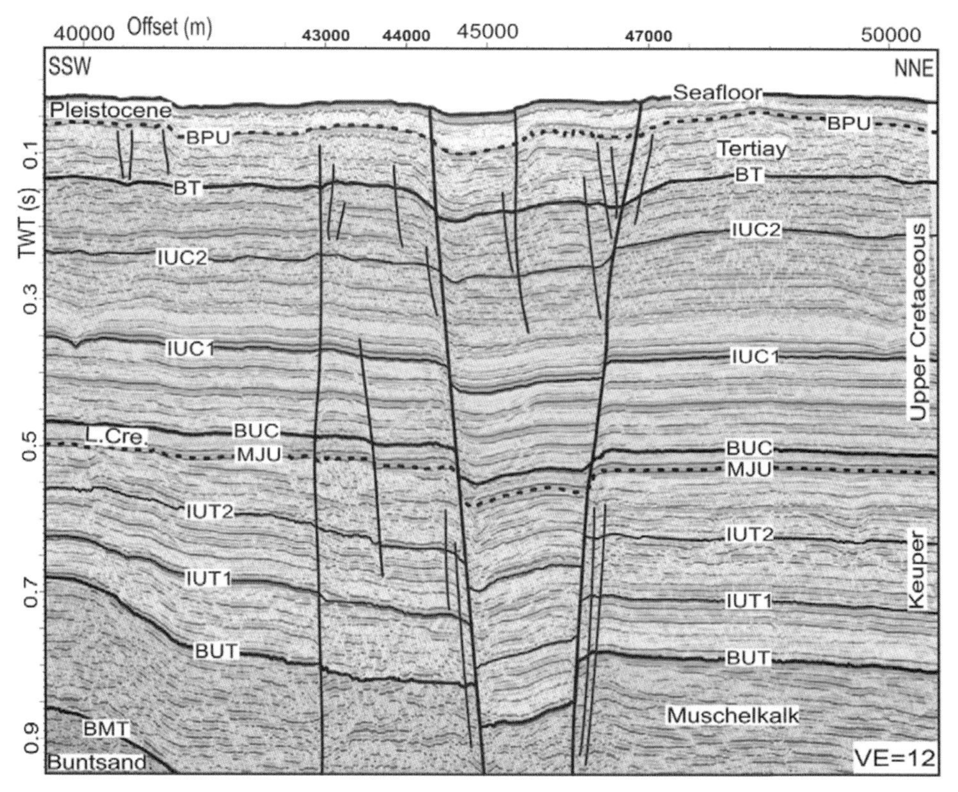

Figure 15.7 Seismic section Langeland. An interpreted SSW-NNE-time-migrated seismic section offshore north-east of Langeland. BMT – Base Middle Triassic; BPU – Base Pleistocene Unconformity; BT – Base Tertiary; BUC – Base Upper Cretaceous; BUT – Base Upper Triassic; IUC 1 and 2 – Internal Upper Cretaceous 1 and 2; IUT 1 and 2 – Internal Upper Triassic 1 and 2; and MJU – Mid Jurassic Unconformity. For location of seismic line, see Figure 15.4 (modified from Al Hseinat & Hübscher, 2017).

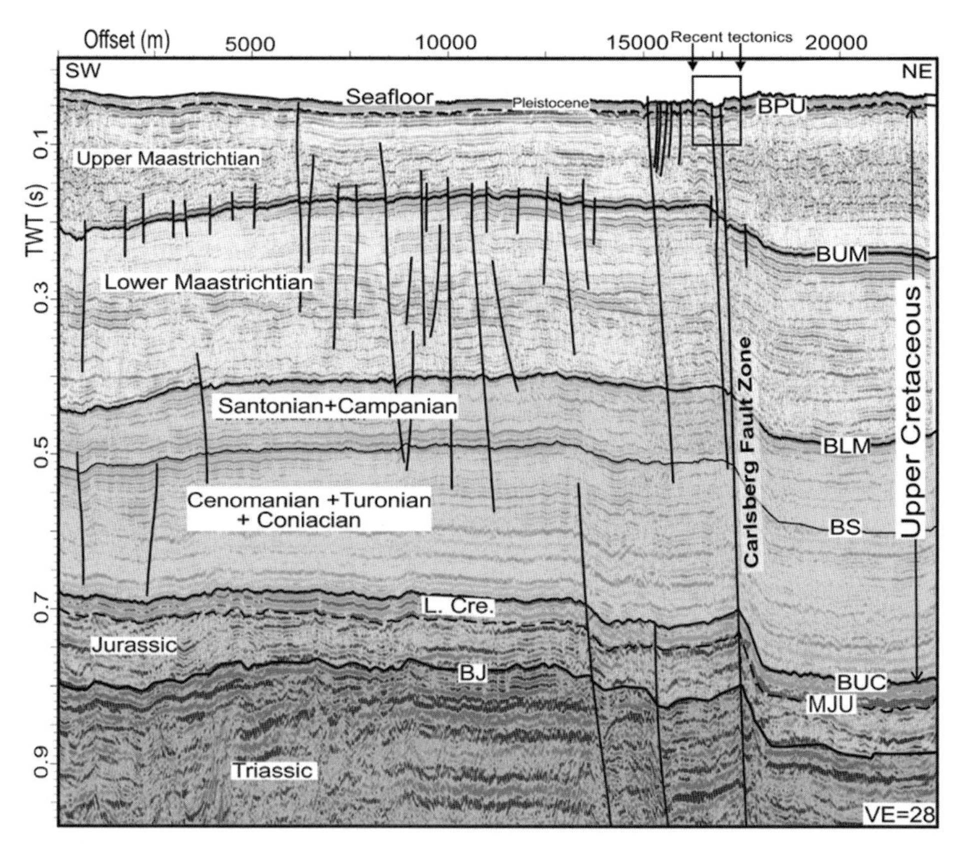

Figure 15.8 Seismic section Copenhagen. An interpreted SW-NE-time-migrated seismic section offshore just south of Copenhagen. The Carlsberg Fault cuts the entire section and pierces the seafloor. BJ – Base Jurassic; BLM – Base Lower Maastrichtian; BPU – Base Pleistocene Unconformity; BS – Base Santonian; BUC – Base Upper Cretaceous; BUM – Base Upper Maastrichtian; MJU – Mid Jurassic Unconformity. For location of seismic line, see Figure 15.4. Modified from Al-Hseinat & Hübscher (2017).

measurements presumably are the result of surficial motion of the geodetic point foundations (Gregersen & Voss, 2014).

The extension of the Carlsberg Fault Zone offshore south of Copenhagen has been investigated by Al-Hseinat and Hübscher (2017), see Figures 15.4 and 15.8. The authors found that faults in the Carlsberg Fault Zone penetrated the Pleistocene sediments all the way to the sea floor, suggesting recent tectonics.

The role of salt kinematics on deformation of near-surface sediments in Denmark is poorly understood. A few descriptions of deformations of Quaternary sediments above salt diapirs in the Norwegian–Danish Basin pointed to a combination of halokinetics, fault reactivation and dissolution of the salt (e.g. Madirazza, 1968; Madirazza & Jacobsen, 1998). Detailed investigations of deformations of sediments and topography above salt diapirs have not been made since. It is however likely that the loading and unloading of ice sheets in the Danish area have had an influence on salt movement during the Quaternary in the same way as described and modelled in German examples (Sirocko et al., 2008; Al Hseinat et al., 2016; Lang et al., 2014).

15.5 Discussion

The main cause of historic intraplate earthquakes in Denmark is believed to be compression from the lithospheric plate motion with only a minor contribution of stress induced by the still ongoing glacial isostatic adjustment. The present-day seismicity is rather small and is – as mentioned earlier – only marginally related to the geologically well-known older fault zones in Denmark.

The different types of examples described in the preceding text indicate that although presumably not seismically active today some fault zones in Denmark appear to have experienced periods of reactivation during the Lateglacial and the Holocene. Lithosphere modelling performed by Brandes et al. (2018) for the STZ has likewise pointed to fault instability following the deglaciation. The modelling results supported field observations of seismically disturbed Lateglacial sediments. In the south-western part of Denmark, Sandersen and Jørgensen (2015) interpreted depressions on the Late Weichselian outwash plain to be the result of fault reactivation of the Tønder Graben in the early Holocene. The findings were supported by dated lake sediments. Descriptions of deformations of a Late Weichselian outwash plain in the middle part of Jutland by Lykke–Andersen et al. (1996) pointed in the same direction, although dating of sediment samples was not performed. These three examples support the change in the stress field of Scandinavia following the deglaciation from a domination by the weight-relief from the ice sheet to the present-day domination by plate motion as described by Gregersen and Voss (2009). So just like in other parts of Northern Europe, the Danish area seems to have had a more seismotectonically active period around the time of deglaciation. The other examples mentioned in the preceding pages may be interpreted in much the same way, but difficulties constraining the time interval of the inferred tectonic events make the connection with deglaciation tectonics uncertain. However, the tectonic events responsible for the surface deformations are in all the shown examples related to the Quaternary.

In the examples mentioned above the tectonic activity seems to be highest in the Lateglacial and the early Holocene, which is also supported by lithosphere modelling. However, seismic events in the interval from the early Holocene and up to the first records of historical seismicity are practically unknown in the Danish area. Only the observations by Hansen (1980) on the island of Læsø suggest tectonic movements as late as around 4,000 years ago. For northern Central Europe and based on numerical simulations, Brandes et al. (2015) suggest that the historical seismicity during the past 1,200 years might be explained by the decay of the Scandinavian ice sheet after the Weichselian glaciation. For Sweden, Mörner (2005) lists several palaeoseismic events and sees both an expected maximum between 9.0 ka BP to 11.0 ka BP and a number of seismic events after this time interval. The historical seismicity in Denmark is well known (Figure 15.1), but bridging the gap between the early Holocene and today will require more data and further scientific investigation. Precise geodetic levelling has in a few places shown significant irregularities over known fault zones, pointing to vertical movements even today (Lykke-Andersen et al., 1996; Lykke-Andersen & Borre, 2000; Gregersen & Schmidt, 2001). However, with no accompanying recent earthquakes in these areas (GEUS, 2020) movements of this type appear to be aseismic.

15.6 Conclusions

Details of the presented faults and tectonic activity may be up for discussion now and in future investigations. Nevertheless, they are part of a generalized pattern, which is convincing to the present authors. The pattern shows tectonic activity in Denmark with a maximum intensity at the time of deglaciation caused by the weight-relief from the ice sheet. Afterwards, there was a transition to the modern stress pattern, dominated by the plate movements.

In the same way as for the other parts of Scandinavia, the stress changes induced by the weight-relief from the ice sheets appear to have been able to create short-lived tectonic instability in the Danish area that is normally considered tectonically stable. The findings in the three examples mentioned above point to reactivations of major fault zones. It is therefore likely that other fault zones in the Danish area would be equally susceptible to reactivation and that tectonic events would be of a recurrent nature during the glaciation-deglaciation cycles (Sandersen & Jørgensen, 2015).

Examples of neotectonic events supported by different datasets are still relatively few in Denmark. But new geophysical data acquisition methods focused on the uppermost parts of the subsurface combined with LiDAR and dating of sediment samples will undoubtedly bring forward new examples. A thorough understanding of the tectonic and seismic history of the Danish area during the Quaternary has great importance for investigations related to e.g. groundwater vulnerability, climate adaptation in low-lying areas, radioactive waste repositories and Carbon Capture and Storage (CCS).

Based on the conclusions above and in line with Gregersen (2014), we suggest the following future activities in order to gain more insight into neotectonics in general and glacially induced faulting specifically in the Danish area:

- Palaeoseismological investigations performed close to fault zones where earthquakes are expected to have occurred. Joint interpretation of data from boreholes, trenches and geophysical surveys in combination with high-detailed digital elevation models.
- Establishment of an improved network of seismographs for location of small earthquakes. Better geographical coverage and more certain information on mechanisms can be obtained.
- Further coverage with permanent geodetic points.
- Interpretation of satellite (interferometric synthetic aperture radar, InSAR) data for detecting recent deformation.
- Comparison of sea-level curves with uplift curves.

Acknowledgements

The authors are grateful for the comments provided by Mathilde B. Sørensen and an anonymous reviewer. Their comments and suggestions greatly improved the manuscript.

References

Al Hseinat, M., Hübscher, C., Lang, J. et al. (2016). Triassic to recent tectonic evolution of a crestal collapse graben above a salt-cored anticline in the Glückstadt Graben/North German Basin. *Tectonophysics*, **680**, 50–66, doi.org/10.1016/j.tecto.2016.05.008.

Al Hseinat, M. and Hübscher, C. (2017). Late Cretaceous to recent tectonic evolution of the north German Basin and the transition zone to the Baltic shield/southwest Baltic Sea. *Tectonophysics*, **708**, 28–55, doi.org/10.1016/j.tecto.2017.04.021.

Andersen, T. R., Westergaard, J. H. and Pytlich, A. (2016). Delineation of fault systems on Langeland, Denmark, based on AEM data and boreholes. *ASEG-PESA-AIG 2016*, August 21–24, 2016, Adelaide, Australia. Extended abstract, 6 pp.

Arvidsson, R., Gregersen, S., Kulhanek, O. and Wahlström, R. (1991). Recent Kattegat earthquakes – evidence of active intraplate tectonics in southern Scandinavia. *Physics of the Earth and Planetary Interiors*, **67**, 275–287, doi.org/10.1016/0031-9201(91)90024-C.

Bendixen, C., Jensen, J. B., Bennike, O. and Boldreel, L. O. (2013). Late glacial to early Holocene development of southern Kattegat. *Geological Survey of Denmark and Greenland Bulletin,* **28**, 21–24, doi.org/10.34194/geusb.v28.4712.

Blem, H. (2002). The Carlsberg Fault – history, location and importance. *Danish Geotechnical Society Bulletin,* **19**, 61–82 (in Danish).

Brandes, C., Steffen, H., Steffen, R. and Wu, P. (2015). Intraplate seismicity in northern Central Europe is induced by the last glaciation. *Geology*, **43**, 611–614, doi.org/10.1130/G36710.1.

Brandes, C., Steffen, H., Sandersen, P. B. E., Wu, P. and Winsemann, J. (2018). Glacially induced faulting along the NW segment of the Sorgenfrei–Tornquist Zone, northern Denmark: implications for neotectonics and Lateglacial fault-bound basin formation. *Quaternary Science Reviews,* **189**, 149–168, doi.org/10.1016/j.quascirev.2018.03.036.

Britze, P. and Japsen, P. (1991). Geological map of Denmark 1:400 000: the Danish Basin: «Top Zechstein» and the Triassic (two-way travel time and depth, thickness and interval velocity). *Geological Survey of Denmark, Map Series,* **31**, 1–4.

Bruun-Petersen, J. (1987). Prækvartæroverfladen i Ribe amt, dens højdeforhold og dannelse samt indflydelse på vandindvindingsmulighederne [The land surface in Ribe county, its height level conditions and formation including influence on water extraction possibilities]. *Dansk Geologisk Forening Årsskrift for 1986,* Copenhagen, 1 June 1987, pp. 35–40 (in Danish).

FENCAT (2020). Fennoscandian earthquake catalogue for 1375–2014, www.seismo.helsinki.fi/bulletin/list/catalog/FENCAT.html.

Fischer, A., Clemmensen, L. B., Donahue, R. et al. (2013). Late Paleolithic Nørre Lyngby – a northern outpost close to the west coast of Europe. *Quartär*, **60**, 137–162, doi.org/10.7485/QU60_07.

Frost, R.T.C. (1977). Tectonic patterns in the Danish Region (as deduced from a comparative analysis of magnetic, landsat, bathymetric and gravity lineaments). *Geologie en Mijnbouw*, **56**, 351–362.

GEUS (2020). Registered earthquakes in Denmark, www.geus.dk/natur-og-klima/jordskaelv-og-seismologi/registrerede-jordskaelv-i-danmark/ (in Danish).
Gregersen, S. (1990). Crustal stress in northern Europe. In R. Freeman and S. Müller, eds., *Proceedings of the VI workshop on the European Geotraverse Project*. ESF Strasbourg, pp. 357–360.
Gregersen, S. (1991). Crustal structure across the Tornquist Zone (southwestern edge of the Baltic Shield): a review of the EUGENO-S geophysical results. *Tectonophysics*, **189**, 165–182, doi.org/10.1016/0040-1951(91)90494-D.
Gregersen, S. (2014). Jordskælvsrisiko i Danmark? – forslag til fremtidens studier [Earthquake risk in Denmark? – Suggestions for future studies]. *Kvant*, June 2014, Danish Geophysical Society.
Gregersen, S., Leth, J., Lind, G. and Lykke-Andersen, H. (1996). Earthquake activity and its relationship with geologically recent motion in Denmark. *Tectonophysics*, **257**, 265–273, doi.org/10.1016/0040-1951(95)00193-X.
Gregersen, S. and Schmidt, K. (2001). Tektonik i Danmark – Sorgenfrei–Tornquist Zonen [Tectonics in Denmark – The Sorgenfrei–Tornquist Zone]. *Geologisk Nyt*, February 2001, 16–17.
Gregersen, S., Voss, P., Nielsen, L. V. et al. (2009). Uniqueness of modelling results from teleseismic P-wave tomography in Project Tor. *Tectonophysics*, **481**, 99–107, doi .org/10.1016/j.tecto2009.01.020.
Gregersen, S. and Voss, P. (2009). Stress change over short geological time: the case of Scandinavia over 9000 years since the Ice Age. In K. Reicherter, A. M. Michetti and P. G. Silva Barroso, eds., *Palaeoseismology: Historical and Prehistorical Records of Earthquake Ground Effects for Seismic Hazard Assessment*. Geological Society, London, Special Publication, Vol. 316, pp. 173–178, doi.org/10.1144/SP316.10.
Gregersen, S. and Voss, P. (2010). Irregularities in Scandinavian postglacial uplift/subsidence in time scales tens, hundreds, thousands of years. *Journal of Geodynamics*, **50**, 27–31, doi.org/10.1016/j.jog.2009.11.004.
Gregersen, S. and Voss, P. H. (2012). Efforts to include geological and geodetic observations in the assessment of earthquake activity in Denmark. *Geological Survey of Denmark and Greenland Bulletin*, **26**, 41–44, doi.org/10.34194/geusb .v26.4747.
Gregersen, S. and Voss, P. H. (2014). Review of some significant claimed irregularities in Scandinavian postglacial uplift on timescales of tens to thousands of years – earthquakes in Denmark? *Solid Earth*, **5**, 109–119, doi.org/10.5194/se-5-109-2014.
Hansen, D. L., Nielsen, S. B. and Lykke-Andersen, H. (2000). The post-Triassic evolution of the Sorgenfrei–Tornquist Zone – results from thermo-mechanical modelling. *Tectonophysics*, **328**, 245–267, doi.org/10.1016/S0040-1951(00)00216-X.
Hansen, J. M. (1980). Læsøs postglaciale udvikling i relation til den Fennoskandiske randzone [Læsø's postglacial development in relation to the Fennoscandian Border Zone]. *Dansk Geologisk Forening Årsskrift for 1979*, 23–30.
Hansen, J. M., Aagaard, T., Stockmarr, J. et al. (2016). Continuous record of Holocene sea-level changes and coastal development of the Kattegat island Læsø (4900 years BP to present). *Bulletin of the Geological Society of Denmark*, **64**, 1–55, doi.org/10.37570/bgsd-2016-64-01.
Heilmann-Clausen, C. (1995). Palæogene aflejringer over danskekalken [Palaeogene deposits above the Danian limestone]. In O. B. Nielsen, ed., *Danmarks geologi fra Øvre Kridt til i dag*. *Århus Geokompendier*, **1**, 79–113, Århus Universitet.
Houmark-Nielsen, M. (1987). Pleistocene stratigraphy and glacial history of the central part of Denmark. *Bulletin of the Geological Society of Denmark*, **36**, 189 pp.

Jakobsen, P. R. and Pedersen, S. A. S. (2009). Fracture valleys in central Jylland – a neotectonic feature. *Geological Survey of Denmark and Greenland Bulletin*, **17**, 33–36, doi.org/10.34194/geusb.v17.5008.

Jakobsen, P. R., Fallesen, J. and Knudsen, C. (2013a). Strukturer i den Københavnske undergrund – folder, forkastninger og sprækker [Structures in Copenhagen's subsurface – folds, faults and cracks]. *Danish Geotechnical Society Bulletin*, **19**, December 2002, 19–29.

Jakobsen, P. R., Wegmuller, U., Capes, R. and Pedersen, S. A. S. (2013b). Terrain subsidence detected by satellite radar scanning of the Copenhagen area, Denmark, and its relation to the tectonic framework. *Geological Survey of Denmark and Greenland Bulletin*, **28**, 25–28, doi.org/10.34194/geusb.v28.4713.

Japsen, P. and Langtofte, C. (1991a). Geologisk kort over Danmark. Det danske Bassin. "Basis Kalk" og Kalk Gruppen [Geological map of Denmark. The Danish Basin. "Basis Lime" and the Lime Group]. *Danmarks Geologiske Undersøgelse Kortserie*, **29**.

Japsen, P. and Langtofte, C. (1991b). Geologisk kort over Danmark, 1:400.000. Det danske Bassin. "Top Trias" og Jura-Nedre Kridt [Geological map of Denmark, 1:400,000. The Danish Basin. "Top Triassic" and Jurassic–Lower Cretaceous]. *Danmarks Geologiske Undersøgelse Kortserie*, **30**.

Jensen, J. B., Petersen, K. S., Konradi, P. et al. (2002). Neotectonics, sea-level changes and biological evolution in the Fennoscandian Border Zone of the southern Kattegat Sea. *Boreas*, **31**, 133–150, doi.org/10.1111/j.1502-3885.2002.tb01062.x.

Jessen, A. and Nordmann, V. (1915). Ferskvandslagene ved Nr. Lyngby. English summary: the fresh-water deposits at Nørre Lyngby. *Geological Survey of Denmark, II Series*, **29**, 66 pp.

Kammann, J., Hübscher, C., Boldreel, L. O. and Nielsen, L. (2016). High-resolution shear-wave seismics across the Carlsberg Fault zone south of Copenhagen – implications for linking Mesozoic and late Pleistocene structures. *Tectonophysics*, **682**, 56–64, doi .org/10.1016/j.tecto.2016.05.043.

Kronborg, C., Bender, H. and Larsen, G. (1978). Tektonik som en mulig medvirkende årsag til daldannelsen i Midtjylland [Tectonics as a possible contributing factor to the valley formation in Mid Jutland]. *Danmarks Geologiske Undersøgelser, Årbog 1977*, 63–76.

Lang, J., Hampel, A., Brandes, C. and Winsemann, J. (2014). Response of salt structures to ice-sheet loading: implications for ice-marginal and subglacial processes. *Quaternary Science Reviews*, **101**, 217–233, doi.org/10.1016/j.quascirev.2014.07.022.

Lykke-Andersen, H. (1979). Nogle undergrundstektoniske elementer i det danske Kvartær [Some subsurficial tectonic elements in the Danish Quaternary]. *Dansk Geologisk Forening*, Årsskrift for 1978, 1–6 (in Danish).

Lykke-Andersen, H. (1981). Indications of neotectonic features in Denmark. *Zeitschrift für Geomorphologie Neue Folge*, **Suppl.-Bd. 40**, 43–54.

Lykke-Andersen, H. (1992). Massebevægelser i Vendsyssels og Kattegats kvartære Aflejringer [Mass movements in the Quaternary deposits of Vendsyssel and the Kattegat]. *Dansk Geologisk Forening Årsskrift for 1990–91*, 93–97 (in Danish).

Lykke-Andersen, H. and Borre, K. (2000). Aktiv tektonik i Danmark – der er liv i Sorgenfrei–Tornquist Zonen [Active tectonics in Denmark – there is life in the Sorgenfrei–Tornquist Zone]. *Geologisk Nyt*, December 2000, 12–13.

Lykke-Andersen, H., Madirazza, I. and Sandersen, P. B. E. (1996). Tektonik og landskabsdannelse i Midtjylland [Tectonics and landscape formation in Mid-Jutland]. *Geologisk Tidsskrift 1996,* **3**, 1–32.

Madirazza, I. (1968). An interpretation of the Quaternary morphology in the Paarup Salt Dome area. *Meddelande Dansk Geologisk Forening*, **18**, København.

Madirazza, I. (2002). The influence of tectonics on the land forms in west Jutland, Denmark. *Bulletin of the Geological Society of Denmark*, **49**, 63–77.

Madirazza, I. and Jacobsen, B. H. (1998). Nøvling: An unusual salt structure on the southern margin of the Danish Zechstein Basin. *Bulletin of the Geological Society of Denmark*, **44**, 139–149.

Markussen, L. M. (2002). Grundvandsforhold i København [Groundwater conditions in Copenhagen]. *DGF-Bulletin*, **19**, 165–182.

Milthers, V. (1916). Spaltedale i Jylland [Spaltedale in Jutland]. *Danmarks Geologiske Undersøgelse IV*, 16 pp.

Mörner, N.-A. (2005). An investigation and catalogue of paleoseismology in Sweden. *Tectonophysics*, **408**, 265–307, doi.org/10.1016/j.tecto.2005.05.039.

Nielsen L. and Thybo, H. (2004). Location of the Carlsberg Fault zone from seismic controlled-source fan recordings. *Geophysical Research Letters*, **31**, doi.org/10.1029/2004GL019603.

Nordmann, V. (1958). Kortbladet Fredericia. A: Kvartære aflejringer [Map sheet of Fredericia. A: Quaternary deposits]. *Danmarks Geologiske Undersøgelse*, **1**, No. 22. English summary, 125 pp.

Ovesen N. K., Blem, H., Gregersen, S., Møller, H. M. F. and Frederiksen, J. K. (2002). Recent terrain-movements in Copenhagen. *Bulletin Geotechnical Society*, **19**, 183–192 (in Danish).

Pedersen, S. A. S. and Gravesen, P. (2016). Tectonic control on the formation of Roskilde Fjord, central Sjælland, Denmark. *Geological Survey of Denmark and Greenland Bulletin*, **35**, 35–38, doi.org/10.34194/geusb.v35.4904.

Pedersen, S. A. S., Rasmussen, L. Aa. and Fredericia, J. (2015). Kortbladsbeskrivelse til Geologisk kort over Danmark, 1:50 000, Sakskøbing 1411 I og 1412 II Syd [Map sheet description for the Geological Map of Denmark, 1:50,000, Sakskøbing 1411 I and 1412 II South] (with a summary in English). *Geological Survey Denmark and Greenland Map Series*, **6**, 42 pp., doi.org/10.34194/geusb.v6.4564.

Pedersen, T., Gregersen, S. and TOR working group (1999). Project Tor: deep lithospheric variation across the Sorgenfrei–Tornquist Zone, Southern Scandinavia. *Bulletin of the Geological Society Denmark*, **46**, 13–24.

Rasmussen, E. S., Dybkjær, K. and Piasecki, S. (2010). Lithostratigraphy of the Upper Oligocene – Miocene succession of Denmark. *Geological Survey of Denmark Greenland Bulletin*, **22**, 92 pp, doi.org/10.34194/geusb.v22.4733.

Rosenkrantz, A. (1937). Bemærkninger om det østsjællandske Daniens stratigrafi og tektonik [Remarks on East Zealand's Danian stratigraphy and tectonics]. *Meddr dansk geologisk Forening*, **9**, 199–212.

Sandersen, P. and Jørgensen, F. (2002). Kortlægning af begravede dale i Jylland og på Fyn Opdatering 2001–2002 [Mapping of Buried Valleys in Jutland and Funen. Update 2001–2002]. *De jysk-fynske amters grundvandssamarbejde*, Vejle Amt, WaterTech a/s, 189 pp. (in Danish).

Sandersen, P. B. E. and Jørgensen, F. (2015). Neotectonic deformation of a Late Weichselian outwash plain by deglaciation-induced fault reactivation of a deep-seated graben structure. *Boreas*, **44**(2), 413–431, doi.org/10.1111/bor.12103.

Sandersen, P. B. E. and Jørgensen, F. (2016). Kortlægning af begravede dale i Danmark. Opdatering 2015 [Mapping of Buried Valleys in Denmark. Update 2015]. *GEUS Special Publication*, December 2016, Vols. 1 & 2 (in Danish).

Sirocko, F., Reicherter, K., Lehne, R. W. et al. (2008). Glaciation, salt and the present landscape. In R. Littke et al., eds., *Dynamics of Complex Intracontinental Basins: The Central European Basin System.* Springer Verlag, Heidelberg, pp. 234–245.

Stewart, I. S., Sauber, J. and Rose, J. (2000). Glacio-seismotectonics: ice sheets, crustal deformation and seismicity. *Quaternary Science Reviews*, **19**, 1367–1389, doi.org/10.1016/S0277-3791(00)00094-9.

Stockmarr, J. (1978). Den prækvartære overflade ved Juelsminde, Danmark [The Pre-Quaternary surface at Juelsminde, Denmark]. *DGU Årbog 1976*, 49–52.

Ter-Borch, N. (1991). Geological map of Denmark, 1:500.000. Structural map of the top chalk group. *Geological Survey of Denmark Map Series*, **7,** 4 pp. Copenhagen.

Ussing, N. V. (1903). Om Jyllands Hedesletter og Teorierne for Deres Dannelse [About Jutland's Hedesletter and the theories of their formation]. *Oversigt over det KGL. danske Videnskabernes Selskabs Forhandlinger*, 1903, **2**.

Vejbæk, O. V. and Britze, P. (1994). Geological map of Denmark, 1:750.000. Top Pre-Zechstein (two-way travel time and depth). *Geological Survey of Denmark Map Series*, **45**, 8 pp. Copenhagen.

Zoback, M. L., Zoback, M. D., Adams, J. et al. (1989). Global patterns of tectonic stress. *Nature*, **341**, 291–298, doi.org/10.1038/341291a0.

16

Glacially Induced Faults in Germany

KATHARINA MÜLLER, JUTTA WINSEMANN, DAVID C. TANNER, THOMAS LEGE, THOMAS SPIES AND CHRISTIAN BRANDES

ABSTRACT

Germany is a geologically diverse, intraplate setting affected by several tectonic phases, which caused a complex fault pattern. Despite the intraplate setting, significant palaeo-, historical and recent seismicity has been observed on many faults, especially in three zones of crustal weakness: the Rhine Rift Valley, the Swabian Alp, and eastern Thuringia/western Saxony. Recent studies have shown that the low seismicity of northern Germany is characterized by fault activity caused by the decay of the Late Pleistocene (Weichselian) ice sheet. Several faults and fault systems show evidence of neotectonic activity, such as the Aller Valley Fault System, Halle Fault System, Harz Boundary Fault, Steinhuder Meer Fault and Osning Thrust, all of which are oriented parallel to the margin of the Pleistocene ice sheets. The timing of fault movements implies that seismicity in northern Germany is likely induced by varying lithospheric stress conditions related to glacial isostatic adjustment (GIA), and faults can be thus classified as glacially induced faults (GIFs). For the Osning Thrust, the Harz Boundary Fault and the Schaabe Fault, this is supported by numerical simulation of GIA-related stress field changes. GIA processes are also a likely driver for the historical and parts of the recent fault activity. The southern extent of GIA-induced fault reactivations caused by the decay of the Fennoscandian ice sheet is not clear. Modelling results imply the influence of GIA reached up to 230 km south of the former Weichselian ice sheet. GIA processes are also described for the Alps, but it is difficult to clearly distinguish between reactivation of faults in the foreland of the Alps due to the Alpine collision and GIA processes.

16.1 Introduction

The geological structure of Germany was shaped by terrane accretion during the Caledonian and Variscan orogenies in the Palaeozoic and subsequently overprinted by Late Palaeozoic, Mesozoic and Cenozoic tectonic phases. The latter include the inversion tectonics in Central Europe, the Alpine collision and the rift phase in the Rhine Valley. This polyphase tectonic history has created a complex fault pattern (Figure 16.1A). The majority of today's seismically active faults are concentrated in the Rhine Rift Valley, the Swabian

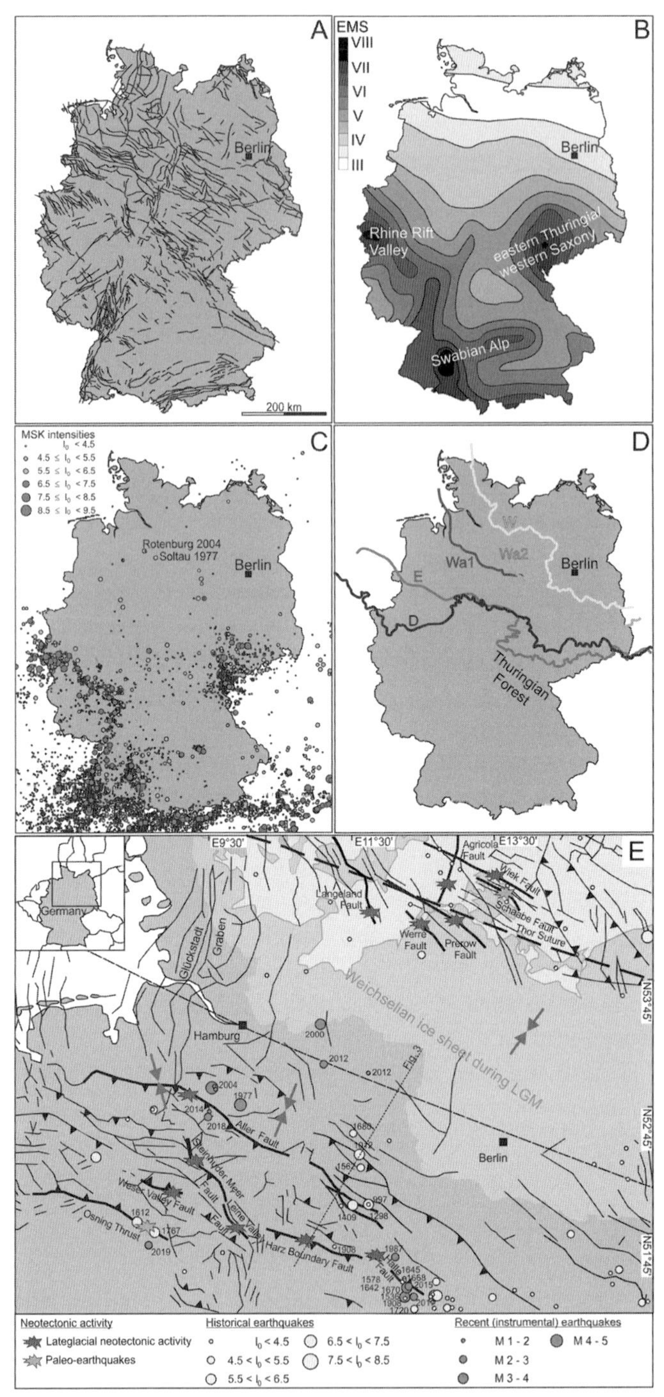

Figure 16.1 Map of Germany. A) Major faults and fault systems in Germany (modified after Schulz, 2013). B) Intensity-based earthquake hazard map. The greyscale displays the macro seismic intensities using the European Macroseismic Scale (EMS) (modified after

Alp and eastern Thuringia/western Saxony, including the earthquake swarm area of the Vogtland (Figure 16.1B) (Grünthal et al., 1998).

Historical earthquakes, with epicentral intensities of up to IX, have occurred in Germany during the last 1,200 years (Figure 16.1C) (Leydecker, 2011), indicating repeated fault activity. In the last 100 years, seismic activity has been instrumentally detected. Besides natural tectonic earthquakes, there have been several recent earthquakes in northern Germany, with epicentres near active gas fields that have been interpreted as the consequence of hydrocarbon exploitation (Dahm et al., 2007, 2015).

Recent studies have shown that northern Germany is characterized by Lateglacial fault activity that is attributed to fault reactivation caused by the decay of the Late Pleistocene Weichselian ice sheet (Figure 16.1D,E) (Brandes et al., 2012). A study on fault reactivation in intraplate areas controlled by ice sheet movements was published by Reicherter et al. (2005). In addition, along salt structures that reach near the Earth's surface, e.g. the Schlieven and Marnitz salt pillows in south-western Mecklenburg (Müller & Obst, 2008) and the Sperenberg and Rambow salt diapirs in Brandenburg (Stackebrandt, 2005, 2015), there is evidence of reactivated basement structures caused by glacial isostatic adjustment (GIA). See also the modelling study of Lang et al. (2014).

Numerical modelling studies imply that many faults in northern Germany became unstable due to GIA-related stress-field changes (Brandes et al., 2012, 2015; Pisarska-Jamroży et al., 2019; Müller et al., 2020b). These findings raise the question of whether there are other glacially induced faults (GIFs) in Germany.

GIFs have been documented in Fennoscandia over the last six decades (e.g. Kujansuu, 1964; Mörner, 1978), a region that was covered by large ice sheets during the Pleistocene. The occurrence of GIFs outside the former glaciated areas is a novelty and has a significant impact on the distribution of neotectonic activity in Germany (and other areas), as shown by the modelling results of Grollimund and Zoback (2001), Hampel et al. (2009), Brandes et al. (2015), Pisarska-Jamroży et al. (2019), Müller et al. (2020b), and therefore requires re-evaluation. Potential GIFs can occur in northern and southern Germany, related to the Fennoscandian ice sheet (Figure 16.1D) and the Alpine ice cover, respectively.

The extent of the area prone to fault reactivation is not clear. The results of Brandes et al. (2015) imply a GIA influence as far south as the Thuringian Forest, located up to 230 km south of the former Weichselian ice sheet. Grützner et al. (2016), Vanneste et al. (2018) and

---

Figure 16.1 (*cont.*) Grünthal et al., 1998). C) Map of earthquake epicentres and adjacent areas for the period CE 800–2008 (modified after Leydecker, 2011). D) Maximum extents of the Elsterian (E), Saalian Drenthe (D), Warthe 1 (Wa1), Warthe 2 (Wa2) and Weichselian (W) ice sheets (modified after Winsemann et al., 2020). E) Tectonic activity along major basement faults in northern Germany. Shown are neotectonic activity, historical earthquakes and recent earthquakes (modified after Brandes et al., 2015; faults from Kley & Voigt, 2008; Seidel et al., 2018; Elbe line from Scheck-Wenderoth & Lamarch, 2005). The red arrows show the recent stress field orientation taken from Marotta et al. (2000, 2002). The numbers represent the years of the earthquakes, which are listed in Table 16.1. (A black and white version of this figure will appear in some formats. For the colour version, please refer to the plate section.)

van Balen et al. (2019) have discussed a potential GIA effect on the faults in the Lower Rhine Rift Valley. However, these faults were already active before the Pleistocene, and a clear reactivation due to GIA processes, e.g. supported by modelling studies, has not yet been shown. Therefore, this study focuses on northern Germany.

16.2 Geological Setting

16.2.1 Main Geological Structures

The geological evolution of Germany was significantly influenced by the Cadomian, Caledonian, Variscan and Alpine orogenies (McCann, 2008). The area consists of a terrane assemblage that was formed progressively during the closure of the Tornquist Sea, the Rhenohercynian Ocean and the Saxothuringian Ocean (Torsvik & Cocks, 2017). From north to south, Germany consists of the East Avalonian Terrane, the Saxothuringian Terrane (composed of Franconia and Thuringia) and the Moldanubian Terrane (Franke, 2000; Franke et al., 2017). The subdivision of the Variscan orogenic belt into three zones, as originally defined by Kossmat (1927), the so-called Rhenohercynian, Saxothuringian and Moldanubian zone, reflects this terrane assemblage (Figure 16.2A).

16.2.2 The Central European Basin System

Northern Germany forms part of the Central European Basin System (CEBS) (Figure 16.2B). After the Variscan Orogeny, the CEBS evolved on top of a Carboniferous Variscan foreland

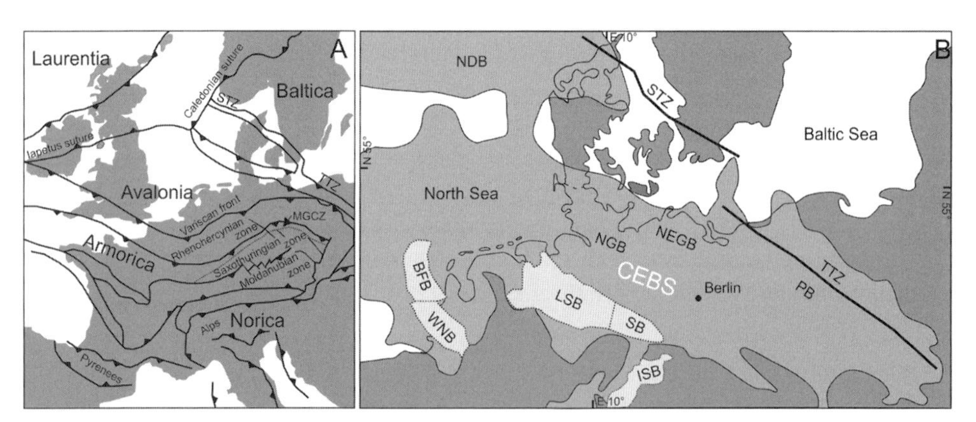

Figure 16.2 A) Tectonostratigraphic units of Germany. TTZ – Tornquist-Teisseyre Zone; STZ – Sorgenfrei-Tornquist Zone; MGCZ – Mid-German Crystalline Zone (modified after Meschede, 2015). B) Overview of the sedimentary basins in the study area; CEBS – Central European Basin System; NGB – North German Basin; NEGB – Northeast German Basin; NDB – Norwegian Danish Basin; PB – Polish Basin; BFB: Broad Fourteens Basin; LSB: Lower Saxony Basin; SB – Subhercynian Basin; WNB – West Netherlands Basin, ISB – intramontane Saale Basin (modified after Scheck-Wenderoth and Lamarch, 2005; Pharaoh et al., 2010).

basin (Littke et al., 2008). During Permian times, after the filling of the peripheral foreland basin, a rift system developed (e.g. Gast & Gundlach, 2006). N-S-striking normal faults were the major structural elements. Strong subsidence in this area during the Mesozoic (e.g. van Wees et al., 2000) was followed by a tectonic inversion phase in Late Cretaceous to Early Palaeocene times (e.g. Mazur et al., 2005; Kley & Voigt, 2008). Within the CEBS, an elongated zone of subsidence (the so-called Mid-German subsidence zone) has shown neotectonic activity since the Oligocene (Stackebrandt, 2004, 2015).

Today, the CEBS consists of several sub-basins such as the North German Basin, the Northeast German Basin, the Lower Saxony Basin and the Subhercynian Basin (Figure 16.2B). As a consequence of different tectonic phases mentioned above, northern Germany is characterized by a dense fault array with several major faults (striking NW-SE) and fault zones with minor faults striking NE-SW or NNE-SSW (Reicherter et al., 2005, see Figure 16.1A,E).

16.2.3 Pleistocene Glaciations

From the Middle Pleistocene onwards, northern Germany was repeatedly influenced by glaciations (Ehlers et al., 2011; Figure 16.1D) and was therefore affected by stresses caused by glacial loading and unloading. The oldest ice advance may have reached the study area during the Cromerian (Marine Isotope Stage (MIS) 16), or even earlier, during the Menapian (MIS 34) (Ehlers et al., 2011; Winsemann et al., 2020).

The following Middle Pleistocene glaciations are termed 'Elsterian' and 'Saalian', where the latter is the younger. The Elsterian glaciation was characterized by three ice advances that are recorded in northern Germany (Eissmann, 2002; Ehlers et al., 2011). It probably can be correlated with MIS 12 and/or MIS 10 (Roskosch et al., 2015; Lauer & Weiss, 2018). Three major ice advances with several sub-phases occurred during the Saalian glaciation (Eissmann, 2002; Ehlers et al., 2011; Lang et al., 2018; Winsemann et al., 2020). They are commonly correlated with MIS 6 and are referred to as the Drenthe and Warthe ice advances (Figure 16.1D; Ehlers et al., 2011; Lang et al., 2018; Winsemann et al., 2020). However, there is evidence for another ice advance during MIS 8 (Roskosch et al., 2015).

The Late Pleistocene Weichselian ice sheet did not cross the Elbe River, and thus periglacial conditions prevailed in northern Germany (Eissmann, 2002; Meinsen et al., 2014; Lehmkuhl et al., 2016). Three main ice marginal positions are known in north-eastern Germany from the Weichselian glaciation. The peak of the Eurasian ice sheet in terms of global ice volume, often referred to as the last glacial maximum (LGM), occurred at around 21–20 ka (Hughes et al., 2016), but in north-eastern Germany the maximum extent was reached at around 30 ka (Hardt et al., 2016), while further to the west, in Mecklenburg and Schleswig-Holstein, it was reached at around 24 ka (Böse et al., 2012) (Figure 16.1D,E). Subsequently, the ice rapidly melted and retreated. A standstill phase during the down-wasting period occurred, which was later followed by the third minor re-advance (Lüthgens & Böse, 2011; Hardt et al., 2016; Hardt & Böse, 2018).

16.2.4 The Stress Field

The present-day stress field in the CEBS has a fan-shaped pattern (Marotta et al., 2002, 2004). The North German Basin is characterized by NW-SE compression and NE-SW extension (Reinecker et al., 2004; Heidbach et al., 2016), whereas the area of the Northeast German Basin shows a NE-SW compression. The major controls on the stress field, which have an effect on the faults in the CEBS, are the NW-SE-directed force induced by the North Atlantic ridge push and the N-S-directed force caused by the Alpine collision (Marotta et al., 2002; Kaiser et al., 2005).

During and after the Cretaceous inversion phase, the compressional stress field in northern Germany was NE-SW and rotated counterclockwise from NE-SW to NW-SE in the Neogene (Kley & Voigt, 2008). From the Pleistocene onwards to the present day, the main factor that influenced the stress field was the postglacial rebound of Fennoscandia (Kaiser et al., 2005).

The growth and decay of the Fennoscandian ice sheet induced a temporal stress field that matched the orientation of the palaeostress field in northern Germany. This NNE-SSW-oriented major stress direction led to the reactivation of pre-existing mainly Mesozoic faults.

Knowledge of the southern extent of GIA processes caused by the Fennoscandian ice sheet in Germany is limited. Coulomb failure stress (CFS) modelling results show that GIA processes play a major role in the area between the Magdeburger Börde and the Sorgenfrei–Tornquist Zone (Figure 16.2), while the area south of the Thuringian Forest experienced minor influence by GIA processes (Figure 16.1D) (Brandes et al., 2015). However, GIA processes are also described for the Alps (e.g. Norton & Hampel, 2010; Mey et al., 2016), but it is difficult to clearly distinguish GIA fault activity in the foreland of the Alps due to the ongoing Alpine collision.

16.3 Neotectonic Activity and Seismicity in Northern Germany

16.3.1 Palaeoseismological Studies and GIA-Induced Movements

Evidence of palaeoearthquakes is shown by Brandes et al. (2012) and Brandes and Winsemann (2013) for the Osning Thrust. Hoffmann and Reicherter (2012) and Pisarska-Jamroży et al. (2018, 2019) found evidence of seismic activity in Pleistocene sediments along the south-western coast of the Baltic Sea in the area of Usedom and Rügen Island. In addition, studies by Hübscher et al. (2004), Lehné and Sirocko (2007, 2010), Brandes et al. (2011), Ludwig (2011), Brandes and Tanner (2012), Al Hseinat and Hübscher (2014), Brandes et al. (2015), Al Hseinat et al. (2016), Grützner et al. (2016), Al Hseinat and Hübscher (2017), Brandes et al. (2018), Grube (2019a,b) and Huster et al. (2020) also point to higher GIA-induced neotectonic activity in northern Germany than previously thought.

A palaeoseismological study in north-western Germany was presented by Grützner et al. (2016) for the Rurrand Fault in the Lower Rhine Embayment, which is part of an intraplate rift system (Roer Valley Rift System). The Peel Boundary fault zone, located in the Netherlands, and the Geelen Fault in Belgium are part of the rift system and were analyzed by Vanneste et al. (2018) and van Balen et al. (2019). They found evidence of seismic fault activity in the Late Pleistocene (Vanneste et al., 2018; van Balen et al., 2019). Grützner et al. (2016) identified two surface rupturing events at the Rurrand Fault in Holocene sediments.

Furthermore, historical earthquakes in this region occurred near Düren in 1755/1756, with epicentral intensities of IV–VI (6.5) (Leydecker, 2011). Grützner et al. (2016) mentioned that surface rupturing events can occur after peak deglaciation on locked faults. However, the fault is part of an intraplate rift system and GIA processes probably play a minor role in this area. Therefore, it is not yet clear if this is a GIF in the proper sense.

Evidence of neotectonic movements in eastern Germany in the Lausitz and Niederlausitz areas are described from several fault zones and graben structures. Above these deep-seated faults, sediment thickness anomalies, facies changes and displaced Pleistocene deposits have been mapped (e.g. Stackebrandt, 2008; Krentz et al., 2010; Kühner, 2010). At these structures repeated phases of reactivation are documented from Cretaceous until Middle Pleistocene times. Evidence of neotectonic activity during the Pleistocene were found at the Erpitzer Fault Zone (Kühner, 2010), the Stradow–Buckower Fault System, the Pritzener Fault (Kühner, 2010) and the Kauscher Graben (Krentz et al., 2010; Kühner, 2009, 2010). Furthermore, evidence was found at the Weisswasser Graben, the Nochtener Graben, the Graben of Calau and the Zinnitzer Graben (Krentz et al., 2010). Stackebrandt (2008) described that some of the NW-SE-oriented graben structures displaced Pleistocene sediments and assumed GIA as a possible trigger. However, only one earthquake with local magnitude (M_L) of 3.2 was registered in this area, in the year 1920 near Cottbus (Leydecker, 2011). The absence of Lateglacial and historical earthquakes in this area and the connection to long-lasting Late Mesozoic–Cenozoic extensional structures that were active for more than 65 Ma make a GIA-driven reactivation of the fault and graben structures unlikely. Therefore, the above-mentioned structures are not considered as GIFs, and GIA may have played a subordinate role in their reactivation in Pleistocene times.

Evidence of neotectonic movements in the Glückstadt Graben, in the North German Basin, are indicated by two supra-salt faults that form a wide crestal-collapse graben and pierce Pleistocene sediments. Holocene growth strata within the graben indicate young tectonic activity (Al Hseinat et al., 2016). A clear relationship exists between the location of the faults and the location of the NNW-SSE-trending basement faults, implying that basement faults control salt tectonics. The recent salt tectonics and upward propagation of the faults may have resulted from differential ice-sheet loading (Al Hseinat et al., 2016).

In the Glückstadt Graben, recent activity due to halokinetic movements and tectonic activity was identified by Lehné and Sirocko (2007). At three locations, recent uplift and subsidence rates in areas with near-surface faults indicate tectonic activity. Possible trigger mechanisms are loading and unloading stresses caused by the Fennoscandian ice sheet or subsidence within the Glückstadt Graben.

Previous studies on palaeoseismicity and potential GIFs or GIA processes are largely based on soft-sediment deformation structures (SSDS) (Hoffmann & Reicherter, 2012; Brandes & Winsemann, 2013; Pisarska-Jamroży et al., 2018; Müller et al., Chapter 4) and numerical simulations of fault reactivation potential (Brandes et al., 2012, 2015; Pisarska-Jamroży et al., 2019). Some palaeoseismological studies are based on SSDS and cannot be connected to a particular fault or fault systems (e.g. Ludwig, 2011; Hoffmann & Reicherter,

2012; Grube, 2019a,b). See Müller et al., Chapter 4, for a review of the use of SSDS in palaeoseismology.

16.3.2 Recent Seismicity

Knowledge of tectonic activity in northern Germany is incomplete due to the rather infrequent earthquakes and the brief period of instrumental observation, which began around 1900. The work of Leydecker (2011) summarizes historical earthquakes in Germany since 800 CE. Historical seismicity, probably connected with basement faults, is shown in Table 16.1.

In recent decades, more than 70 earthquakes (excluding quarry blasts) affected northern Germany. Some of these earthquakes are located near natural gas fields (Dahm et al., 2007, 2015; Müller et al., 2020a).

Two outstanding seismic events that were instrumentally recorded were the Soltau Earthquake, in 1977, with M_L = 4.0, and the Rotenburg Earthquake, in 2004, with a moment magnitude (M_w) of 4.4 (Figure 16.1E) (Leydecker et al., 1980, 2011; Dahm et al., 2007, 2015; Uta et al., 2018). Due to their relatively high magnitude, the Soltau and Rotenburg earthquakes are key events that play a major role in the seismic hazard assessment of northern Germany. A clear classification (whether purely tectonic or triggered by gas recovery) of these earthquakes was investigated, e.g. in Uta et al. (2018) or Müller et al. (2020a), but has not been possible until now.

In addition, there were also natural earthquakes, such as the Wittenburg 2000 event, with $M_L = 3.3$, the Rostock 2001 Earthquake, with $M_L = 3.4$, and two earthquakes in the Halle/Leipzig area, in 2015 and 2017, with $M_w = 3.2$ and 2.8, respectively (Leydecker, 2011; Dahm et al., 2018). From 2000 to 2018, seven deep earthquakes with hypocentre depths of 17.0–31.4 km were registered in northern Germany with $M_L = 1.7-3.1$ (Brandes et al., 2019).

Recent work has shown that northern Germany is a seismically active intraplate region (Brandes et al., 2015). Typical for earthquakes in intraplate areas are long intervals between the earthquakes (Gangopadhyay & Talwani, 2003) and wide spatial and temporal distribution. Seismic activity concentrates for a period of time on one fault and then shifts to the next fault (McKenna et al., 2007). Intraplate earthquakes can be thus characterized as episodic, clustered and migrating.

16.4 Potential GIFs in Northern Germany

In northern Germany, the WNW-ESE-trending faults have the highest reactivation potential because they trend parallel to the former ice margin (Figure 16.1E, Table 16.1) (Brandes et al., 2011, 2012, 2015). For these faults, the maximum horizontal components of the ice-sheet-induced stress are in line with the palaeostress field (cf. Stewart et al., 2000).

Al Hseinat and Hübscher (2017) identified neotectonic movements along several faults in the south-western Baltic Sea, based on a dense grid of seismic reflection profiles. The

Table 16.1. *Potential GIFs in northern Germany. Strike, fault type and earthquakes that were registered in the vicinity of the faults (~20 km)* are listed. *Note, the 2014 and 2018 earthquakes near Walsrode and Rethem were probably too deep to be related to the Aller Valley Fault.*

GIFs	Strike	Fault type	Earthquakes in the vicinity (~20 km)	Reference
Langeland Fault	NNW-SSE	Normal	1888 I_o = IV Baltic Sea (Leydecker, 2011) 2011 M_L = 2.2 Baltic Sea (BGR, 2019)	Al Hseinat et al. (2017)
Prerow Fault and Werre Fault	NW-SE	Crestal collapse graben (strike-slip?)	1997 M_L = 2.6 Zingst (BGR, 2019) 1998 M_L = 2.7 Baltic Sea (BGR, 2019)	Al Hseinat et al. (2017)
Agricola Fault	NNE-SSW	Normal	1981 M_L = 3.1 Baltic Sea (Leydecker, 2011) 1997 M_L = 2.6 Zingst (BGR, 2019) 1997 M_L = 2.8 Baltic Sea (BGR, 2019)	Al Hseinat et al. (2017)
Wiek Fault	WNW-ESE	Normal	1981 M_L = 3.1 Baltic Sea (Leydecker, 2011) 1990 M_L = 2.3 Binz (Leydecker, 2011) 1997 M_L = 2.8 Baltic Sea (BGR, 2019) 2000 M_L = 2.6 Hagen (BGR, 2019)	Al Hseinat et al. (2017)
Schaabe Fault	WNW-ESE	Normal, strike-slip?	1990 M_L = 2.3 Binz (Leydecker, 2011) 2000 M_L = 2.6 Hagen (BGR, 2019)	Pisarska-Jamroży et al. (2018, 2019)
Aller Valley Fault System	NW-SE	Reverse, normal, strike-slip	997 I_o = VI Altmark (Leydecker, 2011) 1298 I_o = IV Magdeburg (Leydecker, 2011) 1409 I_o = VI Magdeburg (Leydecker, 2011) 1576 I_o = IV Magdeburg (questioned) (Leydecker, 2011) 1977 M_L = 4.0 Soltau (Dahm et al., 2007) 2004 M_L = 4.4 Rotenburg (Dahm et al., 2007; Uta et al., 2018) 2014 M_L = 1.3 Walsrode (Brandes et al., 2019) 2018 M_L = 2.0 Rethem (Brandes et al., 2019)	Winsemann et al. (2011)
Halle Fault System	NW-SE	Reverse	1536 I_o = III (3.5) Merseburg (Leydecker, 2011) 1578 I_o = III (3.5) Halle (Leydecker, 2011)	This study

Table 16.1. (*cont.*)

GIFs	Strike	Fault type	Earthquakes in the vicinity (~20 km)	Reference
			1642 I_o = III (3.5) Halle (Leydecker, 2011) 1645 I_o = IV Halle/Weißenfels (Leydecker, 2011) 1668 I_o = IV Halle/Weißenfels (Leydecker, 2011) 1670 I_o = IV Halle/Weißenfels (Leydecker, 2011) 1720 I_o = IV (4.5) Halle/Weißenfels (Leydecker, 2011) 1908 I_o = III Halle/Weißenfels (Leydecker, 2011) 1987 M_L = 2.4 Gröbzig (BGR, 2019) 2015 M_w = 3.2 Röglitz (Dahm et al., 2018) 2015 M_L = 2.0 Röglitz (BGR, 2019) 2017 M_w = 2.8 Schkeuditz (Dahm et al., 2018)	
Harz Boundary Fault	WNW-ESE	Reverse	1908 I_o = III Quedlinburg (Leydecker, 2011)	Müller et al. (2020b)
Leine Valley Fault System	NNW-SSE	Normal/reverse	none	Brandes et al. (2018)
Osning Thrust	WNW-ESE	Reverse	1612 I_o = VI Bielefeld (Leydecker, 2011) 1767 I_o = VI Oerlinghausen (Leydecker, 2011) 2019 M_L = 2.1 NE of Paderborn (BGR, 2019)	Brandes et al. (2012); Brandes & Winsemann (2013)
Weser Valley Fault System	WNW-ESE	Reverse	none	Brandes et al. (2011)
Thor Suture	NW-SE	Reverse	2012 M_L = 2.0 Hitzacker (Brandes et al., 2019) 2014 M_L = 2.2 Rögnitz (Brandes et al., 2019)	Brandes et al. (2019)
Steinhuder Meer Fault	NNW-SSE	Reverse	none	This study

M_w can be calculated using the equation $M_w = 0.682I_0 + 0.16$, according to Grünthal et al. (2009).

Agricola Fault, Langeland Fault, Prerow Fault, Werre Fault and Wiek Fault (Figure 16.1E) cut through unconsolidated Pleistocene sediments and were most likely reactivated due to the combination of the present-day stress field and GIA-induced stress variations. Two historical earthquakes occurred in the years 1888 and 1905, with epicentral intensities of III–IV in the Mecklenburg Bay of the Baltic Sea (Figure 16.1E). In the area of Fehmarn Island, two earthquakes occurred in 1906 and 1907, with epicentral intensities of III–IV (Leydecker, 2011). Further, seven earthquakes with $M_L = 2.3–3.1$ were detected in the Baltic Sea and on Rügen Island (Leydecker, 2011; BGR, 2019).

Close to the **Schaabe Fault**, located near the sea cliff at Dwasieden, in the SW Baltic Sea (Rügen Island), Pisarska-Jamroży et al. (2018, 2019) found evidence of palaeoearthquakes derived from SSDS in Weichselian sediments. The two deformed beds were deposited between 22.7 ± 1.9 ka and 19.0 ± 1.8 ka (Pisarska-Jamroży et al., 2018) and must have formed in front of the advancing Weichselian ice sheet before the LGM (Pisarska-Jamroży et al., 2019). CFS modelling results supported a glacially induced origin of the seismites in this time window. Pisarska-Jamroży et al. (2019) point to GIA-related fault activity at the nearby faults on Rügen Island, e.g. the Parchow, Lietow, Nord-Jasmund, Boldewitz and Wiek faults. Two earthquakes with $M_L = 2.3-2.6$ were detected close to Rügen Island (NW Baltic Sea) (Leydecker, 2011; BGR, 2019).

Evidence of neotectonic movements at the **Aller Valley Fault** are derived from analysis of normal-displaced, shear-deformation bands in Middle Pleistocene Saalian sediments that developed close to the fault and imply that the Aller Valley Fault was active since Saalian times and possibly in historical times. In the Magdeburg area, historical earthquakes occurred with epicentral intensities of IV–VI (Leydecker, 2011, Figure 16.1E). These events are probably related to the Aller Valley Fault. Recent earthquakes were detected in 1977 near Soltau, with $M_L = 4.0$; in 2004 near Rotenburg, with $M_L = 4.4$ (Dahm et al., 2007; Uta et al., 2018); in 2014 close to Walsrode, with $M_L = 1.3$ at a depth of 25.5 km; and in 2018 close to Rethem, with $M_L = 2.0$ at a depth of 28.5 km (Figure 16.1E) (Brandes et al., 2019). However, some of these earthquakes are most likely too deep to be related to the Aller Valley Fault.

At the **Halle Fault System**, evidence of neotectonic movements is indicated by shear-deformation bands with normal displacement, above a blind fault, to the North of the Halle Fault. The neotectonic movements occurred in post Saalian times because the sediments have a Middle Pleistocene Saalian age (Knoth, 1992). The Halle Fault and related faults were active since Middle Pleistocene Saalian times and in historical times. Historical earthquakes with epicentral intensities of III–IV (3.5–4.5) occurred near Merseburg, Halle and Halle/Weißenfels (Leydecker, 2011, Figure 16.1E). Recent earthquakes were detected in 1908 near Halle/Weißenfels, with an epicentral intensity of III; in 1987 near Gröbzig, with $M_L = 2.4$; in 2015 near Röglitz, with $M_L = 3.2$ and $M_L = 2.0$ (Dahm et al., 2018; BGR, 2019); and 2017 near Schkeuditz with $M_L = 2.8$ (Dahm et al., 2018; Figure 16.1E). Historical earthquakes with epicentral intensities of III–IV (3.0–4.5) and recent earthquakes with epicentral intensities of III–VI and magnitudes (M_L) of 2.8 to 4.1 occurred near Leipzig but are most likely linked to the Leipzig Fault System.

At the **Harz Boundary Fault**, evidence of neotectonic movements is indicated by a fault that is exposed in a nearby sinkhole (Franzke et al., 2015; Müller et al., 2020b). This fault shows a polyphase tectonic evolution with initial normal fault movements and a later reactivation as oblique reverse fault with strike-slip component. Luminescence dating of the surrounding deposits indicates that fault movement occurred after ~15 ka (Müller et al., 2020b). CFS modelling results support this reactivation time. A recent earthquake with an epicentral intensity of IV occurred near Quedlinburg, several kilometres north of the Harz Boundary Fault (Leydecker, 2011, Figure 16.1E).

At the **Leine Valley Fault System**, evidence of neotectonic movements are indicated by shear-deformation bands with normal displacement (Figure 16.1E). The shear-deformation bands were analyzed using ground-penetrating radar profiles and can be connected to movement on one of the basement faults (Brandes & Tanner, 2012; Brandes et al., 2018; Winsemann et al., 2018).

At the **Osning Thrust**, evidence of neotectonic movements associated with palaeoearthquakes was described by Brandes et al. (2012). Indicators are shear-deformation bands with normal displacement and SSDS that include clastic dykes and a sand volcano. The neotectonic movements occurred in the Lateglacial between 16 to 13 ka (Brandes et al., 2012; Brandes & Winsemann, 2013). CFS modelling results also support this reactivation time. Historical earthquakes occurred in the nearby area of Bielefeld and near Oerlinghausen, with an epicentral intensity of VI (Leydecker, 2011, Figure 16.1E). A recent earthquake was detected in 2019, NE of Paderborn, with $M_L = 2.1$ (BGR, 2019).

At the **Steinhuder Meer Fault**, evidence of neotectonic movements is indicated by shear-deformation bands with normal displacement (Figure 16.1E). Growth strata point to Saalian neotectonic movements.

Brandes et al. (2019) showed that the major faults of the **Thor Suture** are under reactivation, triggered by stress changes due to GIA processes (Figure 16.1E). Indicators are deep crustal earthquakes at a depth of 17.0–31.4 km.

At the **Weser Valley Fault System**, evidence of neotectonic movements is indicated by a fault system that developed within the glaciolacustrine Emme delta that can be connected to the Mesozoic basement fault (Figure 16.1E; Brandes et al., 2011; Winsemann et al., 2011). The fault system shows syn-sedimentary activity. According to Brandes et al. (2011) and Winsemann et al. (2011), these basement faults were reactivated during the Middle Pleistocene by the advancing Saalian ice sheet, water and sediment loading.

16.5 Discussion

The neotectonic activity in northern Germany is a consequence of the regional lithospheric stress field, which is controlled by the push of the Mid-Atlantic Ridge and the ongoing Alpine Orogeny (Reicherter et al., 2005). Brandes et al. (2011, 2012), Brandes and Winsemann (2013) and Brandes et al. (2015) have shown that additional stress field changes due to glacial isostatic adjustment may have induced the Pleistocene and historical seismicity. In northern Germany, the WNW-ESE-striking faults have the highest reactivation potential because they trend parallel to the former ice margin (Brandes et al., 2015). Glacially induced fault reactivation is largely controlled by deglaciation processes in the

Late Pleistocene. The best evidence available points to the decay of the Weichselian ice sheet as the reason for potential GIFs. However, studies of Brandes et al. (2011), Winsemann et al. (2011) and Pisarska-Jamroży et al. (2018, 2019) also point to possible fault reactivation during the previous ice advances.

16.5.1 Tectonic Structures and the Distribution of Fault Activity

The tectonic structure of northern Germany is the result of the Palaeozoic Caledonian and Variscan orogenies (Krawczyk et al., 2008), Late Palaeozoic to Mesozoic lithospheric extension (Betz et al., 1987) and a distinct Late Mesozoic to Palaeogene inversion phase (Kley & Voigt, 2008). Different studies have shown that the inherited structural grain is an important controlling factor for young tectonic activity (e.g. Sykes, 1978). Brandes et al. (2012) showed that neotectonic activity occurred at the Osning Thrust, and the study of Müller et al. (2020b) found evidence of neotectonic activity on the Harz Boundary Fault System. Both the Osning Thrust and the Harz Boundary Fault were active during the Late Cretaceous inversion phase. Historical seismicity with epicentral intensities of up to VI is also concentrated on Late Cretaceous reverse faults such as the Osning Thrust, the Gardelegen Fault and the Haldensleben Fault.

Recent natural earthquakes detected over the last 18 years point to deep-seated seismicity in northern Germany. These earthquakes reach a magnitude of 3.1 and are partly concentrated on the Thor Suture, which is the remnant of a Silurian subduction zone. Numerical simulations imply that GIA processes play a role in triggering these events (Brandes et al., 2019) and that reverse faults in northern Germany have a high potential to be reactivated as GIFs. For these faults, the CFS became positive with increasing time since deglaciation (Brandes et al., 2015). Lehné and Sirocko (2007), Al Hseinat et al. (2016), Huster et al. (2020) and Christian Hübscher (personal communication, 2020) described evidence of neotectonic activity from the Glückstadt Graben. It is difficult to decide if this activity is the result of GIA processes. Numerical simulations of GIA-related stress-field changes in the southern part of northern Germany imply that normal fault activity was possible in the last 25 ka, but the fact that the Coulomb stress has decreased over the last 20 ka implies that the likelihood of GIA-related normal fault activity will decrease, whereas the likelihood for movements on reverse faults has increased over the last 15 Ma. Further deep earthquakes are located at the Moho at a depth of up to 30 km. This seismicity implies that the Moho may act as a regional detachment on which recent shortening could be compensated (Figure 16.3; Brandes et al., 2019).

16.6 Conclusions

Key parameters to classify faults in northern Germany as GIFs are the timing of reactivation, their orientation to the former ice margins and numerical simulations that show the increase of the CFS over the time since the onset of deglaciation.

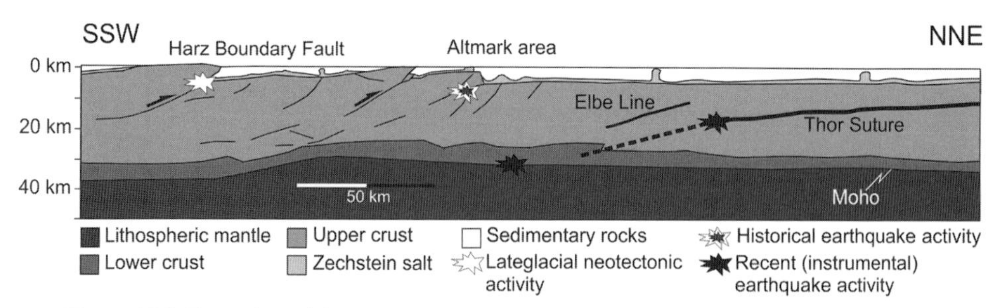

Figure 16.3 Tectonic activity in northern Germany. The cross-section shows Lateglacial neotectonic activity, historical earthquake activity and recent (instrumental) earthquake activity at several fault systems (based on Leydecker, 2011; Brandes et al., 2019; Müller et al., 2020b). Unlike the other faults, the Elbe Line and the Thor Suture are terrane boundaries. The dashed line represents the uncertain location of the Thor Suture. Cross-section is modified after Kley & Voigt (2008).

In the south-western Baltic Sea, the Langeland Fault, the Prerow and Werre faults, the Agricola Fault and the Wiek Fault all show evidence of neotectonic activity induced by GIA processes.

In the south-western Baltic Sea at Rügen Island evidence of palaeoearthquakes were found near the Schaabe Fault. The estimated age of the deformation and modelling results indicate that the reactivation of the fault is glacially induced.

At the Osning Thrust evidence of palaeoearthquakes occurs.

The Thor Suture shows ongoing seismicity that might be related to GIA processes.

At the Weser Valley Fault System, a fault system in the glaciolacustrine Emme delta can be connected to Mesozoic basement faults. These basement faults were probably reactivated during the Middle Pleistocene by the advancing Saalian ice sheet and by water and sediment loading.

At the Aller Valley Fault, the Halle Fault System, the Leine Valley Fault System and the Steinhuder Meer Fault, neotectonic movements are indicated by shear-deformation bands in Middle Pleistocene sediments, which developed in the process zone of the faults.

Neotectonic activity occurred at the Harz Boundary Fault. Lateglacial fault activity (post 15 ka) corresponds with the timing of fault reactivation at the Osning Thrust and the Sorgenfrei–Tornquist Zone.

At the Harz Boundary Fault, the Osning Thrust and Schaabe Fault, field evidence, modelling results and luminescence ages indicate that these faults were reactivated due to GIA-related stress field changes.

Studies in the Lower Rhine Rift show a possible link between GIA processes and fault activity in this area.

The results support the interpretation that northern Germany is a tectonically active intraplate area, with clustered and migrating seismic activity that mainly occurred on major Late Cretaceous reverse faults. The timing of fault movements implies that the seismicity is most likely induced by varying lithospheric stress conditions related to GIA, and some faults are therefore classified as GIFs. This is supported by numerical simulations of GIA-related stress-field changes. Lateglacial, historical and recent earthquakes occurred in this region. Thus, GIA processes are also a likely driver for historical and parts of recent fault activity.

Acknowledgements

We thank Christian Hübscher and Werner Stackebrandt for their constructive reviews, which helped to improve the manuscript. The work has been financially supported by project *Paläoseismische Untersuchung Norddeutschlands* from the BGR (No. 201-10079313).

References

Al Hseinat, M. and Hübscher, C. (2014). Ice-load induced tectonics controlled tunnel valley evolution – instances from the southwestern Baltic Sea. *Quaternary Science Reviews*, **97**, 121–135, doi.org/10.1016/j.quascirev.2014.05.011.

Al Hseinat, M., Hübscher, C., Lang, J. et al. (2016). Triassic to recent tectonic evolution of a crestal collapse graben above a salt-cored anticline in the Glückstadt Graben/North German Basin. *Tectonophysics*, **680**, 50–66, doi.org/10.1016/j.tecto.2016.05.008.

Al Hseinat, M. and Hübscher, C. (2017). Late Cretaceous to recent tectonic evolution of the North German Basin and the transition zone to the Baltic Shield/southwest Baltic Sea. *Tectonophysics*, **708**, 28–55, doi.org/10.1016/j.tecto.2017.04.021.

Betz, D., Führer, F., Greiner, G. and Plein, E. (1987). Evolution of the Lower Saxony Basin. *Tectonophysics*, **137**, 127–170, doi.org/10.1016/0040-1951(87)90319-2.

BGR (2019). Der Geodatendienst GERSEIS innerhalb der interaktiven Kartenanwendung Geoviewer der BGR [The geodata service GERSEIS within the interactive map application Geoviewer of the BGR]. biturl.top/eYFnYn.

Böse, M., Lüthgens, C., Lee, J. R. and Rose, J. (2012). Quaternary glaciations of northern Europe. *Quaternary Science Reviews*, **44**, 1–25, doi.org/10.1016/j.quascirev.2012.04.017.

Brandes, C., Polom, U. and Winsemann, J. (2011). Reactivation of basement faults: interplay of ice-sheet advance, glacial lake formation and sediment loading. *Basin Research*, **23**, 53–64, doi.org/10.1111/j.1365-2117.2010.00468.x.

Brandes, C. and Tanner, D. C. (2012). Three-dimensional geometry and fabric of shear deformation-bands in unconsolidated Pleistocene sediments. *Tectonophysics*, **518**, 84–92, doi.org/10.1016/j.tecto.2011.11.012.

Brandes, C., Winsemann, J., Roskosch, J. et al.(2012). Activity along the Osning Thrust in Central Europe during the Lateglacial: ice-sheet and lithosphere interactions. *Quaternary Science Reviews*, **38**, 49–62, doi.org/10.1016/j.quascirev.2012.01.021.

Brandes, C. and Winsemann, J. (2013). Soft-sediment deformation structures in NW Germany caused by Late Pleistocene seismicity. *International Journal of Earth Sciences*, **102**, 2255–2274, doi.org/10.1007/s00531-013-0914-4.

Brandes, C., Steffen, H., Steffen, R. and Wu, P. (2015). Intraplate seismicity in northern Central Europe is induced by the last glaciation. *Geology*, **43**, 611–614, doi.org/10.1130/G36710.1.

Brandes, C., Igel, J., Loewer, M. et al. (2018). Visualisation and analysis of shear-deformation bands in unconsolidated Pleistocene sand using ground-penetrating radar: implications for palaeoseismological studies. *Sedimentary Geology*, **367**, 135–145, doi.org/10.1016/j.sedgeo.2018.02.005.

Brandes, C., Plenefisch, T., Tanner, D. C., Gestermann, N. and Steffen, H. (2019). Evaluation of deep crustal earthquakes in northern Germany – possible tectonic causes. *Terra Nova*, **31**, 83–93, doi.org/10.1111/ter.12372.

Dahm, T., Cesca, S., Hainzl, S., Braun, T. and Krüger, F. (2015). Discrimination between induced, triggered and natural earthquakes close to hydrocarbon reservoirs: a probabilistic approach based on the modeling of depletion-induced stress changes and seismological source parameters. *Journal of Geophysical Research*, **120**, 2491–2509, doi.org/10.1002/2014JB011778.

Dahm, T., Heimann, S., Funke, S. et al. (2018). Seismicity in the block mountains between Halle and Leipzig, Central Germany: centroid moment tensors, ground motion simulation, and felt intensities of two $M \approx 3$ earthquakes in 2015 and 2017. *Journal of Seismology*, **22**, 985–1003, doi.org/10.1007/s10950-018-9746-9.

Dahm, T., Krüger, F., Stammler, K. et al. (2007). The 2004 M_w 4.4 Rotenburg, northern Germany, earthquake and its possible relationship with gas recovery. *Bulletin of the Seismological Society of America*, **97**, 691–704, doi.org/10.1785/0120050149.

Ehlers, J., Grube, A., Stephan, H. J. and Wansa, S. (2011). Pleistocene glaciations of North Germany-new results. In J. Ehlers, P. L. Gibbard and P. D. Hughes, eds., *Quaternary Glaciations: Extent and Chronology – A Closer Look. Developments in Quaternary Science*, **15**, pp. 149–162, doi.org/10.1016/B978-0-444-53447-7.00013-1.

Eissmann, L. (2002). Quaternary geology of eastern Germany (Saxony, Saxon-Anhalt, South Brandenburg, Thüringia), type area of the Elsterian and Saalian stages in Europe. *Quaternary Science Reviews*, **21**, 1275–1346, doi.org/10.1016/S0277-3791(01)00075-0.

Franke, W. (2000). The mid-European segment of the Variscides: tectonostratigraphic units, terrane boundaries and plate tectonic evolution. In W. Franke, V. Haak, O. Onken and D. Tanner, eds., *Orogenic Processes: Quantification and Modelling in the Variscan Belt.* Geological Society, London, Special Publication, Vol. 179, pp. 35–61, doi.org/10.1144/GSL.SP.2000.179.01.05.

Franke, W., Cocks, L. R. M. and Torsvik, T. H. (2017). The Palaeozoic Variscan oceans revisited. *Gondwana Research*, **48**, 257–284, doi.org/10.1016/j.gr.2017.03.005.

Franzke, H. J., Hauschke, N. and Hellmund, M. (2015). Spätpleistozäne bis frühholozäne Tektonik in einem Karsttrichter im Bereich der Störungszone des Harznordrandes nahe Benzingerode (Sachsen-Anhalt) [Late Pleistocene to Early Holocene tectonics in a karst sinkhole in the area of the northern Harz boundary fault zone near Benzingerode (Saxony-Anhalt)]. *Hallesches Jahrbuch für Geowissenschaften*, **37**, 1–10.

Gangopadhyay, A. and Talwani, P. (2003). Symptomatic features of intraplate earthquakes. *Seismological Research Letters*, **74**, 863–883, doi.org/10.1785/gssrl.74.6.863.

Gast, R. and Gundlach, T. (2006). Permian strike slip and extensional tectonics in Lower Saxony, Germany. *Zeitschrift der Deutschen Gesellschaft für Geowissenschaften*, **157**, 41–56, doi.org/10.1127/1860-1804/2006/0157-0041.

Grollimund, B. and Zoback, M. D. (2001). Did deglaciation trigger intraplate seismicity in the New Madrid seismic zone? *Geology*, **29**, 175–178, doi.org/10.1130/0091-7613(2001)029<0175:DDTISI>2.0.CO;2.

Grube, A. (2019a). Palaeoseismic structures in Quaternary sediments of Hamburg (NW Germany), earthquakes evidence during the younger Weichselian and Holocene. *International Journal of Earth Sciences*, **108**, 845–861, doi.org/10.1007/s00531-019-01681-2.

Grube, A. (2019b). Palaeoseismic structures in Quaternary sediments, related to an assumed fault zone north of the Permian Peissen-Gnutz salt structure (NW Germany) – neotectonic activity and earthquakes from the Saalian to the Holocene. *Geomorphology*, **328**, 15–27, doi.org/10.1016/j.geomorph.2018.12.004.

Grünthal, G., Mayer-Rosa, D. and Lenhardt, W. A. (1998). Abschätzung der Erdbebengefährdung für die D-A-CH-Staaten-Deutschland, Österreich, Schweiz [Estimation of the earthquake hazard for the D-A-CH-countries – Germany, Austria, Switzerland]. *Bautechnik*, **75**, 753–767, doi.org/10.1002/bate.199805380.

Grünthal, G., Stromeyer, D. and Wahlström, R. (2009). Harmonization check of M_w within the central, northern, and northwestern European earthquake catalogue (CENEC). *Journal of Seismology*, **13**, 613–632, doi.org/10.1007/s10950-009-9154-2.

Grützner, C., Fischer, P. and Reicherter, K. (2016). Holocene surface ruptures of the Rurrand Fault, Germany-insights from palaeoseismology, remote sensing and shallow geophysics. *Geophysical Journal International*, **204**, 1662–1677, doi.org/10.1093/gji/ggv558.

Hampel, A., Hetzel, R., Maniatis, G. and Karow, T. (2009). Three-dimensional numerical modeling of slip rate variations on normal and thrust fault arrays during ice cap growth and melting. *Journal of Geophysical Research*, **114**, B08406, doi.org/10.1029/2008JB006113.

Hardt, J., Lüthgens, C., Hebenstreit, R. and Böse, M. (2016). Geochronological (OSL) and geomorphological investigations at the presumed Frankfurt ice-marginal position in northeast Germany. *Quaternary Science Reviews*, **154**, 85–99, doi.org/10.1016/j.quascirev.2016.10.015.

Hardt, J. and Böse, M. (2018). The timing of the Weichselian Pomeranian ice marginal position south of the Baltic Sea: a critical review of morphological and geochronological results. *Quaternary International*, **478**, 51–58, doi.org/10.1016/j.quaint.2016.07.044.

Heidbach, O., Rajabi, M., Reiter, K. and Ziegler, M. (2016). World Stress Map 2016. *GFZ Data Services*, doi.org/10.5880/WSM.2016.002.

Hoffmann, G. and Reicherter, K. (2012). Soft-sediment deformation of Late Pleistocene sediments along the southwestern coast of the Baltic Sea (NE Germany). *International Journal of Earth Sciences*, **101**, 351–363, doi.org/10.1007/s00531-010-0633-z.

Hughes, A. L. C., Gyllencreutz, R., Lohne, Ø. S., Mangerud, J. and Svendsen, J. I. (2016). The last Eurasian ice sheets-a chronological database and time-slice reconstruction, DATED-1. *Boreas*, **45**, 1–45, doi.org/10.1111/bor.12142.

Huster, H., Hübscher, C. and Seidel, E. (2020). Impact of Late Cretaceous to Neogene plate tectonics and Quaternary ice loads on supra-salt deposits at Eastern Glückstadt Graben, North German Basin. *International Journal of Earth Sciences*, **109**, 1029–1050, doi.org/10.1007/s00531-020-01850-8.

Hübscher, C., Lykke-Andersen, H., Hansen, M. B. and Reicherter, K. (2004). Investigating the structural evolution of the western Baltic. *Eos, Transactions American Geophysical Union*, **85**, 115–115, doi.org/10.1029/2004EO120006.

Kaiser, A., Reicherter, K., Hübscher, C. and Gajewski, D. (2005). Variation of the present-day stress field within the North German Basin – insights from thin shell FE modeling based on residual GPS velocities. *Tectonophysics*, **397**, 55–72, doi.org/10.1016/j.tecto.2004.10.009.

Kley, J. and Voigt, T. (2008). Late Cretaceous intraplate thrusting in central Europe: effect of Africa-Iberia-Europe convergence, not Alpine collision. *Geology*, **36**, 839–842, doi.org/10.1130/G24930A.1.

Knoth, W. (1992). Geologische Übersichtskarte von Sachsen-Anhalt 1:400000 [Geological overview map of Saxony-Anhalt 1:400,000]. *Geologisches Landesamt Sachsen-Anhalt*, 1st ed., Halle (Saale).

Kossmat, F. (1927). Gliederung des varistischen Gebirgsbaues [Structure of the Variscan mountains]. *Abhandlungen des Sächsischen Geologischen Landesamts*, **1**, 1–39.

Krawczyk, C. M., McCann, T., Cocks, L. R. M. et al. (2008). Caledonian tectonics. In T. McCann, ed., *The Geology of Central Europe. Precambrian and Paleozoic, Vol. 1*, Geological Society London, pp. 303–381, doi.org/10.1144/CEV1P.7.

Krentz, O., Lapp, M., Seibel, B. and Bahrt, W. (2010). Bruchtektonik [Fracture tectonics]. In Autorenkollektiv, eds., *Die geologische Entwicklung der Lausitz* [The Geological Development of the Lausitz]. Vattenfall Europe Mining AG, Cottbus, pp. 137–160.

Kujansuu, R. (1964). Nuorista siirroksista Lapissa [English summary: Recent faults in Lapland]. *Geologi*, **16**, 30–36 (in Finnish).

Kühner, R. (2009). Neue Ergebnisse zum Nachweis neotektonischer Aktivitäten im Quartär des Tagebaus Welzow-Süd, Südbrandenburg [New results for the detection of neotectonic activities in the Quaternary of the Welzow-Süd opencast mine, southern Brandenburg.]. *Brandenburgische Geowissenschaftliche Beiträge*, **16**, 87–93.

Kühner, R. (2010). Quartär [The Quaternary]. In Autorenkollektiv, eds., *Die geologische Entwicklung der Lausitz* [The Geological Development of the Lausitz]. Vattenfall Europe Mining AG, Cottbus, pp. 97–134.

Lang, J., Hampel, A., Brandes, C. and Winsemann, J. (2014). Response of salt structures to ice-sheet loading: implications for ice-marginal and subglacial processes. *Quaternary Science Reviews*, **101**, 217–233, doi.org/10.1016/j.quascirev.2014.07.022.

Lang, J., Lauer, T. and Winsemann, J. (2018). New age constraints for the Saalian glaciation in northern central Europe: implications for the extent of ice sheets and related proglacial lake systems. *Quaternary Science Reviews*, **180**, 240–259, doi.org/10.1016/j.quascirev.2017.11.029.

Lauer, T. and Weiss, M. (2018). Timing of the Saalian- and Elsterian glacial cycles and the implications for Middle Pleistocene hominin presence in central Europe. *Scientific Reports*, **8**, 5111, doi.org/10.1038/s41598-018-23541-w.

Lehmkuhl, F., Zens, J., Krauß, L., Schulte, P. and Kels, H. (2016). Loess-paleosol sequences at the northern European loess belt in Germany: distribution, geomorphology and stratigraphy. *Quaternary Science Reviews*, **153**, 11–30, doi.org/10.1016/j.quascirev.2016.10.008.

Lehné, R. J. and Sirocko, F. (2007). Rezente Bodenbewegungspotenziale in Schleswig-Holstein (Deutschland) – Ursachen und ihr Einfluss auf die Entwicklung der rezenten Topographie [Recent land motion potentials in Schleswig-Holstein (Germany) – causes and their influence on the development of the recent topography]. *Zeitschrift der Deutschen Gesellschaft für Geowissenschaften*, **158**, 329–347.

Lehné, R. J. and Sirocko, F. (2010). Recent vertical crustal movements and resulting surface deformation within the North German Basin (Schleswig-Holstein) derived by GIS-based analysis of repeated precise leveling data. *Zeitschrift der Deutschen Gesellschaft für Geowissenschaften*, **161**, 175–188.

Leydecker, G., Steinwachs, M., Seidl, D. et al. (1980). Das Erdbeben vom 2. Juni 1977 in der Norddeutschen Tiefebene bei Soltau [The earthquake of June 2, 1977, in the North German Plain near Soltau]. *Geologisches Jahrbuch Reihe E*, **18**, 3–18.

Leydecker, G. (2011). Erdbebenkatalog für Deutschland mit Randgebieten für die Jahre 800 bis 2008 [Earthquake catalog for Germany and peripheral areas for the years 800 to 2008]. *Geologisches Jahrbuch Reihe E*, **59**, 1–198.

Littke, R., Scheck-Wenderoth, M., Brix, M. R. and Nelskamp, S. (2008). Subsidence, inversion and evolution of the thermal field. In R. Littke, U. Bayer, D. Gajewski and S. Nelskamp, eds., *Dynamics of Complex Intracontinental Basins – The Central European Basin System*. Springer-Verlag, Berlin-Heidelberg, pp. 125–141.

Ludwig, A. O. (2011). Zwei markante Stauchmoränen: Peski/Belorussland und Jasmund, Ostseeinsel Rügen/Nordostdeutschland – Gemeinsame Merkmale und Unterschiede [Two distinctive push moraines: Peski/Belarus and Jasmund, Rügen Island/Northeast Germany – common features and difference]. *E&G–Quaternary Science Journal*, **60**, 464–487, doi.org/10.3285/eg.60.4.06.

Lüthgens, C. and Böse, M. (2011). Chronology of Weichselian main ice marginal positions in north-eastern Germany. *E&G – Quaternary Science Journal*, **60**, 236–247, doi .org/10.3285/eg.60.2-3.02.

Marotta, A. M., Bayer, U. and Thybo, H. (2000). The legacy of the NE German Basin-Reactivation by compressional buckling. *Terra Nova*, **12**, 132–140, doi.org/10.1046/j .1365-3121.2000.123296.x.

Marotta, A. M., Bayer, U., Thybo, H. and Scheck, M. (2002). Origin of regional stress in the North German basin: results from numerical modeling. *Tectonophysics*, **360**, 245–264, doi.org/10.1016/S0040-1951(02)00358-X.

Marotta, A. M., Mitrovica, J. X., Sabadini, R. and Milne, G. (2004). Combined effects of tectonics and glacial isostatic adjustment on intraplate deformation in central and northern Europe: applications to geodetic baseline analyses. *Journal of Geophysical Research*, **109**, B01413, doi.org/10.1029/2002JB002337.

Mazur, S., Scheck-Wenderoth, M. and Krzywiec, P. (2005). Different modes of the Late Cretaceous – Early Tertiary inversion in the North German and Polish basins. *International Journal of Earth Sciences*, **94**, 782–798, doi.org/10.1007/s00531-005-0016-z.

McCann, T. (2008). Introduction and overview. In T. McCann, ed., *The Geology of Central Europe: Precambrian and Palaeozoic*. The Geological Society of London, pp. 1–20.

McKenna, J., Stein, S. and Stein, C. A. (2007). Is the New Madrid seismic zone hotter and weaker than its surroundings? In S. Stein and S. Mazzotti, eds., *Continental Intraplate Earthquakes: Science, Hazard, and Policy Issues*. Geological Society of America, Special Paper 425, pp. 167–175, doi.org/10.1130/2007.2425(12).

Meinsen, J., Winsemann, J., Roskosch, J. et.al. (2014). Climate control on the evolution of Late Pleistocene alluvial-fan and aeolian sand-sheet systems in NW Germany. *Boreas*, **43**, 42–66, doi.org/10.1111/bor.12021.

Meschede, M. (2015). *Geologie Deutschlands: Ein prozessorientierter Ansatz* [*Geology of Germany: A Process-Oriented Approach*]. Springer, Berlin/Heidelberg, 249 pp.

Mey, J., Scherler, D., Wickert, A. D. et al. (2016). Glacial isostatic uplift of the European Alps. *Nature Communications*, **7**, 13382, doi.org/10.1038/ncomms13382.

Mörner, N. A. (1978). Faulting, fracturing, and seismicity as functions of glacio-isostasy in Fennoscandia. *Geology*, **6**, 41–45, doi.org/10.1130/0091-7613(1978)6<41:FFASAF>2.0.CO;2.

Müller, U. and Obst, K. (2008). Junge halokinetische Bewegungen im Bereich der Salzkissen Schlieven und Marnitz in Südwest-Mecklenburg [Young halokinetic movements in the area of the salt pillows Schlieven and Marnitz in south-west Mecklenburg]. *Brandenburgische Geowissenschaftliche Beiträge*, **15**, 147–154.

Müller, B., Scheffzük, F., Schilling, M. et al. (2020a). *Reservoir-Management and Seismicity – Strategies to Reduce Induces Seismicity*. DGMK-Research Report 776.

Müller, K., Polom, U., Winsemann, J. et al. (2020b). Structural style and neotectonic activity along the Harz Boundary Fault, northern Germany: a multimethod approach integrating geophysics, outcrop data and numerical simulations. *International Journal of Earth Sciences*, **109**, 1811–1835, doi.org/10.1007/s00531-020-01874-0.

Norton, K. P. and Hampel, A. (2010). Postglacial rebound promotes glacial re-advances – a case study from the European Alps. *Terra Nova*, **22**, 297–302, doi.org/10.1111/j.1365-3121.2010.00946.x.

Pharaoh, T. C., Dusar, M., Geluk, M. C. et al. (2010). Tectonic evolution. In J. C. Doornenbal and A. G. Stevenson, eds., *Petroleum Geological Atlas of the Southern Permian Basin Area*. EAGE Publications, Houten, pp. 25–57.

Pisarska-Jamroży, M., Belzyt, S., Börner, A. et al. (2018). Evidence from seismites for glacio-isostatically induced crustal faulting in front of an advancing land-ice mass (Rügen Island, SW Baltic Sea). *Tectonophysics*, **745**, 338–348, doi.org/10.1016/j.tecto.2018.08.004.

Pisarska-Jamroży, M., Belzyt, S., Börner, A. et al. (2019). The sea cliff at Dwasieden: soft-sediment deformation structures triggered by glacial isostatic adjustment in front of the advancing Scandinavian Ice Sheet. *DEUQUA Special Publications*, **2**, 61–67, doi.org/10.5194/deuquasp-2-61-2019.

Reicherter, K., Kaiser, A. and Stackebrandt, W. (2005). The Post-Glacial landscape evolution of the North German Basin: morphology, neotectonics and crustal deformation. *International Journal of Earth Science*, **94**, 1083–1093, doi.org/10.1007/s00531-005-0007-0.

Reinecker, J., Heidbach, O. and Müller, B. (2004). *World Stress Map* (2004 release). www.world-stress-map.org.

Roskosch, J., Winsemann, J., Polom, U. et al. (2015). Luminescence dating of ice-marginal deposits in northern Germany: evidence for repeated glaciations during the Middle Pleistocene (MIS 12 to MIS 6). *Boreas*, **44**, 103–126, doi.org/10.1111/bor.12083.

Scheck-Wenderoth, M. and Lamarche, J. (2005). Crustal memory and basin evolution in the Central European Basin System-new insights from a 3D structural model. *Tectonophysics*, **397**, 143–165, doi.org/10.1016/j.tecto.2004.10.007.

Schulz, R., Suchi, E., Öhlschläger, D. et al. (2013). Geothermieatlas zur Darstellung möglicher Nutzungskonkurrenzen zwischen CCS und Tiefer Geothermie [Geothermal atlas for illustration of possible competing usage between CCS and deep geothermal energy]. *Leibniz-Institut für Angewandte Geophysik und Bundesanstalt für Geowissenschaften und Rohstoffe*, Hannover, p. 107.

Seidel, E., Meschede, M. and Obst, K. (2018). The Wiek Fault System east of Rügen Island: origin, tectonic phases and its relationship to the Trans-European Suture Zone. In B. Kilhams, P. A. Kukla, S. Mazur et al., eds., *Mesozoic Resource Potential in the Southern Permian Basin*. Geological Society, London, Special Publication, Vol. 469, pp. 59–82, doi.org/10.1144/SP469.10.

Stackebrandt, W. (2004). Zur Neotektonik in Norddeutschland [On neotectonics in Northern Germany]. *Zeitschrift für geologische Wissenschaften*, **32**, 85–95.

Stackebrandt, W. (2005). Neotektonische Aktivitätsgebiete in Brandenburg (Norddeutschland) [Areas of neotectonic activity in Brandenburg (Northern Germany)]. *Brandenburgische Geowissenschaftliche Beiträge*, **12**, 165–172.

Stackebrandt, W. (2008). Zur Neotektonik der Niederlausitz [On neotectonics of the Niederlausitz]. *Zeitschrift der Deutschen Gesellschaft für Geowissenschaften*, **159**, 117–122, doi.org/10.1127/1860-1804/2008/0159-0117.

Stackebrandt, W. (2015). Neotektonische Beanspruchung [Neotonic stress]. In W. Stackebrandt and D. Franke, eds., *Geologie von Brandenburg*. Schweizerbart, Stuttgart, pp. 480–487.

Stewart, I. S., Sauber, J. and Rose, J. (2000). Glacio-seismotectonics: ice sheets, crustal deformation and seismicity. *Quaternary Science Reviews*, **19**, 1367–1389, doi.org/10.1016/S0277-3791(00)00094-9.

Sykes, L. R. (1978). Intraplate seismicity, reactivation of pre-existing zones of weakness, alkaline magmatism, and other tectonism postdating continental fragmentation. *Reviews of Geophysics and Space Physics*, **16**, 621–688, doi.org/10.1029/RG016i004p00621.

Torsvik, T. H. and Cocks, L. R. M. (2017). The integration of palaeomagnetism, the geological record and mantle tomography in the location of ancient continents. *Geological Magazine*, **156**, 242–260, doi.org/10.1017/S001675681700098X.

Uta, P., Brandes, C., Bönnemann, C., Gestermann, N., Kaiser, D., Plenefisch, T. and Winsemann, J. (2018). *Re-evaluation of the Rotenburg mainshock 2004*. DGMK-Project 806, Final Report, 85 pp.

van Balen, R. T., Bakker, M. A. J., Kasse, C., Wallinga, J. and Woolderink, H. A. G. (2019). A Late Glacial surface rupturing earthquake at the Peel Boundary fault zone, Roer Valley Rift System, the Netherlands. *Quaternary Science Reviews*, **218**, 254–266, doi.org/10.1016/j.quascirev.2019.06.033.

Vanneste, K., Camelbeeck, T., Verbeeck, K. and Demoulin, A. (2018). Morphotectonics and past large earthquakes in Eastern Belgium. In A. Demoulin, ed., *Landscapes and Landforms of Belgium and Luxembourg, World Geomorphological Landscapes*. Springer, Cham, pp. 215–236, doi.org/10.1007/978-3-319-58239-9_13.

van Wees, J.-D., Stephenson, R. A., Ziegler, P. A. et al. (2000). On the origin of the Southern Permian basin, central Europe. *Marine and Petroleum Geology*, **17**, 43–59, doi.org/10.1016/S0264-8172(99)00052-5.

Winsemann, J., Brandes, C. and Polom, U. (2011). Response of a proglacial delta to rapid high-amplitude lake-level change: an integration of outcrop data and high-resolution shear wave seismics. *Basin Research*, **23**, 22–52, doi.org/10.1111/j.1365-2117.2010.00465.x.

Winsemann, J., Lang, J., Polom, U. et al. (2018). Ice-marginal forced regressive deltas in glacial lake basins: geomorphology, facies variability and large-scale depositional architecture. *Boreas*, **47**, 973–1002, doi.org/10.1111/bor.12317.

Winsemann, J., Koopmann, H., Tanner, D.C. et al. (2020). Seismic interpretation and structural restoration of the Heligoland glaciotectonic thrust-fault complex: implications for multiple deformation during (pre-)Elsterian to Warthian ice advances into the southern North Sea Basin. *Quaternary Science Reviews*, **227**, 106068, doi.org/10.1016/j.quascirev.2019.106068.

17

Glacially Induced Faulting in Poland

MAŁGORZATA (GOSIA) PISARSKA-JAMROŻY, PIOTR PAWEŁ WOŹNIAK
AND A. J. (TOM) VAN LOON

ABSTRACT

Poland is located in an intraplate area characterized almost everywhere by low recent tectonic activity, which does not imply, however, that it has been unaffected by earthquakes, even in the – geologically speaking – recent geological past. This earthquake activity is due to the Pleistocene glaciations, which left traces in the form of earthquake-induced deformed layers. The strongly deformed layers (seismites) as well as some fault zones with significant offsets crossing also Quaternary sediments can indicate fault (re) activation due to glacial isostatic adjustment. We inventory and describe five sites/areas in the intraplate northern and central parts of Poland, where traces of glacial isostatic adjustment occur. We do not deal, however, with the mountain areas of southern Poland, because Alpine pressure as well as glacial isostatic adjustment may each, possibly jointly, have acted there as a trigger; distinguishing between traces left by them is not yet viable.

17.1 Introduction

The present-day area of Poland was affected during the Middle and Late Pleistocene by three glaciations: the Elsterian (Marine Isotope Stage (MIS) 12), Saalian (MIS 10-6) and Weichselian (MIS 5-2). It is well known from other areas that ice-sheet loading/unloading cycles may induce faulting of the Earth's crust. The alternating phases of ice advance and retreat of the Fennoscandian ice sheet caused glacial isostatic adjustment (GIA) of the lithosphere and seismites (layers with abundant soft-sediment deformation structures (SSDS), attributable to the passage of commonly earthquake-induced seismic waves), which have been described from studies of countries in the vicinity of the Baltic Sea (Mörner, 1991; Brandes et al., 2012; Hoffmann & Reicherter, 2012; van Loon & Pisarska-Jamroży, 2014; van Loon et al., 2016; Pisarska-Jamroży et al., 2019a,b; Belzyt et al., 2021). Most of the significant vertical displacements in northern Poland have been ascribed to GIA during the Cromerian (MIS 21-17), Holsteinian (MIS 11) and Eemian (MIS 5e) interglacials, following melting and retreat of the ice sheet that had formed during the proceeding glacials (Kurzawa, 2003). Recent studies suggest that GIA may also take place in front of an advancing ice sheet (see Brandes et al., 2011; Pisarska-Jamroży et al., 2018, 2019b), but up till now no evidence of such a GIA has been found in Poland.

In the present contribution, we show sites/areas in Poland where traces have been found that are probably due to GIA. The neotectonic activity in the mountain areas of southern Poland are, however, not dealt with because a relationship with GIA is uncertain (Alpine pressure as well as GIA might be responsible). A map of faults that were (re)activated during the Quaternary in SW Poland, including the Sudetes, has been presented by Badura and Przybylski (2000), while case studies from this region have been published by Migoń and Łach (1998), Migoń et al. (1998), Przybylski (1998), Badura et al. (2007), Štěpančíková et al. (2008), Štěpančíková et al. (2010) and Różycka and Migoń (2018). In the Carpathians and their foreland, numerous case studies on Quaternary fault activity of faults have been carried out (e.g. Birkenmajer, 1976; Laskowska-Wysoczańska, 1995; Zuchiewicz, 1995; Wójcik, 2003; Tokarski et al., 2007), but there still is a need of more detailed maps of young faults in the whole region.

17.2 Geological Setting of Poland

Poland is located in the middle of the Eurasian plate, at the contact of two platforms, viz. (1) the East-European Craton, which includes NE Poland, and (2) the Palaeozoic Platform, which includes west and south-west Poland. Moreover, the area of Alpine folding (the Carpathians) forms a third mega-unit in the south of the country (see inset of Figure 17.1). Within these mega-units, second-order tectonic units can be distinguished in a belt-like arrangement (Figure 17.1). This is a consequence of successive stages (the Caledonian, the Variscan and the Alpine) of continental accretion along the south-western border of the East-European Craton (Żelaźniewicz et al., 2011).

Due to the reactivation of Laramide or older structures, tectonic faults and grabens developed in Poland during the Neogene and Quaternary (Figure 17.1). According to Rühle (1973), the Quaternary uplift exceeded 100 m in the north-eastern part of Poland, whereas subsidence ranges from 50 to >100 m in the lower Vistula River valley and in north-west Poland. Additionally, numerous salt diapirs occur in the area of the Mid-Polish Anticlinorium and neighbouring syneclises. The throw of the Quaternary faults is from 40–50 m to more than 100 m in the Sudetes, in the area east of Lublin and in the Inner Carpathians, while it ranges from several to several tens of metres in the Outer Carpathians (Zuchiewicz et al., 2007). Reactivation of some thrusts in the Carpathians is confirmed by studies of fractured clasts in Quaternary fluvial successions (e.g. Tokarski & Świerczewska, 2005; Tokarski et al., 2007).

The average rate of Quaternary faulting in Poland is low: 0.02–0.05 mm per year (Zuchiewicz et al., 2007). The NNE-SSW-oriented compression in the eastern Outer Carpathians, induced by the push from the hinterland, is transmitted farther to the north and changes gradually to NNW-SSE, close to the Baltic Sea (Figure 17.2). A strike-slip fault stress regime dominates in south-eastern Poland; it is less dominant in western Poland, where an extensional stress regime prevails (Jarosiński, 2006).

Epicentres of historical earthquakes are concentrated in southern Poland, along the Sudetes and Carpathians, with some smaller concentrations in central and north-western

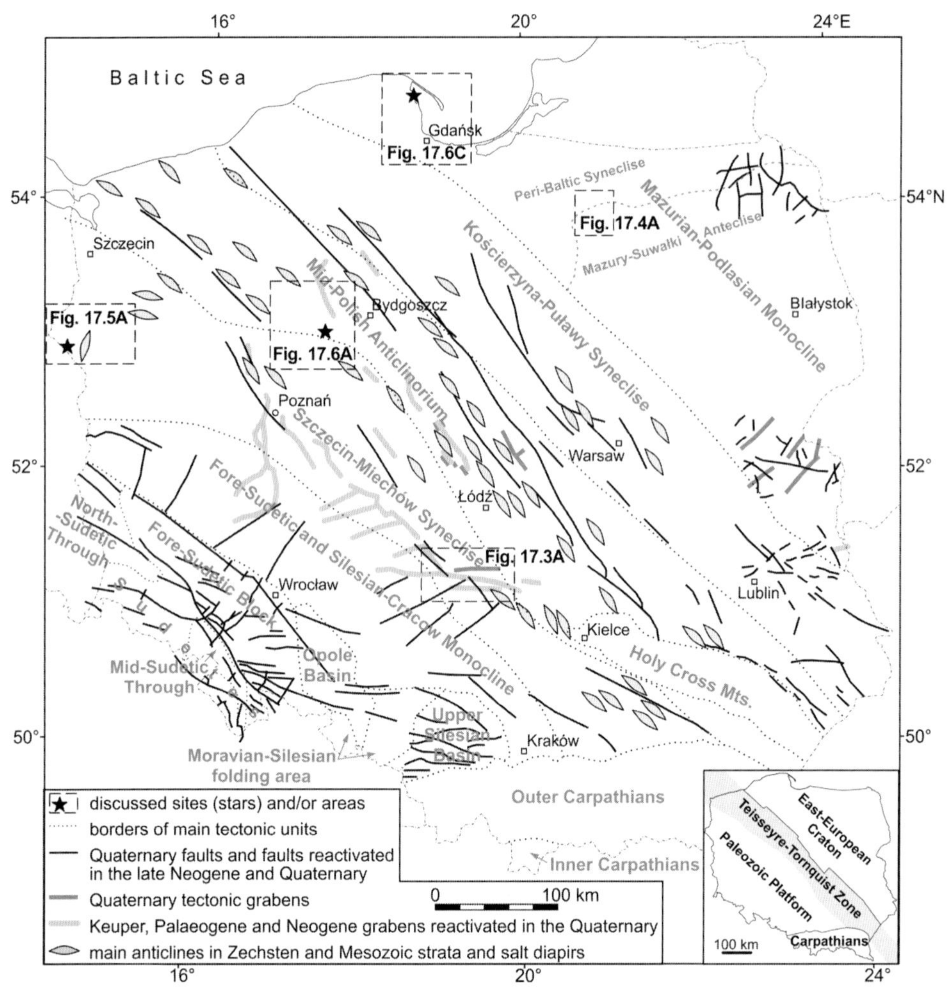

Figure 17.1 Tectonic division of Poland at the sub-Cainozoic surface according to Żelaźniewicz et al. (2011), Peri-Baltic Syneclise and Mazury-Suwałki Anteclise distinguished at the sub-Permian surface, as well as main Quaternary tectonic features according to Zuchiewicz et al. (2007). The areas with possible traces of the GIA discussed in the text are indicated by a star. The inset (lower right corner) shows the three tectonic mega-units and the Teisseyre-Tornquist Zone (grey area).

Poland (Figure 17.2), along the Teisseyre–Tornquist Zone (see the inset in Figure 17.1). Historical seismic activity is often related to strike-slip faults (Zuchiewicz et al., 2007), and the intensities of the earthquakes reach 8° MCS (Mercalli–Cancani–Sieberg scale) in the Sudetes (Pagaczewski, 1972) and 7.5° MCS in the Carpathians (Guterch et al., 2005). The magnitudes are up to 5.4 (Guterch, 2009). In northern Poland, earthquakes are rare, and peak ground acceleration is definitely less than in the south, but single earthquakes with magnitude greater than 4.0 were recently recorded there (Figure 17.2), including

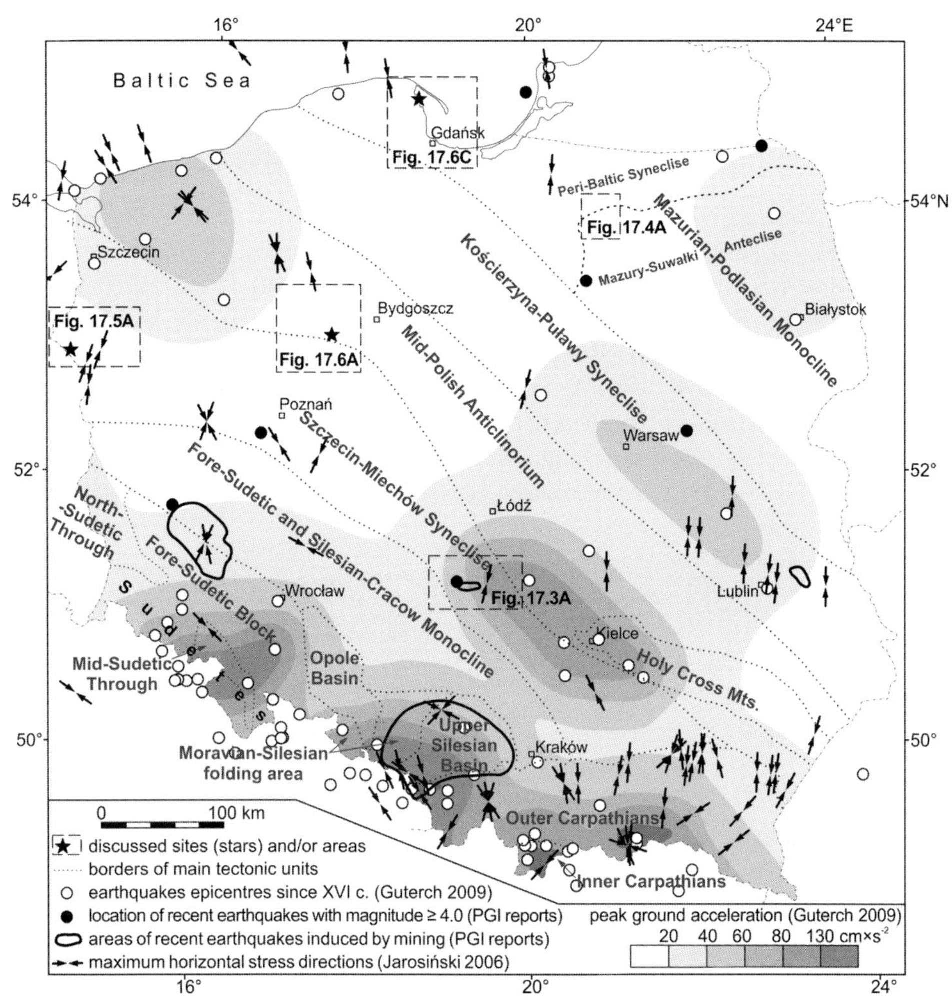

Figure 17.2 Seismotectonic risk in Poland according to Guterch (2009) and according to Polish Geological Institute (PGI) – National Research Institute reports on recent (January 2013–August 2019) geodynamic activity of Poland (www.pgi.gov.pl/mogepl-home/raporty-komunikaty/raporty.html) with maximum horizontal stress directions according to Jarosiński (2006). The areas with possible traces of the GIA discussed in the text are marked with a star.

earthquakes with magnitudes up to 5.2 which occurred in the Gdańsk Bay area (Wiejacz & Dębski, 2001; Gregersen et al., 2007).

17.3 Traces of Glacial Isostatic Adjustment in Poland

Poland was covered by the Fennoscandian ice sheet several times during the Pleistocene, so traces of loading/unloading cycles must have been left in the area. Only a few sites with possible traces of GIA have, however, been described thus far. These sites/areas are (1) the graben area near Bełchatów in central Poland, (2) the zone between the Peri-Baltic Syneclise and the Mazury–Suwałki Anteclise in north-eastern Poland, (3) the Siekierki site in north-western Poland, (4) the Ujście site in western Poland and (5) the Rzucewo site near the Gdańsk Bay (northern Poland). These occurrences are detailed in the following paragraphs.

(1) In the Kleszczów Graben (Figure 17.3A), near Bełchatów (central Poland), the Quaternary deposits locally reach up to over 200 m in thickness. The tills range in age from Elsterian (MIS 12, Mojski, 1985) to latest Saalian (MIS 6) and are commonly intercalated by glaciolacustrine deposits (Brodzikowski et al., 1997), which tend to show liquefaction-induced SSDS (Figure 17.3B; see Brodzikowski et al., 1987a,b). There are also layers (more rarely sets of layers) that form breccias (Gruszka & van Loon, 2007) pointing to significant consolidation before deformation or that show abundant SSDS; both point at a seismic origin. Moreover, probably shock-induced, gravity-flow deposits (Figure 17.3C) are intercalated between glaciolacustrine deposits settled from

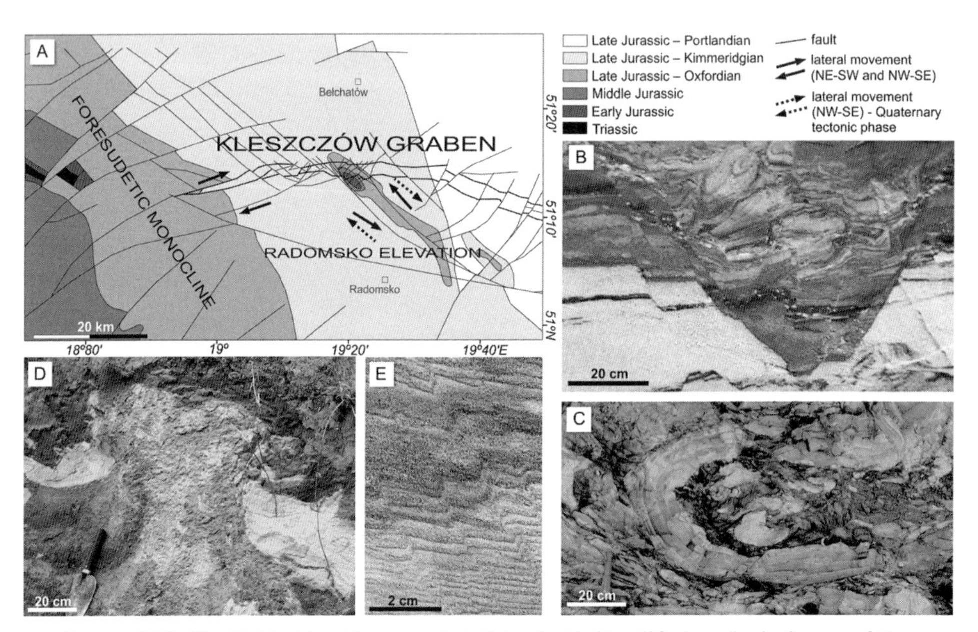

Figure 17.3 The Bełchatów site in central Poland. A) Simplified geological map of the Kleszczów Graben and surroundings without the Cainozoic and Cretaceous cover (modified after Rühle, 1978; Gotowała & Hałuszczak, 2002). B) Glaciolacustrine sandy sediments in the Kleszczów Graben, with a small graben (dead-ice structure) that is not due to endogenic tectonics but from the melting of buried glacial ice below. The resulting depression was filled up with glaciolacustrine silty sediments that became deformed by liquefaction, caused by an earthquake-related shock. C) Head of a complex slump mass in the Kleszczów Graben. The slumping was probably triggered by a seismic shock. D) Small diapir in the Quaternary overburden of the Kleszczów Graben, possibly initiated by a seismic shock. E) Kink structures in glaciofluvial sands within the Kleszczów Graben, representing a (temporary) compressional setting due to glacial push.

suspension (van Loon et al., 1995; Gruszka & Zieliński, 1996). The earthquakes are also commonly considered to have triggered both large (dozens of metres) and small (decimetres to metres) diapirs in the graben at Bełchatów (Figure 17.3D).

The Kleszczów Graben (Figure 17.3A) developed during the Late Oligocene–Early Miocene due to transtensional reactivation of basement faults, oriented WSW-ENE to NW-SE (Gotowała & Hałuszczak, 2002), and it is still active nowadays. The recent activity is expressed as reactivation of some of the pre-existing faults and is ascribed partly to anthropogenic activity (brown coal exploitation), partly to endogenic seismicity resulting in earthquakes with magnitudes up to 4.6 and epicentres aligned parallel to one of the regional basement faults (Zuchiewicz et al., 2007). In addition to these endogenic tectonics, glaciotectonics have affected the Quaternary sediments, as expressed by the presence of SSDS such as kink structures (Figure 17.3E) that indicate a compressional setting (due to glacial push) within the overall tensional setting. The interplay between graben activity and glaciations (Gruszka, 2007) has made the Kleszczów Graben one of the most thoroughly studied Late Cainozoic structures of the Polish Lowlands (Brodzikowski, 1995; Gotowała & Hałuszczak, 2002; Zuchiewicz et al., 2007).

It must be assumed that the Kleszczów Graben had, at least occasionally, a thicker ice cover than its immediate surroundings. At least during retreat of the ice, this will have led to differential loading, which makes it highly probable that GIA occurred, and that at least some (many?) of the seismites must be ascribed to fault activity triggered by GIA.

(2) The provinces of Mazury and Warmia, in north-east Poland were considered by Słodkowska (2009) to have been tectonically stable from the Late Cretaceous to at least the latest Miocene. However, according to Morawski (2009a,b), differential vertical movements of the crystalline basement occurred during the Pleistocene along both the craton slope and the border zone between the two just-mentioned provinces (Figure 17.4A). These movements, which resulted in a block pattern, exceeded 160 m. Two generations of N-S-running vertical fault zones are present (Morawski, 2009a): the older generation consists of reactivated faults along a zone of weakness, running from a depth of approximately 700 m down to the crystalline basement (see the cross-section in Figure 17.4B; see also Figure 5 in Morawski, 2009a), whereas the younger one is represented by faults that displace sediments down to a depth of 300–600 m (Figure 17.4B,C). Both fault generations occur superimposed in a single zone and result in a tectonic graben with an orientation similar to the Mazury–Warmia border zone. This zone had an ice-interlobe position during the Late Weichselian (MIS 2), resulting in a belt of crevasse-infill landforms (Figure 17.4A).

The younger generation of faults has been shown by Morawski to be directly related to GIA. Tectonic activity in the zone during successive Pleistocene glaciations is well documented to have resulted in considerable variations of the relief of the Neogene top, in vertical discontinuities within the Pleistocene and the Neogene, in thick Pleistocene glaciolacustrine successions and in stratigraphic gaps within the Pleistocene succession (Morawski, 2009b).

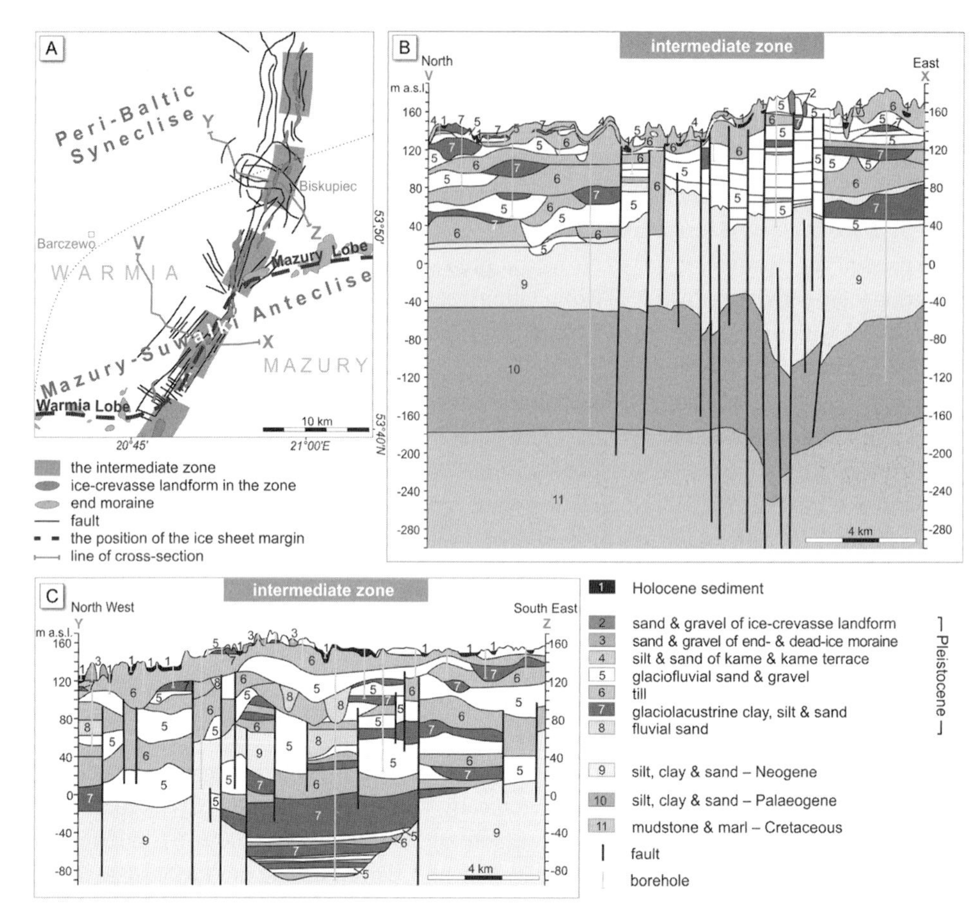

Figure 17.4 The border zone between the provinces of Warmia and Mazury in north-east Poland. A) The situation of the zone following Morawski (2009a,b), with locations of the cross-sections (B and C) and the position of the ice-sheet margin during the Pomeranian Phase of the last glaciation (~17.5 ka according to Tylmann et al., 2019a). The position of the ice-sheet margin during the Last Glacial Maximum (~22 ka according to Tylmann et al., 2019b) is ~20 km further to the south. Note the numerous crevasse-derived landforms. B) Cross-section V-X based on geophysical and geological analyses, according to Morawski (2009a). Note the numerous faults in the zone and their absence in the neighbouring areas. C) Cross-section Y-Z according to Morawski (2009b). Note the thick glaciolacustrine succession in the graben. (A black and white version of this figure will appear in some formats. For the colour version, please refer to the plate section.)

(3) At the Siekierki site (52°48′32″N and 14°14′40″E) in north-western Poland (Figure 17.5A), two strongly deformed layers are present (Figure 17.5B–D). They occur in the fine-grained lacustrine sediments with identical lithology deposited at the end of the Warthanian (MIS 6) Phase (second ice advance of the Saalian glaciation) and at the beginning of Eemian Interglacial (120 ± 1.8 ka BP, MIS 5e, Piotrowski et al., 2012). Both deformed layers show evidence of fluidization and liquefaction, i.e. both well-defined layers contain numerous SSDS such as load casts, pseudonodules,

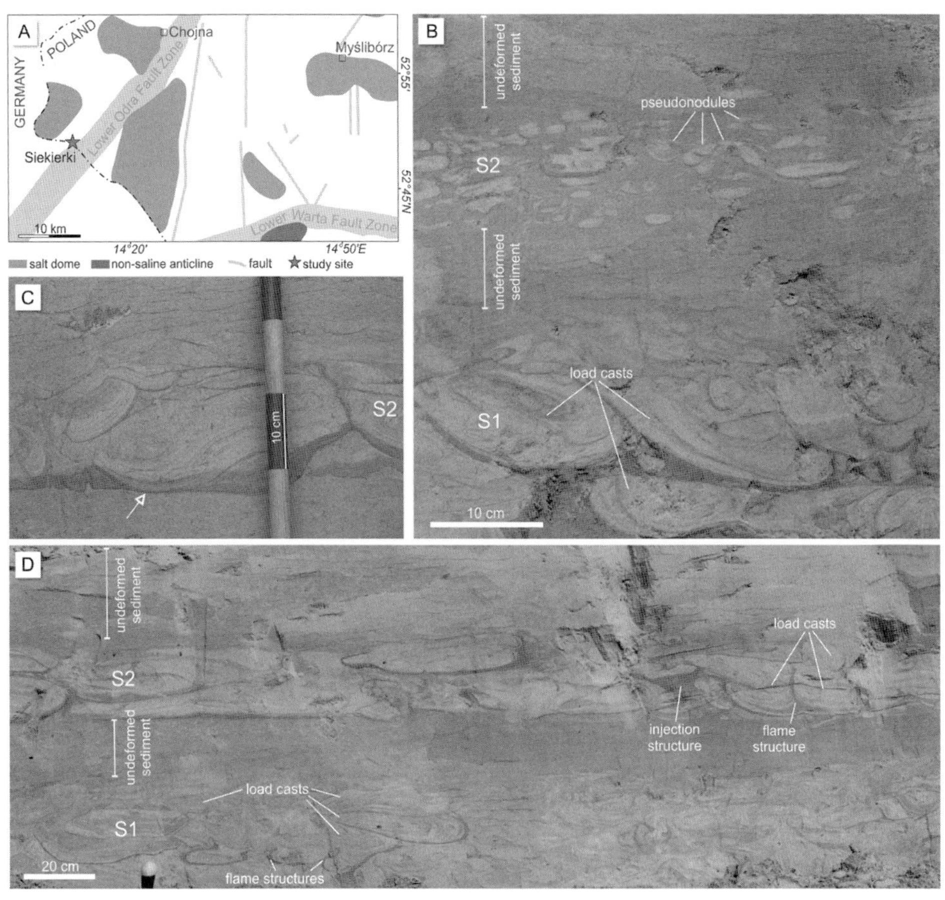

Figure 17.5 The Siekierki site in NW Poland (cf. van Loon & Pisarska-Jamroży, 2014). A) Tectonic setting of the Siekierki site, modified from Dadlez (1979). B) Load casts in the lower seismite (S1) and pseudonodules in the upper one (S2). C) Load casts of different size and different internal complexity in the upper seismite (S2). Note the sharp lower boundary of the layer (see arrow). D) The two seismites (S1 and S2) with abundant SSDS interbedded between undeformed sediments. The seismites were triggered by GIA. (A black and white version of this figure will appear in some formats. For the colour version, please refer to the plate section.)

ball-and-pillow structures, flame structures and fluid-escape structures, which are chaotically dispersed within the layers. The two deformed layers are interbedded between only slightly consolidated sediments that do not show any significant deformations (Figure 17.5B,D). According to van Loon and Pisarska-Jamroży (2014), all above features suggest that the deformed layers are seismites and owe their origin to earthquakes. These earthquakes might be linked to (re)activation of a fault in the western margin of the Lower Odra Fault Zone (Figure 17.5A).

The Permo-Mesozoic Lower Odra Fault Zone (Figure 17.5A) is characterized by differences in the thicknesses of the Quaternary sediments; these differences are explained by fault (re)activation (Kurzawa, 2003), which is also expressed by the occurrence of

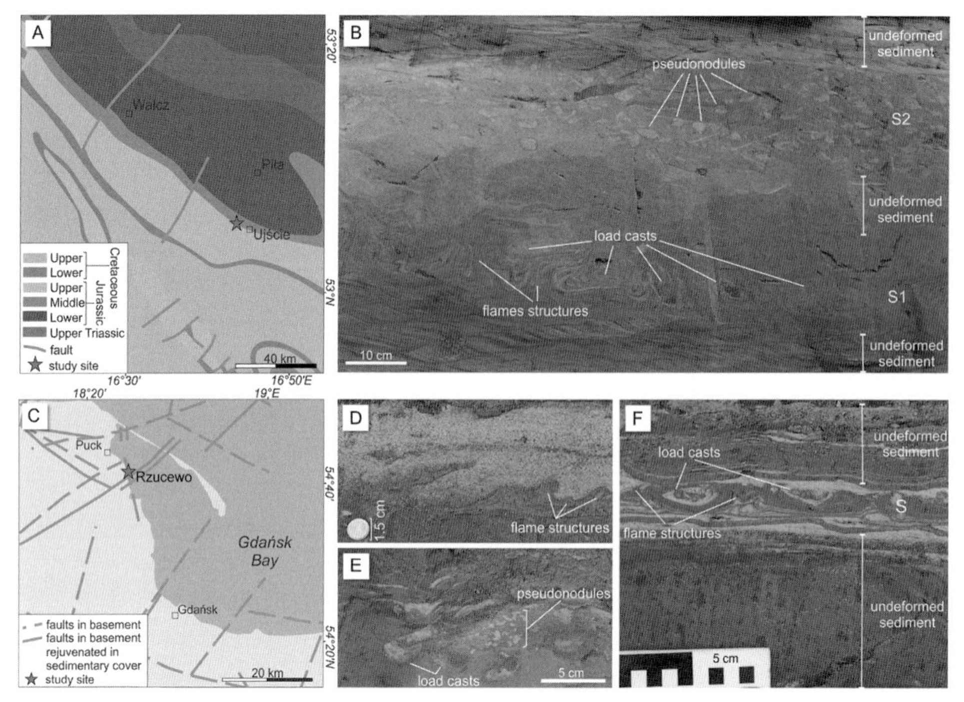

Figure 17.6 The Ujście site in W Poland (A and B, from Pisarska-Jamroży et al., 2019a) and the Rzucewo site in N Poland (C–F, modified from Woźniak & Pisarska-Jamroży, 2016, 2018; Pisarska-Jamroży & Woźniak, 2019) with deformed sediments. A) Pre-Palaeogene map of the surroundings of Ujście with faults that may have been reactivated by Pleistocene GIA (modified from Dadlez & Dembowska, 1965). B) Two strongly deformed layers (S1 and S2) with load casts, pseudonodules and flame structures at Ujście site triggered probably by GIA. C) Main basement faults in the surroundings of Rzucewo (according to Znosko 1998, supplemented by Ryka & Dadlez, 1995; Skompski, 2001). D–F) SSDS (flame structures, load casts) in seismite (S) interbedded between undeformed layers at Rzucewo site. (A black and white version of this figure will appear in some formats. For the colour version, please refer to the plate section.)

elongated depressions in the sub-Cainozoic and sub-Quaternary surfaces. The seismites at the Siekierki site must have been caused by sudden GIA-induced shocks, probably during the Warthanian deglaciation (van Loon & Pisarska-Jamroży, 2014).

(4) In the Ujście site (53°02′40″N, 16°41′42″E) in western Poland (Figure 17.6A), two internally deformed layers with sedimentological evidences of fluidization and liquefaction processes were recognized (Figure 17.6B). The two deformed layers occur in glaciolacustrine sediments that were deposited between 173 ± 1.6 and 140 ± 1.1 ka BP (MIS 6, Pisarska-Jamroży et al., 2019a). These layers are interbedded between undeformed ones of similar lithology and extend laterally over the entire outcrop of some 350 m wide. The lower deformed layer, 20 cm thick, contains complex, sometimes even chaotic, SSDS; most are load casts and flame structures (Figure 17.6B). The upper deformed layer shows almost exclusively pseudonodules occurring at different depths within the deformed layer. The fairly complex structure of the load casts and

chaotic occurrence of pseudonodules in the deformed layer under study here indicates that liquefaction must have been induced by a trigger that acted frequently (Pisarska-Jamroży et al., 2019a).

The pre-Cainozoic surface here is composed of Lower, Middle and Upper Jurassic and, in the SW part, of Upper and Lower Cretaceous (Dzierżek, 1997), so one possibility for the seismite development is Quaternary reactivation of Jurassic faults (Figure 17.6A) running NW of the Ujście site (Pisarska-Jamroży et al., 2019a). In addition to these endogenic tectonics, glaciotectonics have affected the Quaternary sediments. However, no reports of historical earthquakes in the study region are known, nor are there any indications of significant endogenic tectonic activity during the Neogene (Chmal, 2006).

(5) The glaciolacustrine sediments at the Rzucewo site (54°40′58″N, 18°27′57″E) in the Gdańsk Bay area (northern Poland; Figure 17.6C) show abundant SSDS within layers intercalated between undeformed layers. The sediments date most probably from MIS 4-3, as indicated by the set of luminescence dates ranging from 58.2 ± 8.7 to 27.2 ± 4.1 ka BP (Olszak, 1999). Deformed layers show several types of SSDS closely together (Figure 17.6D–F), indicating brittle behaviour, liquefaction and fluidization, but the most common deformations are load structures, flame structures and brittle deformations. The combination of different SSDS could be caused by earthquakes that commonly have different magnitudes, and the resulting shock waves may cause different degrees of liquefaction.

These deformed layers are interpreted by Pisarska-Jamroży and Woźniak (2019) as seismites caused by shocks triggered by reactivation of Palaeozoic fault/faults in the vicinity of the study site (Figure 17.6C). The Palaeozoic faults in the Gdańsk Bay and its vicinity have been interpreted by Ryka (1989), Witkowski (1989), Ryka and Dadlez (1995), Znosko (1998), Skompski (2001) and Pikies (2007) as reactivated during the Quaternary, affecting the overlying rocks, including the Quaternary cover (Figure 17.6C). It is noteworthy in this context that earthquakes with magnitudes up to 5.2 occurred in the Gdańsk Bay area in 2004 (e.g. Gregersen et al., 2007).

Additionally, also in the Rzucewo site other types of SSDS linked to gravity flow were noted and widely described by Woźniak and Pisarska-Jamroży (2018) and Woźniak et al. (2018).

17.4 Discussion and Final Remarks

In almost all cases described above, evidence in the form of specific concentrations and types of SSDS support GIA as the trigger mechanism for the development of seismites. The passage of seismic shock waves results in traces within the sedimentary record in the form of seismites only, however, if the two of the following conditions are met: (1) the magnitude of the earthquake must be sufficiently large ($M_w > 4.5-5$; see Rodríguez-Pascua et al., 2000) and (2) the sediments through which the seismic waves pass must be susceptible to liquefaction. On this basis, seismites interpreted to have been triggered by

GIA have been convincingly shown to exist in Poland. This explains the Quaternary tectonic activity in northern and central Poland, despite its intraplate – and consequently tectonically quiet – position. Sites with a GIA record have only rarely been described from Poland, but this might be due to lack of recognition or to misinterpretation.

The above-detailed sites with evidence of GIA are scattered throughout Poland, being located in different tectonic units; their position commonly cannot be related to the epicentres of historical earthquakes. However, some rare earthquakes have occurred in the present-day area of northern Poland and its vicinity (see Figure 17.2; Wiejacz & Dębski, 2001; Gregersen et al., 2007).

Despite the small number of sites described above, GIA may be recorded in the Quaternary sediments of Poland more often than what is currently known; many GIA-related seismites may not yet have been recognized, and their trigger mechanism may be misinterpreted (many SSDS appear nowadays to have been erroneously ascribed to periglacial processes, cf. van Loon et al., 2020). Moreover, it should be recognized that not all GIA-induced earthquakes left traces in the sediments. Additionally, in the southern and central parts of the country, endogenic tectonics caused by Alpine pressure and graben activity, respectively, have been – and still are – active, so that it has not yet been possible to distinguish here between endogenic tectonics and, e.g. GIA as the responsible trigger for the developments of seismites.

Finally, we want to emphasize that the presence of GIA-induced seismites in intraplate regions such as Poland should be an eye-opener for researchers who study seismites or other tectonic features elsewhere in the world. Since it is not yet possible to distinguish between seismites triggered by GIA or by endogenic tectonics, it seems worthwhile to start investigations that might result in diagnostic criteria.

Acknowledgements

We thank all friends and co-workers of the GREBAL project from Poland, Germany, Lithuania, Latvia and Estonia for many interesting joint fieldworks as well as for fruitful discussions. The suggestions by the reviewers Petra Štěpančíková and Antek Tokarski have greatly helped to improve the manuscript, for which we are most grateful. The work has been financially supported by the National Science Centre Poland (grant No. 2015/19/B/ST10/00661) in the context of the GREBAL project.

References

Badura, J. and Przybylski, B. (2000). *Mapa neotektoniczna Dolnego Śląska* [Neotectonic Map of Lower Silesia]. Unpublished report, Archive of Polish Geological Institute, Lower Silesia Division, Wrocław, map and 43 pp. (in Polish).

Badura J., Zuchiewicz W., Štěpančíková, P. et al. (2007). The Sudetic Marginal Fault: a young morphotectonic feature of central Europe. *Acta Geodynamica et Geomaterialia*, **148**, 7–29.

Belzyt, S., Pisarska-Jamroży, M., Bitinas, A. et al. (2021). Repetitive Late Pleistocene soft-sediment deformation by seismicity-induced liquefaction in north-western Lithuania. *Sedimentology*, doi.org/10.1111/sed.12883.

Birkenmajer, K. (1976). Plejstoceńskie deformacje tektoniczne w Szaflarach na Podhalu [Pleistocene tectonic deformations in Szaflary, Podhale region]. *Rocznik Polskiego Towarzystwa Geologicznego*, **46**, 309–324 (in Polish).

Brandes, C., Polom, U. and Winsemann, J. (2011). Reactivation of basement faults: interplay of ice-sheet advance, glacial lake formation and sediment loading. *Basin Research*, **23**, 53–64, doi.org/10.1111/j.1365-2117.2010.00468.x.

Brandes, C., Winsemann, J., Roskosch, J. et al. (2012). Activity along the Osning Thrust in central Europe during the Lateglacial: ice-sheet and lithosphere interactions. *Quaternary Science Reviews*, **38**, 49–62, doi.org/10.1016/j.quascirev.2012.01.021.

Brodzikowski, K. (1995). Pleistocene glacigenic deposition in a tectonically active, subsiding zone: the Kleszczów Graben, central Poland. In J. Ehlers, S. Kozarski and P. Gibbard, eds., *Glacial Deposits in North-East Europe*. Brookfield Balkema, A. A. Rotterdam, pp. 361–385.

Brodzikowski, K., Gotowała, R., Kasza, L. and van Loon, A. J. (1987a). The Kleszczów Graben (central Poland): reconstruction of the deformational history and inventory of the resulting soft-sediment deformation structures. In M. E. Jones and R. M. F. Preston, eds., *Deformation of Sediments and Sedimentary Rocks*. Geological Society, London, Special Publication, Vol. 29, pp. 241–254, doi.org/10.1144/GSL.SP.1987.029.01.18.

Brodzikowski, K., Hałuszczak, A., Krzyszkowski, D. and van Loon, A. J. (1987b). Genesis and diagnostic value of large-scale gravity-induced penecontemporaneous deformation horizons in Quaternary sediments of the Kleszczów Graben (central Poland). In M. E. Jones and R. M. F. Preston, eds., *Deformation of Sediments and Sedimentary Rocks*. Geological Society, London, Special Publication, Vol. 29, pp. 287–298, doi.org/10.1144/GSL.SP.1987.029.01.22.

Brodzikowski, K., van Loon, A. J. and Zieliński, T. (1997). Development of a lake in a subsiding basin in front of a Saalian ice sheet (Kleszczów Graben, central Poland). *Sedimentary Geology*, **113**, 55–80, doi.org/10.1016/S0037-0738(97)00046-8.

Chmal, R. (2006). *Objaśnienia do szczegółowej mapy geologicznej Polski w skali 1:50000. Arkusz Piła* [Explanations to the Detailed Geological Map of Poland, Scale 1:50000. Sheet Piła]. Polish Geological Institute Press, Warsaw (in Polish).

Dadlez, R. and Dembowska, J. (1965). Budowa geologiczna parantyklinorium pomorskiego [The geology of the Pomeranian para-anticlinorium]. *Prace Instytutu Geologicznego* 40, 175 pp. (in Polish).

Dadlez, R. (1979). Tektonika Kompleksu Cechsztyńsko–Mezozoicznego [Tectonics of the Zechstein–Mesozoic Complex]. In M. Jaskowiak-Schoeneich, ed., *Budowa geologiczna niecki szczecińskiej i bloku Gorzowa* [The Geological Structure of the Szczecin Trough and the Gorzow Block]. Prace Instytutu Geologicznego, **96**, pp. 108–121 (in Polish).

Dzierżek, J. (1997). Geology of sub-Quaternary basement and stratigraphy of Quaternary sediments in the middle Noteć river valley, western Poland. *Annales Societatis Geologorum Poloniae*, **67**, 57–81.

Gotowała, R. and Hałuszczak, A. (2002). The Late Alpine structural development of the Kleszczow Graben (Central Poland) as a result of reactivation of the pre-existing, regional dislocations. *EGU Stephan Mueller Special Publication*, **1**, 137–150, doi.org/10.5194/smsps-1-137-2002.

Gregersen, S., Wiejacz, P., Dębski, W. et al. (2007). The exceptional earthquakes in Kaliningrad district, Russia on September 21, 2004. *Physics of the Earth and Planetary Interiors*, **164**, 63–74, doi.org/10.1016/j.pepi.2007.06.005.

Gruszka, B. (2007). The Pleistocene glaciolacustrine sediments in the Bełchatów mine (central Poland): endogenic and exogenic controls. *Sedimentary Geology*, **193**, 149–166, doi.org/10.1016/j.sedgeo.2006.01.008.

Gruszka, B. and van Loon, A. J. (2007). Pleistocene glaciolacustrine breccias of seismic origin in an active graben (central Poland). *Sedimentary Geology*, **193**, 93–104, doi .org/10.1016/j.sedgeo.2006.01.009.

Gruszka, B. and Zieliński, T. (1996). Gravity flow origin of glaciolacustrine sediments in a tectonically active basin (Pleistocene, central Poland). *Annales Societatis Geologorum Poloniae*, **66**, 59–81.

Guterch, B. (2009). Sejsmiczność Polski w świetle danych histycznych [Seismicity of Poland in the light of historical records]. *Przegląd Geologiczny,* **57**, 513–520. (in Polish)

Guterch, B., Lewandowska-Marciniak, H. and Niewiadomski, J. (2005). Earthquakes recorded in Poland along the Pieniny Klippen Belt, Western Carpathians. *Acta Geologica Polonica*, **53**, 28–44.

Hoffmann, G. and Reicherter, K. (2012). Soft-sediment deformation of Late Pleistocene sediments along the southwestern coast of the Baltic Sea (NE Germany). *International Journal of Earth Sciences*, **101**, 351–363, doi.org/10.1007/s00531-010-0633-z.

Jarosiński, M. (2006). Recent tectonic stress field investigations in Poland: a state of the art. *Geological Quarterly*, **50**, 303–321.

Kurzawa, M. (2003). The sedimentary record and rates of Quaternary vertical tectonic movements in NW Poland. *Quaternary International*, **101**, 137–148, doi.org/10 .1016/S1040-6182(02)00096-4.

Laskowska-Wysoczańska, W. (1995). Neotectonic and glacial control on geomorphic development of middle and eastern parts of the Sandomierz Basin and the Carpathian margin. *Folia Quaternaria*, **66**, 105–122.

Migoń, P. and Łach, J. (1998). Geomorphological evidence of neotectonics in the Kaczawa sector of the Sudetic marginal fault, southwestern Poland. *Geologica Sudetica*, **31**, 307–316.

Migoń, P., Krzyszkowski, D. and Gogół, K. (1998). Geomorphic evolution of the front of the Sudetes between Dobromierz and Paszowice and adjacent areas, with particular reference to the fluvial systems. *Geologica Sudetica*, **31**, 289–305.

Mojski, J. E. (1985). *Geology of Poland. Volume I Stratigraphy, Part 3b Cainozoic*. Geological Press, Warsaw, 248 pp.

Morawski, W. (2009a). Neotectonics induced by ice-sheet advances in NE Poland. *Geologos*, **15**, 199–217, doi.org/10.2478/v10118-009-0004-z.

Morawski, W. (2009b). Differences in the regional stratigraphy of NE Poland caused by vertical movements due to glacioisostasy. *Geologos*, **15**, 235–250, doi.org/10.2478/ v10118-009-0006-x.

Mörner, N. A. (1991). Intense earthquakes and seismotectonics as a function of glacial isostasy. *Tectonophysics*, **188**(3–4), 407–410, doi.org/10.1016/0040-1951(91)90471-4.

Olszak, I. (1999). Chronostratigraphy of the western part of the cliff of Kępa Swarzewska near Jastrzębia Góra (Baltic Coast). *Peribalticum*, **7**, 41–63.

Pagaczewski, J. (1972). Catalogue of earthquakes in Poland in 1000–1972 years. *Publications of the Institute of Geophysics Polish Academy of Sciences*, **51**, 3–36.

Pikies, R. (2007). Influence of tectonic processes on the relief of sub-Quaternary surface and influence of these processes on formation of Quaternary in deep-water part of the southern Baltic Sea. *MELA Conference, Living Morphotectonics of the European Lowland, Cedynia, Poland*, pp. 60–70.

Piotrowski, A., Brose, F., Sydor, P., Seidler, J. and Pisarska-Jamroży, M. (2012). Stanowisko 3 – Siekierki. Osady Interglacjału Eemskiego w Siekierkach [Eemian interglacial deposits at Siekierki]. In M. Błaszkiewicz and F. Brose, eds., *Korelacja osadów plejstocenu na pograniczu polsko-niemieckim w Dolinie Dolnej Odry* [Pleistocene Sediment Correlations in the Lower Odra Region]. Polish Geological Institute Press, Warsaw, pp. 161–163 (in Polish).

Pisarska-Jamroży, M., Belzyt, S., Börner, A. et al. (2018). Evidence from seismites for glacio-isostatically induced crustal faulting in front of an advancing land-ice mass (Rügen Island, SW Baltic Sea). *Tectonophysics*, **745**, 338–348, doi.org/10.1016/j.tecto.2018.08.004.

Pisarska-Jamroży, M. and Woźniak, P.P. (2019). Debris flow and glacioisostatic-induced soft-sediment deformation structures in a Pleistocene glaciolacustrine fan: the southern Baltic Sea coast, Poland. *Geomorphology*, **326**, 225–238, doi.org/10.1016/j.geomorph.2018.01.015.

Pisarska-Jamroży, M., van Loon, A. J., Roman, M. and Mleczak, M. (2019a). Enigmatic gravity-flow deposits at Ujście (western Poland), triggered by earthquakes (as evidenced by seismites) caused by Saalian glacioisostatic crustal rebound. *Geomorphology*, **326**, 239–251, doi.org/10.1016/j.geomorph.2018.01.010.

Pisarska-Jamroży, M., Belzyt, S., Bitinas, A., Jusienė, A. and Woronko, B. (2019b). Seismic shocks, periglacial conditions and glacitectonics as causes of the deformation of a Pleistocene meandering river succession in central Lithuania. *Baltica*, **32**, 63–77, doi.org/10.5200/baltica.2019.1.6.

Przybylski, B. (1998). Late Quaternary evolution of the Nysa Kłodzka river valley in the Sudetic Foreland, southwestern Poland. *Geologia Sudetica,* **31**, 197–211.

Rodríguez-Pascua, M. A., Calvo, J. P., De Vicente, G. and Gomez-Gras, D. (2000). Soft-sediment deformation structures interpreted as seismites in lacustrine sediments of the Prebetic Zone, SE Spain, and their potential use as indicators of earthquake magnitudes during the Late Miocene. *Sedimentary Geology*, **135**, 117–135, doi.org/10.1016/S0037-0738(00)00067-1.

Różycka, M. and Migoń, P. (2018). Tectonic geomorphology of the Sudetes Mountains (central Europe) – a review and re-appraisal. *Annales Societatis Geologorum Poloniae*, **87**, 275–300.

Rühle, E. (1973). Ruchy neotektoniczne w Polsce [Neotectonic movements in Poland]. In E. Rühle, ed., *Metodyka badań osadów czwartorzędowych* [Methodology of research of Quaternary sediments]. Geological Press, Warsaw, pp. 13–31 (in Polish).

Rühle, E. (1978). Mapa geologiczna Polski bez utworów kenozoicznych i kredowych 1: 500 000 [Geological map of Poland without Cenozoic and Cretaceous strata 1: 500,000]. Instytut Geologiczny, Warsaw (in Polish).

Ryka, W. (1989). Podłoże krystaliczne polskiej części południowego Bałtyku [Crystaline basement of the South Baltic Sea]. *Kwartalnik Geologiczny,* **34**, 21–36 (in Polish).

Ryka, W. and Dadlez, R. (1995). Podłoże krystaliczne [Crystalline basement]. In J. E. Mojski, ed., *Atlas geologiczny południowego Bałtyku* [Geological Atlas of the South Baltic Sea]. Polish Geological Institute Press, Warsaw, (in Polish).

Skompski, S. (2001). *Objaśnienia do szczegółowej mapy geologicznej Polski w skali 1:50000, Arkusz Puck* [Explanations to the detailed Geological Map of Poland, Scale 1:50000. Sheet Puck]. Polish Geological Institute Press, Warsaw, 40 pp. (in Polish).

Słodkowska, B. (2009). Palynology of the Palaeogene and Neogene from Warmia and Mazury areas (NE Poland). *Geologos*, **15**, 219–234, doi.org/10.2478/v10118-009-0005-y.

Štěpančíková, P., Stemberk, J., Vilímek, V. and Košťák, B. (2008). Neotectonic development of drainage networks in the East Sudeten Mountains and monitoring of recent fault displacements (Czech Republic). *Geomorphology*, **102**, 68–80, doi.org/10.1016/j.geomorph.2007.06.016.

Štěpančíková, P., Hók, J., Nývlt, D. et al. (2010). Active tectonics research using trenching technique on the south-eastern section of the Sudetic Marginal Fault (NE Bohemian Massif, central Europe). *Tectonophysics*, **485**, 269–282, doi.org/10.1016/j.tecto.2010.01.004.

Tokarski, A. K. and Świerczewska, E. (2005). Neofractures versus inherited fractures in structural analysis: a case study from Quaternary fluvial gravels. *Annales Societatis Geologorum Poloniae*, **75**, 95–104.

Tokarski, A. K., Świerczewska, E. and Zuchiewicz, W. (2007). Fractured clasts in neotectonics reconstructions: an example from Nowy Sącz Basin, western Outer Carpathians, Poland. *Studia Quaternaria*, **24**, 47–52.

Tylmann, K., Rinterknecht V. R., Woźniak, P. P. et al. (2019a). Retreat of the southern front of the last Scandinavian Ice Sheet: dates and rates. *20th Congress of the International Union for Quaternary Research (INQUA). Book of Abstracts*, app.oxfordabstracts.com/events/ 574/ program-app/titles.

Tylmann, K., Rinterknecht, V. R., Woźniak, P. P. et al.(2019b). The local last glacial maximum of the southern Scandinavian Ice Sheet front: cosmogenic nuclide dating of erratics in northern Poland. *Quaternary Science Reviews*, **219**, 36–46, doi.org/10.1016/j.quascirev.2019.07.004.

van Loon, A. J., Brodzikowski, K. and Zieliński, T. (1995). Shock-induced resuspension deposits from a Pleistocene proglacial lake (Kleszczów Graben, central Poland). *Journal of Sedimentary Research*, **A65**, 417–422, doi.org/10.1306/D42680DB-2B26-11D7-8648000102C1865D.

van Loon, A. J. and Pisarska-Jamroży, M. (2014). Sedimentological evidence of Pleistocene earthquakes in NW Poland induced by glacioisostatic rebound. *Sedimentary Geology*, **300**, 1–10, doi.org/10.1016/j.sedgeo.2013.11.006.

van Loon, A. J., Pisarska-Jamroży, M., Nartišs, M., Krievāns, M. and Soms, J. (2016). Seismites resulting from high-frequency, high-magnitude earthquakes in Latvia caused by Late Glacial glacio-isostatic uplift. *Journal of Palaeogeography*, **5**, 363–380, doi.org/10.1016/j.jop.2016.05.002.

van Loon, A. J., Pisarska-Jamroży, M. and Woronko, B. (2020). Sedimentological distinction in glacigenic sediments between load casts induced by periglacial processes from those induced by seismic shocks. *Geological Quarterly*, **64**, 626–640, doi.org/10.7306/gq.1546.

Wiejacz, P. and Dębski, W. (2001). New observation of Gulf of Gdansk seismic events. *Physics of the Earth and Planetary Interiors*, **123**, 233–245, doi.org/10.1016/S0031-9201(00)00212-0.

Witkowski, A. (1989). Ewolucja i tektonika staropaleozoicznego kompleksu strukturalnego południowego Bałtyku [Evolution and tectonics of the Lower Palaeozoic structural complex in the southern Baltic Sea]. *Kwartalnik Geologiczny*, **34**, 51–66 (in Polish with English summary).

Woźniak, P. P. and Pisarska-Jamroży, M. (2016). Rzucewo-soft-sediment deformation structures in glaciolimnic sediments-different trigger mechanisms. In R. Sokołowski and D. Moskalewicz, eds., *Quaternary Geology of North-Central Poland from the Baltic Coast to the LGM Limit*, University of Gdańsk, pp. 53–67.

Woźniak, P. P. and Pisarska-Jamroży, M. (2018). Debris flows with soft-sediment clasts in a Pleistocene glaciolacustrine fan (Gdańsk Bay, Poland). *Catena*, **165**, 178–191, doi.org/10.1016/j.catena.2018.01.022.

Woźniak, P. P., Pisarska-Jamroży, M. and Elwirski, Ł. (2018). Orientation of gravels and soft-sediment clasts in subaqueous debrites – implications for palaeodirection reconstruction: case study from Puck Bay, northern Poland. *Bulletin of the Geological Society of Finland,* **90**, 161–174, doi.org/10.17741/bgsf/90.2.002.

Wójcik, A. (2003). Czwartorzęd zachodniej części Dołów Jasielsko-Sanockich (polskie Karpaty Zewnętrzne) [The Quaternary of the western part of Jasielsko-Sanockie Doły (Polish Outer Carpathians)]. *Prace Państwowego Instytutu Geologicznego*, **178**, 148 pp.

Żelaźniewicz, A., Aleksandrowski, P., Buł, Z. et al. (2011). *Regionalizacja tektoniczna Polski* [Tectonic division of Poland]. Committee of Geological Sciences, Polish Academy of Sciences Press, Wrocław, 60 pp.

Znosko, J. (1998). *Mapa tektoniczna Polski* [Tectonic map of Poland]. Polish Geological Institute Press, Warsaw (in Polish).

Zuchiewicz, W. (1995). Selected aspects of neotectonics of the Polish Carpathians. *Folia Quaternaria*, **66**, 145–204.

Zuchiewicz, W., Badura, J., Jarosiński, M. and Commission on Neotectonics, Committee for Quaternary Research, Polish Academy of Sciences (2007). Neotectonics of Poland: an overview of active faulting. *Studia Quaternaria*, **24**, 5–20.

18

Soft-Sediment Deformation Structures in the Eastern Baltic Region

Implication in Seismicity and Glacially Triggered Faulting

ALBERTAS BITINAS, JURGA LAZAUSKIENĖ AND MAŁGORZATA (GOSIA) PISARSKA-JAMROŻY

ABSTRACT

Recent studies of soft-sediment deformation structures in the Eastern Baltic Region (EBR) demonstrated that their formation could have been caused by fluidization and liquefaction of sediments possibly triggered by palaeoearthquakes; thus, they could be interpreted as seismites. The EBR is characterized by still rather low seismic activity, although from the year 1375 several tens (~40) of both historical and instrumental earthquakes are described and recorded in the region and adjacent territories. The distribution of seismic activity in the EBR is possibly attributed to several factors, predominantly related to the lithospheric geodynamic processes and variations of rheological properties of lithosphere within the study area. The seismic activity has an irregular distribution – the northern part of the region is more seismically active than the southern one, but still is higher compared with more inland cratonic areas. Overall, a major portion of registered earthquakes within the EBR are associated with mapped faults and might be related to the reactivation of pre-existing fault systems. Still, in some cases it is difficult to associate single earthquakes with certain faults unambiguously due to significant errors of location of seismic events and location of faults. Moreover, not all earthquakes in the EBR correspond unambiguously to mapped fault zones, as it is complicated to associate some earthquakes with certain faults due to errors in identifying the location of low-magnitude seismic events. An identification of the seismogenic faults might also be complicated due to the rather small scale of the tectonic dislocations in the intracratonic area, with up to 2.5-km-thick Phanerozoic sedimentary cover. Nevertheless, a portion of the soft-sediment deformation structures can be interpreted as seismites and attributed to the seismic events triggered by glacial isostatic adjustment of the lithosphere during the Last (Weichselian) glacial advance and subsequent deglaciation. The detailed overview of soft-sediment deformation structures of possibly seismic origin discovered in several sites in Estonia, Latvia, Lithuania, Belarus and the Kaliningrad District of Russia is presented.

18.1 Introduction

In the second decade of the twenty-first century, based on results of detailed field analysis of soft-sediment deformation structures (SSDS), geological mapping and careful analysis of published scientific data, an assumption of possibly seismically induced SSDS owing to palaeoseismic events in the Eastern Baltic Region (EBR) was developed and published (Rogozhin et al., 2010, 2014; Bitinas & Lazauskienė, 2011; Bitinas, 2012). In a number of previous publications most of the deformation features in sediments are attributed to sedimentary structures triggered by different geological processes – permafrost, slope or glaciotectonic processes, while the possibly seismic origin of these structures is omitted. The more detailed studies of SSDS demonstrated that their formation could have been caused by liquefaction of sediments possibly triggered by palaeoearthquakes due to glacial isostatic adjustment (GIA) of the lithosphere (e.g. van Loon et al., 2016; Pisarska-Jamroży et al., 2018a, 2019a,b,c; Pisarska-Jamroży & Woźniak, 2019; van Loon et al., 2020).

For many years the entire EBR has been considered a low seismicity area. There are two main reasons that seismic activity has not been adequately assessed: (1) a priori prevailing opinion that the region, as a stable part on the East European Craton, is of low seismic activity and (2) lack of experience and skills to identify reliably traces of palaeoearthquakes. But this opinion began to change essentially after a few earthquakes that occurred at the juncture of the twentieth and twenty-first centuries: after the earthquake of 4.7 magnitude in the Estonian island of Osmussaare, in 1976, and especially after two successive 5.0- and 5.2-magnitude earthquakes in the Kaliningrad District of the Russian Federation in 2004 (Nikonov, 2005; Gregersen et al., 2007). These events stimulated palaeoseismological investigations not only in the Kaliningrad District (Rogozhin et al., 2010, 2014; Nikonov, 2011; Druzhinina et al., 2017) but in the wider territory of the EBR as well – an international project 'Recognition of traces left by earthquakes in Pleistocene sediments affected by GIA in the Baltic Sea Basin (GREBAL)', financed from the National Science Centre Poland, began in 2015 and covered, among others, the territories of Estonia, Latvia and Lithuania (Pisarska-Jamroży & Bitinas, 2018).

18.2 Geological Structure and Tectonics

The territory of the EBR covers the central part of the Baltic Sedimentary Basin – the largest sedimentary basin located on the western margin of the East European Craton. The structure of the basin is defined by features within the underlying Precambrian crystalline basement that occurs at a depth of 500–1,000 m in the northern and eastern parts of the basin, increasing to a depth of 2,300 m in Western Lithuania and to 3,000–5,000 m to the south-western part, close to the Teisseyre–Tornquist Zone (Paškevičius, 1997). The sedimentary cover of the Baltic Sedimentary Basin is represented by the Vendian and all the geological systems of the Phanerozoic to Quaternary. The present extent of the Baltic Sedimentary Basin represents only a part of an initial Early Palaeozoic basin that has been considerably eroded in succeeding periods by denudation processes, especially during Carboniferous–Early Permian times.

The geodynamic evolution of the intracratonic Baltic Sedimentary Basin was closely related to the location of the basin on the western edge of the East European Craton and specific mechanical properties of the basin's lithosphere. The area was not subjected to intense folding, and rather weak faulting was related to the flexure of the Baltica plate in front of the Caledonian orogen (Lazauskiene et al., 2002). The seismo-tectonic framework of the areas under study has been outlined several times in recent decades. A number of faults and fault zones have been distinguished in the EBR and adjacent territories based on geological and geophysical data (Figure 18.1). The crystalline basement of the Baltic Sedimentary Basin is strongly dissected by tectonic faulting. Two major types of faults

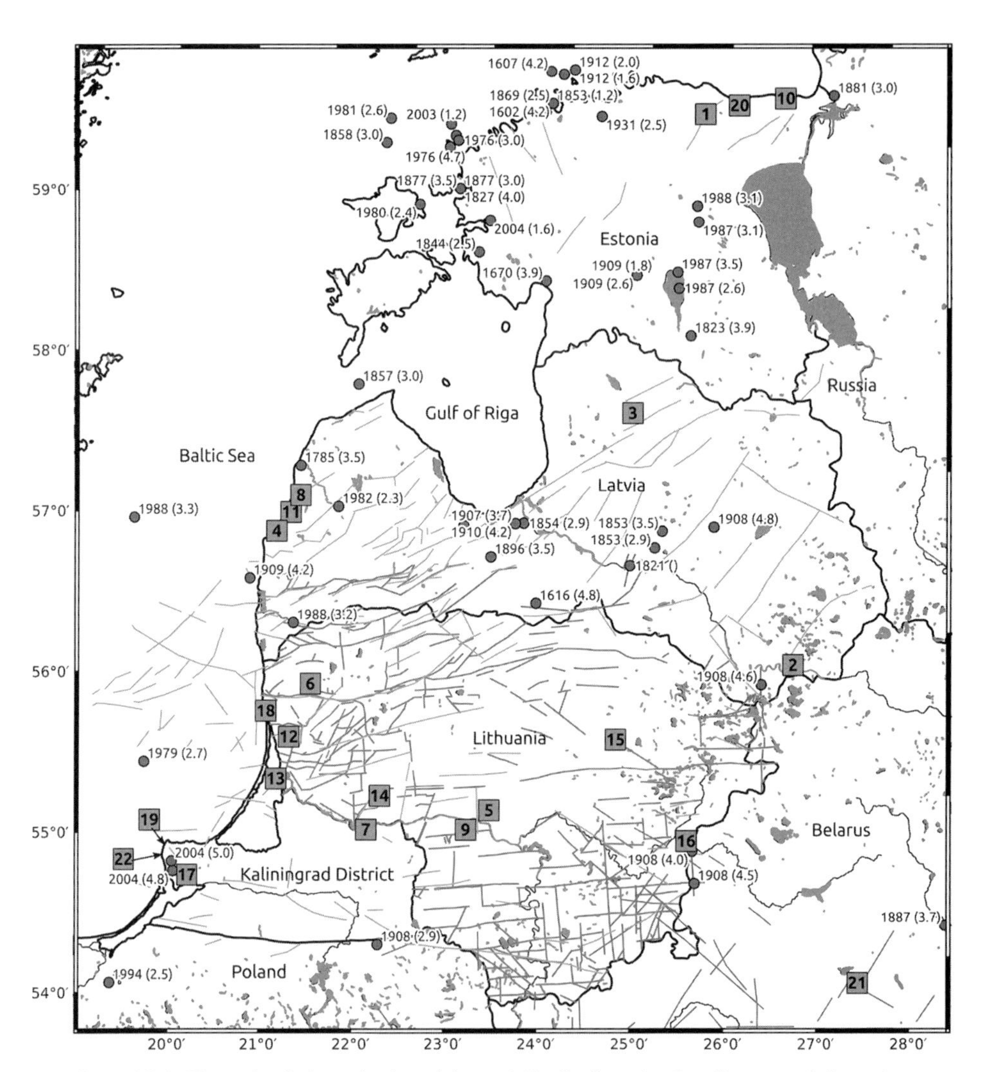

Figure 18.1 The major faults, seismic activity and distribution of soft-sedimentary deformation structures of possibly seismic origin in the Eastern Baltic Region. Coloured straight lines – network of tectonic faults compiled from separate sources/authors; red circles – location/ epicentre of historical and instrumentally recorded earthquake (numbers side by side indicates year and magnitude of seismic event); green tetragonals – sites with SSDS of possibly seismic origin (number of sites correspond to the numbering presented in Table 18.1). (A black and white version of this figure will appear in some formats. For the colour version, please refer to the plate section.)

prevail in the region, i.e. the oldest pre-platform and younger platform features. The former is defined in the crystalline basement and does not dissect the sedimentary cover, whereas the latter penetrates the sediments overlying the crystalline basement. The faults dissecting the sedimentary cover are oriented N-S, W-E, NW-SE and NE-SW, predominantly.

The distribution of faults is rather variable in the territory of Lithuania (Figure 18.1). In the western part of Lithuania faults of latitudinal direction dominate; the faults slightly deflect to east while going from south to north. In south-western Lithuania and the Kaliningrad District of Russia faults of WSW-NNE direction prevail. In western Latvia and north-western Lithuania faults trending W-E direction dominate, then deflect slightly to NE direction. The density of the faults here is higher than in the rest of the study region. In eastern Latvia the faults mainly tend in SW-NE direction. Faults of meridional direction dominate in northern Latvia, whereas in Estonia the most significant faults strike in SW-NE and NNW-SSE directions. The majority of potentially seismogenic faults in the studied area are of N-S and SW-NE orientations. According to preliminary measurements of tectonic stresses in Lithuania (Grünthal & Stromeyer, 1995; Šliaupa & Zakarevičius, 2000) and experiments in Fennoscandia (Gregersen & Basham, 1989) it was assumed that approximately half of the earthquakes might be related to the thrust faults and another majority to the strike-slip movements, and only ~10 per cent are confined to normal faults (Andrius Pačėsa, personal communication, 2020).

18.3 Seismic Activity

The EBR is characterized by rather low seismic activity: several tens (> 40) of historical earthquakes with intensities of VI–VII (MSK-64 scale) and instrumental earthquakes (with local magnitudes up to $M_L = 5$) have been recorded in the region and adjacent territories since the year 1303 (Grünthal & Riedel, 2007; Pačėsa et al., 2005; Lazauskienė et al., 2012). The majority of seismic events in the EBR that so far were identified and included in seismic catalogues are historical ones, and the primary sources of information do not provide evaluations of errors of epicentre locations for historical events. The historical seismic activity in the EBR is also significantly lower compared to seismicity of the Fennoscandian Shield. Palaeoseismic activity of the Fennoscandian Shield and adjacent Baltic Sea territories during the Lateglacial and Holocene (the last 13,000 years) is well documented by numerous palaeoseismic investigations (Mörner, 2005, 2008; Jakobsson et al., 2016; Brandes et al., 2018; Mattila et al., 2019; Grube, 2019). Despite this, until recently for the eastern part of the Baltic Region the palaeoseismic activity has not been adequately assessed due to the lack of experience and of skills among researchers to identify reliably traces of seismically induced SSDS. Still, despite rather low seismic activity, the EBR is more seismically active compared with the more 'inland' territories of the East European Craton (Grünthal, 1999; Sharov et al., 2007). The most significant earthquakes in the study region were: (1) Osmussare (Estonia) in 1976 with a maximal magnitude up to $M_L = 4.75$, (2) Kokneses (Estonia) – 22 February 1821, $M_L = 4.5$, (3) Gudogai (Belarus) – 28 December 1908, $M_L = 4.5$, (4) Madona (Latvia) – 29 December 1908, $M_L = 4.5$ (Boborikin et al., 1993). The magnitude of the strongest one recorded in the EBR, the Kaliningrad earthquake in 2004, was assessed as high as $M_W = 5.2$ (Gregersen et al., 2007) and M_b as high as 5.4 (data of the Russian Academy of Sciences). These earthquakes are presumably related with the network of the fault systems of different levels that dissect the territory of Kaliningrad District – with the Prieglius Fault Zone (Pačėsa et al., 2005), or with more W-NW-located Jantarnenskaya and Bakalinskaya faults (Rogozhin et al., 2010, 2014). As more than one earthquake is related to these faults, they could be regarded as potentially seismically active. Moreover, the earthquakes in the Kaliningrad District resulted in surface disturbances and fractures mostly of the superficial character that emerged partially (or completely) as delayed consequences in the proximal zone of the Kaliningrad earthquake on 21 September 2004 (Nikonov, 2011). They also caused ground shaking throughout Lithuania and, moreover, up to a distance of ~800 km, thus in the whole Baltic Sea Region (Gregersen et al., 2007). These earthquakes are very important in reassessing the seismicity and seismic hazard potential of the EBR, as they had the highest magnitudes ever instrumentally recorded in this region and occurred in the stable intracratonic setting.

18.4 Soft-Sediment Deformation Structures and Their Potential Relation to Glacially Triggered Faulting

Investigations of most possibly seismically triggered SSDS in the EBR started relatively late. At a current stage of palaeoseismological investigations all the localities with probably SSDS driven by seismicity could possibly be subdivided into four groups (Table 18.1, Figure 18.2):

(1) seismic origin of SSDS is confirmed after detailed investigations; the results were published, or the publications are in preparation (Table 18.1, sites no 1–7);
(2) SSDS were investigated in detail, but the seismic/non-seismic origin is still discussed (Table 18.1, sites no. 8–9);
(3) SSDS discovered previously but more detailed sedimentological studies are needed (Table 18.1, sites no. 10–19);
(4) SSDS selected from the previously published papers (Raukas & Teedumäe, 1997; Karabanov & Yelovicheva, 1997, etc.) where they have been interpreted as structures of non-seismic origin, but detailed investigations are needed to confirm or refute the seismic origin (Table 18.1, sites no. 20–22).

SSDS caused by fluidization and liquefaction processes induced by seismic tremor and already interpreted as seismites were found in several sites of Estonia, Latvia, Lithuania, Belarus and the Kaliningrad District (Figure 18.1). The mentioned structures were developed in sediments of different genesis – generally of glaciolacustrine/lacustrine or fine-grained glaciofluvial/fluvial origin. The SSDS are attributed to sediments accumulated over a relatively long time period, i.e. starting possibly from the Middle or Early Pleistocene and ending in middle Holocene. According to results of the stress assessment

Table 18.1. *Soft-sediment deformation structures (SSDS) in the Eastern Baltic Region (possibly triggered by seismic events). WES = water escape structures.*

	Locality	Coordinates	Type of SSDS	Deformed sediments	Time of deformation	GIA	Reference
1	Rakvere, sand quarry, Estonia	~59°N ~26°E	Load structures, flame structures, WES	Late Weichselian glaciolacustrine sediments, ~1 ka	Postglacial	Yes	Unpublished data
2	Rakuti, the Dauguva River outcrop, Latvia	55°53′51″ N 27°05′30″ E	Load structures, flame structures, WES	Late Weichselian glaciofluvial, glaciolacustrine sediments, ~16–17 ka	Postglacial	Yes	van Loon et al., 2016
3	Valmiera, the Gauja River outcrop, Latvia	57°32′45″ N 25°26′41″ E	Load structures, flame structures, WES	Late Weichselian glaciofluvial and glaciolacustrine sediments, ~14.5 ka	Postglacial	Yes	van Loon et al., 2016
4	Baltmuiža, the Baltic Sea cliff, Latvia	56°55′31″ N 21°15′23″ E	Load structures, flame structures, WES	Proglacial glaciolacustrine sediments, ~26 ka	Before the Last Glacial advance; later affected by glaciotectonic processes	No	Belzyt et al., 2018a;
5	Slinkis, the Dubysa River outcrop, Lithuania	55°05′43″ N 23°26′51″ E	Load structures, flame structures	Proglacial, alluvial sediments, ~21–24 ka	Before the Last Glacial advance	No	Belzyt et al., 2018b; Pisarska-Jamroży et al., 2019c
6	Dyburiai, the Minija River outcrop, Lithuania	55°56′21″ N 21°35′40″ E	Load structures, flame structures, ball-and-pillows, injection structures	Intramorainic sediments, ~98–111 ka	Early/Middle Weichselian	Yes	Pisarska-Jamroży et al., 2018b; Belzyt et al., 2021

Table 18.1. (*cont.*)

	Locality	Coordinates	Type of SSDS	Deformed sediments	Time of deformation	GIA	Reference
7	Riadino, the Šesupė River terrace, Kaliningrad District of Russia	55°01′32″ N 22°11′56″ E	Diapir-like liquefaction structures	Second overbank terrace, ~61–44 ka	After 8.1 ka BP	Yes	Druzhinina et al., 2017
8	Sarnate, the Baltic Sea cliff, Latvia	57°03′54″ N 21°24′55″ E	Load casts, WES	Late Weichselian glaciolacustrine sediments	Postglacial	Yes	Saulīte et al., 2007; Nartišs et al., 2018
9	Kumečiai, sand quarry, Lithuania	55°03′39″ N 23°24′59″ E	Load structures, flame, ball-and-pillows, injection structures	Intermorainic glaciolacustrine sediments, ~46–76 ka	Before the Last Glacial advance	?	Pisarska-Jamroży et al., 2018c
10	Voka, the Baltic Sea (Gulf of Finland) cliff, Estonia	59°24.9′ N 27°35.9′ E	Load casts	Middle Weichselian limnic (?) sediments	Before the Last Glacial advance; later affected by glaciotec-tonic processes (?)	Yes	Unpublished data
11	Jurkalne, the Baltic Sea cliff, Latvia	~57°00′30″ N ~21°22′45″ E	WES	Lateglacial glaciolacustrine sediments	Postglacial	Yes	Bitinas & Lazauskienė, 2011
12	Juodikiai, sand quarry, Lithuania	55°36′39″ N 21°20′39″ E	Load structures	Postglacial glaciofluvial sandur-deltaic sediments	Postglacial	Yes	Bitinas & Damušytė, 2018
13	Ventės Ragas, Curonian Lagoon outcrop, Lithuania	55°20′57″ N 21°11′50″ E	Diapires, 'sinked' boulders	Postglacial lacustrine sediments	Postglacial	Yes	Bitinas & Damušytė, 2018
14	Šakiai (Tauragė district), clay quarry (currently re-cultivated), Lithuania	55°14′05″ N 22°21′21″ E	Pseudonodules, WES	Postglacial glaciolacustrine sediments	Postglacial	?	Bitinas & Lazauskienė, 2011

15	Anykščiai, Neogene quartz sand quarry, Lithuania	55°31′33″ N 25°01′27″ E	Pseudonodules	Intermorainic, Early vs Middle Pleistocene sediments	Middle Pleistocene	?	Unpublished data
16	Buivydžiai (Punžionys), the Neris River outcrop, Lithuania	54°51′58″ N 25°44′36″ E	Pseudonodules	Intermorainic Middle Weichselian sediments	Before the Last Glacial advance	?	biturl.top/73aE7b (Accessed: 3 March 2017)
17	Logvino-Cherepanovo area, two excavations, Kaliningrad District	~54°45′ N ~20°12′ E	Diapir-like structures, pseudonodules (?)	Lateglacial glaciolacustrine sediments	Late Holocene	No	Rogozhin et al., 2010
18	Giruliai, the relict Litorina Sea cliff, Lithuania	55°46′22″ N 21°05′01″ E	Mega-landslide	Lateglacial glaciolacustrine sediments	Holocene (after the maximal Litorina Sea transgression)	Yes	Damušytė & Bitinas 2018
19	Mys Taran, the Baltic Sea cliff, Kaliningrad District	~54°57′10″ N ~19°58′25″ E	Mega-landslide	Pleistocene and Neogene (?) sediments	Late Holocene	No	Rogozhin et al., 2014
20	Purtse River outcrop, Estonia	~59°24′ N ~27°01′ E	WES	Postglacial (Younger Dryas) sediments	Postglacial or Holocene	?	Raukas & Teedumäe, 1997
21	Zaslavl, sand/gravel quarry, Belarus	~54°02.0′ N ~27°14″ E	WES, flame structures	Eemian Interglacial and Early Weichselian sediments	Early/ Middle Weichselian	?	Karabanov & Yelovicheva, 1997; Yelovicheva & Drozd, 2005
22	Bakalino, the Baltic Sea cliff, Kaliningrad District	~54°55′30″ N ~19°56′45″ E	Diapir, flexure	Palaeogene, Neogene, Pleistocene and late Holocene sediments	Late Holocene	No	Rogozhin et al., 2014

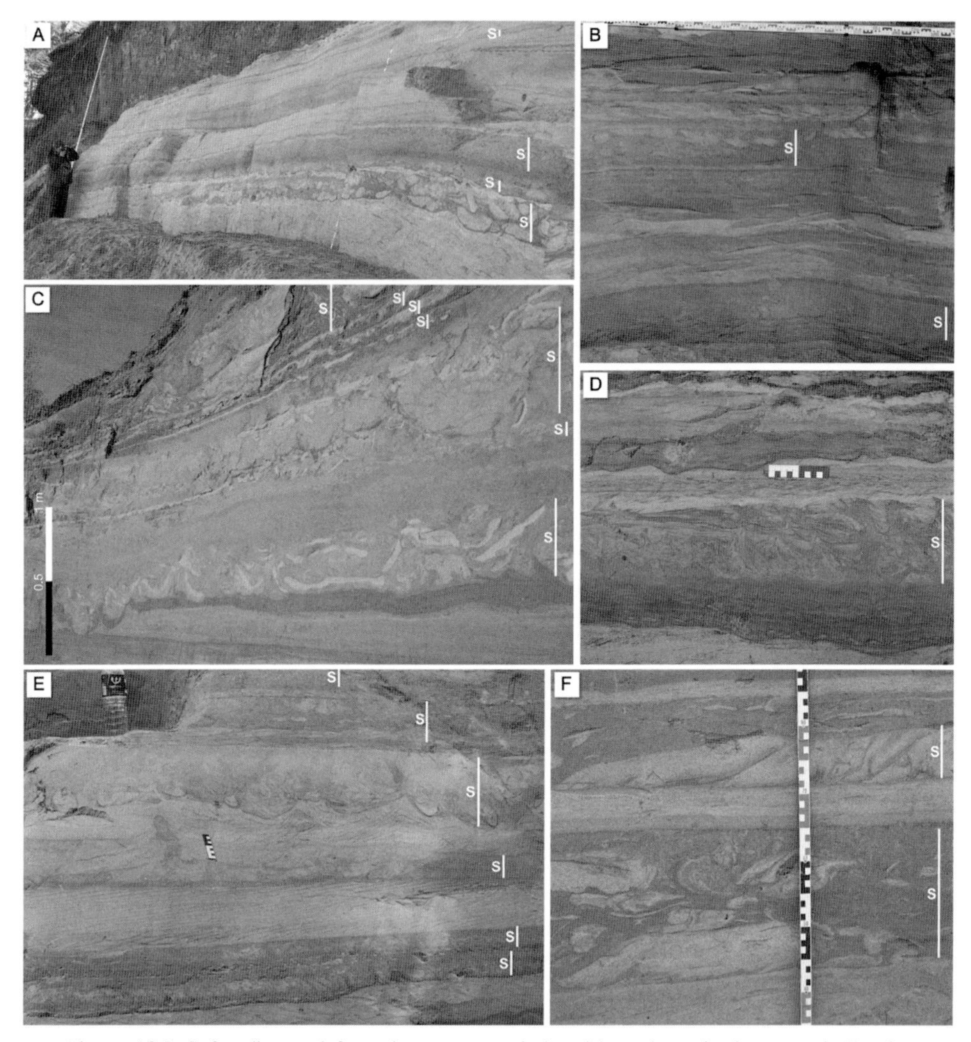

Figure 18.2 Soft-sediment deformation structures induced by palaeoseismic events in Latvia, Lithuania and Estonia. The 'S' letter indicates occurrence of internally deformed layer (seismite). Note that the lower and upper boundaries of seismites in most cases are sharp and well visible. A) Deformed fine-grained silty and sandy layers with load casts, flame structures and water/sediment escape-structures occurring along a cliff of the Baltic Sea coast in Baltmuiža outcrop in Western Latvia: five seismites were recognized (Belzyt et al., 2018a). B) Two from seven seismites recognized in the Valmiera outcrop occurring along the bank of the Gauja River (north-eastern Latvia) with load casts, pseudonodules and flame structures (cf. van Loon et al., 2016). C) Seven from twelve seismites in the Rakuti outcrop (south-eastern Latvia). These seismites contain load casts, pseudonodules, flame structures, water/sediment escape-structures as well as fragments of broken-up beds (see the lowermost deformed layer). The deformed layers occur sandwiched between undeformed layers (cf. van Loon et al., 2016). D) One of two deformed layers from the Slinkis outcrop located along the steeply incised Dubysa River valley in Central Lithuania (Pisarska-Jamroży et al., 2019). E) Several deformed layers in the Dyburiai outcrop along the Minija River in North-western Lithuania. The seismites, with load casts, pseudonodules, flame structures, water/sediment escape-structures, are separated by undeformed layers. Note that the size of deformation structures in deformed layers in most cases is limited to the thickness of these layer (Belzyt et al., 2018b, 2021). F) Two internally deformed layers with soft-sediment deformation structures near Rakvere in northern Estonia. (A black and white version of this figure will appear in some formats. For the colour version, please refer to the plate section.)

obtained by Steffen et al. (2019), SSDS in some of them could be triggered by GIA. The stress regime and subsidence of the EBR were strongly affected by the Last Weichselian glaciation during the Late Pleistocene. The Weichselian ice sheet has significantly changed the stress regime both during and after glaciations (Šliaupa, 2003). It caused a reactivation of pre-existing basement and Palaeozoic faults. During glaciation the stability of faults has increased in the central-eastern part of the Baltic Sedimentary Basin (Estonia and Latvia), while faults were destabilized in the Lithuanian territory under an extensional regime. This situation changed after ice sheet retreat. The faults should have been strongly activated in Estonia and, essentially, Latvia under a compressional regime, while in the Lithuanian territory the extension was still prevailing (Šliaupa, 2003).

In Estonia the seismites were recognized in glaciotectonically deformed glaciolacustrine sediments deposited ~13 ka near Rakvere. The beds with SSDS are gently tilted from the depositional position (Figure 18.2). Activation of the nearby occurring Aseri Fault System due to GIA could be among the most likely trigger mechanisms of these layers with SSDS development (cf. Pisarska-Jamroży & Belzyt, 2019).

In Latvia the seismites are identified at three sites: on the Baltic Sea coastal cliff at Baltmuiža site (Belzyt et al., 2018a), at Valmiera site in the north-eastern part of Latvia, and at Rakuti site in the Daugava River in the south-eastern part of the country (van Loon et al., 2016). In all these three sites the beds with SSDS (load structures like load cast, pseudonodules, flame structures, water/sediment escape structures and fragments of broken-up layers) occurred mainly separated by undeformed layers; they formed sandwich-like architecture and extended laterally some hundreds of metres (Figure 18.2). At the Baltmuiža site, five layers with 'trapped' SSDS sandwiched between undeformed layers were recognized in glaciolacustrine sediments deposited ~26 ka BP, which afterwards were gently tilted/reoriented by glaciotectonic processes during Late Weichselian. According to Steffen et al. (2019) calculations of stress fields in the vicinity of this site, the reactivation of faults in the basement was not possible, so the most probable triggering mechanisms responsible for the origin of the SSDS was a local seismic tremor of moderate magnitude caused, e.g. by glacial earthquakes, as large-scale, stick-slip motion nowadays can cause earthquakes with magnitude up to 5.5 (cf. Ekström et al., 2003, 2006; Podolskiy & Walter, 2016, and references therein). At the Valmiera site, seven seismites were found in the section of glaciofluvial and glaciolacustrine sands and silts deposited in an ice-dammed lake where a stream could still run through (Nartišs, 2014; Krievāns, 2015). The lake was formed between the Linkuva and Valdemārpils ice marginal zones (Nartišs, 2014), which means somewhat before 14.5 ka BP, so the SSDS cannot be older (van Loon et al., 2016). The seismites at the Valmiera site were linked to the reactivation of Caledonian and Hercynian faults (Nikulin, 2011) caused by local changes in the weight of the retreating ice sheet. GIA was also responsible for the reactivation of some faults in southern Latvia near Rakuti (Lukševičs et al., 2012; Popovs et al., 2015; Steffen et al., 2019), where twelve seismites in glaciolacustrine and glaciofluvial sands and silts, deposited 17–16 ka BP, were found (van Loon et al., 2016).

In Lithuania the seismites were recognized at two sites (Figure 18.2), both in steeply incised river valleys: at Slinkis site in Dubysa River valley in the Central Lithuania (Belzyt et al., 2018b; Pisarska-Jamroży et al., 2019c) and at Dyburiai site in Minija River valley in north-western Lithuania (Pisarska-Jamroży et al., 2018b; Belzyt et al., 2021). At Slinkis site two seismites with water/sediment escape structures, load and flame structures were described in the Pleistocene meandering river succession deposited between 22.4 ± 1.2 ka BP and 22.6 ± 1.4 ka BP (Pisarska-Jamroży et al., 2019c). Recent modelling of the thrust-faulting mechanism in the vicinity of the Slinkis site (Steffen et al., 2019) does not support the GIA interpretation for this time period. However, Pisarska-Jamroży et al. (2019) suggest also the second trigger mechanism of seismite development – large-scale, stick-slip motion of ice sheet (cf. Ekström et al., 2003, 2006; Podolskiy & Walter, 2016, and references therein). At the second site in Lithuania–Dyburiai, ten internally deformed layers were recognized. Within these layers' load, flame structures and water/sediment escape structures (e.g. silty and sandy volcanoes), fragments of a broken-up laminae were recognized. These deformed sediments as well as sediments in between were deposited in a glaciolacustrine environment ~111–98 ka BP (Pisarska-Jamroży et al., 2018b). These sediments were deformed shortly after their deposition but before the Last Glaciation ice-sheet advance. There are two possibilities of triggering mechanism: GIA (cf. Steffen et al., 2019) or glacial earthquake.

The recent palaeoseismological studies of the liquefaction-induced SSDS in Ryadino site in the north-eastern part of the Kaliningrad District concluded that formation of seismites identified in the site was triggered by earthquake-induced shaking in middle or late Holocene times and the resulting sedimentary structures might be related to the movements of the Earth's crust blocks (Druzhinina et al., 2017).

Two investigated sites with identified SSDS (Table 18.1, Sarnate, Kumečiai) need more detailed studies, thus, assumed seismic origin of these SSDS is still under discussion. SSDS in a number of other sites – Voka, Jurkalne, Juodikiai, Buivydžiai (Punžionys), Anykščiai, Logvino-Cherepanovo, etc. – also could be potentially linked with palaeoseismic events, but they, too, need a more detailed investigation of their origin mechanisms. There is an assumption that a palaeoseismic event could also trigger the mega-scale landslide (360 m long) developed on the palaeocliff of the Litorina Sea near Giruliai vicinity, Western Lithuania (Damušytė & Bitinas, 2018). A similar mega-landslide, about 700 m long, is identified on the recent Baltic Sea cliff close to the Mys Taran Peninsula, Kaliningrad District. According to single radiocarbon dating of a deformed layer of palaeosoil developed inside this structure, the landslide could have occurred due to a palaeoseismic event less than 1.2–1.3 ka BP (Rogozhin et al., 2014).

There are many published papers or monographs as well as unpublished geological reports, descriptions, photos or hand-made sketches with SSDS of probably seismic origin. However, such assumptions should be confirmed by detailed field investigations, e.g. deformation structures in the Zaslavl quarry in the western part of Belarus (Yelovicheva & Drozd, 2005; Figure 18.3) or deformations in the Purtse River Outcrop (Raukas & Teedumäe, 1997) in north-eastern Estonia (Table 18.1). The SSDS in both sites have been initially interpreted as permafrost structures, but such interpretation seems somewhat debatable – the aforementioned SSDS could also have been formed during liquefaction triggered by seismic events. The next example is a diapir fold of Palaeogene–Neogene

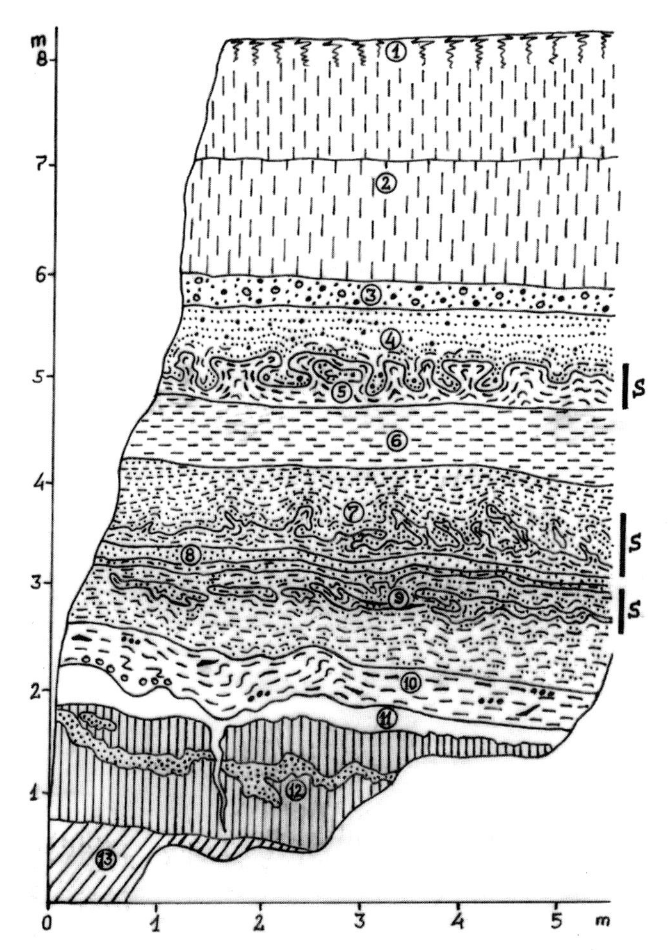

Figure 18.3 Sedimentary succession near Zaslavl (western part of Belarus) with Eemian Interglacial (lowermost part of section, below the 2.5-m marker) and Early Weichselian sediments; redrawn and generalized by authors from figure published in Yelovicheva & Drozd (2005). Lithology of layers (numbers in circle): 1 – soil; 2 – loess-like silty sediments; 3 – sand with gravel; 4 – silty sand; 5 – sandy silt; 6 – silt; 7 – sandy silt and silty sand; 8 – fine-grained sand; 9 – sandy silt and silty sand, 10 – silt with inserts of peat and single gravel; 11 – peat, brownish black, with plant remnants; 12 – peat, black, with sand interlayers; 13 – till. A few layers with well visible soft-sedimentary deformation structures (marked by vertical bold dash and 'S') were attributed by Karabanov & Yelovicheva (1997) to permafrost processes, but they are more similar to the water escape-structures and flame structures formed due to liquefaction triggered by palaeoseismic event.

sediments incorporated into Pleistocene glacial sediments and initially interpreted as glaciotectonic structure (Dodonov et al., 1976); however, in later publications they were reinterpreted as tectonic structure possibly related to the latter earthquakes in the Kaliningrad District (site no. 22 in Table 18.1; Rogozhin et al., 2014).

18.5 Discussion and Final Remarks

One of the major features in recognizing SSDS of seismic origin is in sediments where there can be found evidence of upward directed structures (features) caused by fluidization and liquefaction processes like water/sediment escape-structures as sand volcanoes or flame structures linked to load structures. The other challenge concerns identification of deformation-triggering mechanisms, as liquefaction can be induced in many ways including, e.g. meteorite impact, tsunami, sediment loading, etc. However, there are no specific criteria to distinguish liquefaction induced by earthquake from the other previously mentioned triggers (van Loon et al., 2020). Moreover, multi-triggering mechanisms could be crucial for formation of SSDS in sediments. Finally, careful insight into the apparent discrepancy between the tectonic deformations of the Earth's crust during ice advance and ice retreat is required coupled with identification of faults in the bedrock succession that might have been reactivated during glaciation and deglaciation phases (Brandes et al., 2015; Pisarska-Jamroży et al., 2018a). Most of the contemporary studies only consider the isostatic response from loading and unloading of ice, thereby ignoring the complex geomorphic evolution associated with ice-stream carving and the isostatic rebound caused by this sediment unloading. It is necessary to consider contribution of individual factors such as incoming sediment thickness, rheological contrast and climatic changes in resulting deformation geological processes such as isostatic rebound and glacial sediment transfer.

Overall, the major part of the registered earthquakes within the EBR are associated with mapped faults or fault systems – e.g. the Daugavpils earthquake of 1908 (Latvia) was related to a block shift along the N-S-trending Anisimovitshi Fault; some earthquakes in Latvia are associated with the Liepaja–Saldus ridge formed during the latest Silurian–earliest Devonian (Pačėsa & Šliaupa, 2011). Moreover, the Prieglius, Jantarnenskaya and Bakalinskaya fault zones in the Kaliningrad District are clearly identified as hosting the earthquakes of year 2004 (Rogozhin et al., 2010, 2014). Still, in the central part of the Baltic Sedimentary Basin, it is sometimes difficult to unambiguously associate these single earthquakes with certain faults because of significant errors in locating weak seismic events and due to uncertain identification of some fault locations (Lazauskienė et al., 2012). The identification of the seismogenic faults is also rather complicated due to the small scale of the tectonic dislocations in the intracratonic area with up to a 2.5-km-thick Phanerozoic sedimentary cover (Lazauskienė et al., 2012), especially in determining the potential for fault displacement at or near the surface. The seismic activity in the EBR has an irregular distribution – the northern part of the region is more seismically active than the southern one. The maximum activity is recorded in Latvia, characterized by most intense faulting of the sediment layers. Seismic activity is slightly lower in Estonia, while the territory of Lithuania seems to be the quietest (Lazauskienė et al., 2012). A very uneven distribution of seismic activity in the EBR could be attributed to several factors, predominantly related to the lithospheric-scale geodynamic processes. GIA is still apparent in Estonia (Rosentau et al., 2012) and to a slightly less extent in Latvia, as indicated by the distribution of the recent earthquakes. Vertical ground movements are defined by precise levelling and by changes of the gravity field (Pačėsa & Šliaupa, 2011). It is implied that possible seismic activity of the faults is related to the regional stress field that affects the lithosphere of the entire Baltic Region, but no uniform stress pattern can be found for the EBR. Considering results of GIA modelling made by Steffen et al. (2019), some SSDS identified in the EBR could be attributed to seismic events triggered by GIA during the Weichselian.

Acknowledgements

The study was financially supported by a grant for the GREBAL project no. 2015/19/B/ST10/00661) from the National Science Centre Poland. We express our sincere gratitude to Olga Druzhinina, Christian Brandes and one anonymous reviewer for valuable remarks and suggestions.

References

Belzyt, S., Nartišs, M., Pisarska-Jamroży, M., Woronko, B. and Bitinas, A. (2018a). Large-scale glaciotectonically-deformed Pleistocene sediments with deformed layers sandwiched between undeformed layers, Baltmuiža site, Western Latvia. In M. Pisarska-Jamroży and A. Bitinas, eds., *Soft-sediment Deformation Structures and Palaeoseismic Phenomena in the South-eastern Baltic Region. Excursion Guide of International Palaeoseismological Field Workshop, 17–21 September 2018*. Lithuanian Geological Survey, Lithuanian Geological Society, Vilnius, pp. 38–42.

Belzyt, S., Pisarska-Jamroży, M., Bitinas, A., Damušytė, A. and Woronko, B. (2018b). Soft-sediment deformation structures in the Pleistocene meandering-river floodplain (Slinkis Outcrop, Central Lithuania). In M. Pisarska-Jamroży and A. Bitinas, eds., *Soft-sediment Deformation Structures and Palaeoseismic Phenomena in the South-eastern Baltic Region. Excursion Guide of International Palaeoseismological Field Workshop, 17–21 September 2018*. Lithuanian Geological Survey, Lithuanian Geological Society, Vilnius, pp. 16–20.

Belzyt, S., Pisarska-Jamroży, M., Bitinas, A. et al. (2021). Repetitive Late Pleistocene soft-sediment deformation by seismicity-induced liquefaction in north-western Lithuania. *Sedimentology*, doi.org/10.1111/sed.12883.

Bitinas, A. (2012). Implications of the paleoseismicity of the Eastern Baltic Sea Region. *Quaternary International*, **52**, 279–280, doi.org/10.1016/j.quaint.2012.07.230.

Bitinas, A. and Lazauskienė J. (2011). Implications of palaeoseismic events based on the analysis of the structures of the Quaternary deposits. *Baltica*, **24**, 127–130 (in Lithuanian).

Bitinas, A. and Damušytė, A. (2018). Ventės Ragas outcrop and Juodikiai quarry: soft-sediment deformation structures of enigmatic genesis in the Lithuanian Maritime Region. In M. Pisarska-Jamroży and A. Bitinas, eds., *Soft-sediment Deformation Structures and Palaeoseismic phenomena in the South-eastern Baltic Region. Excursion Guide of International Palaeoseismological Field Workshop, 17–21 September 2018*. Lithuanian Geological Survey, Lithuanian Geological Society, Vilnius, 26–30.

Boborikin, A. M., Avotinia I. Y., Yemelianov A. P., Sildvee A. and Suveizdis P. (1993). Catalogue of historical earthquakes of Belarus and the Baltic Region. *Seismological Report of Seismic Stations of Minsk–Pleshchenitsi and Naroch for 1988*. Minsk, pp. 126–137.

Brandes, C., Steffen, H., Steffen, R. and Wu, P. (2015). Intraplate seismicity in northern central Europe is induced by the last glaciation. *Geology*, **43**, 611–614, doi.org/10.1130/G36710.1.

Brandes, C., Steffen, H., Sandersen P. B. E., Wu, P. and Winsemann, J. (2018). Glacially induced faulting along the NW segment of the Sorgenfrei–Tornquist Zone, northern Denmark: implications for neotectonics and Lateglacial fault-bound basin formation. *Quaternary Science Reviews*, **189**, 149–168, doi.org/10.1016/j.quascirev.2018.03.036.

Damušytė, A. and Bitinas, A. (2018). Giruliai mega landslide (Litorina Sea palaeo cliff): possible relict of palaeoseismic event. In M. Pisarska-Jamroży and A. Bitinas, eds., *Soft-sediment Deformation Structures and Palaeoseismic Phenomena in the South-eastern Baltic Region. Excursion Guide of International Palaeoseismological Field Workshop, 17–21 September 2018*. Lithuanian Geological Survey, Lithuanian Geological Society, Vilnius, p. 31.

Dodonov A. E., Namestnikov Y. G. and Jakushova A. F. (1976). *Noveyshaya tektonika yugo-vostochnoy chasti Baltiyskoy sinekliz* [The Latest Tectonics of the South-eastern Part of the Baltic Syneclise]. Moscow University Press, Moscow (in Russian).

Druzhinina, O., Bitinas, A., Molodkov, A. and Kolesnik, T. (2017). Palaeoseismic deformations in the Eastern Baltic region (Kaliningrad District of Russia). *Estonian Journal of Earth Sciences*, **66**, 119–129, doi.org/10.3176/earth.2017.09.

Ekström, G., Nettles, M. and Abers, G. A. (2003). Glacial earthquake. *Science*, **302**, 622–624, doi.org/10.1126/science.1088057.

Ekström, G., Nettles, M. and Tsai, V. C. (2006). Seasonality and increasing frequency of Greenland glacial earthquakes. *Science*, **311**, 1756–1758, 10.1126/science.1122112.

Gregersen, S. and Basham, P. V. (1989). *Earthquakes at North Atlantic Margins: Neotectonics and Postglacial Rebound*. NATO ASI Series, Vol. 266. Kluwer Academic Publishers, Dordrecht. doi.org/10.1007/978-94-009-2311-9.

Gregersen, S., Wiejacz, P., Dębski, W. et al. (2007). The exceptional earthquakes in Kaliningrad district, Russia on September 21, 2004. *Physics of the Earth and Planetary Interiors*, **164**, 63–74, doi.org/10.1016/j.pepi.2007.06.005.

Grube, A. (2019). Palaeoseismic structures in Quaternary sediments of Hamburg (NW Germany), earthquake evidence during the younger Weichselian and Holocene. *International Journal of Earth Sciences*, **108**(3), 845–861, doi.org/10.1007/s00531-019-01681-2.

Grünthal, G. and GSHAP Region 3 Working Group (1999). Seismic hazard assessment for Central, North and Northwest Europe: GSHAP Region 3. *Annali di Geofisica*, **42**(6), 999–1011, doi.org/10.4401/ag-3783.

Grünthal, G. and Riedel, P. (2007). Zwei angebliche Erdbeben in den Jahren 1303 und 1328 im heutigen Raum Kaliningrad [Two supposed earthquakes in 1303 and 1328 in what is now Kaliningrad region]. *Zeitschrift für Geologische Wissenschaften*, **35**(3), 157–163.

Grünthal, G. and Stromeyer, D. (1995). Rezentes Spannungsfeld und Seismizitat des baltischen Raumes und angrenzender Gebiete – ein Ausdruck aktueller geodynamischer Prozesse [Recent stress field and seismicity of the Baltic region and adjacent areas – an expression of current geodynamic processes]. *Brandenburgische Geowissenschaftliche Beitrage*, **2**, 69–76.

Jakobsson, M., O'Regan, G. M., Greenwood, S. L. et al. (2016). Postglacial tectonic structures and mass wasting in Lake Vättern, southern Sweden. *Geological Society Memoir,* **46**(1), 119–120, doi.org/10.1144/M46.58.

Karabanov, A. K. and Yelovicheva, Ya. K. (1997). Geological objects of excursions. "Zaslavl" quarry. In *Excursion Guidebook 'Quaternary Deposits and Neotectonics in the Area of Pleistocene Glaciations', May 12–16, 1997*. Minsk, pp. 15–18.

Krievāns, M. (2015). *Formation of the Hydrographic Network in the Lower Gauja Spillway Valley Adjoining Area during the Late Weichselian Deglacial*. Summary of PhD thesis, University of Latvia, Riga.

Lazauskienė, J., Stephenson, R. A, Šliaupa, S. and van Wees, J.-D. (2002). 3D flexural model of the Silurian Baltic Basin. *Tectonophysics*, **346**, 115–135, 10.1016/S0040-1951(01)00231-1.

Lazauskienė, J., Pačėsa, A. and Satkūnas, J. (2012). Seismotectonic and seismic hazard maps of Lithuania – recent implications of intracratonic seismicity in the Eastern Baltic Region. *Geologija*, **54**, 1–9, doi.org/10.6001/geologija.v54i1.2364.

Lukševičs, E., Stinkulis, G., Mūrnieks, A. and Popovs, K. (2012). Geological evolution of the Baltic Artesian Basin. In A. Delina, ed., *Highlights of Groundwater Research in the Baltic Artesian Basin*. University of Latvia, Riga, pp. 7–53.

Mattila, J., Ojala, A. E. K. et al.(2019). Evidence of multiple slip events on postglacial faults in northern Fennoscandia. *Quaternary Science Reviews*, **215**, 242–252, doi.org/10.1016/j.quascirev.2019.05.022.

Mörner, N. A. (2005). An interpretation and catalogue of Paleoseismicity in Sweden. *Tectonophysics*, **408**, 265–307, doi.org/10.1016/j.tecto.2005.05.039.

Mörner, N. A. (2008). Tsunami events within Baltic. *Polish Geological Institute Special Papers*, **23**, 71–76.

Nartišs, M. (2014). *Ice Meltwater Lakes of Northern Vidzeme and Middle Gauja Lowlands during the Late Weichselian Deglaciation*. Summary of PhD thesis, University of Latvia, Riga.

Nartišs, M., Woronko, B., Pisarska-Jamroży, M., Belzyt, S. and Bitinas, A. (2018). Injection structures and load casts in lagoon sediments (Sārnate outcrop, W Latvia). In M. Pisarska-Jamroży and A. Bitinas, eds., *Soft-sediment Deformation Structures and Palaeoseismic Phenomena in the South-eastern Baltic Region. Excursion Guide of International Palaeoseismological Field Workshop, 17–21 September 2018*. Lithuanian Geological Survey, Lithuanian Geological Society, Vilnius, pp. 32–37.

Nikonov, A. A. (2005). Following the Kaliningrad Earthquake. *Priroda*, **3**, 47–53.

Nikonov, A. A. (2011). Surface disturbances connected with the Kaliningrad earthquake of September 21, 2004, and their correlation with macroseismic scales. *Seismic Instruments*, **47**, 148–157, doi.org/10.3103/S074792391102006X.

Nikulin, V. (2011). Assessment of the seismic hazard in Latvia. Version of 2007 year. *Material Science and Applied Chemistry*, **24**, 110–115.

Pačėsa, A. and Šliaupa, S. (2011). Seismic activity and seismic catalogue of the East Baltic region. *Geologija*, **53**(3), 134–146, doi.org/10.6001/geologija.v53i3.1894.

Pačėsa, A., Šliaupa, S. and Satkūnas, J. (2005). Recent earthquake activity in the Baltic region and seismological monitoring in Lithuania. *Geologija*, **50**, 8–18 (in Lithuanian).

Paškevičius, J. (1997). *The Geology of the Baltic Republics*. Lithuanian Geological Survey, Vilnius.

Pisarska-Jamroży, M. and Bitinas, A. (2018). *Soft-sediment Deformation Structures and Palaeoseismic Phenomena in the South-Eastern Baltic Region. Excursion Guide of International Palaeoseismological Field Workshop, 17–21 September 2018*. Lithuanian Geological Survey, Lithuanian Geological Society, Vilnius.

Pisarska-Jamroży, M. and Belzyt, S. (2019). Is there any relationship between occurrence of soft-sediment deformation structures caused by Pleistocene seismicity and induced by glaciotectonics? *ICHEPS-19 7th International Colloquium on Historical Earthquakes & Paleoseismology Studies*, 4–6 November 2019, Barcelona, 32.

Pisarska-Jamroży, M., Belzyt, S., Börner, A. et al. (2018a). Evidence from seismites for glacio-isostatically induced crustal faulting in front of an advancing land-ice mass (Rügen Island, SW Baltic Sea). *Tectonophysics*, **745**, 338–348, doi.org/10.1016/j.tecto.2018.08.004.

Pisarska-Jamroży, M., Belzyt, S., Bitinas, A. et al. (2018b). A glaciolacustrine succession (Dyburiai outcrop, NW Lithuania) with numerous deformed layers sandwiched between undeformed layers. In M. Pisarska-Jamroży and A. Bitinas, eds., *Soft-sediment Deformation Structures and Palaeoseismic Phenomena in the South-Eastern Baltic Region. Excursion Guide of International Palaeoseismological Field Workshop, 17–21 September 2018*. Lithuanian Geological Survey, Lithuanian Geological Society, Vilnius, pp. 43–48.

Pisarska-Jamroży, M. and Woźniak, P. P. (2019). Debris flow and glacioisostatic-induced soft-sediment deformation structures in a Pleistocene glaciolacustrine fan: the southern Baltic Sea coast, Poland. *Geomorphology*, **326**, 225–238, doi.org/10.1016/j.geomorph.2018.01.015.

Pisarska-Jamroży, M., Belzyt, S, Börner, A. et al.(2019a). The sea cliff at Dwasieden: soft-sediment deformation structures triggered by glacial isostatic adjustment in front of the advancing Scandinavian Ice Sheet. *DEUQUA Special Publications*, **2**, 61–67, doi.org/10.5194/deuquasp-2-1-2019.

Pisarska-Jamroży, M., van Loon, A. J., Roman, M. and Mleczak, M. (2019b). Enigmatic gravity-flow deposits at Ujście (western Poland), triggered by earthquakes (as evidenced by seismites) caused by Saalian glacioisostatic crustal rebound. *Geomorphology*, **326**, 239–251, doi.org/10.1016/j.geomorph.2018.01.010.

Pisarska-Jamroży, M., Belzyt, S., Bitinas, A., Jusienė, A. and Woronko, B. (2019c). Seismic shocks, periglacial conditions and glaciotectonics as causes of the deformation of a Pleistocene meandering river succession in central Lithuania. *Baltica*, **32**, 63–77, doi.org/10.5200/baltica.2019.1.6.

Podolskiy, E. A. and Walter, F. (2016). Cryoseismology. *Reviews of Geophysics*, **54**, 708–758, doi.org/10.1002/2016RG000526.

Popovs, K., Saks, T. and Jātnieks, J. (2015). A comprehensive approach to the 3D geological modelling of sedimentary basins: example of Latvia, the central part of the Baltic Basin. *Estonian Journal of Earth Sciences*, **64**, 173–188.

Rogozhin, E. A., Ovsyuchenko, A. N., Novikov, S. S. and Marakhanov A. V. (2010). Active tectonic of the 21 September 2004 Kaliningrad earthquake's region. *Voprosy inzhinernoj seismologii*, **37**(3), 5–20 (in Russian).

Rogozhin, E. A., Ovsyuchenko, A. N., Gorbatikov, A. V. et al. (2014). Detal'naya otsenka seysmicheskoy opasnosti territorii Kaliningrada i tektonicheskiy analiz zemletryaseniy 2004 g [Detailed seismic hazard assessment of the Kaliningrad territory and the tectonic position of the earthquakes occurred in 2004]. *Inzhinernyje iziskanija*, **12**, 26–38 (in Russian).

Raukas, A. and Teedumäe, A. (1997). *Geology and Mineral Resources of Estonia*. Estonian Academy Publishers, Tallinn.

Rosentau, A., Harff, J., Oja, T. and Meyer, M. (2012). Postglacial rebound and relative sea level changes in the Baltic Sea since the Litorina transgression. *Baltica*, **25**, 113–120, doi.org/10.5200/baltica.2012.25.11.

Saulīte, A., Kalnina, L. Stinkulis, G. and Cerina, A. (2007). A new data from the outcrop at the coastal cliff of the Baltic Sea near to Sarnate. *Proceedings of the Field Symposium 'The Quaternary of Western Lithuania: From the Pleistocene*

Glaciations to the Evolution of the Baltic Sea'. May 27–June 02, 2007, Plateliai, Lithuania, pp. 73–74.
Sharov, N. V., Malovichko, A. A. and Shukin, Yu. K. (2007). *Earthquakes and Microseismicity of East European Platform – Recent Geodynamic Approach. Book 1 Earthquakes*. KarNC RAN, Petrozavodsk (in Russian).
Šliaupa, S. (2003). *Geodynamic Evolution of the Baltic Sedimentary Basin*. Habilitation thesis. Institute of Geology, Vilnius.
Šliaupa, S. and Zakarevičius, A. (2000). *Recent Stress Pattern in the Eastern Part of the Baltic Basin, Lithuania, Joint Meeting of EUROPROBE (TESZ) and PACE Projects, Zakopane*. Workshop Abstracts Volume, Warsaw, 79 pp.
Steffen, H., Steffen, R. and Tarasov, L. (2019). Modelling of glacially-induced stress changes in Latvia, Lithuania and the Kaliningrad District of Russia. *Baltica*, **32**, 78–90, doi.org/10.5200/baltica.2019.1.7.
van Loon, A. J., Pisarska-Jamroży, M., Nartišs, M. and Krievâns, M. (2016). Seismites resulting from high-frequency, high-magnitude earthquakes in Latvia caused by Late Glacial glacio-isostatic uplift. *Journal of Palaeogeography*, **5**, 363–380, doi.org/10.1016/j.jop.2016.05.002.
van Loon, A. J., Pisarska-Jamroży, M. and Woronko, B. (2020). Sedimentological distinction in glacigenic sediments between load casts induced by periglacial processes from those induced by seismic shocks. *Geological Quarterly,* **64**, 626–640, doi.org/10.7306/gq.1546.
Yelovicheva, Ya. K. and Drozd, Ye. N. (2005). Zaslavl' – opornyy razrez muravinskogo mezhlednikov'ya Belarusi. [Zaslavlj – the main section of the Murava interglacial in Belarus]. Monografiya deponirovannyye v BelISA, Belarusian State University, Minsk, 81 pp. (in Russian).

Part V

Glacially Triggered Faulting Outside Europe

The following chapters summarize findings, suggestions and indications of glacially triggered faulting outside Europe. This concerns formerly and presently glaciated areas in North America and the polar areas on both hemispheres.

19

The Search for Glacially Induced Faults in Eastern Canada

JOHN ADAMS AND GREGORY R. BROOKS

ABSTRACT

There is abundant evidence of high levels of seismic activity during deglaciation of Eastern Canada, suggesting that the seismic response of Eastern Canada to deglaciation is analogous to Fennoscandia, where numerous glacially induced faults (GIF) have been confirmed. However, the Canadian record of GIFs is scant. The two probable GIFs that are described are few compared to the statistically expected amount of 100+ surface ruptures. Alternative explanations to account for the small number of known ruptures are provided together with an interpretation of certain normal faulting that has been observed in glaciolacustrine sediments. It is recommended that interpretation of prospective GIF features should utilize a sceptical approach employing judgemental scales that reflect data limitations and associated uncertainties.

19.1 Introduction

Eastern Canada (Figure 19.1) represents a large area of glaciated terrain, much of which is underlain by Precambrian Shield bedrock. It was unloaded of 2–3 km of ice mostly between 17,000 and 8,000 years ago (Dyke, 2004), a process marked by an increase in earthquake shaking events immediately after the local deglaciation (Brooks, 2018). Northern Fennoscandia is a much smaller region (about 900,000 km^2) that otherwise appears equivalent in terms of crustal age, crustal thickness, contemporary compressive tectonic stress regime, recent ice load and deglacial process. There, over a dozen deglacial fault scarps have been identified (e.g. Olesen et al., 2004; Lagerbäck & Sundh, 2008; Sutinen et al., 2014; Mikko et al., 2015).

As yet there has been no systematic survey to identify candidate glacially induced faults (GIFs) in Canada. In a recent review, Brooks and Adams (2020) examined GIFs in Eastern Canada that they defined as occurring within an interval of 'several' thousand years after ice-sheet meltback at a given location. For their purposes, Eastern Canada was considered to encompass the Atlantic Canada provinces of Nova Scotia, New Brunswick, Prince Edward Island, Newfoundland and Labrador, the central Canada provinces of Ontario and Quebec, and Manitoba, the easternmost province of Western Canada, an area of about 4 million km^2 (Figure 19.1). They examined the evidence for eleven features from an area

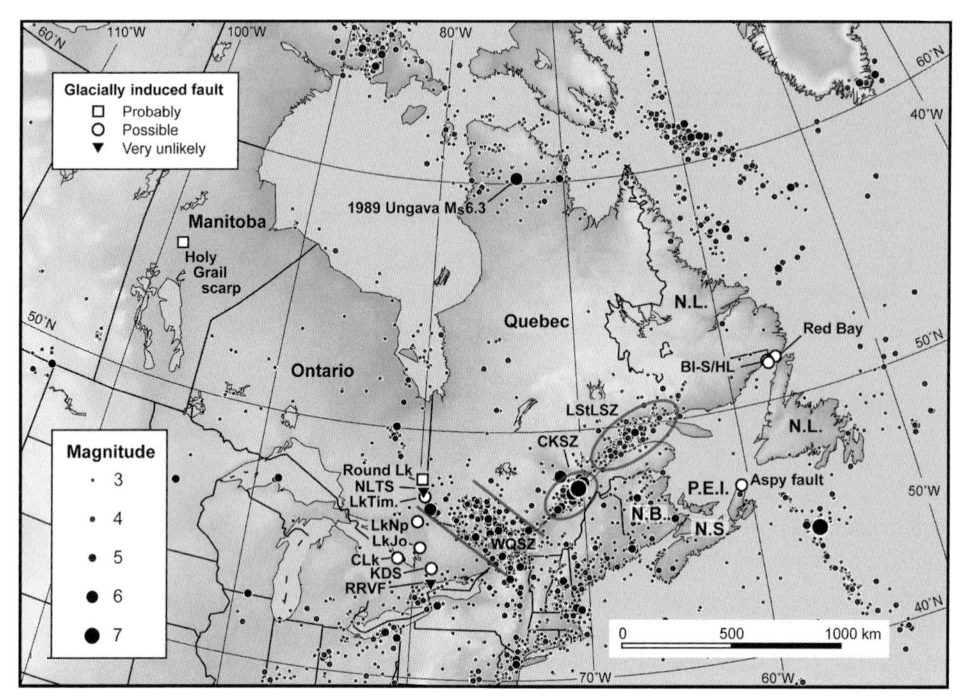

Figure 19.1 Map of Eastern Canada showing the location of historical earthquake epicentres, major seismic zones and the fault features mentioned in the text (modified from Brooks & Adams, 2020). Bl-S/HL – Blanc-Sablon and Hammone Lake; CKSZ – Charlevoix-Kamouraska Seismic Zone; CLk – Charles Lake; LkJo – Lake Joseph; LkNp – Lake Nipissing; LkTim. – Lake Timiskaming; LStLSZ – Lower St. Lawrence Seismic Zone; N.B. – New Brunswick; N.L. – Newfoundland and Labrador; NLTS – New Liskeard-Thornloe scarp; N.S. – Nova Scotia; P.E.I. – Prince Edward Island; RRVF – Rouge River Valley Faults; TESF –Timiskaming East Shore Fault; KDS – Kirkfield deformed shorelines; WQSZ – Western Quebec Seismic Zone (base earthquake map courtesy of the Canadian Hazards Information Service, NRCan).

about four times larger than Fennoscandia and concluded that just the two discussed herein are probable early postglacial GIFs (Figure 19.1; Table 19.1). Neither feature has been documented and confirmed to the degree of the Fennoscandian faults.

This chapter provides context for the occurrence of GIFs in Eastern Canada and reviews briefly the two identified probable examples of GIFs. It discusses the apparent contradiction of observed normal faulting within a region where the contemporary crustal stress conditions produce reverse or strike-slip faulting and provides several alternative explanations to account for the small number of known ruptures.

19.2 Expected Numbers of Glacially Induced Faults

Fennoscandia has been the basis for a model that suggests: (i) seismic strain energy is accumulated during major glaciations because the weight of the ice inhibits reverse faulting

Table 19.1 *Listing of candidate glacially induced faults ranked as 'probable' by Brooks & Adams (2020).*

Location	Latitude/ Longitude	Comments
Holy Grail scarp, Manitoba	55.197° N 99.026° W	Low-angled curvilinear to linear scarp, ~30 km long and 5–8 m high, that forms the east side of a slightly elevated cuesta.
Round Lake, Ontario	48.015° N 80.041° W	A series of faults, some aligned or parallel, with overall ~2-km extent, buried in lake sub-bottom. Faults are normal, 60–545 m long, and offset glaciolacustrine sediments. Offsets are on low-angle fault planes (as low as ~30°) and locally up to several metres. Position and alignment of faults coincides approximately with the mapped Long Lake Fault, which is ~21 km long.

of shield regions; and (ii) the accumulation is released as a burst of large-magnitude earthquakes when the ice load is removed during deglaciation (Johnston, 1987; Adams, 2005; Steffen et al., 2014; Craig et al., 2016). This appears a plausible mechanism for Eastern Canada and is supported by the intense shaking record of early postglacial earthquakes from western Quebec. That record indicates seismicity rates about two orders of magnitude higher than today's rate (Brooks, 2018), even though none of the causative faults have been identified.

Unlike other deglaciated shield regions, Eastern Canada has a documented historical surface rupture, showing that today's crustal stress conditions are favourable for earthquakes to create surface scarps. The 1989 Ungava earthquake (Figure 19.1) was magnitude M_s 6.3, rupturing the top 5–6 km of the crust and forming a 10-km-long fault rupture with a scarp up to 1.3 m high (average height 0.8 m; Adams et al., 1991). Modelling of the surface deformation suggests the rupture occurred on a chiefly reverse fault dipping at 70°, with 1.8 m of slip and with the slip mainly in the top 3 km. Large areas of the Canadian Shield appear to have seismicity confined to the top 5–7 km of crust (Ma et al., 2008), meaning that a significant fraction of their $M > 6$ earthquakes should generate surface ruptures. A statistical analysis (Fenton et al., 2006) suggests that if the historical rates had persisted since deglaciation, 28–160 surface ruptures should have occurred in the Eastern Canada region of Figure 19.1. To these should be added perhaps four times as many from the hypothesized burst of GIFs (by analogy to Sweden; Adams, 2005), for an expected number in the 100–600 range. The additional assumptions to get the factor-of-four increase are that the strain build-up is constant with time, that the last glacial persisted for about four times longer than the time since deglaciation and that the ratio of small to large earthquake numbers (*b*-value) has not changed.

19.3 Probable Glacially Induced Faults

19.3.1 Holy Grail Scarp, Manitoba

The strongest candidate for a GIF in Eastern Canada is the Holy Grail scarp, located in north-central Manitoba about 600 km north of Winnipeg (Figure 19.1). The ~30-km-long

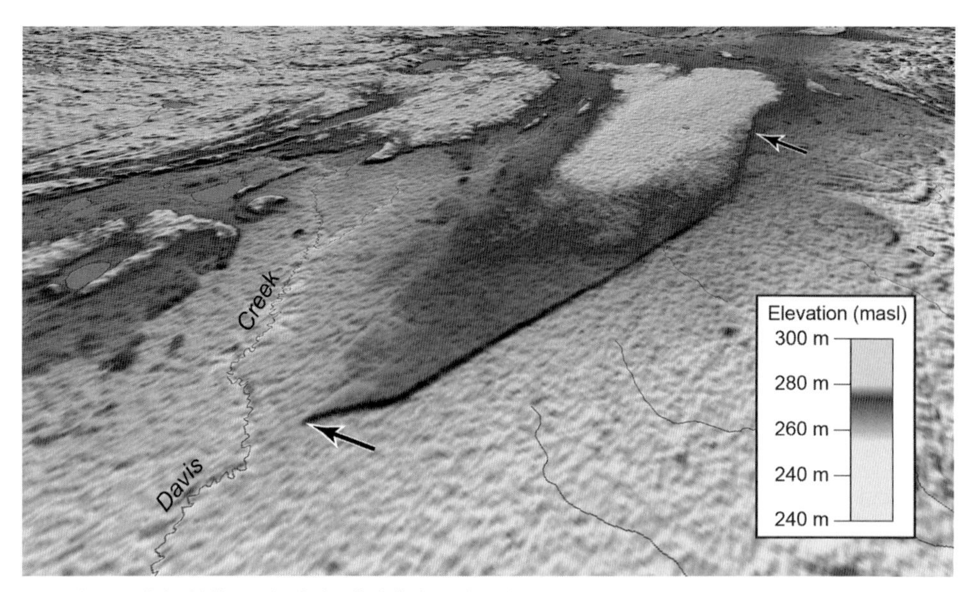

Figure 19.2 Oblique shaded relief digital elevation model (DEM) showing the Holy Grail scarp, north-central Manitoba (see Figure 19.1 for location; V.E. 6X; DEM derived from Shuttle Radar Topography Mission 90 m DEM data courtesy of NASA/JPL-Caltech). Two arrows spaced ~29 km apart mark the extent of the possible fault scarp along the east side of a slightly elevated cuesta. View is approximately northwards. The irregular texture over much of the ground surface probably is an artifact of the low-resolution SRTM DEM. (A black and white version of this figure will appear in some formats. For the colour version, please refer to the plate section.)

scarp is evident on satellite imagery and aerial photographs and forms the east side of a slightly elevated cuesta that is 10–15 m above the surrounding terrain (Figure 19.2). The scarp itself is 5–8 m high, faces east and slopes at less than 5°. Given the low angle of the scarp and the reported geology of the area (see Manitoba Energy and Mines, 1989; McMartin, 1997), the scarp is probably composed of surficial materials.

Remote interpretation of the scarp is complicated because a recessional shoreline of glacial Lake Agassiz was positioned against at least part of the scarp during the Early Holocene (McMartin, 2000; Trommelen, 2014). However, Brooks and Adams (2020) consider it unlikely that the scarp is exclusively the product of shoreline erosion because over its length it has a different longitudinal slope than the shoreline's tilt.

Towards the north, the scarp becomes indistinct. Towards the south, it exhibits a sharp, 26° bend for the last 2 km north of its termination. The feature may extend further to the south of the termination, but the ground surface there is at a lower elevation (light grey in Figure 19.2), and any extension of the rupture may be buried beneath subsequent glacial Lake Agassiz sediments. Burial of a low-level extension by glacial lake deposits is consistent with an early postglacial age for the Holy Grail scarp. Where Davis Creek crosses the possible extension, its channel exhibits changes in course and sinuosity (Figure 19.2; but better shown by Figure 3 of Brooks and Adams, 2020), which may reflect an alteration of the valley slope caused by a buried rupture.

The inferred creek valley slope change and the topographic cuesta are consistent with the scarp caused by slip on a west-dipping thrust fault. With an overall strike of 030 degrees, such a fault would be favourably oriented for reactivation in the generally E-W compressive stress field of Eastern Canada. At a minimum, the length of the sharpest part of the scarp (~15 km) and the scarp height (~5 m) suggest the causative earthquake might have been larger than the Ungava earthquake by a 0.5-magnitude unit (see below), i.e. $M_w = 6.8+$. The Holy Grail scarp is assessed as 'probably' representing a GIF, but trenching or other on-site investigations is required to conclusively establish its earthquake origin.

19.3.2 Round Lake Faulting, Ontario

Based on the preliminary interpretation of sub-bottom acoustic profiling data, Brooks and Adams (2020) reported that under a 1.8 × 0.5-km part of Round Lake are ten short, normal faults that offset just the lower half to two-thirds of the lake's glaciolacustrine deposits. The faults have dips as low as ~30° (Brooks, unpublished data) and have vertical offsets up to several metres. Immediately overlying the faulted sediment is a mass transport deposit, locally up to 2 m thick, that is about 9,100 years old (distinct from, and about 20 varve years older than, the earthquake discussed by Brooks (2020)). The mass transport deposit implies synchronicity for all the fault offsets and (based on experience in other lakes; see Brooks, 2016, 2018) their generation during a shaking event. As the low rigidity of the sediments and the short length of the faults preclude the observed faults being the source of the shaking, the causative earthquake is presumed to have occurred in the underlying Precambrian bedrock.

The occurrence of normal faults like these (similar ones are inferred from profiles in Lake Vättern, for example; see Jakobsson et al., 2014) poses a problem for those studying the seismotectonics of glaciated continents. Firstly, the dominant faulting style for contemporary earthquakes in stable continental regions is reverse or strike-slip faulting – normal faulting mechanisms are unusual. Secondly, there is no strong contender for a normal-faulting GIF, as the Nordmannvikdalen feature in Norway has been now assessed 'very unlikely' to be neotectonic (Redfield & Hermanns, 2016). Brooks and Adams (2020) indicate that the normal faults that offset sediments in Round Lake are not glaciotectonic melt-out structures and that they occurred during the early postglacial period because of their stratigraphic position within the glaciolacustrine deposits of a large glacial lake. The nearby Long Lake Fault, a well-defined bedrock lineament, ~21 km long, extends NNE from the Round Lake area (Figure 19.3) and is an attractive candidate for the causative earthquake rupture, even though there is, as yet, no evidence for postglacial slip on the fault.

How can we reconcile the expected reverse slip on the causative fault with the normal fault offsets of the lake sediments? Below, we suggest a model to link low-dip extensional faulting in deglacial sediments to inferred high-dip reverse faulting in the underlying bedrock; the model might be useful where the bedrock offset has not been imaged.

Consider the effect of a suddenly formed reverse-fault bedrock displacement on a thick overlying sequence of cohesive glaciolacustrine sediments. At its simplest, the originally undeformed horizontal beds extending across the fault would then occur at different

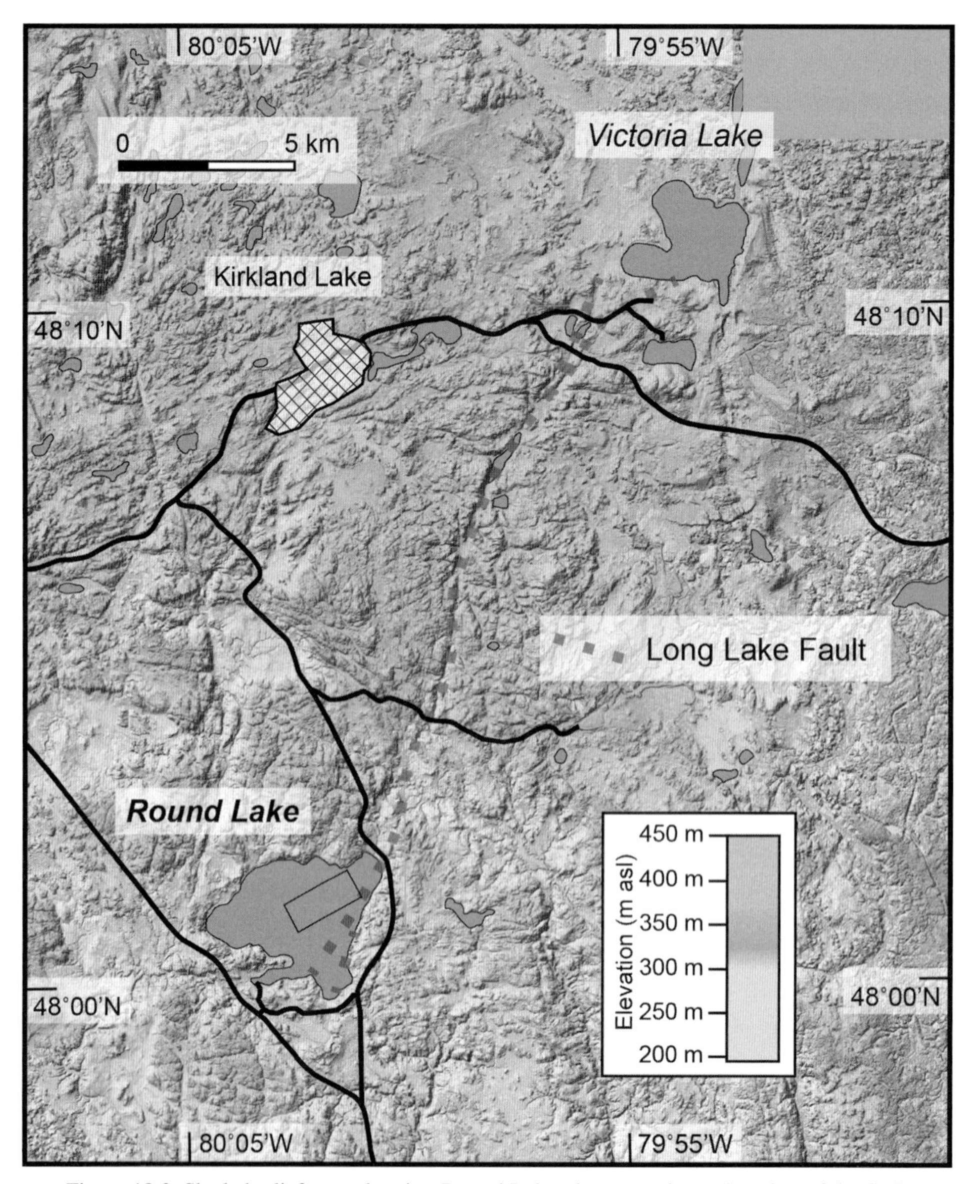

Figure 19.3 Shaded relief map showing Round Lake, the approximate location of the fault features discussed in the text (delineated by a rectangle in Round Lake) and the mapped location of Long Lake Fault and associated bedrock lineament. Location of Long Lake Fault after Ager & Trowell (2000) (V.E. ~6X; shaded relief map derived from a DEM provided by the Ontario Ministry of Natural Resources and Forestry and used under the Open Government License Ontario). (A black and white version of this figure will appear in some formats. For the colour version, please refer to the plate section.)

elevations, established by the uplift on the fault. In some cases, the elevation transition might be modelled as a monoclinal fold with the beds deforming plastically as a trishear (e.g. Erslev, 1991, his Figure 4B). Erslev notes that 'beds adjacent to the anticlinal hinge are commonly thinned and elongated, particularly for higher angle faulting, due to material flux into the footwall'. However, we observe that some glaciolacustrine sediments behave in a brittle fashion (i.e. macro-faulting of beds in seismic reflection profiles and in subaerial exposures and micro-faulting of rhythmic laminations/beds within core) and so assume in those cases none of the thinning is accommodated by plastic deformation.

As a semi-quantitative example, a prototypical GIF reverse fault with 5 m of throw on a 70° dip causes 1.7 m of shortening of the bedrock and sediments, but 4.7-m uplift on the scarp (Figure 19.4A). The reverse fault, in turn, would cause 4.7-m lengthening to an overlying sediment bed (this assumes the sediment bends closely over the fault scarp), and thus a net 3-m lengthening of the thinned and deformed sedimentary layers across the fault (Figure 19.4B). Steeper bedrock fault dips give even more lengthening, equalling the fault

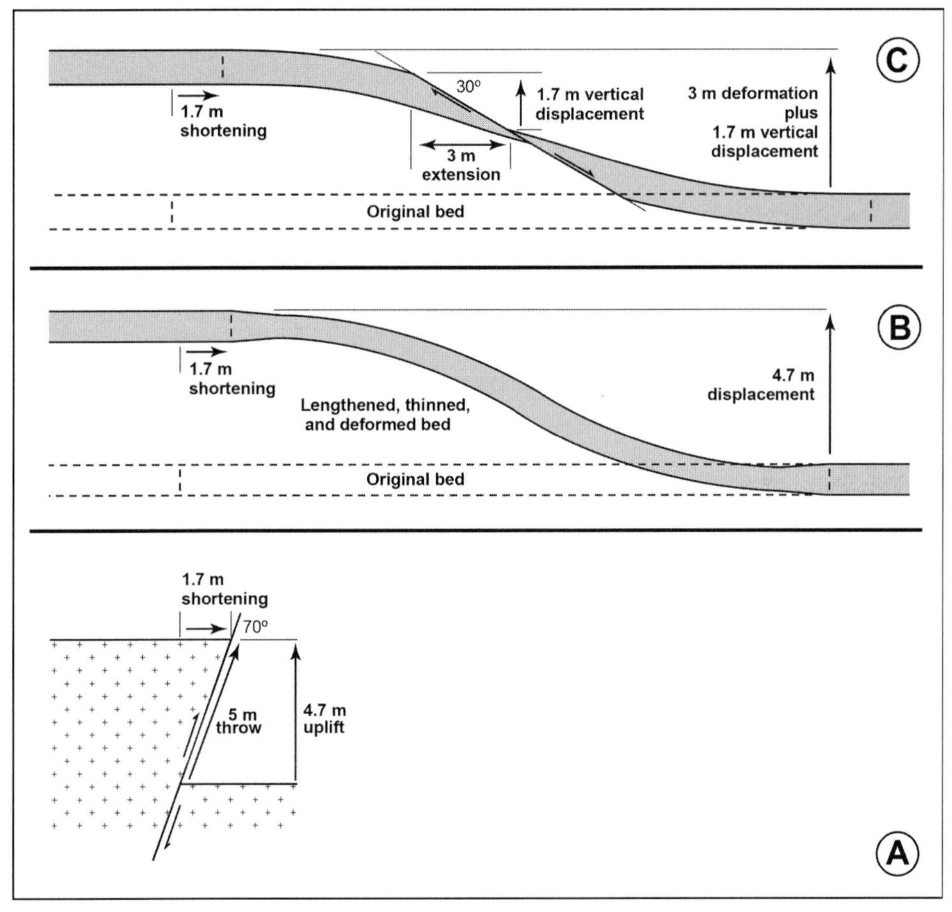

Figure 19.4 Cartoon of example in text (to approximate scale), showing (from bottom) how 5 m of reverse slip on a bedrock fault (A) can cause either (B) plastic bed thinning (deformed bed is 75 per cent of its original thickness) or (C) low-angle normal faulting (as observed at Round Lake) in an overlying sedimentary bed. Both cases would probably not co-exist in a sedimentary sequence above a reverse fault. In this illustration, the footwall is taken as fixed. Thickening of the total sedimentary sequence on the footwall might be expected but is not shown in the figure for clarity.

throw for 90°, while a 45° dip gives no net lengthening. For brittle materials, lengthening is accommodated by extension on normal faults. If the extension takes place on, for example, a 30° dipping normal fault within the sediments (like the Round Lake faults, where we believe we are seeing the true dip), then 3 m of extension could result in about 1.7 m of vertical offset of sedimentary layers (Figure 19.4C). The total extension in the sediments (with corresponding vertical offsets) might be distributed on a number of lesser conjugate normal faults, both synthetic and antithetic, but the sum of the orthogonal extension should relate to the bedrock displacement in the manner above. Furthermore, the normal faults in the sediments should be rooted near the bedrock fault and might dip towards it. 'Near' in this context is within several sediment thicknesses. In this way, a minimum rupture slip (minimum because any horizontal slip component is ignored) might be evaluated from profiles across the fault, then used together with slip-to-magnitude relations to estimate a magnitude for the causative earthquake and hence the likely length of the rupture. Such estimates would provide a basis for reasonableness of the interpretation and could predict the height of exposed bedrock scarps that might be searched for nearby and the distance over which they might be found.

19.4 Other Candidates for Glacially Induced Faults

Brooks and Adams (2020) assessed nine features as 'possible' GIFs (Figure 19.1). The evidence for these is varied: some are probably neotectonic but of the wrong age (Aspy Fault), whereas others are probably the right age but have weak evidence of being neotectonic (sub-aqueous faulting in glaciolacustrine deposits within lakes Timiskaming, Nipissing and Joseph, Ontario; irregularities to the shoreline of a large glacial lake near Kirkfield, Ontario). In addition, there are features with uncertain ages and weak neotectonic evidence (i.e. sinuous ridge feature near Charles Lake, Ontario; examples of reported offset glaciated bedrock surface near Blanc-Sablon and near Hablon Lake, Quebec, and near Red Bay, Newfoundland and Labrador). In every case, more investigation is needed to better evaluate the potential for a GIF origin. In demonstrations of rigorous science, two putative neotectonic faults, the Lower Rouge River Valley faulting and the New Liskeard–Thornloe scarp (Figure 19.1), have been investigated sufficiently to assess them as 'very unlikely' to be GIFs. The former is ascribed to glaciotectonic deformation (ice push; Godin et al., 2002) and the latter is an erosional scarp (Brooks & Pugin, 2019).

19.5 Discussion

The large number of expected surface ruptures (>100) relative to the few known ruptures (the Ungava plus the two probable GIF examples) suggests four possible alternatives.

1. GIFs and surface ruptures never existed in Eastern Canada. However, the region has contemporary earthquakes, with the 1989 Ungava earthquake providing an excellent counterexample of a (late Holocene) surface rupture. Furthermore, the record of early

postglacial earthquake shaking indicates that earthquake activity was widespread and intense during the deglacial period, even if the causative faults have not been identified (Brooks, 2018; Brooks, 2020; Brooks & Adams, 2020). We consider this alternative to be unlikely.

2. GIFs previously existed in Eastern Canada, but almost all scarps formed inside the ice margin and were destroyed by forward movement of the warm-based ice sheet during its last stages of meltback. The erosive nature of the ice was much less in Fennoscandia, which was a cold-based ice sheet, so that any fault scarps formed under its ice could emerge without much modification as the ice melted back. We find the alternative of destroyed scarps unlikely for Canada, as firstly, the large amount of shaking evidence within the postglacial sediments of proglacial lakes in Eastern Canada (see Brooks, 2018; Brooks, 2020; Brooks & Adams, 2020) would only be consistent if the many shaking episodes represented only very large earthquakes far away (so as to be inside the contemporary ice front). For the sequence of earthquakes in the proglacial lakes, this would mean that the required magnitudes would need to be ever larger with time, as the ice front moved ever farther away. Secondly, there are GIFs in Fennoscandia that clearly postdate the ice meltback (e.g. Smith et al., 2014; Sutinen et al., 2014). We recognize that the possibility of preservation through a glacial cycle implies that some Fennoscandian GIFs may be multi-event features and thus more evident in the landscape than a single-event Canadian GIF, but we do not know if this is a contributing factor to the recognition of GIFs there.
3. GIFs still exist but are, and will remain, unrecognized. We note that the long-term discernibility of the Ungava rupture is low because it ruptured through hummocky terrain without clear markers to show offset, extended under lakes for more than half of its length, seldom exposed offset bedrock, and much of its discernibility was due to torn peat and newly exposed lichen-free boulders that will re-vegetate in a few tens to hundreds of years. Even north of the treeline the visual signature is cryptic; in southern Canada widespread tree cover and thick glacial sediment in valleys, lowlands and underlying clay plains additionally reduce the visibility of such faults. We reject this alternative because new data imagery sources, such as Google Earth satellite mosaics with seasonal images, Shuttle Radar Topography Mission DEMs, and LiDAR- and aerial-photograph-derived DEMs, increase the chances that prospective faults will be found. Overall, these new imagery sources – especially bare-earth DEMs – provide good reason to anticipate that additional GIFs will be recognized from a widespread, systematic search for such features.
4. GIFs exist and will be clearly evident once identified. This is the most likely alternative, in our view. However, even if a potential GIF is identified via improved imagery, there are two additional factors that will hinder its confirmation as a GIF. Firstly, GIF study has not been a research priority, meaning that there are few researchers with the skills, experience and resources active in Eastern Canada. Secondly, large areas of northern Quebec, Ontario, Manitoba and Labrador have no road access, thus greatly increasing the cost of the site investigations needed to confirm candidates identified by imagery. As an illustration of both factors, even though the Holy Grail scarp was discovered in the 1970s, there has been no follow-up work.

Although we consider that good potential exists for GIFs in Eastern Canada, we urge an approach that employs scepticism to the interpretation of features as GIFs. Certain topographic features or geological phenomena (examples might be centimetre-scale postglacial faults, superficial stress-release 'pop-ups' in bedrock, river sinuosity changes, sediment drape over drumlins, iceberg furrows, blocky talus along linear features, sackungen and deformed/displaced shorelines attributable to ice remnants) may give a similar appearance to, but lack the continuity and scale expected for, actual fault scarps. The interpretation of a feature as seismogenic (whether glacially induced or otherwise) should be qualified appropriately using a judgement scale that reflects data limitations and associated uncertainties, as was applied by Brooks and Adams (2020). This scale follows Muir Wood (1993) and expresses the interpretation of potential seismogenic faults as 'almost certainly', 'probably', 'possibly', 'probably not' and 'very unlikely' to reflect available information and associated uncertainties. Brooks and Adams (2020) also recognized that the classification of a fault or site-specific evidence of shaking as glacial-induced can be hampered by the feature or event age being poorly constrained. They therefore applied a similar judgement scale for the interpretation of a feature as being glacially induced (i.e. early postglacial in age), applying the terms 'almost certainly', 'probably', 'possibly', 'probably not' and 'very unlikely' as data permit.

19.6 Conclusions

Only two features have been assessed as 'probable' GIFs in Eastern Canada: the Holy Grail scarp, Manitoba, and faulted glaciolacustrine deposits within Round Lake, Ontario. Additional investigation is required to conclusively establish an earthquake origin for the features at both locations. Nine features are considered 'possible' GIFs, but in every case there is much uncertainty about feature age and/or a neotectonic origin.

Additional examples of GIFs should be present in Eastern Canada. The historical background rate of seismicity predicts that 28–160 surface ruptures should have occurred since deglaciation. Perhaps four times as many ruptures can be added to this, arising from the hypothesized burst of seismicity during the early postglacial period.

There is good potential for new prospective GIFs to be identified in Eastern Canada, especially because of the increasing availability of high-resolution, bare earth DEMs. Confirmation that such features are GIFs, however, will be hampered, since GIF study is not a research priority and because of the cost of accessing what is largely remote terrain.

The interpretation of a prospective GIF feature should utilize an approach employing scepticism, qualified appropriately by using judgement scales that reflect data limitations and associated uncertainties.

Acknowledgements

Work on this paper was supported by the Canadian Hazards Information Service and the Public Safety Geoscience Program, Lands and Minerals Sector, Natural Resources Canada.

We thank Dan Clark, Andrée Blais-Stevens and Monica Giona Bucci for their thorough and constructive reviews. This paper represents NRCan Contribution 20190637.

References

Adams, J. (2005). On the probable rate of magnitude $\geq$ 6 earthquakes close to a Swedish site during a glacial cycle. Appendix 5. In S. Hora and J. Mikael, eds., *Expert Panel Elicitation of Seismicity Following Glaciation in Sweden*. Swedish Radiation Protection Authority, No. SSI–2005-20, pp. 33–60.

Adams, J., Wetmiller, R. J., Hasegawa, H. S. and Drysdale, J. (1991). The first surface faulting from a historical intraplate earthquake in North America. *Nature*, **352**, 617–619, doi.org/10.1038/352617a0.

Ager, J. A. and Trowell, N. F. (2000). Geological compilation of the Kirkland Lake area, Abitibi greenstone belt. Ontario Geological Survey, Preliminary Map Series, scale 1:100,000, P3425.

Brooks, G. R. (2016). Evidence of late glacial paleoseismicity from mass transport deposits within Lac Dasserat, northwestern Quebec, Canada. *Quaternary Research*, **86**, 184–199, doi.org/10.1016/j.yqres.2016.06.005.

Brooks, G. R. (2018). Deglacial record of paleoearthquakes interpreted from mass transport deposits at three lakes near Rouyn-Noranda, northwestern Quebec, Canada. *Sedimentology*, **65**, 2439–2467, doi.org/10.1111/sed.12473.

Brooks, G. R. (2020). Evidence of a strong paleoearthquake in $\sim$9.1 ka cal BP interpreted from mass transport deposits, northeastern Ontario – western Quebec, Canada. *Quaternary Science Reviews*, **234**, doi.org/10.1016/j.quascirev.2020.106250.

Brooks, G. R. and Adams, J. (2020). A review of evidence of glacially-induced faulting and seismic shaking in southeastern Canada. *Quaternary Science Reviews*, **228**, doi.org/ 10.1016/j.quascirev.2019.106070.

Brooks, G. R. and Pugin, A. J.-M. (2019). Assessment of a seismo-neotectonic origin for the New Liskeard–Thornloe scarp, Timiskaming graben, northeastern Ontario. *Canadian Journal of Earth Sciences*, **57**(2), 267–274, doi.org/10.1139/cjes-2019-0036.

Craig, T. J., Calais, E., Fleitout, L., Bollinger, L. and Scotti, O. (2016). Evidence for the release of long-term tectonic strain stored in continental interiors through intraplate earthquakes. *Geophysical Research Letters*, **43**, doi.org/10.1002/ 2016GL069359.

Dyke, A. S. (2004). An outline of North American deglaciation with emphasis on central and northern Canada. In J. Ehlers and P. L. Gibbard, eds., *Quaternary Glaciations – Extent and Chronology, Part II: North America. Developments in Quaternary Science*, Vol. 2, Elsevier, Amsterdam, pp. 373–424, doi.org/10.1016/S1571-0866 (04)80209-4.

Erslev, E. A. (1991). Trishear fault-propagation folding. *Geology*, **19**(6), 617–620, doi.org/ 10.1130/0091-7613(1991)019<0617:TFPF>2.3.CO;2.

Fenton, C. H., Adams, J. and Halchuk, S. (2006). Seismic hazards assessment for radio-active waste disposal sites in regions of low seismic activity. *Geotechnical and Geological Engineering*, **24**, 579–592, doi.org/10.1007/s10706-005-1148-4.

Godin, L., Brown, R. L., Dreimanis, A., Atkinson, G. M. and Armstrong, D. K. (2002). Analysis and reinterpretation of deformation features in the Rouge River valley, Scarborough, Ontario. *Canadian Journal of Earth Sciences*, **39**, 1373–1391, doi .org/10.1139/e02-059.

Jakobsson, M., Björck, S., O'Regan, M. et al.(2014). Major earthquake at the Pleistocene–Holocene transition in Lake Vättern, southern Sweden. *Geology*, **42**, 379–382. Data Repository item 2014142, doi.org/10.1130/G35499.1.
Johnston, A.C. (1987). Suppression of earthquakes by large continental ice sheets. *Nature*, **330**, 467–469, doi.org/10.1038/330467a0.
Lagerbäck, R. and Sundh, M. (2008). *Early Holocene Faulting and Paleoseismicity in Northern Sweden*. Geological Survey of Sweden Research Paper Series C, Volume 836, 80 pp.
Ma, S., Eaton, D. W. and Adams, J. (2008). Intraplate seismicity of a recently deglaciated shield terrane: a case study from Northern Ontario, Canada. *Bulletin of the Seismological Society of America*, **98**, 2828–2848, doi.org/10.1785/0120080134.
Manitoba Energy and Mines (1989). Bedrock Geology Compilation Map Series, preliminary edition, Nelson House, NTS 63-O.
McMartin, I. (1997). Surficial geology, Wuskatasko River area, Manitoba. *Geological Survey of Canada Open File*, **3324**, doi.org/10.4095/208906.
McMartin, I. (2000). Paleogeography of Lake Agassiz and regional post-glacial uplift history of the Flin Flon region, central Manitoba and Saskatchewan. *Journal of Paleolimnology*, **24**, 293–315, doi.org/10.1023/A:1008127123310.
Mikko, H., Smith, C. A., Lund, B., Ask, M. V. S. and Munier, R. (2015). LiDAR-derived inventory of post-glacial fault scarps in Sweden. *GFF*, **137**, 334–338, doi.org/10.1080/11035897.2015.1036360.
Muir Wood, R. (1993). *A Review of Seismotectonics of Sweden*. SKB Technical Report TR 93-13, Stockholm, 243 pp.
Olesen, O., Blikra, L. H., Braathen, A. et al. (2004). Neotectonic deformation in Norway and its implications: a review. *Norwegian Journal of Geology*, **84**, 3–34.
Redfield, T. F. and Hermanns, R. L. (2016). Gravitational slope deformation, not neotectonics: Revisiting the Nordmannvikdalen feature of northern Norway. *Norwegian Journal of Geology*, **96**, 1–29, doi.org/10.17850/njg96-3-05.
Smith, C. A., Sundh, M. and Mikko, H. (2014). Surficial geology indicates early Holocene faulting and seismicity, central Sweden. *International Journal of Earth Sciences*, **103**, 1711–1724, doi.org/10.1007/s00531–014-1025-6.
Steffen, R., Wu, P., Steffen, H. and Eaton, D. W. (2014). The effect of earth rheology and ice-sheet size on fault slip and magnitude of postglacial earthquakes. *Earth and Planetary Science Letters*, **388**, 71–80, doi.org/10.1016/j.epsl.2013.11.058.
Sutinen, R., Hyvönen, E., Middleton, M. and Ruskeeniemi, T. (2014). Airborne LiDAR detection of postglacial faults and Pulju moraine in Palojärvi, Finnish Lapland. *Global and Planetary Change*, **115**, 24–32, doi.org/10.1016/j.gloplacha.2014.01.007.
Trommelen, M. S. (2014). Surficial point and line features of the Nelson House map sheet (NTS 63O), Manitoba. Manitoba Mineral Resources, Manitoba Geological Survey Surficial Geology Compilation Map Series SG-GF2013–63O.

20

Glacially Induced Faulting in Alaska

JEANNE SAUBER, CHRIS ROLLINS, JEFFREY T. FREYMUELLER
AND NATALIA A. RUPPERT

ABSTRACT

Southern Alaska provides an ideal setting to assess how surface mass changes can influence crustal deformation and seismicity amidst rapid tectonic deformation. Since the end of the Little Ice Age, the glaciers of southern Alaska have undergone extensive wastage, retreating by kilometres and thinning by hundreds of metres. Superimposed on this are seasonal mass fluctuations due to snow accumulation and rainfall of up to metres of equivalent water height in fall and winter, followed by melting of gigatons of snow and ice in spring and summer and changes in permafrost. These processes produce stress changes in the solid Earth that modulate seismicity and promote failure on upper crustal faults. Here we quantify and review these effects and how they combine with tectonic loading to influence faulting in the south-east, St. Elias and south-west regions of mainland Alaska.

20.1 Introduction and Tectonic Setting

In southern Alaska, a strong seasonal cycle of snow accumulation and melt is superimposed on a high rate of interannual glacier mass wastage. The magnitudes of both are spatially and temporally variable, as seen in Gravity Recovery and Climate Experiment (GRACE)-derived observations across the region (Figure 20.1). These mass changes occur directly atop the southern Alaska plate boundary zone, which transitions from mainly strike-slip faulting in Southeast Alaska to mainly upper-crustal thrusting in the St. Elias region to subduction along the Alaska–Aleutian megathrust further west.

Southeast Alaska features the largest average seasonal signal of the three regions (~40 cm water equivalent, w.e.), superimposed on ongoing ice wastage with an interannual trend of approximately 14 cm/yr w.e. (2003–2016) (Figure 20.1). This coastal region has been undergoing rapid post-Little Ice Age (LIA) deglaciation since around 1770, particularly in the Glacier Bay region, which alone has lost > 3,000 km^3 of ice (Larsen et al., 2005). At present, the solid-Earth isostatic response to past and ongoing mass loss produces surface uplift rates of up to > 3 cm/yr (among the fastest on Earth) and > 7 mm/yr of horizontal motion centred on Glacier Bay (Larsen et al., 2005; Hu & Freymueller, 2019). Running beneath and alongside this rapid deformation is the strike-slip Fairweather Fault, which is the effective plate boundary in Southeast Alaska, accommodating > 4 cm/yr of

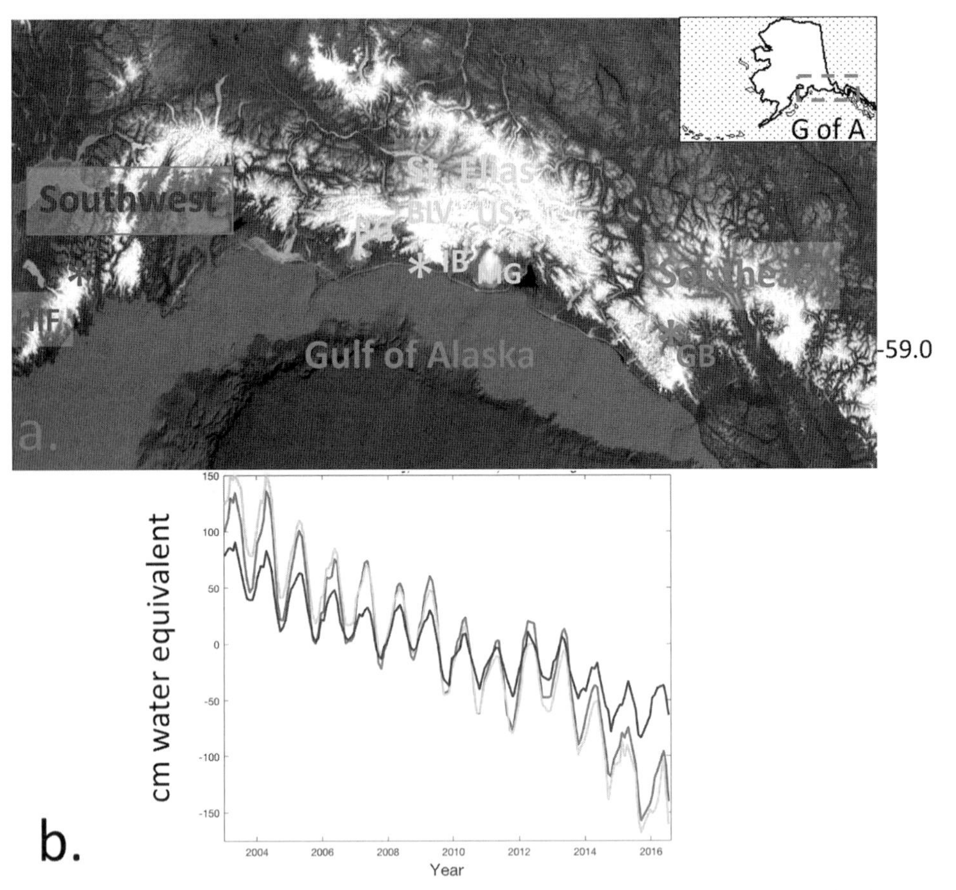

Figure 20.1 a) Overview of study region and locations of major glaciers. BIV – Bagley Ice Valley; BG – Bering Glacier; GB – Glacier Bay; HIF – Harding Icefield; IB – Icy Bay; MG – Malaspina Glacier; US – Upper Seward Glacier, location of 2014 M = 6 oblique-slip earthquake. Background image: Moderate Resolution Imaging Spectroradiometer (MODIS) image from 9 August 2003. Coloured stars: Locations of mass changes, plotted in (b). b) Recent cryospheric changes in cm water equivalent (w.e.) estimated from Gravity Recovery and Climate Experiment (GRACE)-derived mass change (mascons) for the south-eastern, St. Elias and south-west regions of coastal Alaska. The 2003–2016 data used are available at neptune .gsfc.nasa.gov/gngphys/index.php?section=470 (Luthcke et al., 2013; Loomis & Luthcke, 2014). (A black and white version of this figure will appear in some formats. For the colour version, please refer to the plate section.)

right-lateral motion (Fletcher & Freymueller, 2003; Elliott et al., 2010; Koehler & Carver, 2018). The Fairweather Fault has ruptured in four $M \geq 7$ earthquakes since 1927 (Doser & Lomas, 2000) including an $M = 7.8$ earthquake in 1958, which nucleated south-west of Glacier Bay (Plafker et al., 1978; Doser, 2010) (Figure 20.2). The north-west Fairweather Fault connects to the St. Elias compressional margin, the site of two $M \approx 8.1$ thrust earthquakes in 1899 (Plafker & Thatcher, 2008) and an $M_s = 7.4$ event in 1979. Here we evaluated how post-1770 ice loss may have influenced these and other recent earthquakes.

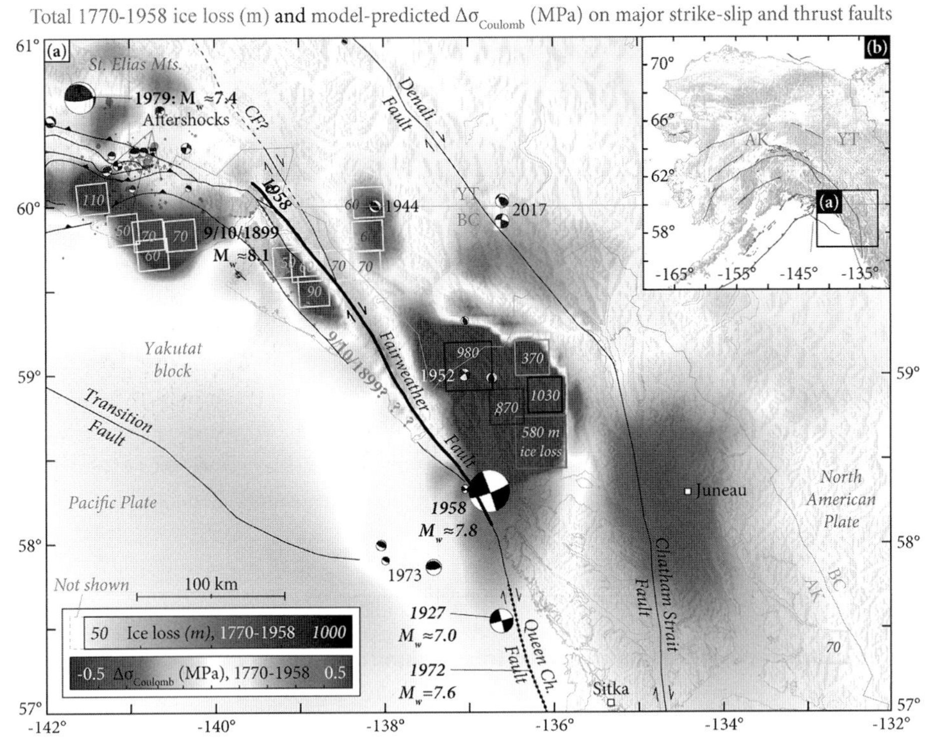

Figure 20.2 a) Post-1770 ice loss, related stress changes and recent earthquakes in Southeast Alaska. Greyscale squares: regions of $\geq$ 50 m total 1770–1958 ice loss in the model of Hu & Freymueller (2019, see references within for specifics of ice loss history), coloured and annotated (italics) by total ice loss. Areas of $<$50 m total ice loss are omitted for readability. Coloured surface: Total 1770–1958 Coulomb stress changes imparted to faults in Southeast Alaska by the integrated elastic and viscoelastic responses to the ice loss, as computed with the Hu & Freymueller (2019) ice and Earth structure model. The full stress change tensor field was computed at 10-km depth, then Coulomb stress changes were calculated by projecting onto a grid of variable receiver fault plane orientations and slip directions spatially interpolated from those of major faults (black lines), capturing the spatial variation in faulting type. We assumed an effective friction of 0.4. Beach balls: Focal mechanisms of earthquakes since 1899 (Doser & Lomas, 2000; Johnson et al., 2020). b) Location of study region. c–f) For physical intuition, we use a reference model of a 100-km-wide ice mass losing 1 m/yr of equivalent water height alongside a vertical right-lateral strike-slip fault, within the same Hu & Freymueller (2019) Earth structure. c) Model setup. d) Cumulative shear stress changes for right-lateral slip after 200 yr of melting and viscoelastic rebound (MPa). (Colours would be reversed for a left-lateral strike-slip fault.) e) Normal stress changes (MPa), positive for unclamping (reduction in normal stress). f) Coulomb stress changes (MPa) assuming an effective friction of 0.4. (A black and white version of this figure will appear in some formats. For the colour version, please refer to the plate section.)

The St. Elias margin has the most rapid present-day ice wastage of the three regions (17 cm/yr w.e.) as well as an average seasonal load amplitude of ~35 cm w.e. (Figure 20.1). Post-LIA deglaciation in this region began around 1880 between the Malaspina and Bering glaciers and likely induces up to 10–20 mm/yr of viscoelastic uplift at present day (Sauber et al., 1997, 2000; Elliott et al., 2013). The St. Elias region is undergoing rapid NNW-SSE tectonic compression and uplift as the Pacific plate carries the thick Yakutat block into and beneath interior Alaska (Sauber et al., 1997; Chapman et al., 2008; Bruhn et al., 2012; Elliott et al., 2013). Here we reviewed and evaluated the influence of seasonal and interdecadal mass changes on the thrust and oblique-slip faulting that pervades this area.

The eastern Alaska–Aleutian subduction zone features more modest interannual (~9 cm/yr w.e.) and seasonal (22 cm w.e.) cryospheric changes (Figure 20.1, Southwest), but also ruptures in great earthquakes including the 1964 M = 9.2 Alaska earthquake, as well as multi-annual slow-slip events (e.g. Li & Freymueller, 2018). Here we explored the influence of mass changes on the initiation of slow-slip events beneath the Harding Icefield on the Kenai Peninsula.

20.2 Cryosphere-Solid Earth Interactions and Large Earthquakes in Southeast Alaska

To estimate the history of cryosphere-related stress changes in Southeast Alaska (Figure 20.2b), we used the model of Hu and Freymueller (2019) and references therein, which couples an ice history model with an elastic-viscoelastic Earth structure. The Earth model comprises a 55-km-thick elastic layer overlying a 230-km-thick linear viscoelastic asthenosphere with Maxwell viscosity $3 \cdot 10^{19}$ Pa·s. In a three-dimensional search over lithosphere thickness, asthenosphere thickness and viscosity, Hu and Freymueller (2019) found that this combination of the three parameters (coupled with the ice model) produced the best fit to interannual uplift rates extracted from 20 years of Global Positioning System (GPS) data in the region. Their original model and parameter search used TABOO (Spada et al., 2003), which does not calculate stress, so we converted the model to Relax (Barbot & Fialko, 2010a,b) to calculate stress changes on faults. To compute stress changes on faults in a way that accounts for the change in dominant faulting type between Southeast Alaska and the St. Elias, we computed the full history of the stress change tensor at a 10-km depth, then calculated the changes in Coulomb failure stress (see discussion in Steffen et al., Chapter 2) on a grid of variable receiver fault plane orientations and slip directions spatially interpolated from those of major faults in the Alaska Division of Geological and Geophysical Survey's Quaternary Faults and Folds map, augmented with the block boundaries in the Elliott et al. (2013) St. Elias block model.

Our main finding is that pre-1958 ice loss and viscoelastic rebound loaded the future epicentre of the 1958 M = 7.8 Fairweather Fault earthquake with a 0.2–0.6-MPa (2–3-bar) increase in Coulomb stress (Figure 20.2a). These stress changes can be understood physically via a simple reference model of a vertical strike-slip fault with a thinning surface load abutting the left side of the fault in the fault's strike direction (Figure 20.2c–f), within the same Hu and Freymueller (2019) Earth structure. This reference model shows that cryosphere-induced deformation increases the shear stress for right-lateral slip ahead of the load in the strike direction (Figure 20.2d) (the opposite for left-lateral slip) and unclamps (reduces the confining stress on) the fault alongside the thinning load (Figure 20.2e), summing to a Coulomb stress increase for right-lateral slip (promoting fault instability), whose maximum is near the corner of the load (Figure 20.2f). In Southeast Alaska, the location of the Glacier Bay unloading was such that the 1958 epicentre was located near this maximum, on the only section of the right-lateral Fairweather Fault on which both the shear and normal stress were increased (Figure 20.2a). Although a 0.2–0.6-MPa stress change is equivalent to just $\sim 2-3$ years of tectonic loading on the fast-slipping Fairweather Fault, these results suggest that cryosphere-solid Earth interactions might have affected the nucleation and characteristics of the 1958 earthquake (Rollins et al., 2021).

Like the earlier calculations given in Sauber et al. (2000), Sauber and Molnia (2004) and Sauber and Ruppert (2008), our results using the Hu and Freymueller (2019) model in the St. Elias region indicate that ice loss and viscoelastic rebound increased the Coulomb stress on thrust faults there, likely promoting the 1979 ($M_s = 7.4$) St. Elias earthquake and/or its rich aftershock sequence (Figure 20.2a). In fact, the majority of known $M \geq 6$ twentieth-century earthquakes in Southeast Alaska (Doser & Lomas, 2000; Johnson et al., 2020) appear to have occurred at locations where the post-LIA cryosphere-solid Earth interactions would have increased rather than decreased Coulomb stress on faults (Figure 20.2a). This correlation should not be oversold as all of these earthquakes also occurred within the rapidly deforming plate boundary zone; nevertheless, it suggests that stress changes from cryosphere-solid Earth interactions are worth considering in regional seismic hazard models.

20.3 Ice Mass Fluctuations and Earthquakes in the St. Elias Region

Our previous work in the St. Elias region comprised two main approaches. In the first, we estimated the possible stress influence of post-LIA ice loss on the 1979 St. Elias earthquake (Sauber et al., 2000; Sauber & Ruppert, 2008, and references therein). We used a two-dimensional finite element model with a simple geology-based representation of ice loss to calculate the incremental stresses and change in fault stability along the shallow-dipping main thrust zone (MTZ) and at the surface. Along the MTZ, our results indicated that the region between the coast and the 1979 aftershock zone had been brought closer to failure by 0.2–1.2 MPa by 1979. In the second approach, we used the strong seasonal variations and the modern-day seismic catalogue to vet the hypothesis that these Coulomb stress changes can influence earthquake behaviour. We found that background seismicity rates vary seasonally and also appeared to vary interannually in a seasonally dependent way (between 1988 and 2006.9), with stronger seasonal fluctuations in warmer-than-average years than in colder years.

Recently, we extended this analysis by (1) calculating the *annual* 3D stresses and Coulomb stress changes associated with tectonic loading to place the stresses from cryospheric fluctuations in context; (2) improving the estimate of cryosphere-related stress changes by incorporating GRACE and other decadal-scale mass change data; (3) using updated and relocated seismicity data to evaluate the modulation of seismic energy release and testing for a cryospheric influence on hypocentre depth; and (4) using well-constrained focal mechanisms to estimate stress orientations as a function of time.

To calculate the first-order features of annual tectonic stresses and surface displacements, we used a 3D finite element model (FEM) of a subduction zone in PyLith (Aagaard et al., 2017). Our model consisted of a slab and a subduction interface of which the shallow portion (<40 km) is locked and the deeper portion creeps at the rate of Pacific–North American plate convergence (5 cm/yr). Our results indicate up to 4 cm/yr of ~N-S horizontal contraction onshore and tectonic uplift rates that range from <1 mm/yr to 1 cm/yr further inland. The corresponding annual tectonic stressing rate has a maximum principal stress (σ_1, directed ~North) of up to 20 kPa/yr, whereas the vertical stress rate (σ_V) is <4 kPa/yr. These stress changes promoted Coulomb failure on faults by ~9 kPa/year.

To compare with the stresses associated with cryospheric changes, we used PyLith to estimate the incremental crustal displacements and stresses due to seasonal mass variations and annual wastage during Water Year 2005 (WY05; 2004.8–2005.8; this year represents an average year with net mass loss). We used a layered rheological model with a 40-km (effective) elastic layer over a viscoelastic layer (Sauber et al., 2016). The magnitude and timing of the fluctuations were estimated using GRACE $1^\circ \times 1^\circ$ mascon results (Figure 20.1) validated qualitatively with GPS reflectometry (K. Larson, cGPS site, AB09), glacier studies (Muskett et al., 2008a,b, 2009) and snow depth change as a function of elevation estimated from another mountainous region (Kirchner et al., 2014). Figure 20.3 shows how incremental stresses and Coulomb stress changes vary between fall/winter and spring/summer and between the Gulf of Alaska coast and the Bagley Ice Valley ~50 km inland (BIV, Figure 20.1). In winter the more abundant snow north of the coast at higher altitudes causes subsidence and tilting towards the north (near coast), moving the faults away from failure on the thrust faults that pervade this environment. In contrast, from late March until the early fall, the melting of this snow and ice produces uplift and tilting towards the south in the coastal region, promoting Coulomb failure there by 5 kPa. However, the maximum Coulomb stress increase is inland in the Bagley Ice Valley (26 kPa) due to both uplift and north-directed horizontal motions that are similar in orientation to those from tectonic loading. At GPS site AB35, along the coast between Icy Bay and the Bering Glacier, we observed seasonal oscillations in the detrended position time series comparable to those predicted by these calculations (Fu et al., 2012; Sauber et al., 2016).

To best evaluate the influence of seasonal cryospheric change on seismicity, we estimated temporal variations in the seismic energy release per month and examined relocated hypocentral depths for the Icy Bay region given in Sauber and Ruppert (2008). We examined seasonal fluctuations in seismic energy release rates (Figure 20.4) over four time periods: (1) July 2005–June 2010, which featured the lowest detection threshold thanks to the additional coverage provided by the STEEP project (Bruhn et al., 2012), providing

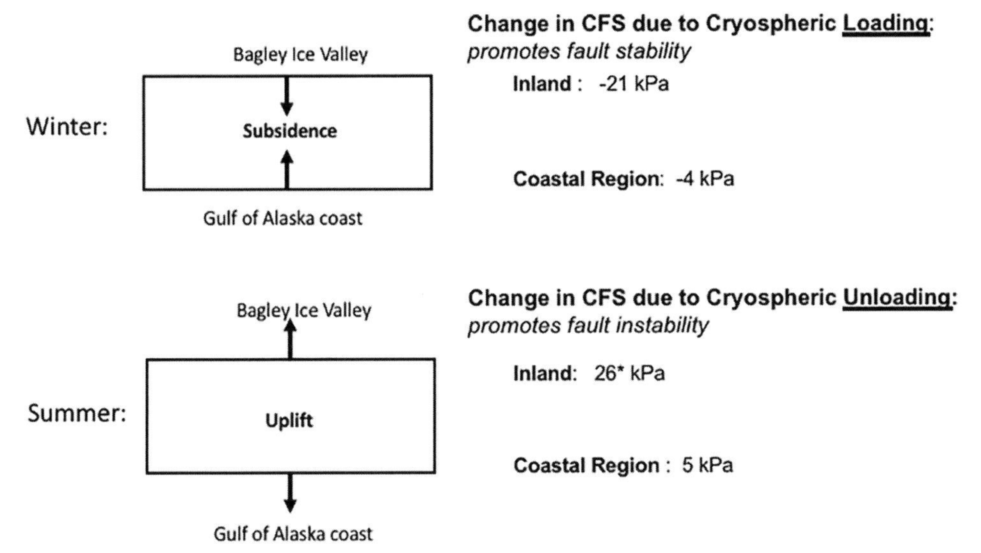

Figure 20.3 St. Elias region: Schematic diagram of calculated seasonal surface displacements (left) and changes in Coulomb stress for WY05 assuming an optimal, pre-existing fault orientation (see Steffen et al., Chapter 2) for regions close to the coast and ~50 km inland in Bagley Ice Valley (BIV in Figure 20.1). Top: Modelled loading in late fall to early spring induces mm of horizontal displacement (black arrows, left) and ~cm of subsidence, promoting fault stability (top right). Bottom: The same but for the surface unloading scenario during spring to early fall that leads to fault instability. *Calculation assumes most of winter snow melts in summer; represents upper bound for displacements and fault stability changes.

the best baseline for background seismicity, (2) November 2009–October 2012, (3) November 2012–October 2015, and (4) November 2015–September 2018. Seismic energy release rates notably increased in late fall in all time periods. Although our earlier study suggested that this seasonality may be stronger during warmer-than-average years, here all time periods showed a similar seasonal dependence. In parallel, tests for seasonal variability in the hypocentral depths of relocated earthquakes with $M \geq 1.4$ (the overall magnitude of completeness over the 2005–2018 period) revealed no systematic depth variations between seasons or between the four time periods. The largest earthquake in the St. Elias region between 2005 and 2018 was a $M_w = 6.0$ event below the Upper Seward Glacier (Figure 20.1) on 17 July 2014. This shallow event (6 km depth) occurred at the end of the high-altitude melt season during a period of maximum fault instability (Figure 20.3).

Although the predicted seasonal stresses are small relative to total stress levels (Sauber et al., 2000, and references therein), we tested whether well-constrained stress tensors from the study region changed between the summer and winter months, especially in the vertical component (σ_V). We used P-wave first motion focal mechanisms to compute best-fitting stress tensors in the Icy Bay region following the approach given in Ruppert (2008). We divided the dataset into 4-month-long seasons (summer and winter periods) and compared

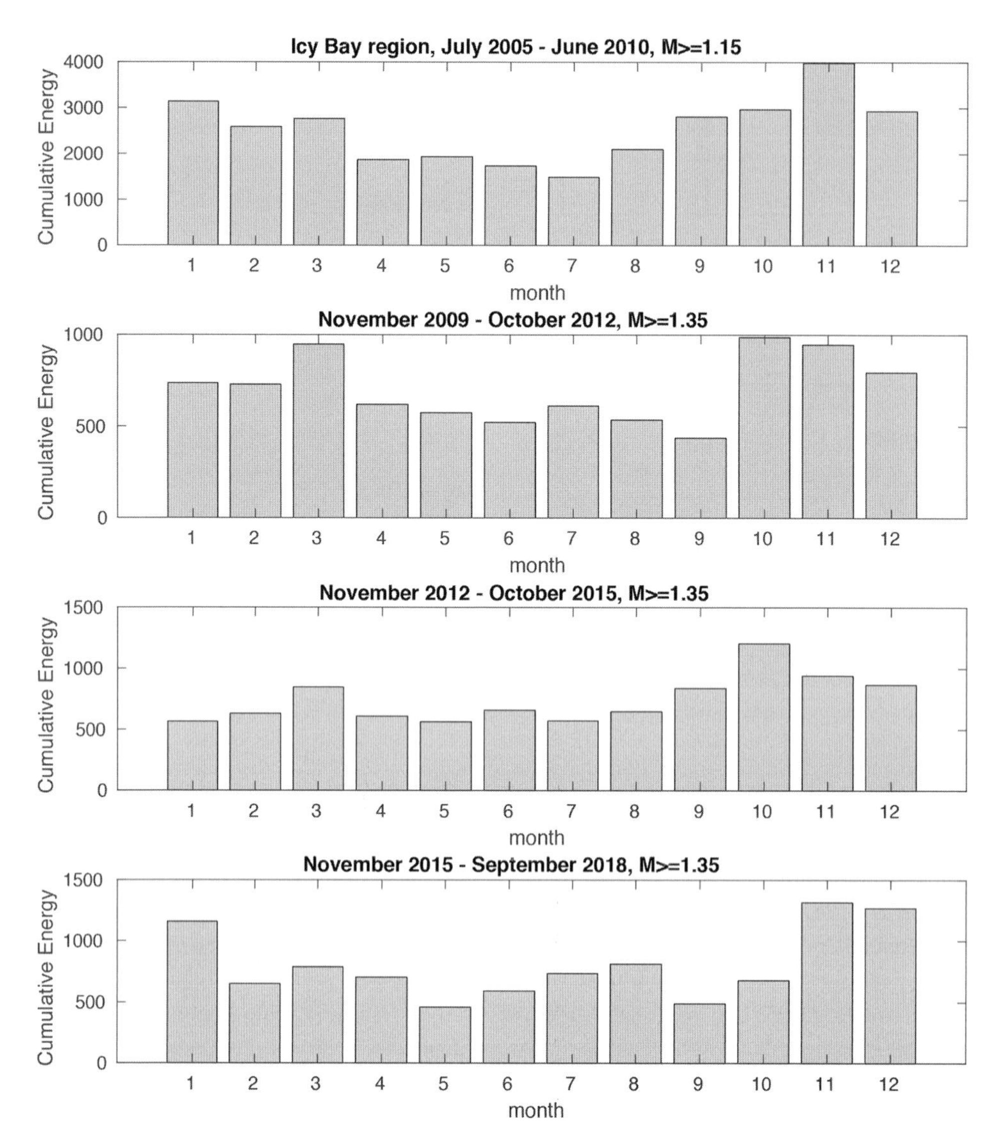

Figure 20.4 Total energy released by tectonic events per month over four different time periods. Energy of an earthquake is defined as $\text{LogE} = 11.8 + 1.5 \cdot \text{M}$. Earthquake data is from the Alaska Earthquake Center Catalog supplemented with the STEEP dataset (Bruhn et al., 2012). Icy Bay region is defined as 140.4–142.2°W and 59.7–60.5°N. Only tectonic earthquakes were included (9,135 within the Icy Bay region). Note the increase of seismic energy release in fall during all time periods.

them to the stress tensor orientations from the full year (Figure 20.5). The resulting stress tensors for the total dataset versus summer or winter were within statistical uncertainties of each other; however, there is a hint that the vertical component (σ_3) migrates to be more vertical in summer than the year-round average.

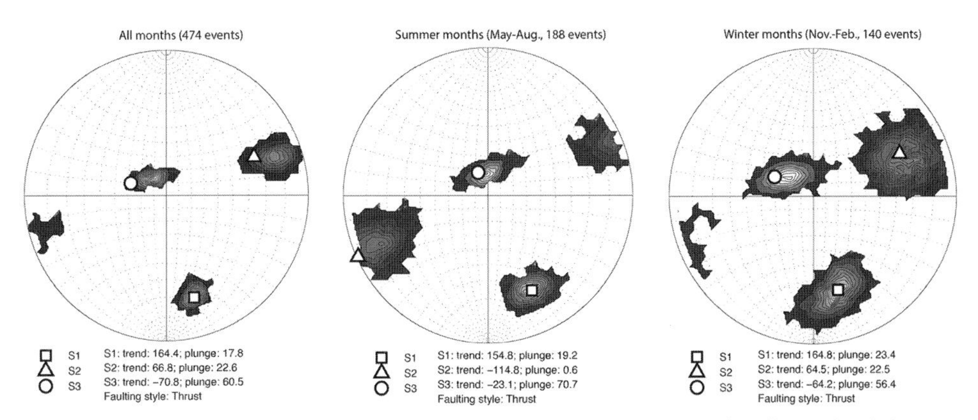

Figure 20.5 Best-fitting stress tensors for events in the Icy Bay area for all months (left), May–August only (middle) and November–February only (right). Open squares, triangles and circles represent best-fitting maximum, intermediate and minimum stress directions, respectively; parameters are given at the bottom of plots. Shaded areas show uncertainty levels as computed by a bootstrapping technique. (A black and white version of this figure will appear in some formats. For the colour version, please refer to the plate section.)

20.4 Evaluating the Influence of Seasonal Fluctuations on Slow-Slip below the Harding Icefield

Although the seasonal and interannual mass changes in the eastern Alaska–Aleutian subduction zone (Figure 20.1) are milder than in the other two regions, we explored their influence on the onset of slow-slip events (effectively very slow earthquakes, similarly involving fault slip). In Cook Inlet, slow slip events occur on the interface between the subducting Pacific plate and the overriding plate and tend to last several years (e.g. Li & Freymueller, 2018). They occur at depths of $\sim 30-50$ km, deeper than large 1964-style thrust earthquakes and near the transition from unstable (stick-slip) to stable friction. Lowry (2006) found a periodicity in slow-slip events in Guerrero, Mexico, and proposed that they may occur in response to climate-driven stress perturbations. Although the lower Cook Inlet slip events do not show an obvious periodicity, they do tend to begin in the late fall/ early winter, similar to the peak in seismicity rate.

Here we compared documented slow-slip events (Li & Freymueller, 2018) to the predicted changes in Coulomb stress in the region of slow-slip events. To calculate the seasonal stress changes prior to one of the slow-slip events (2010.8–2011.0), we ran a PyLith model with seasonal snow loading and unloading plus interannual ice wastage on the Harding Icefield. The location of maximum predicted fault stability change roughly corresponds to the upper portion of the 2004 slow-slip event ($\sim$35 km depth). We calculated that snow accumulation between November 2009 and March 2010 promoted fault stability there by 1.2 kPa, whereas melting and ongoing wastage in the summer months preceding the event (April–early October 2010) produced uplift and tilting in the approximate direction of plate convergence, promoting fault instability by 1.3 kPa. These minor changes may have had a small influence on promoting in time the initiation of the 2010.8 slow-slip event.

20.5 Summary and Future Work

In this study we summarized previous and new work evaluating the influence of cryospheric fluctuations on fault slip in three tectonically distinct regions of Southern Alaska: the strike-slip plate boundary in the south-east, the St. Elias crustal collision zone and the eastern Alaska–Aleutian subduction zone. In the south-east, long-term ice wastage may promote major earthquakes in time, though it represents a second-order perturbation on top of the rapid tectonic loading. For the dominantly thrust-faulting environment of the St. Elias, and to a lesser extent slow-slip events below the Harding Icefield region, seasonal unloading can lead to significant changes in the fault stability relative to annual tectonic loading rates, but these changes are quite localized.

As noted in Hardebeck (2004) and Parsons (2005), a sudden change in stress due to an earthquake or rapid cryospheric change (seasonal ablation or a glacial surge) is only likely to change earthquake probabilities over a time interval that is short compared to the repeat time of large earthquakes (Sauber et al., 1995; Sauber et al., 2000; Sauber & Ruppert, 2008). These localized, transient stresses might thus have a more testable effect on small and moderate earthquakes. An additional complexity was highlighted by Johnson et al. (2019), who conducted a systematic study of shallow, $M > 2.0$ earthquakes across southern Alaska to test for seasonal variations during the annual hydrological cycle. Their results indicated a general increase in regional seismicity rates in the late fall that lags behind the optimal conditions for promoted fault slip by $\sim$3 months, similar to their findings in some regions of California (Johnson et al., 2017). To date, the effect of recent regional, time-dependent stresses due to glacial wastage on faulting potential has not been evaluated in any Alaska-wide earthquake hazard assessment (Mueller et al., 2015).

Acknowledgements

We thank Holger Steffen and two anonymous reviewers for their constructive reviews of our submission and Rebekka Steffen for providing Chapter 2 of the present book in advance of publication. Funding for this study for J. Sauber, J. Freymueller and C. Rollins was provided by NASA's Earth Surface and Interior grant 15-ESI2015–0051 and GRACE-FO science team grant 15-GRACE15–0006 (J. Sauber). This research was supported in part by the Geophysical Institute, University of Alaska Fairbanks (N. A. Ruppert).

References

Aagaard, B., Knepley, M. and Williams, C. (2017). PyLith v2.2.1, Computational Infrastructure for Geodynamics, doi.org/10.5281/zenodo.886600, geodynamics.org/cig/software/pylith/

Barbot, S. and Fialko, Y. (2010a). Fourier-domain Green's function for an elastic semi-infinite solid under gravity, with applications to earthquake and volcano deformation. *Geophysical Journal International*, **182**(2), 568–582, doi.org/10.1111/j.1365-246X.2010.04655.x.

Barbot, S. and Fialko, Y. (2010b). A unified continuum representation of postseismic relaxation mechanisms: semi-analytic models of afterslip, poroelastic rebound and viscoelastic flow. *Geophysical Journal International*, **182**(3), 1124–1140, doi.org/10.1111/j.1365-246X.2010.04678.x.

Bruhn, R. L., Sauber, J., Cotton, M. M. et al. (2012). Plate margin deformation and active tectonics along the northern edge of the Yakutat Terrane in the Saint Elias Orogen, Alaska, and Yukon, Canada. *Geosphere*, **8**(6), 1384–1407, doi.org/10.1130/GES00807.1.

Chapman, J. B. et al. (2008). Neotectonics of the Yakutat Collision: changes in deformation driven by mass redistribution. *American Geophysical Union: Active Tectonics and Seismic Potential of Alaska*. Geophysical Monograph Series 179, doi.org/10.1029/179GM21.

Doser, D. (2010). A re-evaluation of the 1958 Fairweather, Alaska, earthquake sequence. *Bulletin of the Seismological Society of America*, **100**(4), 1792–1799, doi.org/10.1785/0120090343.

Doser, D. I. and Lomas, R. (2000). The transition from strike-slip to oblique subduction in southeastern Alaska from seismological studies. *Tectonophysics*, **316**, 45–65, doi.org/10.1016/S0040-1951(99)00254-1.

Elliott, J., Freymueller, J. T. and Larsen, C. F. (2013). Active tectonics of the St. Elias orogen, Alaska, observed with GPS measurements. *Journal of Geophysical Research Solid Earth*, **118**, 5625–5642, doi.org/10.1002/jgrb.50341.

Elliott, J., Larsen, C. F., Freymueller, J. T. and Motyka, R. J. (2010). Tectonic block motion and glacial isostatic adjustment in southeast Alaska and adjacent Canada constrained by GPS measurements. *Journal of Geophysical Research*, **115**, B09407, doi.org/10.1029/2009JB007139.

Fletcher, H. J. and Freymueller, J. T. (2003). New constraints on the motion of the Fairweather Fault, Alaska, from GPS observations. *Geophysical Research Letters*, **30**(3), 1139, doi.org/10.1029/2002GL016476.

Fu, Y., Freymueller, J. T. and Jensen, T. (2012). Seasonal hydrological loading in southern Alaska observed by GPS and GRACE. *Geophysical Research Letters*, **39**(15), doi.org/10.1029/2012GL052453.

Hardebeck, J. L. (2004). Stress triggering and earthquake probability estimates. *Journal of Geophysical Research*, **109**, B04310, doi.org/10.1029/2003JB002437.

Hu, Y. and Freymueller, J. T. (2019). Geodetic observations of time-variable glacial isostatic adjustment in southeast Alaska and its implications for Earth rheology. *Journal of Geophysical Research*, **124**(9), 9870–9889, doi.org/10.1029/2018JB017028.

Johnson, C. W., Fu, Y. and Bürgmann, R. (2017). Stress models of the annual hydrospheric, atmospheric, thermal, and tidal loading cycles on California faults: Perturbation of background stress and changes in seismicity. *Journal of Geophysical Research: Solid Earth,* **122**(12), 10,605–10,625, doi.org/10.1002/2017JB014778.

Johnson, C. W., Fu, Y. and Bürgmann, R. (2020). Hydrospheric modulation of stress and seismicity on shallow faults in southern Alaska. *Earth and Planetary Science Letters*, **530**, 115904 doi.org/10.1016/j.epsl.2019.115904.

Kirchner, P. B., Bales, R. C., Molotch, N. P, Flanagan, J. and Guo, Q. (2014). LiDAR measurement of seasonal snow accumulation along an elevation gradient in the southern Sierra Nevada, California. *Hydrology and Earth System Sciences*, **18**(10), 4261–4275, doi.org/10.5194/hess-18-4261-2014.

Koehler, R. D. and Carver, G. A. (2018). Active Faulting and Seismic Hazards in Alaska. Alaska Division of Geological and Geophysical Surveys, Miscellaneous Publication 160.

Larsen, C. F., Motyka, R. J., Freymueller, J. T., Echelmeyer, K. A. and Ivins, E. R. (2005). Rapid viscoelastic uplift in southeast Alaska caused by post-Little Ice Age glacial retreat. *Earth and Planetary Science Letters*, **237**(3-4), 548–560, doi.org/10.1016/j.epsl.2005.06.032.

Li, S. and Freymueller, J. T. (2018). Spatial variation of slip behavior beneath the Alaska Peninsula along Alaska–Aleutian subduction zone. *Geophysical Research Letters,* **45**(8), 3453–3460, doi.org/10.1002/2017GL076761.

Loomis, B. D. and Luthcke, S. B. (2014). Optimized signal denoising and adaptive estimation of seasonal timing and mass balance from simulated GRACE-like regional mass variations. *Advances in Adaptive Data Analysis*, **6**(1), 1450003, doi.org/10.1142/S1793536914500034.

Lowry, A. R. (2006). Resonant slow fault slip in subduction zones forced by climatic load stress. *Nature*, **442**, doi.org/10.1038/nature05055.

Luthcke, S. B., Sabaka, T. J., Loomis, B. D. et al. (2013). Antarctica, Greenland and Gulf of Alaska land-ice evolution from an iterated GRACE global mascon solution. *Journal of Glaciology*, **59**(216), 613–631, doi.org/10.3189/2013JoG12J147.

Mueller, C. S., Briggs, R. W., Wesson, R. L. and Petersen, M. D. (2015). Updating the USGS seismic hazard maps for Alaska. *Quaternary Science Reviews*, **113**, 39–47, doi.org/10.1016/j.quascirev.2014.10.006.

Muskett, R. R., Lingle, C. S., Sauber, J. M., Rabus, B. T. and Tangborn, W. V. (2008a). Acceleration of surface lowering on the tidewater glaciers of Icy Bay, Alaska, USA from InSAR DEMs and ICESat altimetry. *Earth and Planetary Science Letters*, **265**(3–4), 345–359, doi.org/10.1016/j.epsl.2007.10.012.

Muskett, R. R., Lingle, C. S., Sauber, J. M. et al. (2008b). Surging, accelerating surface lowering and volume reduction of the Malaspina Glacier system, Alaska, USA, and Yukon, Canada, from 1972 to 2006. *Journal of Glaciology*, **54**(188), 788–800, doi.org/10.3189/002214308787779915.

Muskett, R. R., Lingle, C. S., Sauber, J. M. et al. (2009). Airborne and spaceborne DEM-and laser altimetry-derived surface elevation and volume changes of the Bering Glacier system, Alaska, USA, and Yukon, Canada, 1972–2006. *Journal of Glaciology*, **55**(190), 316–326, doi.org/10.3189/002214309788608750.

Parsons, T. (2005). Significance of stress transfer in time-dependent earthquake probability calculations. *Journal of Geophysical Research*, **110**, B05S02, doi.org/10.1029/2004JB003190.

Plafker, G., Hudson,T., Bruns, T. R. and Rubin, M. (1978). Late Quaternary offsets along the Fairweather faults and crustal plate interactions in southern Alaska. *Canadian Journal of Earth Sciences*, **15**(5), 805–816, doi.org/10.1139/e78-085.

Plafker, G. and Thatcher, W. (2008). Geological and geophysical evaluation of the mechanisms of the great 1899 Yakutat Bay earthquakes. *American Geophysical Union: Active Tectonics and Seismic Potential of Alaska*, Geophysical Monograph Series 179, doi.org/10.1029/179GM21.

Rollins, C., Freymueller, J. T. and Sauber, J. M. (2021). Stress promotion of the 1958 M_w~7.8 Fairweather Fault earthquake and others in southeast Alaska by glacial isostatic adjustment and inter-earthquake stress transfer. *Journal of Geophysical Research Solid Earth*, **126**, e2020JB020411, doi.org/10.1029/2020JB020411.

Ruppert, N. A. (2008). Stress map for Alaska from earthquake focal mechanisms. *American Geophysical Union: Active Tectonics and Seismic Potential of Alaska*, Geophysical Monograph Series 179, doi.org/10.1029/179GM20.

Sauber, J., Plafker, G. and Gipson, J. (1995). Geodetic measurements used to estimate ice transfer during Bering Glacier surge. *Eos, Transactions American Geophysical Union*, **76**(29), 289–290, doi.org/10.1029/95EO00171.
Sauber, J., McClusky, S. and King, R. (1997). Relation of ongoing deformation rates to the subduction process in southern Alaska. *Geophysical Research Letters*, **24**, 2853–2856, doi.org/10.1029/97GL52979.
Sauber, J., Plafker, G., Molnia, B. F. and Bryant, M. A. (2000). Crustal deformation associated with glacial fluctuations in the eastern Chugach Mountains, Alaska. *Journal of Geophysical Research*, **105**, 8055–8077, doi.org/10.1029/1999JB900433.
Sauber, J. M., Freymueller, J. T., Han, S. C., Davis, J. L. and Ruppert, N.A. (2016). Short-term response of the solid Earth to cryosphere fluctuations and the earthquake cycle in south-central Alaska. *American Geophysical Union, Fall Meeting 2016*, Abstract #G11A-1057 (poster available on ResearchGate).
Sauber, J. M. and Molnia, B. F. (2004). Glacier ice mass fluctuations and fault instability in tectonically active Southern Alaska. *Global and Planetary Change*, **42**, 279–293, doi .org/10.1016/j.gloplacha.2003.11.012.
Sauber, J. M. and Ruppert, N. (2008). Rapid ice mass loss: does it have an influence on earthquake occurrence in Southeast Alaska? *American Geophysical Union: Active Tectonics and Seismic Potential of Alaska*, Geophysical Monograph Series 179, doi.org/10.1029/179GM21.
Spada, G., Antonioli, A., Boschi, L. et al. (2003). *TABOO, User Guide*. Samizdat Press, Golden-White River Junction.

21

Indications on Glacially Triggered Faulting in Polar Areas

HOLGER STEFFEN AND REBEKKA STEFFEN

ABSTRACT

The polar region is the area surrounding the Earth's geographical poles (Antarctica and the Arctic). While glacially induced faults (GIFs) are well known in the formerly glaciated areas of Northern Europe, such faults within the Arctic and Antarctica are unidentified, although the theory of their physical mechanism would allow their presence. Mainly, the fact that most of the polar region is covered either by ocean (Arctic) or ice sheets (Antarctica, Greenland) prevents the detailed analysis of those regions with respect to GIFs. However, there are several indications that suggest an existence of GIFs in the polar region. Here, we summarize findings about potential GIFs in Northern Canada, Greenland, Iceland and Svalbard in the northern hemisphere and revisit seismicity in Antarctica.

21.1 Introduction

Northern Europe is regarded as a key area for studying the process of glacial isostatic adjustment (GIA), which describes the Earth's response to the waxing and waning of ice sheets during cold climate periods (Steffen & Wu, 2011). The infrastructure and dedicated research activities going back to the seventeenth century allowed a rich record of documents and observations witnessing the GIA process, such as relative sea-level (RSL) changes and geodetic deformation records. In the twentieth century remarkable faults were found in the most northern parts of Norway, Sweden and Finland (e.g. Kujansuu, 1964; Lagerbäck & Sundh, 2008; Olesen et al., see Chapter 11; Smith et al., see Chapter 12; Sutinen et al., see Chapter 13), which were also related to GIA, i.e. interpreted as stress release during or soon after the deglaciation. They are nowadays termed 'glacially induced faults' (GIFs).

The underlying physical process was introduced by Johnston (1987) and confirmed with modelling by Wu and Hasegawa (1996a,b). The principle can be briefly summarized: in a thrust-faulting tectonic background stress regime as is apparent in northern Europe, the load of the ice sheet increases normal stresses at the fault. Thereby fault slip is prevented. During load removal the accumulated stress can be released if the fault orientation in the stress field allows it (a detailed explanation can be found in Steffen et al., see Chapter 2). One can thus speculate that glaciation of a large area would cause increased earthquake activity at the end of the deglaciation.

Knowing about the GIFs in Northern Europe and the underlying physical process, Johnston (1996) put the question forward vis-à-vis the location of the Canadian and Siberian GIFs. Indeed, potential faults pointing to activity due to glacial influence are suggested for the formerly glaciated regions of Eastern Canada (Fenton, 1994; Brooks & Adams, 2020, and see Chapter 19 herein) and the British Isles (Firth & Stewart, 2000). Moreover, Johnston (1987) argued that both Greenland and Antarctica are currently rather stable in view of earthquake activity as the surface is covered with huge ice sheets, but when the ice melts, stress may be released in form of earthquakes (Gregersen, 1989; Arvidsson, 1996; Ivins et al., 2003). Such behaviour could also be anticipated for several islands in the Arctic, e.g. in Northern Canada.

In the following, we summarize indications in the literature on glacially triggered faulting (GTF) and the current seismicity in the polar areas of the Earth. However, the literature on this topic is sparse. Often, indications are mentioned only in unpublished reports or conference abstracts. Dedicated research is limited, which is likely related to the fact that the focus during expeditions into these remote areas is set to many other, rather urgent scientific investigations in view of climatic changes. Coincidentally, these climatic changes currently cause rapid retreat of the ice caps and glaciers (e.g. Jacob et al., 2012) and thus may lead to reactivation of pre-existing faults as suggested by Arvidsson (1996). We limit ourselves in this chapter to indications found in parts of Northern Canada, Greenland, Iceland, Svalbard and Antarctica, which were discussed in the last 40 years. Arctic areas of continental Northern Europe, e.g. in Norway and Russia, are discussed in Olesen et al., see Chapter 11, and Nikolaeva et al., see Chapter 14, respectively.

21.2 Northern Canada

The Laurentide Ice Sheet covered most parts of the islands in Canada's Arctic (Dyke, 2004). The ice sheet reached its maximum in the area about 21,400 years ago (Dyke, 2004). Most islands became ice-free, except for smaller glaciers, between 14,000 and 10,000 years ago (Dyke et al., 1991). Some stages of ice sheet distribution can be found in Figure 21.1a. The development of the ice sheet and especially its deglaciation are well documented by a wealth of data such as RSL collected mainly by the Geological Survey of Canada (see e.g. Simon et al., 2018).

Seismicity in the area is low (Figure 21.1a) but we also note that seismograph station coverage is sparse. Hence, detection of events is hampered. There were several light earthquakes with magnitudes between 4 and 5, but of any above 5 only few were recorded. The largest recorded earthquake of magnitude 5.9 according to Earthquakes Canada (6.0 according to USGS and 6.1 according to Motazedian & Ma, 2018) happened recently (8 January 2017) north of Somerset Island. Two of the more than 20 aftershocks reached magnitude 5.2 (Motazedian & Ma, 2018). Earthquakes seem to cluster along a line from the Boothia Peninsula towards the north, peaking in number and magnitude between Somerset Island and Devon Island. Small clusters are also apparent in south-eastern Melville Island and between Somerset Island and Baffin Island in the Prince Regent Inlet.

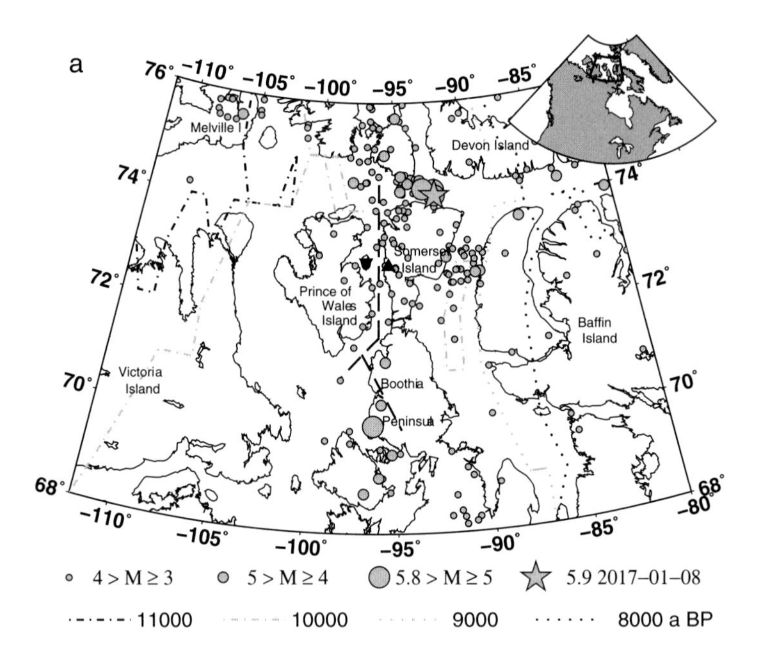

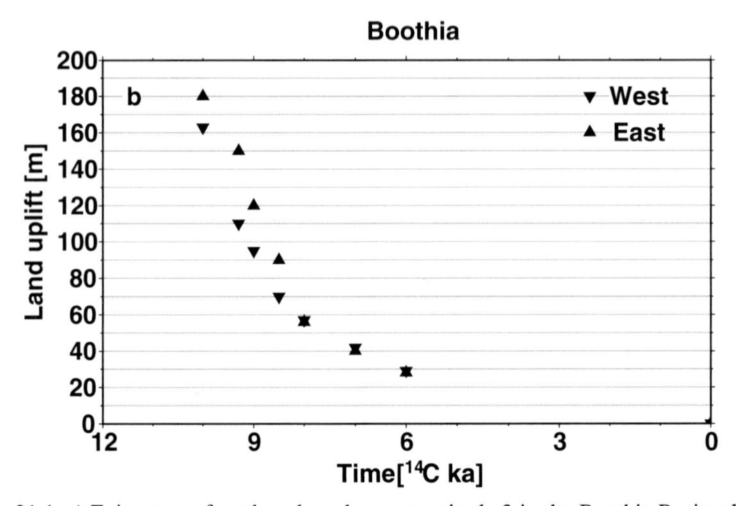

Figure 21.1 a) Epicentres of earthquakes above magnitude 3 in the Boothia Peninsula area in Northern Canada (see inlet for location in North America), for years 1900–2019 (based on the USGS Earthquake Catalog until 1985, earthquake.usgs.gov/earthquakes/search/ and the Earthquakes Canada, GSC, Earthquake Search (On-line Bulletin), biturl.top/2eyQZb, Natural Resources Canada, 2020-03-19, for years 1985–2019). Black dashed lines mark a hypothetical fault zone discussed in Dyke et al. (1991, 1992). The Peel Sound is the area between Prince of Wales Island and Somerset Island. Grey and black dash-dotted and dotted lines mark ice limits at four selected time steps from the ICE-7G_NA ice history model (Roy & Peltier, 2018). The triangles in the centre are locations for extracting land uplift values from Figure 14 in Dyke et al. (1991). b) Corresponding values are plotted over time (not calibrated, see text).

Currently, the most suitable data record to identify any GTF is RSL data. Dyke et al. (1991) analyzed more than 100 RSL data in the Canadian Arctic and presented maps of the GIA behaviour of the archipelago. According to the authors, these maps are drawn conservatively based on RSL data whose height error is at most 5 per cent, usually much less. While most areas in these maps showed the typical exponential emergence curve, the authors found some unexpected 'steep sided, north-south oriented ridge in the region of Boothia Peninsula and western Somerset Island' between 9.3 and 8.0 ka (not calibrated). This ridge leads to differences of locally up to 100 m in the height of neighbouring RSL index points around Peel Sound between Prince of Wales Island and Somerset Island.

We show an example of two selected locations west and east of the sound. They are marked with triangles in Figure 21.1a, one on the east side of Prince of Wales Island and one on the west side of Somerset Island. We have extracted their corresponding land uplift values for each time slice in Figure 14 in Dyke et al. (1991) and plotted them in Figure 21.1b. Note that the time values are not calibrated as in Dyke et al. (1991). While the land uplift values match from 8.0 ka on, there are large discrepancies of at least 20 m before that time. The points are ~50 km apart, thus significant tilting or another process that levelled both areas happened between 8.5 and 8.0 ka. This cannot be explained with GIA that would be due to retreating ice. Dyke et al. (1991) suggested therefore a hypothetical fault zone (Figure 21.1a) that was active between 8.5 and 8.0 ka. To support their hypothesis, they looked for GIFs during mapping, but none could be identified. However, a few minor postglacial lineaments that might have tectonic origin and strike parallel to the Peel Sound were found. These may hint at fault offsets in the Peel Sound area.

Dyke et al. (1991, 1992) add that something similar has not been found in Arctic Canada. In a later study, Dyke (1998) highlights that no new explanation has been found yet for this behaviour in the RSL data. Since then, no other mechanism has been discussed nor investigation undertaken to identify the proposed fault zone to our knowledge.

On another note, surficial faults with apparently young traces of activity were spotted on Ellesmere Island north-west of Greenland during fieldwork, but they were not the target of those expeditions (Bernard Guest, personal communication, 2020). These faults merit future investigation. A fault scarp north of Esayoo Bay in western Ellesmere Island, for example, extends across a U-shaped glacial valley with a maximum relief of about 2 metres on the west side of the valley. Although probably an old fault associated with Palaeogene–Neogene Eurekan Orogen tectonics, a neotectonic reactivation is suggested as it is very linear, fresh looking and perpendicular to two big glaciers nearby.

21.3 Greenland

The initial glaciation of Greenland had already begun 18 million years ago (Thiede et al., 2011), whereas the Greenland ice sheet that we know today formed mainly in the last approximately 122,000 years (Seierstad et al., 2014), although parts may be as old as 1 million years (Yau et al., 2016). Greenland forms part of the stable craton of eastern North America, with the actively spreading Mid-Atlantic Ridge to the east. Moderate to

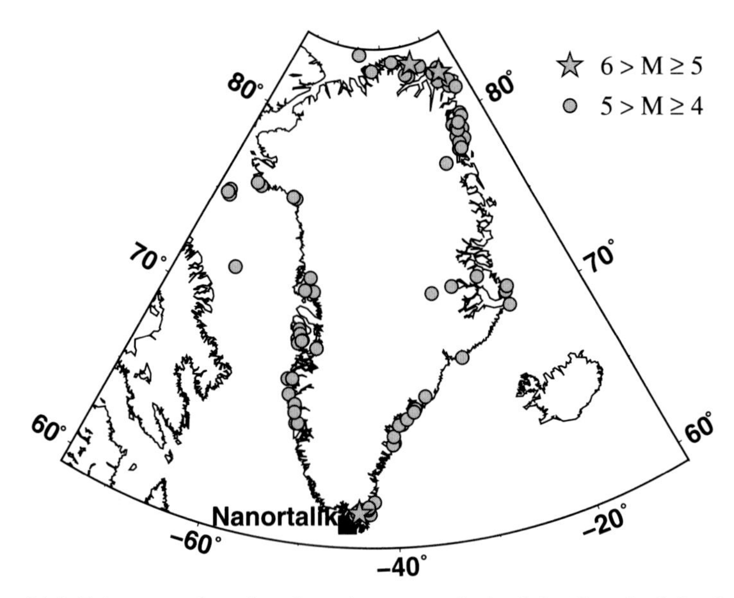

Figure 21.2 Epicentres of earthquakes above magnitude 4 in Greenland in the years 1900–2019 (based on the USGS Earthquake Catalog, earthquake.usgs.gov/earthquakes/search/ and the Greenland earthquakes list available from the Geological Survey of Denmark and Greenland (GEUS), biturl.top/beyuuy). The town of Nanortalik in southern Greenland, where a significant shift in the height of relative sea level index points is observed (see text), is marked.

high seismicity is thus not expected there (Giardini et al., 2003). Investigation of seismicity in Greenland is in general difficult due to the ice coverage, but the number of seismographs within the Greenland Ice Sheet Monitoring Network (GLISN) is increasing steadily (Clinton et al., 2014). Strong earthquakes with a moment magnitude above 6 have not been reported yet (Voss et al., 2007), but light earthquakes with magnitudes between 4 and 5 are observed every year (Figure 21.2). Only three moderate earthquakes of magnitude 5 or larger are listed in the USGS Earthquake Catalog (Figure 21.2).

Since the last glacial maximum (LGM) between 26,000 and 19,500 years ago, the ice sheet lost, with some fluctuations, ~4.7 m in sea-level equivalent (Lecavalier et al., 2014). Stress modelling for Greenland pointed to unstable conditions $\sim 10,615 \pm 250$ years ago at the southern tip of Greenland near the town of Nanortalik (Figure 21.2; Steffen et al., 2020). As RSL index points show significant shifts around this time that cannot be reliably explained with isostatic surface deformation, the authors suggest GTF as the mechanism that led to this shift via a NW-SE-striking fault reactivated in thrust conditions. Since the potential earthquake happened offshore, they also point to corresponding tsunami hazard. This may also play an important role in case of future earthquakes related to current climate-change-induced ice melting. Even though this fault has not yet been identified, (potential) GIF reactivation was suggested within other areas in southern Greenland (Peulvast et al., 2011; David Greene, personal communication, 2020). Moreover, a moderate earthquake with a (moment) magnitude of 5.1 occurred in 1998 north-east of Nanortalik

(Figure 21.2; Chung, 2002). We note though that this earthquake is listed as a minor earthquake with a (Richter) magnitude of only 2.5 in the Danish earthquake catalogue for Greenland.

Earthquakes with a magnitude above 5 also occurred in north-eastern Greenland in 1987 and 1993 (Figure 21.2). Chung and Gao (1997) and Chung (2002) suggested that these moderate earthquakes – both with normal faulting focal mechanism – took place in the peripheral bulge of the ice sheet and were caused by ice cap changes. Gregersen (2006) argued instead that the stress field in Greenland is dominated by plate motion from the Northern Mid-Atlantic Ridge, a view shared by Einarsson (1989) for Iceland. However, the epicentres of these two Greenland earthquakes were too far away from the plate boundary, and Gregersen (2006) could not suggest any alternative explanation for their generation. On the other hand, there are no geologically confirmed GIFs in the area and, interestingly, both earthquakes are not listed in the Danish earthquake catalogue for Greenland.

Due to the rather dramatic ongoing and accelerating ice melt in Greenland (Shepherd et al., 2020) it has been argued that earthquake activity will increase (Arvidsson, 1996). This seems not to have happened yet (Olivieri & Spada, 2015). Such increase is naturally subject to pre-existing geological structures that support the release of accumulated energy and the distribution of seismic stations. Steffen et al. (2020) therefore strongly advocated that appropriate fault mapping on land and on the seafloor should be performed along those rapidly melting areas.

21.4 Iceland

Iceland was covered by an ice cap during the last glaciation, and several smaller ice caps are still present. The geodynamic situation of Iceland is special as it is a stretched and magma-covered continental crust between divergent plate boundaries (Foulger et al., 2020). This results in a normal faulting tectonic regime, which is mainly stable during and after glaciation (Steffen et al., see Chapter 2), and only around the ice cap or at the very edge normal faulting due to glacially induced stresses can occur. One might speculate that even GTF could have happened, but it seems at first glance unlikely.

Hjartardóttir et al. (2009) inspected the approximately 30-km-long Kerlingar Fault of the Northern Volcanic Rift Zone (NVZ) in Northeast Iceland (Figure 21.3). The fault is located at the eastern flank of the rift zone. The authors reported a 2–9-m-fault throw down to the east, thus implying a normal fault. As the fault forms a well distinguishable offset in a flat moraine, the authors concluded that the fault was active in the Holocene, i.e. after the pre-Holocene glaciation. Moreover, according to the authors the NNW-SSE-orientated fault is different from other faults in the NVZ due to the following: (i) at 30 km, it is significantly longer than other faults there, most of which are only 1 km long and also rather straight and continuous; (ii) it is not parallel to the spreading ridge and the throw is down to east (being located at the eastern flank it should normally throw to west); but (iii) the fault is parallel to a major geological boundary between the NVZ and the so-called Eastern Fjords Block; and (iv) several faults in the area have similar/parallel strike directions to the Kerlingar Fault.

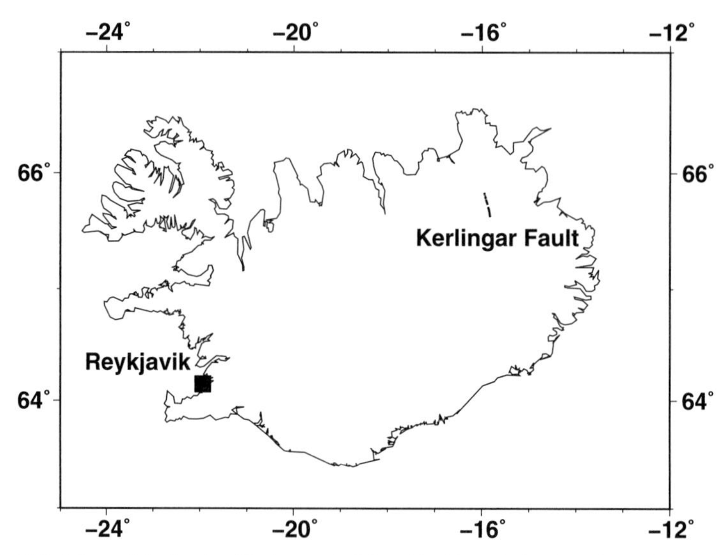

Figure 21.3 Map of Iceland with its capital Reykjavik and the location of the Kerlingar Fault in Northeast Iceland.

Three suggestions were provided for the generation of the fault: a rifting event, a stress transfer from a nearby transform fault or GTF, with the latter favoured by the authors. Several possible explanations were presented for potential GTF. The most plausible include different effective viscosities or densities in the mantle between the two nearby geological provinces. Those differences lead to regionally different uplifts that cause stress and finally fault motion. We note that these explanations are different from the generally discussed principle of GTF as in Johnston (1987) or Steffen et al., see Chapter 2, where stress build-up causing fault activity is mainly due to bending of the lithosphere and stress migration from the mantle to the upper lithosphere. Hence, the Kerlingar Fault should not be considered as GIF if indeed regional rheological differences in the upper mantle cause a different uplift behaviour.

21.5 Svalbard

The Svalbard Archipelago is about midway between mainland Norway and the North Pole at the north-western margin of the Barents Sea. During the last glaciation, the archipelago was covered by the north-western part of the Svalbard–Barents Sea–Kara Sea Ice Sheet and became likely (almost) ice-free 11,000 years ago (Hughes et al., 2016). At least, the glaciers on western Svalbard were smaller than present-day glaciers so the valleys were ice-free (Svendsen & Mangerud, 1997). Today, 55 per cent of the island is covered with glaciers, but these are retreating due to climate change (van den Heuvel et al., 2020).

Svalbard is situated on the passive margin of the slow-spreading Arctic Ocean and the Norwegian–Greenland seas, thus adjacent to two active rift zones (Dörr et al., 2013).

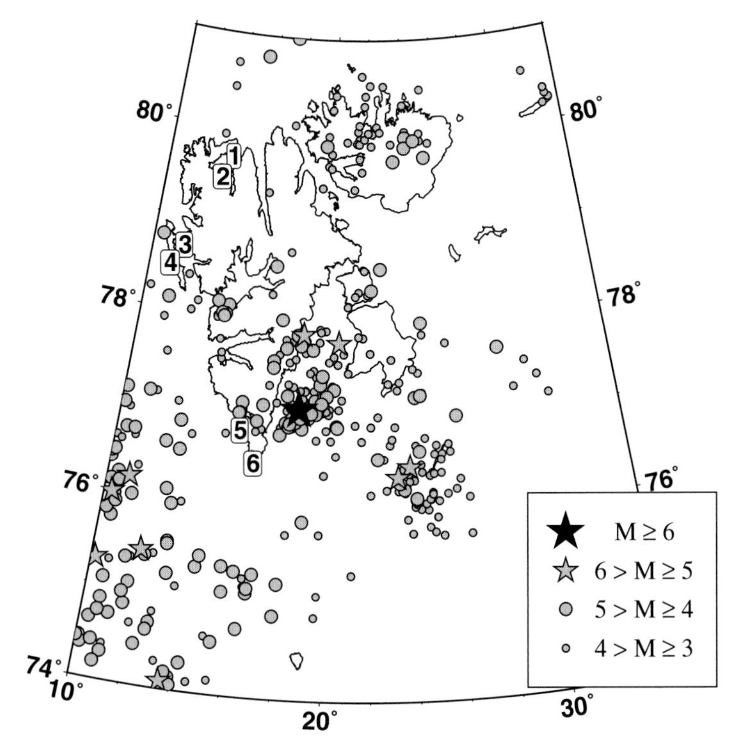

Figure 21.4 Epicentres of earthquakes above magnitude 3 in Svalbard in the years 1900–2019 (based on the USGS Earthquake Catalog, earthquake.usgs.gov/earthquakes/search/ and the FENCAT catalogue www.seismo.helsinki.fi/EQ-search/query.php). 1 – Woodfjord; 2 – Bockfjord; 3 – Forlandsundet; 4 – Geddesfjell on Prins Karls Forland; 5 – Breinesflya; 6 – Sørkapp.

Earthquake activity is consequently larger than in continental areas with minor to moderate earthquakes (Figure 21.4). Many epicentres are in the west in proximity to the Mid-Atlantic Ridge. A strong earthquake with magnitude of 6.1 occurred on 21 February 2008 (black star in Figure 21.4) south-east of the main island.

Olesen et al. (2004) and (2013) summarized a total of six locations with indications of GTF in Svalbard, see Figure 21.4. Four of them (numbers 2–5 in Figure 21.4) were re-investigated by aerial photographs by Olesen et al. (2004), whereas only the Bockfjord location (2), suggested by Piepjohn (1994), was considered as possibly neotectonic. The other locations were ranked as probably not (3 and 4) or unlikely (5) neotectonic because of missing fault scarps or inconsistencies in fault scarp heights (Olesen et al., 2004). Piepjohn (1994) also found differences in the pattern of Holocene shorelines at Woodfjord (1). While shorelines to the west of the fjord dip to the west with an irregular development, the ones to the east dip to the south but with a regular development during continuous Holocene land uplift (Piepjohn, 1994). As geologic mapping by another researcher did not see such differences (Olesen et al., 2004), the finding is ranked as probably not neotectonics by Olesen et al. (2004).

During collection of bathymetric data with the Spanish research vessel 'B.I.O. Hespérides (A-33)', a several km-long, NNW–SSE-trending and WSW-facing escarpment was found ~40 km to the SSE of Sørkapp (6) in a trough mouth fan at a depth of ~300 m of water (Angelo Camerlenghi, personal communication, 2020). Additional sub-bottom profiler data across the fault demonstrated that post-LGM deposits are offset with a height appearing to be consistent. This bathymetric information on the inner shelf has not been published, to the knowledge of the authors.

21.6 Antarctica

The Antarctic continent roughly fills the area south of 66° S (Figure 21.5). It is covered by the world's largest ice sheet of almost 14 million km^2 in area and nearly 27 million km^3 in volume (continental and shelf ice; Fretwell et al., 2013). Icing of Antarctica began at least 34 million years ago (Goldner et al., 2014) and glaciation underwent some fluctuations.

The seismicity was always considered low (e.g. Behrendt, 1999) with only sporadic larger events (e.g. a light earthquake with magnitude 4.5, Adams et al., 1985). Until 1985, the hypocentres of only 5 events could be placed in Antarctica (Adams et al., 1985). South of 70° S, an earthquake of magnitude of 6 or higher has never been recorded. However, one

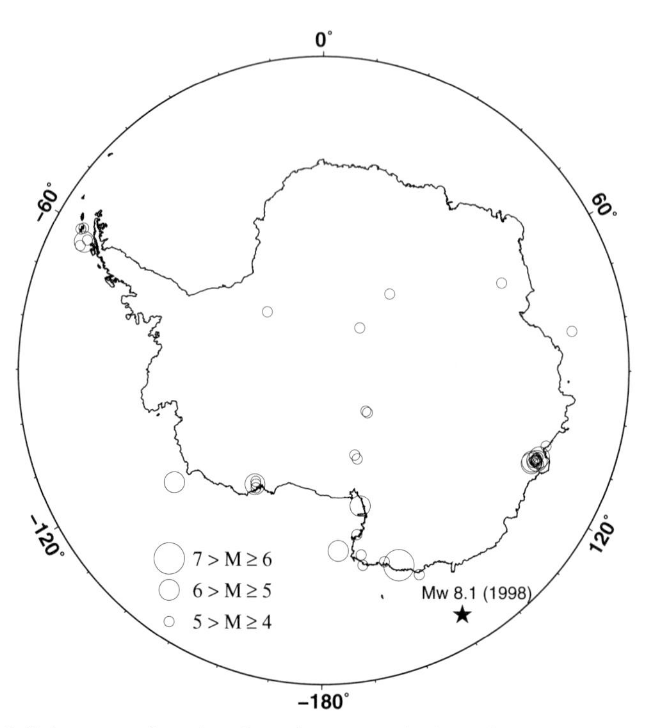

Figure 21.5 Epicentres of earthquakes above magnitude 4 in Antarctica in the years 1900–2019 (based on the USGS Earthquake Catalog, earthquake.usgs.gov/earthquakes/search/). The epicentre of the 1998 M_w 8.1 earthquake near the Balleny Islands is marked with a star.

must note that the distribution of seismographs has always been rather sparse given that the area of Antarctica is larger than Europe. Research programmes in recent years increased the amount of data and thus density and magnitude threshold so that now more earthquakes are observed and analyzed (e.g. Lough et al., 2018). Based on such new data, Lough et al. (2018) suggested that the seismicity of Antarctica could be compared to that of the Canadian Shield.

An unusually great earthquake with magnitude $M_w = 8.1$ occurred on 25 March 1998, 330 km north of the Antarctic Shelf close to the Balleny Islands (Figure 21.5; Henry et al., 2000). It was argued that this was due to GIA (Tsuboi et al., 2000) or, at least, that GIA cannot be ruled out (Kreemer & Holt, 2000). With modelling, both Ivins et al. (2003) and Kaufmann et al. (2005) could show that deglaciation can have an influence this far from the Antarctic continent, thus GIA can indeed be considered as a driver of this event.

Such a great magnitude earthquake is still an exception, and no comparable event has been recorded so far within the glaciated area. Nonetheless, the increased amount of data shows that there are more earthquakes beneath the ice than the anticipated physical process according to Johnston (1987) would allow. Lough et al. (2018) speculated that tectonic intraplate stresses could accumulate long enough over time until the GIA stress that suppresses activity would be overcome. Then, any glaciated area could potentially behave like a typical continental area without ice. This hypothesis should be further investigated.

To our knowledge, there is only one study that discusses geological evidence for GTF in Antarctica. Three small faults (Keyhole Ridge, Lower Hidden Valley and Garwood Valley faults) in South Victoria Land at the margin of the Antarctic Ice Sheet show apparent sinistral strike-slip displacements at some time during the (based on the glaciological history of the area) last 50,000 years and possibly within the last few thousand years (Jones, 1997). Munier and Fenton (2004) note though that their motion is more likely the result of regional tectonics.

21.7 Discussion

The polar areas are still glaciated by large ice sheets and glaciers. There have been fluctuations of their size in the last 25,000 years and there are significant mass decreases ongoing due to climate change. According to meanwhile well-established theory, the ice retreat could lead under certain circumstances to reactivation of pre-existing faults. This is well-known for Northern Europe and the British Isles. No such structure, however, has been clearly identified to date in the polar areas – none that could point to reactivation in the last 25,000 years, nor was such a structure affected in recent years. However, there are indications in records of RSL index points in the Canadian Arctic and at the southern tip of Greenland that such reactivation happened at the end of the last deglaciation.

Dedicated geological and geophysical investigations should be performed to support or reject these hypothetical faults. In addition, these investigations should be accompanied by appropriate modelling, as is needed for the Kerlingar Fault in Iceland, where GTF has been suggested as a driver but is not confirmed and where care must be taken, as regional differences in mantle rheology could be the actual cause. Hjartardóttir et al. (2009) already suggested modelling as a tool to solve the cause for the existence of the Kerlingar Fault.

Antarctica seems not to have any evidence of GTF. Seismicity is low, even at the ice sheet edges, where current melting is ongoing. Geological evidence is lacking, but we note that analysis regarding GTF has not been the target of research in Antarctica (Mike Bentley and Terry Wilson, personal communication, 2019). As mass loss is ongoing especially in West Antarctica (e.g. Barletta et al., 2013), it can be speculated whether earthquake activity is significantly increasing and if pre-existing faults can be reactivated, and moreover, whether evidence of activity can then be found on the surface. Knowledge of such events – past, present or future – is also important for geodetic measurements in Antarctica, which can be affected by (post)seismic deformation (King & Santamaria-Gomez, 2016).

The Barents Sea including Svalbard and parts of the Russian Arctic were glaciated as well, and here also the existence of GIFs can per se not be excluded.

21.8 Conclusions

Clear evidence for glacially triggered faulting in polar areas is missing. This includes Northern Alaska, the Canadian Arctic, Greenland, Iceland, Svalbard and Antarctica. In the last 40 years some indications, mainly from earthquakes or relative sea-level data, have been mentioned, but rarely in the peer-reviewed literature. Quite often reports or conference abstracts point to such indications, but dedicated research is lacking. Hence, especially in view of established theory of glacially triggered faulting and the ongoing ice melt in polar areas, we suggest that scientists investigating bedrock in polar areas with sophisticated geoscientific tools should be (more) sensitized to the problem of glacially triggered faulting and its corresponding causes. This should increase the amount of clear evidence for glacially triggered faulting in polar areas in the future. Also, seismograph distribution in these rather remote areas should be densified to increase detectability and enhance accurate determination of source parameters of future events.

Acknowledgements

We are grateful to Søren Gregersen and an anonymous reviewer as well as the editor Odleiv Olesen for their many suggestions that helped improve the manuscript. We thank Mike Bentley, Angelo Camerlenghi, Arthur Dyke, David Greene, Bernard Guest, Ásta Hjartardóttir, Karsten Piepjohn and Terry Wilson for discussions. Figures were drawn with GMT5 (Wessel et al., 2013).

References

Adams, R. D., Hughes, A. A. and Zhang, B. M. (1985). A confirmed earthquake in continental Antarctica. *Geophysical Journal International*, **81**(2), 489–492, doi.org/10.1111/j.1365-246X.1985.tb06416.x.

Arvidsson, R. (1996). Fennoscandian earthquakes: whole crustal rupturing related to postglacial rebound. *Science*, **274**, 744–746, doi.org/10.1126/science.274.5288.744.

Barletta, V. R., Sørensen, L. S. and Forsberg, R. (2013). Scatter of mass changes estimates at basin scale for Greenland and Antarctica. *The Cryosphere*, **7**, 1411–1432, doi.org/10.5194/tc-7-1411-2013.

Behrendt, J. (1999). Crustal and lithospheric structure of the West Antarctic Rift System from geophysical investigations – a review. *Global and Planetary Change*, **23**, 25–44, doi.org/10.1016/S0921-8181(99)00049-1.

Brooks, G. R. and Adams, J. (2020). A review of evidence of glacially-induced faulting and seismic shaking in southeastern Canada. *Quaternary Science Reviews*, **228**, 106070, doi.org/10.1016/j.quascirev.2019.106070.

Chung, W.-Y. (2002). Earthquakes along the passive margin of Greenland: evidence for postglacial rebound control. *Pure and Applied Geophysics*, **159**, 2567–2584, doi.org/10.1007/s00024-002-8748-1.

Chung, W.-Y. and Gao, H. (1997). The Greenland earthquake of 11 July 1987 and Postglacial Fault reactivation along a passive margin. *Bulletin of the Seismological Society of America*, **87**(4), 1058–1068.

Clinton, J. F., Nettles, M., Walter, F. et al. (2014). Seismic network in Greenland monitors Earth and ice system. *Eos Transactions American Geophysical Union*, **95**, 13–24, doi.org/10.1002/2014EO020001.

Dörr, N., Clift, P. D., Lisker, F. and Spiegel, C. (2013), Why is Svalbard an island? Evidence for two-stage uplift, magmatic underplating, and mantle thermal anomalies. *Tectonics*, **32**, 473–486, doi.org/10.1002/tect.20039.

Dyke, A. S. (1998). Holocene delevelling of Devon Island, Arctic Canada: implications for ice sheet geometry and crustal response. *Canadian Journal of Earth Sciences*, **35**, 885–904, doi.org/10.1139/cjes-35-8-885.

Dyke, A. S. (2004). An outline of North American deglaciation with emphasis on central and northern Canada. In J. Ehlers and P. L. Gibbard, eds., *Quaternary Glaciations – Extent and Chronology, Part II. North America., Developments in Quaternary Science 2*. Elsevier, New York, pp. 373–424, doi.org/10.1016/S1571-0866(04)80209-4.

Dyke, A. S., Morris, T. F. and Green, D. E. C. (1991). *Postglacial Tectonic and Sea Level History of the Central Canadian Arctic.* Geological Survey of Canada Bulletin, **397**, 56 pp.

Dyke, A. S., Morris, T. F., Green, D. E. C. and England, J. H. (1992). *Quaternary Geology of Prince of Wales Island, Arctic Canada.* Geological Survey of Canada Memoir, 433, 142 pp.

Einarsson, P. (1989). Intraplate earthquakes in Iceland. In S. Gregersen and P. W. Basham, eds., *Earthquakes at North-Atlantic Passive Margins: Neotectonics and Postglacial Rebound.* Kluwer Academic Press, Dordrecht/Boston/London, pp. 329–344.

Fenton, C. (1994). *Postglacial Faulting in Eastern Canada.* Geological Survey of Canada Open File, 2774, 94 pp.

Firth, C. R. and Stewart, I. S. (2000). Postglacial tectonics of the Scottish glacio-isostatic uplift centre. *Quaternary Science Reviews*, **19**, 1469–1493, doi.org/10.1016/S0277-3791(00)00074-3.

Foulger, G. R., Doré, T., Emeleus, C. H. et al. (2020). The Iceland microcontinent and a continental Greenland–Iceland–Faroe Ridge. *Earth-Science Reviews*, **206**, 102926, doi.org/10.1016/j.earscirev.2019.102926.

Fretwell, P., Pritchard, H. D., Vaughan, D. G. et al. (2013). Bedmap 2: improved ice bed, surface and thickness datasets for Antarctica. *The Cryosphere*, **7**, 375–393, doi.org/10.5194/tc-7-375-2013.

Giardini, D., Grünthal, G., Shedlock, K. M. and Zhang, P. (2003). The GSHAP Global Seismic Hazard Map. In W. Lee, H. Kanamori, P. Jennings and C. Kisslinger, eds., *International Handbook of Earthquake & Engineering Seismology, International Geophysics Series 81B*. Academic Press, Amsterdam, pp. 1233–1239.
Goldner, A., Herold, N. and Huber, M. (2014). Antarctic glaciation caused ocean circulation changes at the Eocene–Oligocene transition. *Nature*, **511**(7511), 574–577, doi .org/10.1038/nature13597.
Gregersen, S. (1989). The seismicity of Greenland. In S. Gregersen and P.W. Basham, eds., *Earthquakes at North-Atlantic Passive Margins: Neotectonics and Postglacial Rebound*. Kluwer Academic Press, Dordrecht/Boston/London, pp. 345–353.
Gregersen, S. (2006). Intraplate earthquakes in Scandinavia and Greenland. Neotectonics or postglacial uplift. *Journal of the Indian Geophysical Union*, **10**, 25–30.
Henry, C., Das, S. and Woodhouse, J. H. (2000). The March 25, 1998 Mw = 8.1 Antarctic Plate earthquake: moment tensor and rupture history. *Journal of Geophysical Research Solid Earth*, **105**, 16097–16119, doi.org/10.1029/2000JB900077.
Hjartardóttir, Á. R., Einarsson, P. and Brandsdóttir, B. (2011). The Kerlingar fault, Northeast Iceland: a Holocene normal fault east of the divergent plate boundary. *Jökull*, **60**, 103–116.
Hughes, A. L. C., Gyllencreutz, R., Lohne, Ø. S., Mangerud, J. and Svendsen, J. I. (2016). The last Eurasian ice sheets – a chronological database and time-slice reconstruction, DATED-1. *Boreas*, **45**(1), 1–45, doi.org/10.1111/bor.12142.
Ivins, E. R., James, T. S. and Klemann, V. (2003). Glacial isostatic stress shadowing by the Antarctic Ice Sheet. *Journal of Geophysical Research Solid Earth*, **108**(B12), doi.org/10.1029/2002LB002182.
Jacob, T., Wahr, J., Pfeffer, W. T. and Swenson, S. (2012). Recent contributions of glaciers and ice caps to sea level rise. *Nature*, **482**, 514–518, doi.org/10.1038/nature10847.
Johnston, A. C. (1987). Suppression of earthquakes by large continental ice sheets. *Nature*, **330**, 467–469, doi.org/10.1038/330467a0.
Johnston, A. C. (1996). A wave in the Earth. *Science*, **274**, p. 735, 10.1126/science.274 .5288.735.
Jones, S. (1997). Late Quaternary faulting and neotectonics. South Victoria Land, Antarctica. *Journal of the Geological Society of London*, **153**, 645–653, doi.org/10 .1144/gsjgs.154.4.0645.
Kaufmann, G., Wu, P. and Ivins, E. R. (2005). Lateral viscosity variations beneath Antarctica and their implications on regional rebound motions and seismotectonics. *Journal of Geodynamics*, **39**, 165–181, doi.org/10.1016/j.jog.2004.08.009.
King, M. A. and Santamaría-Gómez, A. (2016). Ongoing deformation of Antarctica following recent Great Earthquakes. *Geophysical Research Letters*, **43**, 1918–1927, doi.org/10.1002/2016GL067773.
Kreemer, C. and Holt, W. E. (2000). What caused the March 25, 1998 Antarctic plate earthquake?: inferences from regional stress and strain rate fields. *Geophysical Research Letters*, **27**, 2297–2300, doi.org/10.1029/1999GL011188.
Kujansuu, R. (1964). Nuorista siirroksista Lapissa [English summary: Recent faults in Lapland]. *Geologi*, **16**, 30–36 (in Finnish).
Lagerbäck, R. and Sundh, M. (2008). *Early Holocene Faulting and Paleoseismicity in Northern Sweden*. Geological Survey of Sweden Research Paper, Series C, Vol. 836, 80 pp.
Lecavalier, B. S., Milne, G. A., Simpson, M. J. R. et al. (2014). A model of Greenland ice sheet deglaciation constrained by observations of relative sea level and ice extent. *Quaternary Science Reviews*, **102**, 54–84, doi.org/10.1016/j.quascirev.2014.07.018.

Lough, A. C., Wiens, D. A. and Nyblade, A. (2018). Reactivation of ancient Antarctic rift zones by intraplate seismicity. *Nature Geoscience*, **11**(7), 515–519, doi.org/10.1038/s41561-018-0140-6.

Motazedian, D. and Ma, S. (2018). Source parameter studies on the 8 January 2017 Mw 6.1 Resolute, Nunavut, Canada, Earthquake. *Seismological Research Letters*, **89**, 1030–1039, doi.org/10.1785/0220170260.

Munier, R. and Fenton, C. (2004). Review of postglacial faulting. In R. Munier and H. Hökmark, eds., *Respect Distances*. SKB Technical Report TR-04-17, Swedish Nuclear Fuel and Waste Management Co., Stockholm, pp 157–218.

Olesen, O., Blikra, L. H., Braathen, A. et al.(2004). Neotectonic deformation in Norway and its implications: a review. *Norwegian Journal of Geology*, **84**, 3–34.

Olesen, O., Bungum, H., Dehls, J. et al. (2013). Neotectonics, seismicity and contemporary stress field in Norway – mechanisms and implications. In L. Olsen, O. Fredin and O. Olesen, eds., *Quaternary Geology of Norway, Geological Survey of Norway Special Publication, Vol. 13*. pp. 145–174.

Olivieri, M. and Spada, G. (2015). Ice melting and earthquake suppression in Greenland. *Polar Science*, **9**, 94–106, doi.org/10.1016/j.polar.2014.09.004.

Peulvast, J.-P., Bonow, J. M., Japsen, P., Wilson, R. W. and McCaffrey, K. J. W. (2011). Morphostructural patterns and landform generations in a glaciated passive margin: the Kobberminebugt-Qaqortoq region of South Greenland. *Geodinamica Acta*, **24**(1), 1–19, doi.org/10.3166/ga.24.1-19.

Piepjohn, K. (1994). *Tektonische Evolution der Devongräben (Old Red) in NW-Svalbard* [Tectonic evolution of Devonian graben (Old Red) in NW Svalbard]. Unpublished PhD thesis, Westfälische Wilhelms-Universität, Münster, 170 pp.

Roy, K. and Peltier, W. R. (2018). Relative sea level in the Western Mediterranean basin: a regional test of the ICE-7G_NA (VM7) model and a constraint on Late Holocene Antarctic deglaciation. *Quaternary Science Reviews*, **183**, 76–87, doi.org/10.1016/j.quascirev.2017.12.021.

Seierstadt, I., Abbott, P. M., Bigler, M. et al. (2014). Consistently dated records from the Greenland GRIP, GISP2 and NGRIP ice cores for the past 104 ka reveal regional millennial-scale δ18O gradients with possible Heinrich event imprint. *Quaternary Science Reviews*, **106**, 29–46, doi.org/10.1016/j.quascirev.2014.10.032.

Shepherd, A., Ivins, E., Rignot, E. et al. (2020). Mass balance of the Greenland Ice Sheet from 1992 to 2018. *Nature*, **579**, 233–239 doi.org/10.1038/s41586-019-1855-2.

Simon, K. M., James, T. S., Henton, J. A. and Dyke, A. S. (2016). A glacial isostatic adjustment model for the central and northern Laurentide Ice Sheet based on relative sea-level and GPS measurements. *Geophysical Journal International*, **205**, 1618–1636, doi.org/10.1093/gji/ggw103.

Steffen, H. and Wu, P. (2011). Glacial isostatic adjustment in Fennoscandia – a review of data and modeling. *Journal of Geodynamics*, **52**, 169–204, doi.org/10.1016/j.jog.2011.03.002.

Steffen, R., Steffen, H., Weiss, R. et al. (2020). Early Holocene Greenland-ice mass loss likely triggered earthquakes and tsunami. *Earth and Planetary Science Letters*, **546**, 116443, doi.org/10.1016/j.epsl.2020.116443.

Svendsen, J. I. and Mangerud, J. (1997). Holocene glacial and climatic variations on Spitsbergen, Svalbard. *Holocene*, **7**, 45–57, doi.org/10.1177%2F095968369700700105.

Thiede, J., Jessen, C., Knutz, P. et al. (2011). Millions of years of Greenland Ice Sheet history recorded in ocean sediments. *Polarforschung*, **80**(3), 141–159, doi.org/10.2312/polarforschung.80.3.141.

Tsuboi, S., Kikuchi, M., Yamanaka, Y. and Kanao, M. (2000). The March 25, 1998 Antarctic earthquake: great earthquake caused by postglacial rebound. *Earth Planets Space*, **52**, 133–136, doi.org/10.1186/BF03351621.

van den Heuvel, F., Hübner, C., Błaszczyk, M., Heimann, M. and Lihavainen, H. (2020). *SESS Report 2019 – The State of Environmental Science in Svalbard – An Annual Report*. Svalbard Integrated Arctic Earth Observing System, Longyearbyen.

Voss, P., Kildegaard Poulsen, S., Simonsen, S. and Gregersen, S. (2007). Seismic hazard assessment of Greenland. *GEUS Bulletin*, **13**, 57–60.

Wessel, P., Smith, W. H. F., Scharroo, R., Luis, J. F. and Wobbe, F. (2013). Generic Mapping Tools: improved version released. *Eos Transactions American Geophysical Union*, **94**, 409–410, doi.org/10.1002/2013EO450001.

Wu, P. and Hasegawa, H. S. (1996a). Induced stresses and fault potential in eastern Canada due to a disc load: a preliminary analysis. *Geophysical Journal International*, **125**, 415–430, doi.org/10.1111/j.1365-246X.1996.tb00008.x.

Wu, P. and Hasegawa, H. S. (1996b). Induced stresses and fault potential in Eastern Canada due to a realistic load: a preliminary analysis. *Geophysical Journal International*, **127**, 215–229, doi.org/10.1111/j.1365-246X.1996.tb01546.x.

Yau, A. M., Bender, M. L., Blunier, T. and Jouzel, J. (2016). Setting a chronology for the basal ice at Dye-3 and GRIP: implications for the long-term stability of the Greenland Ice Sheet. *Earth and Planetary Science Letters*, **451**, 1–9, doi.org/10.1016/j.epsl.2016.06.053.

Part VI

Modelling of Glacially Induced Faults and Stress

As glacially induced faults are reactivated due to a combination of tectonic and glacially induced isostatic stresses it is interesting to model the corresponding fault slip with dedicated models. This chapter presents a modelling approach using a well-established model of glacial isostatic adjustment (GIA); this is followed by a review of stresses to be considered in sophisticated future modelling.

22

Glacial Isostatic Adjustment Models for Earthquake Triggering

PATRICK WU, REBEKKA STEFFEN, HOLGER STEFFEN AND BJÖRN LUND

ABSTRACT

In order to model glacial triggering of earthquakes, the spatio-temporal variation of glacial isostatic adjustment (GIA)-induced stress that occurs during a glacial cycle needs to be obtained. This can be computed efficiently using commercial finite-element (FE) codes with appropriate modifications to include the important effects of 'pre-stress advection', 'internal buoyancy' and 'self-gravity'. The modifications described in Wu (2004) are reviewed for incompressible and so-called materially compressible flat-earths. When GIA-induced stress is superimposed on background tectonic stress and overburden pressure, the time variation of earthquake potential at various locations in the Earth can be evaluated for any fault orientation. In order to model more complex slip and fault behaviour over time, the three-stage FE model approach of Steffen et al. (2014) is reviewed. Finally, selected numerical examples and their results from both modelling approaches are shown.

22.1 Introduction

Early studies of the causal relationship between glacial isostatic adjustment (GIA) and intraplate earthquakes are mainly based on the modelled 'maximum stress difference' (e.g. Walcott, 1970; Spada et al., 1991), the 'maximum flexural stress' (e.g. Stein et al., 1979) or the strain-rate (e.g. James & Bent, 1994). To study the change in stability of faults during a glacial cycle, the relative position between the Coulomb failure envelope and the Mohr circle must be examined (Steffen et al., see Chapter 2). By combining background stress (tectonic stress, overburden pressure, topographic stress, sediment loads, etc.) with the GIA-induced flexural stress from thin-plate theory, Quinlan (1984) studied the mode of faulting for recent earthquakes in Eastern Canada using the so-called *fault stability margin* (*FSM*), which represents the shortest distance between the Mohr circle and the Mohr-Coulomb failure envelope. However, the thin plate theory assumes that the mantle is inviscid, thus the important effect of stress migration from the relaxing mantle to the lithosphere is completely ignored. Using Mohr-Coulomb failure theory, Johnston (1987) showed that the presence of glacial load actually suppresses earthquakes beneath it. However, it remained unclear how a fault can become unstable near the end of deglaciation if its original state was stable before the start of the glacial cycle. Using *FSM* with total stress from a combination of GIA stress modelled in a viscoelastic Earth with background stress, Wu and Hasegawa (1996a,b) explained the underlying mechanism that led to rupture, and they showed that the predicted timing of GIA-induced earthquakes is in reasonable agreement with the observed data. However, in their model all faults are assumed to be virtual faults. The concept of a virtual fault means that no physical fault surface is included in the model, and therefore its presence does not affect the stress distribution. Ivins et al. (2003) also investigated the use of the Mogi–von Mises criterion to study fault stability but found that the predicted results just confirm those that use *FSM* as the criterion. The definition of FSM favours the computation of optimally oriented faults but is not restricted to it (Wu & Johnston, 1998). To estimate the stability of non-optimally and optimally oriented faults, we shall replace *FSM* by the change in Coulomb failure stress ΔCFS, which gives the vertical distance between the Mohr circle and the Mohr-Coulomb failure envelope (Lund, 2005, 2006; Lund et al., 2009; Steffen et al., 2019). However, in these studies the fault is still assumed to be a virtual fault, and the fault surface is not included in the model. It is important to note that while negative values of *FSM* point to unstable fault conditions, for Coulomb failure stresses it is the positive ΔCFS values that refer to unstable, whereas negative ΔCFS values point to stable conditions. For our discussion here, GIA models with virtual faults (e.g. Wu & Hasegawa 1996a,b; Lund, 2005, 2006; Lund et al., 2009) are called the Type I model (see Section 22.2 for details). This type of model predicts that GIA-induced stress is available to reactive faults at onset time, but it cannot predict the amount of fault slip nor its slip history.

A second type of model exists wherein the fault structure can release a stress induced by an ice load and produce fault slip. These models, which include the fault surface (including its geometry and frictional properties), are called Type II models. There are two variations of Type II models. The first is described in Hetzel and Hampel (2005) and subsequent papers. However, this approach has several limitations, such as how it incorrectly predicts continuous fault slip over thousands of years (even before glaciation) with rates much higher than are generally observed for glacially triggered faulting; and it fails to properly implement the stress migration from the mantle to the lithosphere (see Steffen et al., 2015, for more details). The Type II model that is adopted here is the one described in Steffen et al. (2014a,b,c), which combines a regular GIA model (whose depth extends down to the core-mantle boundary or CMB) with a fault at the Earth's surface to investigate the fault slip and fault activation time during a glacial cycle (see Section 22.3 for details). In this approach, background stress is input directly into the model instead of applying plate-velocities at the sides of the model (as in Hetzel & Hampel, 2005). So, although the faults were initially stressed critically by the tectonic background stress, the fault slips predicted are entirely triggered by GIA stresses (and not confounded with the effects of ongoing plate velocities) and are thus suitable for studying GIA-induced earthquakes. Also, neither a fault slip nor an earthquake is predicted before glaciation in a thrust-faulting regime, and the predicted fault slips in the model of Steffen et al. (2014a,b,c) are almost instantaneous and thus more consistent with observation. However, normal and strike-slip faulting stress regimes have not been tested, but earthquake triggering is expected to be confined to certain onset time interval(s).

In the following we describe the basic equations of GIA (Section 22.2) and the application of these in commercial finite-element (FE) codes to estimate glacially induced stresses (Section 22.3). In addition, two examples of such models with simple and realistic ice loading histories will be presented. We finish with a description of the Type II model from Steffen et al. (2014a) in Section 22.4. The implementation of background stresses is explained in Gradmann and Steffen, see Chapter 23.

22.2 Basic Equations and Boundary Conditions for GIA Modelling

The Earth is viscoelastic as it behaves like an elastic body initially under an applied load, but it can flow like a viscous body if the applied load is maintained over a long period of time. The simplest viscoelastic model, which can describe GIA as a steady-state process, is that of a Maxwell body. So, the constitutive relation between stress and strain (Cathles, 1975) is given by:

$$\frac{\partial}{\partial t}\sigma_{ij} + \frac{\mu}{\eta}\left(\sigma_{ij} - \frac{1}{3}\,\sigma_{kk}\delta_{ij}\right) = \lambda\frac{\partial}{\partial t}\,\epsilon_{rr}\,\delta_{ij} + 2\mu\,\frac{\partial}{\partial t}\epsilon_{ij}, \qquad (22.1)$$

where σ_{ij} and ϵ_{ij} are stress and strain tensors, respectively, δ_{ij} is diagonal tensor, λ and μ are elastic Lamé parameters, and η is viscosity. For a compressible Earth, $\epsilon_{rr} \equiv \vec{\nabla} \cdot \vec{u} \neq 0$ and the value of λ remains finite. On the other hand, if $\epsilon_{rr} \rightarrow 0$ and $\lambda \rightarrow \infty$, but $\lambda\epsilon_{rr}$ remains a finite constant, then Earth's material is incompressible.

Earth's initial state, before application of any deforming stresses, is assumed to be one of hydrostatic equilibrium (see Chapter II in Cathles, 1975), i.e.

$$\frac{dp_0}{dr} = -\rho_0 g_0, \qquad (22.2)$$

where p_0 is pressure, ρ_0 is density, g_0 is gravity and the subscript 0 denotes the initial state in each. The gravity of the initial field is related to the gravitational potential ϕ_0 by Poisson's equation:

$$\nabla^2\phi_0 = -\nabla \cdot \vec{g}_0 = 4\pi G\rho_0. \qquad (22.3)$$

This equation states that the density distribution determines both the gravitational potential and the gravitational acceleration (or gravity). The Earth model that satisfies Poisson's equation is often called self-gravitating because when deforming stresses are applied the movement of masses alters the local gravity and its potential (via gravitational attraction) and Poisson's equation remains valid (see equations (22.4), (22.5) and (22.6)).

The perturbed state occurs when deforming stresses are applied and there is motion. The deformation gives rise to strain ϵ_{ij} and stress σ_{ij} in addition to the perturbed density ρ_1, gravity g_1 and gravitational potential ϕ_1.

These perturbed quantities must satisfy the linearized version of Newton's law of momentum conservation (Farrell, 1972; Cathles, 1975; Wu & Peltier, 1982):

$$0 = \vec{\nabla} \cdot \bar{\bar{\sigma}} - \vec{\nabla}\left(\vec{u} \cdot \rho_0 g_0 \hat{r}\right) - \rho_0\,\vec{\nabla}\,\phi_1 - \rho_1 g_0 \hat{r}. \qquad (22.4)$$

Note that the acceleration term on the left side of equation (22.4) is neglected because the glacial isostatic process is too slow to cause significant accelerations. The terms on the right side of the equation are the divergence of stress, the advection of pre-stress, the perturbed gravity field (or self-gravity) and the buoyancy force of local density perturbation (or internal buoyancy), respectively. Here, $\hat{r}$ is a unit vector in the radial direction, and $\vec{u}$ is the displacement vector with $\vec{u} \cdot \hat{r} = u_r$.

Cathles (1975) pointed out that (i) the advection of the pre-stress term $\nabla(\rho_0 g_0 u_r)$ exists only for an elastic solid and not for a viscous fluid, and (ii) the buoyancy term $(\rho_0 g_0 u_r)$ exists only in the normal stress boundary condition for a viscous fluid and not for an elastic solid. However, if equation (22.4) and the elastic boundary conditions are used for a viscoelastic Earth at the initial time, then it is unclear what role the advection of the pre-stress term will play when the viscoelastic Earth relaxes and approaches the fluid limit and whether the fluid boundary condition can be satisfied. Wu and Peltier (1982) pointed out that, with the stress transformation in equation (22.9), equation (22.4) becomes the equation for a fluid, and thus the elastic normal stress boundary condition becomes that for a fluid. This means that either the elastic or the fluid form of (22.4) can be used for viscoelastic media as long as the consistent (elastic/fluid) boundary conditions are applied. Wu (1992a) further clarified that the advection of pre-stress term provides the restoring force of isostasy in the fluid and that without it there will be no postglacial rebound. Detlef Wolf (personal communication, 1992) pointed out that $\rho_0 g_0 u_r$ in the advection of the pre-stress term connects the material incremental stress to the local incremental stress – in other words, it translates between Lagrangian and Eulerian forms of the equation. We shall see below that this has led to the development of the stress transformation being used with the FE method (Wu, 2004).

Since mass must be conserved during the deformation process, the third equation used in GIA models is the linearized continuity equation:

$$\rho_1 = -\rho_0 \vec{\nabla} \cdot \vec{u} - \vec{u} \cdot (\partial_r \rho_0) \hat{r}. \tag{22.5}$$

For a self-gravitating Earth, this perturbed density ρ_1 causes a change in the gravitational potential as described by Poisson's equation:

$$\nabla^2 \phi_1 = 4\pi G \rho_1. \tag{22.6}$$

Equations (22.1) to (22.6) form the complete set of 'equations of motion' that must be solved in order to model the deformation of a spherical, self-gravitating Maxwell Earth. The solution that satisfies these equations of motion must also satisfy a specific set of boundary conditions, which depends on the problem, but the ones consistent with the equations of motion for a spherical self-gravitating Earth are:

(i) Continuity of gravitational potential throughout the Earth: $[\phi_1]_-^+ = 0$.
(ii) At the surface of the Earth $r = a : [\bar{\bar{\sigma}} \cdot \hat{r}]_-^+ = 0$, so that for normal stress, $\sigma_{rr}|_{r=a} = -\gamma g_0$, and for shear stress, $\sigma_{r\Theta}|_{r=a} = 0$. Here, γ is the surface mass density

of the applied surface load. Also, $[f]_-^+ = f_+ - f_-$ is the difference between the values above (f_+) and below (f_-) the boundary. In addition, the gradient of gravitational potential satisfies $[\nabla\phi_1 \cdot \hat{r}]_-^+ + 4\pi G\rho_0 u_r = 4\pi G\gamma$.

(iii) At internal solid-solid boundaries, $[\bar{\bar{\sigma}} \cdot \hat{r}]_-^+ = 0$, so that $\sigma_{rr}|_-^+ = \sigma_{r\Theta}|_-^+ = 0$. In addition, there is continuity of displacement, $[\vec{u}]_-^+ = 0$ and $[\nabla\phi_1 \cdot \hat{r} + 4\pi G\rho_0 u_r]_-^+ = 0$.

(iv) At the CMB, $[\bar{\bar{\sigma}} \cdot \hat{r}]^+ = \rho_f g_0 u_r \hat{r}$, where ρ_f is the density at the top of the fluid core and u_r is the radial displacement at the elastic-fluid boundary (e.g. Chinnery, 1975, and equation (48) in Wu & Peltier, 1982). In addition, shear stress vanishes at the CMB, $[\vec{u}]_-^+ = 0$ and $[\nabla\phi_1 \cdot \hat{r}]_-^+ + 4\pi G[\rho_0]_-^+ u_r = 0$.

Given the geometry, structure and material properties of the Earth model and the spatial-temporal evolution of the surface loads, solutions for the equations of motion and the boundary conditions can be obtained.

In the following, our focus is on flat-earth models where Earth's sphericity and the effect of self-gravitation can be neglected. The flat-earth approximation works well for ice loads with characteristic length comparable to, or smaller than the Fennoscandian ice sheet (Amelung & Wolf, 1994), but it also works, under certain conditions, for larger ice loads (see Section 22.5 below). For a non-self-gravitating Earth, gravity $g_0(z)$ remains constant in time and $\phi_1 = 0$. In addition, if the flat-earth model is composed of uniform incompressible layers/elements (so that density and material properties change only between layers/elements), then inside the element/layer equation (22.5) gives: $\rho_1 = 0$, and the only equation of motion that needs to be solved is equation (22.4), which becomes:

$$0 = \vec{\nabla} \cdot \bar{\bar{\sigma}} - \rho_0 \, g_0 \vec{\nabla}(w) \tag{22.7}$$

where $w = \vec{u} \cdot \hat{z}$ is the vertical displacement. The boundary conditions for this flat, incompressible Earth now become (Wu, 2004):

(i) At the surface of the Earth ($z = 0$): normal stress $\sigma_{zz}|_{z=0} = -\gamma g_0$ and shear stress $\sigma_{zx}|_{z=0} = 0$.

(ii) At solid-solid boundaries: $\sigma_{zz}|_-^+ = \sigma_{zx}|_-^+ = 0$, and $[w]_-^+ = [v]_-^+ = 0$, where v is the horizontal displacement.

(iii) At the CMB, $[\sigma_{zz}]^+ = \rho_f g_0 w$, where ρ_f is the density at the top of the fluid core and w is the radial displacement at the elastic-fluid boundary. In addition, shear stress vanishes at the CMB. Note that for ice loads with characteristic length comparable to or less than the Fennoscandian ice sheet, the CMB is not 'felt' by the deformation (Wu & Peltier, 1982), so this boundary condition is not important.

Spatio-temporal evolution of GIA stress can be computed numerically for some simple Earth and ice models based on their analytical solutions in the transformed domain (e.g. Johnston et al., 1998; Klemann & Wolf, 1999). However, for more complicated ice histories or when there is lateral heterogeneity in material properties or non-linear mantle rheology, it is more convenient to use the FE method, which will be described briefly next.

22.3 Modelling Glacially Induced Stress Changes

The FE method can consider strong lateral variation of material properties in the Earth. Unlike the spectral method, which works well only if mantle creep is linear, the FE method can also be used to model non-linear creep flow inside the mantle. However, for the FE method to work well, the spatial scale of the elements must be small enough so that the lateral change in displacement or stress within an element can be properly represented. Otherwise, the gradient in stress and displacements cannot be resolved. Thus, the computation normally involves FE grids with fine spatial-resolution, and this increases the computational memory, time and cost.

There are several commercially available FE packages; however, most of them only solve the equation:

$$0 = \vec{\nabla} \cdot \bar{\bar{\sigma}}. \tag{22.8}$$

Comparing equation (22.8) with equation (22.4) shows that the important contributions of the advection of pre-stress, the perturbed gravity field and the buoyancy force of local density perturbation are not included in commercial FE packages. Wu (2004) shows that by using the stress transformation:

$$\bar{\bar{\sigma}}^V = \bar{\bar{\sigma}} - \rho_0 g_0 w \bar{\bar{I}}, \tag{22.9}$$

equation (22.7) becomes $0 = \vec{\nabla} \cdot \bar{\bar{\sigma}}^V$, which has the same form as equation (22.8). This means that commercial FE packages can be used to model the GIA deformation of a non-self-gravitating incompressible flat Earth. With (22.9), the modified boundary conditions now become:

(i) At the surface of the Earth ($z = 0$): normal stress $\left[\sigma^V_{zz} + \rho_0 g_0 w\right]_{z=0} = -\gamma g_0$ and shear stress $\sigma^V_{zx}\big|_{z=0} = 0$.
(ii) At solid-solid boundaries: $\sigma^V_{zz}\big|^+_- = (\rho_- - \rho_+) g_0 w$, $\sigma^V_{zx}\big|^+_- = 0$ and $[w]^+_- = [v]^+_- = 0$.
(iii) At the CMB (i.e. $z = -H$), $\left[\sigma^V_{zz}\right]_{z=-H+\epsilon} = (\rho_f - \rho_s) g_0 [w]_{z=-H}$, where ρ_s is the density at the bottom of the solid mantle at $z = -H + \epsilon$.

In FE models, the buoyancy terms in the above modified boundary conditions act in the upward direction and $\rho_0 g_0 w$, $(\rho_- - \rho_+) g_0 w$ or $(\rho_f - \rho_s) g_0 w$ can be simulated by Winkler foundations (see Wu, 2004) with spring constants set to be the vertical density contrast across the material interface multiplied by the vertical component of gravity (i.e. $\rho_0 g_0$, $(\rho_- - \rho_+) g_0$ and $(\rho_f - \rho_s) g_0$, respectively), provided that the material interfaces are perpendicular to the direction of gravity. For inclined material interfaces, elastic springs (see Schmidt et al., 2012) should be used. Also, the stress output of the FE method must be converted back to a physical stress through:

$$\bar{\bar{\sigma}} = \bar{\bar{\sigma}}_{FE} + \rho_0 g_0 w \bar{\bar{I}}. \tag{22.10}$$

The GIA stresses obtained from equation (22.10) can be combined with background stresses (e.g. tectonic stresses and overburden pressure) to analyze ΔCFS over time (Steffen et al., see Chapter 2). The inclusion of tectonic stress in the model is important because it can strongly affect the spatio-temporal variation of ΔCFS and the mode of failure outside the ice margin (Wu et al., 1999). The tensor of the total stress is then diagonalized to obtain the principal stresses and their principal orientations from which the normal and shear stress changes are obtained and applied in the estimation of ΔCFS for every time point during a glacial cycle.

A simple numerical example of glacially induced stresses for Type I models is shown in Figure 22.1 for a circular ice load on a flat-earth model, created using the FE software ABAQUS. The elements have a resolution of 50×50 km in the horizontal and decrease in size outside of a box with dimensions of $6{,}000 \times 6{,}000$ km. The model has a depth of 2,891 km and is divided into six material layers (two elastic and four viscoelastic layers; see Table 22.1). The ice model has a parabolic profile with thickness at location (x, y) and time t given by:

$$ice_{\text{height}}(x,y,t) = \begin{cases} \sqrt{1-\left|\dfrac{x}{r_x(t)}\right|}\sqrt{1-\left|\dfrac{y}{r_y(t)}\right|}\,ice_{\text{thickness}}(0,0,t), \text{ for } \left|\dfrac{x}{r_x(t)}\right| \leq 1 \text{ and } \left|\dfrac{y}{r_y(t)}\right| \leq 1 \\ 0, \qquad \text{for } \left|\dfrac{x}{r_x(t)}\right| > 1 \text{ and } \left|\dfrac{y}{r_y(t)}\right| > 1 \end{cases}, \tag{22.11}$$

where $r_x(t)$ and $r_y(t)$ are the radius of the ice sheet in the x and y directions at time t respectively, and $ice_{\text{thickness}}(0,0,t)$ the thickness of the ice at the center of the model ($x = 0, y = 0$). At glacial maximum, the ice load has a diameter of 2,000 km and maximum thickness of 2,000 m at the centre, but its size and height vary linearly with time t during loading and unloading. The glacial cycle has a length of 110 ka, where the loading phase has a duration of 100 ka and the unloading phase is 10 ka. In both phases, the ice thickness changes linearly over time. A thrust-faulting stress regime is applied as background stress with σ_1 (maximum principal stress) in the horizontal x-direction and σ_2 (intermediate principal stress) in the horizontal y-direction. The direction of the maximum horizontal stress S_H is in the x-direction. σ_3 (minimum principal stress) is in the vertical direction and has a value equal to the overburden pressure, $S_V = \rho g h$.

In order that the fault be critically stressed before the onset of glaciation, the maximum horizontal stress is set to $S_H = \left(\sqrt{\mu^2+1}+\mu\right)^2 S_V$ with the coefficient of friction $\mu = 0.6$ (Byerlee, 1978) and the minimum horizontal stress relates to the other stresses via the stress ratio R (Etchecopar et al., 1981; $S_h = (1-R)S_H + RS_V$), which is set to 0.5 in this example. The crust is assumed to be critically stressed, so that $CFS = 0$ MPa before glaciation, and positive ΔCFS is equivalent to unstable conditions, while a negative ΔCFS refers to stable condition. The ΔCFS is calculated at a depth of 5 km for optimally oriented faults, which is equivalent to a dip angle of $\sim 30°$ and a strike angle of $0°$ (measured from the positive y-direction).

ΔCFS for optimally oriented faults is shown for six different time points in Figure 22.1: 25 ka (during loading), 75 ka (during loading), 100 ka (maximum loading), 105 ka (during unloading), 110 ka (end of unloading) and 120 ka (10 ka after the end of the glacial cycle).

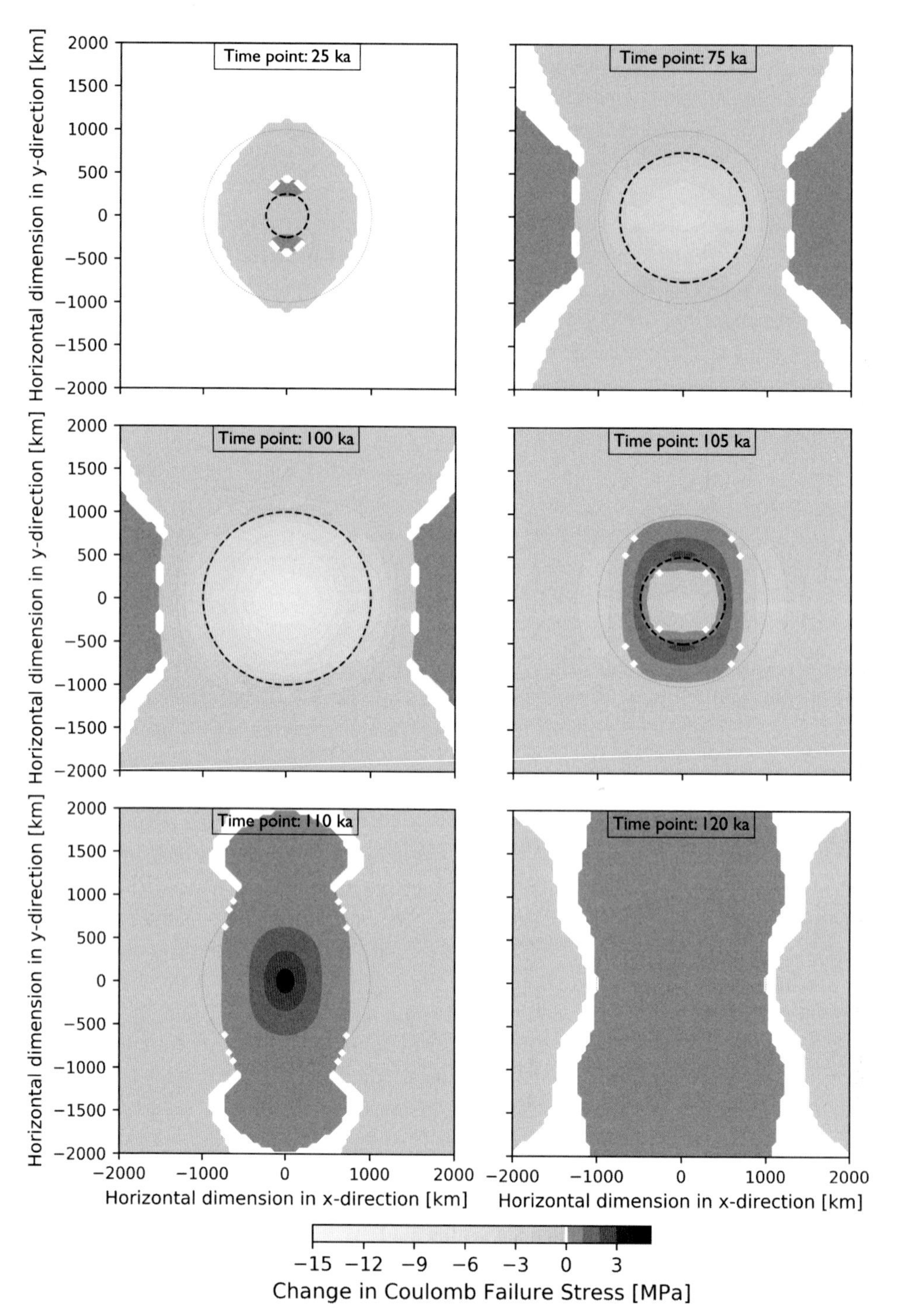

Figure 22.1 Results of Type I Model: Change in Coulomb failure stress (ΔCFS) for a circular ice sheet (diameter: 2,000 km, ice thickness: 2,000 m), which changes in size during

Table 22.1. *Parameters of the simple 1D model in Figure 22.1. Material values are volume-averaged from the Preliminary Reference Earth Model (PREM, Dziewonski & Anderson, 1981).*

Layer name	Layer thickness [km]	Density [kg/m^3]	Young's modulus [Pa]	Poisson's ratio	Viscosity [Pa·s]
Crust	30	3,094	$1.355 \cdot 10^{11}$	0.27	–
Lithosphere	60	3,376	$1.703 \cdot 10^{11}$	0.28	–
Upper mantle 1	320	3,498	$1.955 \cdot 10^{11}$	0.30	$5 \cdot 10^{20}$
Upper mantle 2	250	3,908	$2.849 \cdot 10^{11}$	0.30	$5 \cdot 10^{20}$
Lower mantle 1	1,140	4,798	$5.357 \cdot 10^{11}$	0.28	$2 \cdot 10^{21}$
Lower mantle 2	1,091	5,341	$6960 \cdot 10^{11}$	0.30	$2 \cdot 10^{21}$

The area with positive ΔCFS values (Figure 22.1, dark grey colour = unstable) is located outside of the ice margin during the loading phase.

When loading increases, this unstable area aligns with the stress direction of the maximum horizontal stress (x-direction), while the area beneath the ice sheet becomes negative (Figure 22.1, light grey colour = stable). The stable area increases in size and is larger than the size of the ice sheet until glacial maximum. However, as soon as the ice melts, the area that becomes ice-free shows instability reaching values of up to 5 MPa. In contrast to the alignment of the unstable zone with the maximum horizontal stress direction during loading, the unstable area during unloading and afterwards follows the direction of the minimum horizontal stress (y-direction) and increases in size over the former ice margin. The amplitude of instability reaches its maximum value during the end of unloading beneath the former ice sheet maximum. Additionally, 10 ka after the end of the glacial cycle, the unstable area has increased in size and formerly stable zones outside the former ice margin at glacial maximum become unstable. The $S_{H\max}$ orientation changes during the glacial cycle by about 3°, which results in a variation of the orientation of the optimally oriented fault.

A second numerical example of Type I model shows ΔCFS for Northern Europe when a more realistic ice history model is used. A flat-earth model is created using the FE software ABAQUS with an element resolution of 50 × 50 km in the inner part, and which has a

Figure 22.1 (*cont.*) the glacial cycle (maximum extension is shown by the light grey dashed line, current time-point extension is shown by the black dashed line). The loading phase is 100 ka long and the unloading phase 10 ka. 25 ka and 75 ka show a time point during loading, 100 ka is at maximum loading, 105 ka is during unloading, 110 ka is at the end of unloading and 120 ka is 10 ka after the end of the glacial cycle. The thrust-faulting background stress field has $S_{H\max}$ oriented in the x-direction. Positive values of ΔCFS (dark grey to black colour) refer to unstable condition.

dimension of 4,500 × 4,500 km. The rest of the 60,000 × 60,000-km box is meshed with increased horizontal element dimensions. Like the previous model, this also has a depth of 2,891 km and is divided into two elastic layers and four viscoelastic layers; 18 element layers are used, which results in nearly 218,000 elements. The lithospheric thickness varies laterally and is taken from Wang and Wu (2006). The viscosities of the upper and lower mantle are the same as in the previous example (see Table 22.1). Also, the realistic ice model ICE-6G_C (Peltier et al., 2015) is applied. It can be shown that this GIA model is able to provide a reasonably good fit between GIA observations and predictions (e.g. Huang et al., 2019).

The background stress is prescribed with a thrust-faulting stress regime with a NNW-SSE-orientated maximum horizontal stress (angle of N 150° E) and the background stresses are calculated using the same parameters as in the first example (see above). The obtained ΔCFS is negative (i.e. the faults are stable) for the entire area during glaciation but reaches positive values and therefore unstable conditions soon after ice melting started at 22 ka before present (Figure 22.2). The first areas that become unstable are at the margin of the former ice sheet or even further away in the peripheral bulge, but this area grows in time. The maximum positive changes in ΔCFS are found in northern Fennoscandia where the ice thickness was the highest and where the prominent glacially induced faults are found

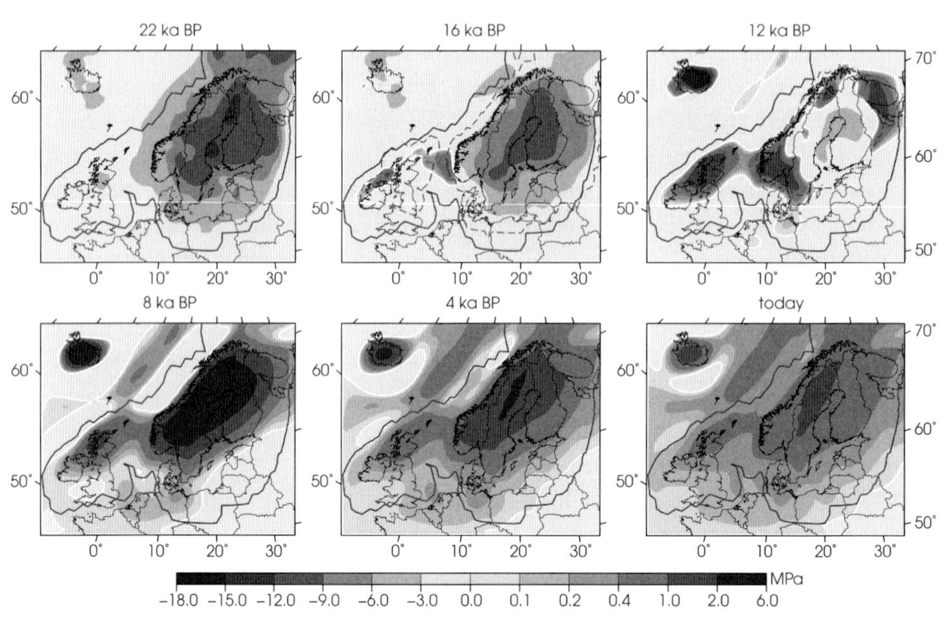

Figure 22.2 Results of Type I Model. Change in Coulomb failure stress (ΔCFS in MPa) at 2.5 km depth for Northern Europe from Last Glacial Maximum until today divided into six time points (in thousand years before present, ka BP). The green solid line shows the maximum ice margin at Last Glacial Maximum and the green dashed line is the ice margin at the respective time shown on top of each subfigure. ICE-6G_C (Peltier et al., 2015) is used as ice model in this numerical simulation. Note that positive ΔCFS refers to unstable condition. Also, the colour scale is not linear for positive values. (A black and white version of this figure will appear in some formats. For the colour version, please refer to the plate section.)

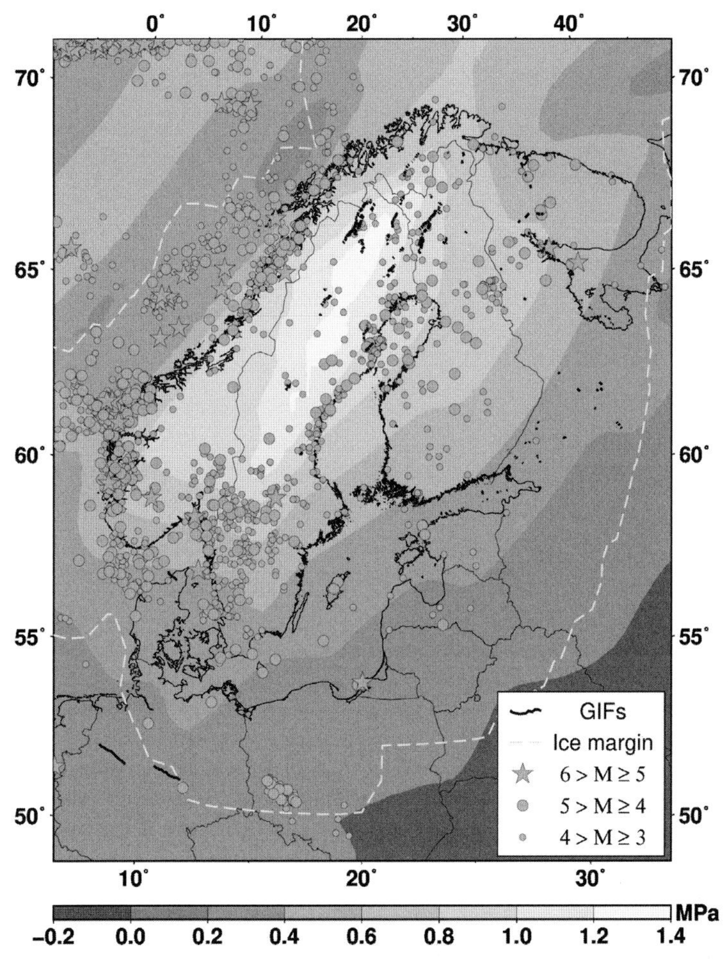

Figure 22.3 Results of Type I Model. Today's change in Coulomb failure stress (ΔCFS in MPa) at 2.5 km depth for Northern Europe (same as Figure 22.2 lower right) in comparison to historic and recent seismicity above magnitude 3 and the location of glacially induced faults (GIFs). Seismicity (stars and circles) taken from FENCAT (2020). GIFs (black lines) according to Munier et al. (2020). The light grey dashed line shows the maximum ice margin at Last Glacial Maximum from model ICE-6G_C (Peltier et al., 2015).

(Figure 22.3). Figures 22.2 and 22.3 also show that not only does the area that was formerly covered by the ice sheet become unstable but so also does the surrounding area, which remains in unstable mode even up to today. This supports several indications for glacially triggered faulting that have been found in this area (see Chapters 11–18). Comparison of ΔCFS to historic and recent seismicity (Figure 22.3), however, does not show a distinct correlation that would imply a GIA influence on today's seismicity. Wu et al. (1999) postulated that the current spatial distribution of seismicity may be largely explained by: (i) coastal weakening that results in the clustering of seismicity along the land-water boundary; (ii) stress release by the reactivation of large faults (e.g. Pärvie Fault) early in

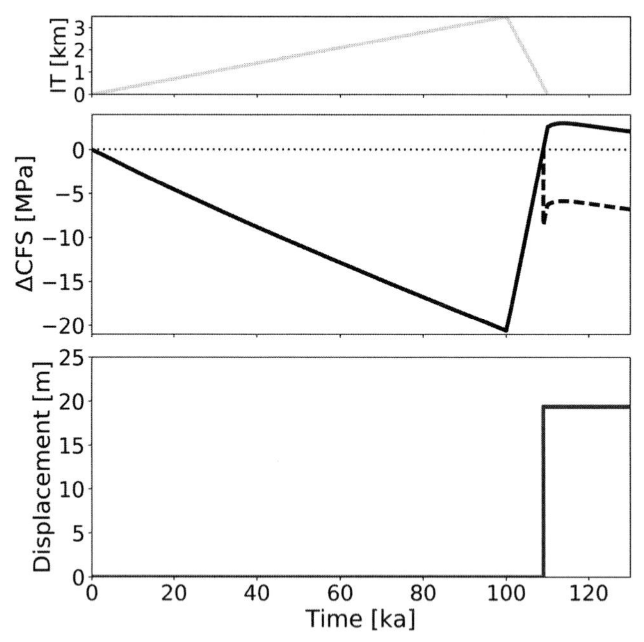

Figure 22.4 Fault slip obtained from a numerical simulation using the Type II model in a thrust-faulting stress field. The upper panel shows the ice-thickness (IT) variation, the central panel shows the change in Coulomb failure stress ΔCFS during the glacial cycle for a point at 4-km depth beneath the ice-sheet centre (solid: ΔCFS without earthquake, dashed: change in ΔCFS due to an earthquake), and lower panel shows the fault slip over time. The fault is reactivated only once at 109 ka, which leads to a change in ΔCFS afterwards (difference between the solid and dashed curve in the central panel).

the deglaciation stage (see the dashed line in the middle panel of Figure 22.4) that results in low seismicity in regions with the largest value of ΔCFS today. However, the low seismicity near the centre of rebound may exist because the faults there were more stable than critically stressed before the onset of glaciation, or it may be that they are not optimally oriented, so that extra stress (more than 1.5 MPa) is needed to bring the pre-existing faults to failure.

22.4 Modelling Glacially Induced Fault Reactivation with the Type II Model

While the previous sections focused on the usage of the commercial FE method for GIA purposes, the applicability of ABAQUS for fault modelling has been shown by several authors (e.g. Ellis et al., 2006; Hergert & Heidbach, 2010). Consequently, it is desirable to combine the well-known fault modelling approaches with the benchmarked GIA-modelling technique described above. However, the problem is that if we directly insert a free fault into the GIA model, then the new boundary condition will cause the stresses near the inserted fault surface to change. In addition, for the computation of the total stress, the stress output from the FE model needs to be recomputed according to equation (22.10) to give the full GIA stress before it can be added to the background stress. Thus, the fault structure cannot be implemented directly in the GIA model, but additional FE models have to be used with the modified GIA stresses. This can be solved by developing a three-stage FE model approach to overcome the difficulties of using a fault structure in the correct GIA stress settings (see Steffen et al., 2014a).

The three-stage FE model approach involves the usage of a GIA model (the first model as described in Section 22.3), another FE model to obtain reaction forces (second model) and a third FE model being the fault model. The software ABAQUS is used for all three models. The GIA model includes the lithosphere and mantle to a depth of 2,891 km. The modelled GIA displacements can be fitted to GIA observations allowing the usage of realistic model structures (lithospheric depth, viscosity structure of the mantle, ice model, etc.). The stresses obtained from this first model are then converted to GIA stresses according to equation (22.10). The second FE model uses only the lithosphere now, but also applies the total stresses as well as the displacements from the first FE model for a specific time step, and it has a fault structure included. The total stresses together with the displacements are not in equilibrium, as additional stresses were added. Therefore, the second FE model alone is not stable and would not reach equilibrium conditions. To obtain a stable model setup, all degrees of freedom are fixed in all directions and reaction forces are determined for each degree of freedom for the nodes in a static analysis from this model using ABAQUS. The fault structure is also tied, allowing no movement. The displacement and stress fields are not changed, since the additional reaction forces hold the model in equilibrium. Those reaction forces are then applied together with the displacement and stress field in the third FE model (the fault model) and the fault structure is untied and therefore is free to move. This results in a static fault movement that lasts for only a few seconds and simulates the occurrence of a glacially triggered earthquake. The displacements and stresses from the third model can now be used to analyze the fault slip behaviour. The second and third FE model are created only for the time points when the analysis of ΔCFS (obtained from the first model) has shown an instability. In addition, when an earthquake occurs, the changed stress and displacement conditions are applied to the stress analysis from the first FE model and ΔCFS is evaluated for the subsequent time steps with the already occurred earthquake in mind. However, the GIA model itself (first model) is not changed. A new set of second and third models would only be required if ΔCFS becomes positive again.

The advantage of using the three-stage FE model approach is six-fold: 1) results from any GIA model can be used and a fault structure is not required in the first GIA model; 2) as background stresses are not used in the first GIA model but rather in the second and third FE model only, different background stress conditions can be tested without the need to run a computationally time-consuming GIA model again; 3) the nonexistence of a fault structure in the GIA model also allows the test of different fault configurations in the third FE model without the need to run a new GIA model again; 4) the fault is released in a stress field, which uses the GIA stresses and boundary conditions from the GIA equations; 5) the applied background stresses are implemented as an additional, time-independent stress field, which is also suitable for formerly glaciated regions; 6) separate events (earthquakes) are obtained as soon as ΔCFS is found to be positive (unstable conditions) within a specific region.

A numerical example of the Type II model is shown in Figure 22.4, where a GIA model is loaded with a 3,000-km wide and 3,500-m thick ice sheet. The glacial cycle has a duration of 110 ka, with a linear ice accumulation phase of 100 ka and a linear melting phase of 10 ka (Figure 22.4). The fault is located beneath the centre of the ice sheet, has a dip of 45° and is subject to a thrust-faulting stress regime. The corresponding ΔCFS shows a decrease during loading followed by an increase during unloading. Just before the end of the melting phase at 109 ka, ΔCFS becomes positive and an earthquake with a vertical displacement of 19.4 m is created. This large displacement is due to the large ice cap and the large tectonic background stresses for a 45° dipping fault (see Steffen et al., 2014a, for details). Due to the slip event, ΔCFS changes indicate that the fault has become stable again (dashed line). No additional earthquake is induced afterwards. If no earthquake were to occur at 109 ka, stresses would increase even further, leading to more unstable conditions (solid line), before they decrease again after 114 ka.

22.5 Discussion

In this paper, the spatio-temporal variation of GIA-induced stress is computed using the flat-earth model. The flat-earth approximation is valid for ice loads with size comparable to or smaller than the Fennoscandian ice sheet. A question is whether the flat-earth approximation also works well for larger ice loads. Amelung and Wolf (1994) showed that the neglect of sphericity is partially compensated by ignoring self-gravitation, so that the deformation of a spherical, self-gravitating Earth within the ice margin can be reasonably approximated by flat-earth models even for ice loads as large as the Laurentide ice sheet. This was confirmed by Wu and Johnston (1998).

In addition, Wu (2004) has shown that the deformation in a spherical, self-gravitating incompressible Earth can be computed by: (i) applying the transformation of equation (22.9) to equation (22.4) and the boundary conditions listed in Section 22.2, and (ii) coupling these with Poisson's equation (22.6), where $\rho_1 = 0$, and solving them iteratively. Even the sea level equation (Farrell & Clark, 1976; Milne & Mitrovica, 1998) can be solved this way using an iterative procedure. This method of Wu (2004) is also called the iterative stress transform (IST) method. The computed deformation and stress of the IST method have been benchmarked with the spectral normal method (Johnston et al., 1998; Wong & Wu, 2019), and excellent agreement between the two methods has been found.

For a compressible Earth, the stress transformation in equation (22.9) no longer works (Bängtsson & Lund, 2008; Wong & Wu, 2019). In that case, the iterative body force (IBF) method (Wong & Wu, 2019) has to be used. This means that all the body forces in equation (22.4) within each element are replaced by their volumetric average and the governing equations of motion (22.4), (22.5) and (22.6) are solved iteratively. However, the IBF method is much more computationally intensive, so an efficient way of obtaining the approximate solution is by using the IST approach with the inclusion of 'material compressibility' (Klemann et al., 2003) – i.e. using the values of λ and μ for a compressible Earth in the first term of equation (22.4) but ignoring the last term of equation (22.4). The difference between a compressible Earth and a 'material compressible' Earth has been shown to be small for relative sea levels and current uplift rates (see supporting information in Li et al., 2020). Actually, Wu et al. (1999) have compared fault instability predictions in Fennoscandia, computed with incompressible and compressible Earths. They found that the main effect of incompressibility is on the amplitude of fault stability in and around Fennoscandia, but it has little effect on the onset time of instability or the mode of failure within the LGM ice margin. However, the effect of incompressibility may impact the onset time of instability in the British Isles and Iceland offshore. Using a compressible, self-gravitating spherical earth, Wu and Johnston (2000) showed that GIA is able to trigger postglacial earthquakes both inside and outside the former ice margin in North America.

Wu et al. (1999) have also studied the effects of (i) mantle viscosity, (ii) lithospheric thickness, (iii) tectonic stress and (iv) ice model on fault instability predictions. They found that the largest effect is due to the ice model, which strongly influences the onset time of fault instability. For example, for isolated, small ice caps with horizontal dimension close to about 12 times the lithospheric thickness, the bending stress can be greatly magnified when compared to the vertical stress so that fault instability can be induced even under the ice load - thus strongly affecting the onset timing (Johnston et al., 1998). Tectonic stress also strongly affects the mode of failure outside the ice margin, although its effect within the ice margin is small. On the other hand, the effects of mantle viscosity are found to be small on the onset timing and have no effect on the mode of failure. Wu and Mazzotti (2007) also investigated the impact of a laterally heterogeneous lithospheric weak zone and found that the effects on fault stability are significant; however, the effect on the onset time inside the ice margin is small.

As for the Type II model, numerical tests of different Earth and ice model configurations showed that fault slip magnitude depends mainly on ice thickness and crustal as well as lithospheric thickness; ice width affects reactivation time (Johnston et al., 1998; Steffen et al., 2014b). The viscosity of the mantle, however, has an effect only on the fault slip, depending on the location of the fault with respect to the ice sheet (beneath the ice sheet centre or beneath the ice closer to the margin). In addition, Steffen et al. (2014b,c) showed that it is possible to reactivate 45° and 60° dipping faults. However, the 60° dipping fault was found to slip twice, which is supported by observations (Smith et al., 2018). A fault slip was only obtained for nearly critically stressed conditions (ΔCFS before glaciation between 0 and –2 MPa, Steffen et al., 2014c). The fault slips obtained are consistent with the observations of surface offsets up to 10 to 20 m.

Finally, it should be noted that the Type II model described in Section 22.4 treated the background stresses as time-independent inputs – since the time scales of tectonics is on the order of millions of years, whereas that of the GIA process is on the order of only a thousand years. So, the tectonic stresses keep the faults critically stressed, while the fault slips are triggered entirely by the time-dependent GIA stress alone. Also, because the stresses have to exceed the Mohr-Coulomb criterion before the faults are reactivated, faults do not slip continuously during a glacial cycle but are confined to certain onset time interval(s), and the fault slips occur almost instantaneously as is observed. On the other hand, if the time variation of tectonics stress or strain rate overlaps with that for GIA, so that these two processes interact significantly with each other, then another model type needs to be developed. However, in that model type, the strain rates must exceed a certain criterion (to be determined) before faults are allowed to slip; otherwise continuous fault slips with unrealistic slip rates before glaciation might be predicted.

22.6 Conclusion

The basic equations and boundary conditions for GIA modelling on a flat Earth are reviewed. The iterative stress transform method of Wu (2004), see equations (22.9) and (22.10), can be used to compute efficiently the spatial-temporal variation of GIA-induced stress and Coulomb failure stress for the Type I model, where virtual faults are assumed. This is illustrated by numerical examples using a simple circular ice model and the more realistic ice model ICE-6G_C for Northern Europe and Iceland. Clearly, application of the Type I model is not limited to Northern Europe, but it can be used to study glacially induced stresses on fault stability in North America (Wu & Hasegawa, 1996a,b; Wu & Johnston, 2000), Antarctica (Kaufmann et al., 2005), Greenland (Steffen et al., 2020), Alaska (Sauber et al., 2000) and other glaciated areas.

To model GIA-induced fault slip with a more realistic fault surface, the Type II model has to be used. It contains a three-stage FE model approach and its predicted fault motion is illustrated with a 2D Type II model that includes a single fault. Clearly it is desirable to (i) extend this to a 3D Type II model and (ii) study the reactivation of a system of faults and their interaction by GIA-induced stress in future work.

Acknowledgements

We are indebted to Volker Klemann and an anonymous reviewer as well as editor Odleiv Olesen for their constructive comments. The FE calculations were performed with the ABAQUS package from Hibbitt, Karlsson and Sorensen Inc.

References

Amelung, F. and Wolf, D. (1994). Viscoelastic perturbations of the earth: significance of the incremental gravitational force in models of glacial isostasy. *Geophysical Journal International*, **117**, 864–879, doi.org/10.1111/j.1365-246X.1994.tb02476.x.

Byerlee, J. D. (1978). Friction of rock. *Pure and Applied Geophysics*, **116**, 615–626, doi .org/10.1007/BF00876528.

Bängtsson, E. and Lund, B. (2008). A comparison between two solution techniques to solve the equations of glacially induced deformation of an elastic earth. *International Journal for Numerical Methods in Engineering*, **75**(4), 479–502, doi.org/10.1002/nme.2268.

Cathles, L. M. III (1975). *The Viscosity of the Earth's Mantle*. Princeton University Press, Princeton.

Chinnery, M. A. (1975). The static deformation of an Earth with a fluid core: a physical approach. *Geophysical Journal of the Royal Astronomical Society*, **42**, 461–475, doi .org/10.1111/j.1365-246X.1975.tb05872.x.

Dziewonski, A. M. and Anderson D. L. (1981). Preliminary reference Earth model. *Physics of the Earth and the Planetary Interiors*, **25**, 297–356, doi.org/10.1016/0031-9201(81)90046-7.

Ellis, S., Beavan, J., Eberhart-Phillips, D. and Stöckhert, B. (2006). Simplified models of the Alpine Fault seismic cycle: stress transfer in the mid-crust. *Geophysical Journal International*, **166**, 386–402, doi.org/10.1111/j.1365-246X.2006.02917.x.

Etchecopar, A., Vasseur, G. and Daignieres, M. (1981). An inverse problem in microtectonics for the determination of stress tensors from fault striation analysis. *Journal of Structural Geology*, **3**(1), 51–65, doi.org/10.1016/0191-8141(81)90056-0.

Farrell, W. E. (1972). Deformation of the earth by surface loads. *Reviews of Geophysics* **10**, 761–797, doi.org/10.1029/RG010i003p00761.

Farrell, W. E. and Clark, J. A. (1976). On postglacial sea level. *Geophysical Journal of the Royal Astronomical Society*, **46**, 647–667, doi.org/10.1111/j.1365-246X.1976.tb01252.x.

FENCAT (2020). Fennoscandian earthquake catalogue for 1375–2014 (available at www.seismo.helsinki.fi/bulletin/list/catalog/FENCAT.html).

Hergert, T. and Heidbach, O. (2010). Slip-rate variability and distributed deformation in the Marmara Sea fault system. *Nature Geoscience*, **3**, 132–135, doi.org/10.1038/NGEO739.

Hetzel, R. and Hampel, A. (2005). Slip rate variations on normal faults during glacial-interglacial changes in surface loads. *Nature*, **435**, 81–84, doi.org/10.1038/nature03562.

Huang, P., Wu, P. and Steffen, H. (2019). In search of an ice history that is consistent with composite rheology in Glacial Isostatic Adjustment modelling. *Earth and Planetary Science Letters*, **517**, 26–37, doi.org/10.1016/j.epsl.2019.04.011.

Ivins, E. R., James, T. S. and Klemann, V. (2003). Glacial isostatic stress shadowing by the Antarctic ice sheet. *Journal of Geophysical Research*, **108**(B12), 2560, doi.org/10.1029/2002JB002182.

James, T. S. and Bent, A. L. (1994). A comparison of eastern North American seismic strain-rates to glacial rebound strain-rates. *Geophysical Research Letters*, **21**, 2127–2130, doi.org/10.1029/94GL01854.

Johnston, A. C. (1987). Suppression of earthquakes by large continental ice sheets. *Nature*, **330**, 467–469, doi.org/10.1038/330467a0.

Johnston, P., Wu, P. and Lambeck, K. (1998). Dependence of horizontal stress magnitude on load dimension in glacial rebound models. *Geophysical Journal International*, **132**, 41–60, doi.org/10.1046/j.1365-246x.1998.00387.x.

Kaufmann, G., Wu, P. and Ivins, E. R. (2005). Lateral viscosity variations beneath Anatarctica and their implications on regional rebound motions and seismotectonics. *Journal of Geodynamics*, **39**, 165–181, doi.org/10.1016/j.jog.2004.08.009.

Klemann, V. and Wolf, D. (1999). Implications of a ductile crustal layer for the deformation caused by the Fennoscandian ice sheet. *Geophysical Journal International*, **139**, 216–226, doi.org/10.1046/j.1365-246X.1999.00936.x.

Klemann, V., Wu, P. and Wolf, D. (2003). Compressible viscoelasticity: stability of solutions for homogeneous plane earth models. *Geophysical Journal International*, **153**, 569–585, doi.org/10.1046/j.1365-246X.2003.01920.x.

Li, T., Wu, P., Wang, H. S. et al. (2020). Uncertainties of Glacial Isostatic Adjustment model predictions in North America associated with 3D structure. *Geophysical Research Letters*, **47**, e2020GL087944, doi.org/10.1029/2020GL087944.

Lund, B. (2005). *Effects of Deglaciation on the Crustal Stress Field and Implications for Endglacial Faulting: A Parametric Study of Simple Earth and Ice Models*. SKB

Technical Report TR-05-04, Swedish Nuclear Fuel and Waste Management Co., Stockholm, 68 pp.

Lund, B. (2006). *Stress Variations during a Glacial Cycle at 500 m Depth in Forsmark and Oskarshamn: Earth Model Effects*. SKB Report R-06-95, Swedish Nuclear Fuel and Waste Management Co., Stockholm, 38 pp.

Lund, B., Schmidt, P. and Hieronymus, C. (2009). *Stress Evolution and Fault Stability during the Weichselian Glacial Cycle*. SKB Technical Report TR-09-15, Swedish Nuclear Fuel and Waste Management Co., Stockholm, Sweden, 106 pp.

Milne, G. A. and Mitrovica, J. X. (1998). Postglacial sea-level change on a rotating Earth. *Geophysical Journal International*, **133**, 1–19, doi.org/10.1046/j.1365-246X.1998.1331455.x.

Munier, R., Adams, J., Brandes, C. et al. (2020). International Database of Glacially-Induced Faults. *PANGAEA*, doi.org/10.1594/PANGAEA.922705.

Peltier, W. R., Argus, D. F. and Drummond, R. (2015). Space geodesy constrains ice age terminal deglaciation: the global ICE-6G_C (VM5a) model. *Journal of Geophysical Research Solid Earth*, **120**(1), 450–487, doi.org/10.1002/2014JB011176.

Quinlan, G. (1984). Postglacial rebound and the focal mechanisms of eastern Canadian earthquakes. *Canadian Journal of Earth Sciences*, **21**, 1018–1023, doi.org/10.1139/e84-106.

Sauber, J., Plafker, G., Molnia, B. F. and Bryant, M. A. (2000). Crustal deformation associated with glacial fluctuations in the eastern Chugach Mountains, Alaska. *Journal of Geophysical Research Solid Earth*, **105**, 8055–8077, doi.org/10.1029/1999JB900433.

Schmidt, P., Lund, B. and Hieronymus, C. (2012). Implementation of the glacial rebound prestress advection correction in general-purpose finite element analysis soft-ware: springs versus foundations. *Computers & Geosciences*, **40**, 97–106, doi.org/10.1016/j.cageo.2011.07.017.

Smith, C. A., Grigull, S. and Mikko, H. (2018). Geomorphic evidence of multiple surface ruptures of the Merasjärvi "postglacial fault", northern Sweden. *GFF*, **140**(4), 318–322, doi.org/10.1080/11035897.2018.1492963.

Spada, G., Yuen, D. A., Sabadini, R. and Boschi, E. (1991). Lower-mantle viscosity constrained by seismicity around deglaciated regions. *Nature*, **351**, 53–55, doi.org/10.1038/351053a0.

Steffen, R., Wu, P., Steffen, H. and Eaton, D. W. (2014a). On the implementation of faults in finite-element glacial isostatic adjustment models. *Computers & Geosciences*, **62**, 150–159, doi.org/10.1016/j.cageo.2013.06.012.

Steffen, R., Wu, P., Steffen, H. and Eaton, D. W. (2014b). The effect of earth rheology and ice-sheet size on fault slip and magnitude of postglacial earthquakes. *Earth and Planetary Science Letters*, **388**, 71–80, doi.org/10.1016/j.epsl.2013.11.058.

Steffen, R., Steffen, H., Wu, P. and Eaton, D. W. (2014c). Stress and fault parameters affecting fault slip magnitude and activation time during a glacial cycle. *Tectonics*, **33**(7), 1461–1476, doi.org/10.1002/2013TC003450.

Steffen, R., Steffen, H., Wu, P. and Eaton, D. W. (2015). Reply to comment by Hampel et al. on "Stress and fault parameters affecting fault slip magnitude and activation time during a glacial cycle". *Tectonics*, **34**(11), 2359–2366, doi.org/10.1002/2015TC003992.

Steffen, H., Steffen, R. and Tarasov, L. (2019). Modelling of glacially-induced stress changes in Latvia, Lithuania and the Kaliningrad District of Russia. *Baltica*, **32**(1), 78–90, doi.org/10.5200/baltica.2019.1.7.

Steffen, R., Steffen, H., Weiss, R. et al. (2020). Early Holocene Greenland-ice mass loss likely triggered earthquakes and tsunami. *Earth and Planetary Science Letters*, **546**, 116443, doi.org/10.1016/j.epsl.2020.116443.

Stein, S., Sleep, N. H., Geller, R. J. et al. (1979). Earthquakes along the passive margin of Eastern Canada. *Geophysical Research Letters*, **6**, 537–540, doi.org/10.1029/GL006i007p00537.

Walcott, R. I. (1970). Isostatic response to loading of the crust in Canada. *Canadian Journal of Earth Sciences*, **7**, 716–727, doi.org/10.1139/e70-070.

Wang, H. and Wu, P. (2006). Effects of lateral variations in lithospheric thickness and mantle viscosity on glacially induced relative sea levels and long wavelength gravity field in a spherical, self-gravitating Maxwell Earth. *Earth Planet Science Letters*, **249**(3), 368–383, doi.org/10.1016/j.epsl.2006.07.011.

Wong, M. CK. and Wu, P. (2019). Using commercial finite-element packages for the study of Glacial Isostatic Adjustment on a compressible self-gravitating spherical earth – 1: harmonic loads. *Geophysical Journal International*, **217**, 1798–1820, doi.org/10.1093/gji/ggz108.

Wu, P. (1992a). Viscoelastic vs. viscous deformation and the advection of pre-stress. *Geophysical Journal International*, **108**, 35–51, doi.org/10.1111/j.1365-246X.1992.tb00844.x.

Wu, P. (1992b). Deformation of an incompressible viscoelastic flat earth with Power Law Creep: a Finite Element approach. *Geophysical Journal International*, **108**, 136–142, doi.org/10.1111/j.1365-246X.1992.tb00837.x.

Wu, P. (2004). Using commerical finite element packages for the study of earth deformations, sea levels and the state of stress. *Geophysical Journal International*, **158**, 401–408, doi.org/10.1111/j.1365-246X.2004.02338.x.

Wu, P. and Hasegawa, H. S. (1996a). Induced stresses and fault potential in Eastern Canada due to a disc load: a preliminary analysis. *Geophysical Journal International*, **125**, 415–430, doi.org/10.1111/j.1365-246X.1996.tb00008.x.

Wu, P. and Hasegawa, H. S. (1996b). Induced stresses and fault potential in Eastern Canada due to a realistic load: a preliminary analysis. *Geophysical Journal International*, **127**, 215–229, doi.org/10.1111/j.1365-246X.1996.tb01546.x.

Wu, P. and Johnston, P. (1998). Validity of using flat-earth finite element models in the study of postglacial rebound. In P. Wu, ed., *Dynamics of the Ice Age Earth: A Modern Perspective*. Trans Tech Publications, Switzerland, pp. 191–202.

Wu, P. and Johnston, P. (2000). Can deglaciation trigger earthquakes in N. America? *Geophysical Research Letters*, **27**, 1323–1326, doi.org/10.1029/1999GL011070.

Wu, P. and Mazzotti, S. (2007). Effects of a lithospheric weak zone on postglacial seismotectonics in Eastern Canada and Northeastern USA. In S. Stein and S. Mazzotti, eds., *Continental Intraplate Earthquakes: Science, Hazard and Policy Issues*. Geological Society of America, Special Paper, Vol. 425, pp. 113–128.

Wu, P. and Peltier, W. R. (1982). Viscous gravitational relaxation. *Geophysical Journal of the Royal Astronomical Society*, **70**, 435–486, doi.org/10.1111/j.1365-246X.1982.tb04976.x.

Wu, P., Johnston, P. and Lambeck, K. (1999). Postglacial rebound and fault instability in Fennoscandia. *Geophysical Journal International*, **139**, 657–670, doi.org/10.1046/j.1365-246x.1999.00963.x.

23

Crustal-Scale Stress Modelling to Investigate Glacially Triggered Faulting

SOFIE GRADMANN AND REBEKKA STEFFEN

ABSTRACT

Modelling of stresses that influence glacially triggered faulting has progressed substantially in the last decades with more complex models and improved modelling techniques, incorporating a variety of relevant processes, better constraints of the ice loading history, higher model resolution and three-dimensional geometries. A number of recent developments are collected in this chapter to portray the scope and variability of numerical modelling relevant to glacially triggered faulting. These range from modelling of the general in situ stress field to studies on the stress field induced by glacial loading and unloading.

An appropriate estimation of the ambient background stress field is crucial for determining the effect of additional ice loading (or unloading) on pre-stressed faults. Contributions from local and far-field stress sources (topography, tectonics) need to be reconciled with in situ measurements from boreholes and fault-plane solutions from earthquakes. We describe the different types of stresses in glaciated regions but focus on Scandinavia together with techniques used to incorporate stresses into numerical models.

23.1 Introduction

The loading and unloading of ice masses changes the existing stress field both in areas covered by ice and in their surroundings. These stresses, named 'glacially induced stresses', have been considered the trigger of major large earthquakes that occurred during and after the ice cap melting. Stress modelling and quantification is important for estimating and understanding the occurrence of these earthquakes. Glacially induced stresses add to the existing in situ stress field (here referred to as background stress), which again comprises several components. Knowing about the different stress field contributions and how they interact is essential to understanding overall stress development during a glacial cycle (Steffen et al., see Chapter 2).

23.2 Stress Field Modelling

Modelling Earth's in situ stress field has been done at various scales – from global tectonic stress models to stresses around single faults. The advance of computational methods and availability of global datasets has widely evolved and improved the stress modelling approaches. Here, we primarily discuss modelling of local-to-regional stress fields with emphasis on Scandinavia.

The first regional stress models made use of a two-dimensional thin-sheet approximation (e.g. Grünthal & Stromeyer, 1992; Jarosiński et al., 2006, and references therein), which is appropriate to capture the main tectonic stresses and requires significantly less computational power than three-dimensional (3D) stress models. The latter have only been developed more recently and focus on somewhat smaller scales (e.g. Lund et al., 2009; Hergert & Heidbach, 2011; Heidbach et al., 2014; Reiter & Heidbach, 2014; Hergert et al., 2015; Gradmann et al., 2018). Yet these different studies use widely varying approaches to implement the various stress field components.

Stress fields can be modelled with different techniques. Spectral codes were used in the past to model glacially induced stresses (Klemann & Wolf, 1998), but the most common method is finite-element modelling. Finite-element methods have several advantages: they include 3D variations of Earth's structure and they use non-linear rheologies. Observations that constrain the numerical models need to be used. Here, in situ measurements of present-day stress orientations as compiled in the World Stress Map database (Zoback, 1992; Heidbach et al., 2018) are primary information. Absolute stress magnitudes are seldomly measured. Fault-plane solutions of earthquakes also provide information about the stress regime (Steffen et al., 2012), in particular at larger depths of several kilometres. Secondary information from land uplift (for example from the Global Navigation Satellite System (GNSS) and interferometric synthetic aperture radar (InSAR), fault offsets, sea-level indicators) as well as changes in the Earth's gravity field are used to constrain the modelling of glacial isostatic adjustment (GIA) and the GIA-induced stresses (e.g. Steffen & Wu, 2011).

23.3 Stress Field Components

The Earth's stress field can be considered as the sum of different components, which in turn are often classified as long-wavelength (regional) and short-wavelength (local) components (Figure 23.1). An exhaustive list may not exist, but the commonly discussed components are as follows. The long-wavelength components are the (1) overburden stress, (2) tectonic stresses and, with much smaller magnitude, stresses induced by other (3) geodynamic motions and (4) large-scale thermal effects. The shorter-wavelength components can be associated with (5) topographic stresses and topology of other density contrast interfaces such as the Moho, (6) loading/unloading of ice or rock masses, (7) changes in water content or (8) lateral changes of lithology including faults. A detailed review is provided in Zoback and Zoback (2015).

The choice of modelling depends on the type of stress and its source and development over time. Large and regional lithospheric stresses have been created over several millions of years and can be calculated with various methods numerically and analytically. In contrast, stresses due to loading/unloading of ice masses – glacially induced stresses – depend on the history of the ice sheets and are modelled on 1- to 1,000-year time scale.

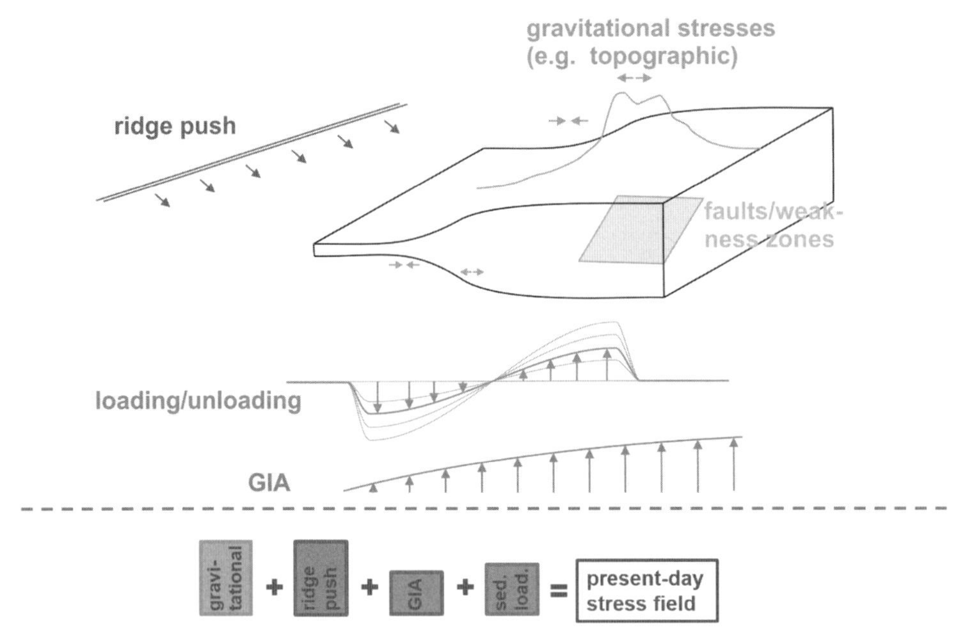

Figure 23.1 Schematic presentation of the stress field components contributing to the present-day stress regime on the Norwegian margin. From Gradmann et al. (2018). (A black and white version of this figure will appear in some formats. For the colour version, please refer to the plate section.)

A compilation of data of the Earth's in situ stress field is shown in Figure 23.2. Here, the ratio of mean horizontal stress to vertical stress ($k = S_{H\text{mean}}/S_V$) from worldwide measurements related to mining and civil engineering projects is shown. The stress ratio k is indicative of the stress regime ($k \gg 1$ for compressional stress regimes, $k \ll 1$ for tensional regimes, $k \approx 1$ for lithostatic or strike-slip regimes). Whereas worldwide stress measurements show that the vertical stress generally corresponds to the weight of the overburden, the horizontal stress exhibits a wide range of values in the upper 5 km. Very high near-surface horizontal stresses have been documented for Scandinavia (ranging up to 10 MPa) as well as Canada (Hast, 1958; Myrvang, 1993; Stephansson, 1993; Bell & Wu, 1997; Pascal et al., 2010). At larger depths the stress field becomes close to lithostatic, meaning vertical and horizontal stresses are equal.

23.3.1 Background Stress

For studies that are particularly interested in a single stress field component (such as models of glacially triggered faulting, GTF), it is often useful to calculate all other stress field components in a comprehensive, ambient state of stress, here termed 'background stress'. This approach has been used for GTF models by, for example, Lund (2005), Lund et al. (2009) and Steffen et al. (2019). The vertical stress is easily calculated as the overburden load ($S_V = \rho g z$) and the horizontal stress is related through $S_H = kS_V$. A common concept is that the Earth's crust is critically stressed, thus everywhere at failure. Evidence for this can be found independently in (1) the widespread seismicity induced by fluid injection or

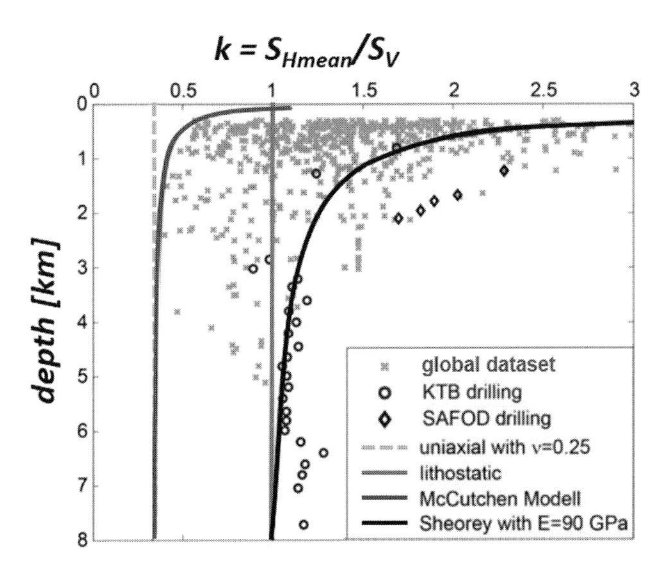

Figure 23.2 Global compilation of k-values from >300 m depth from the global database of Brown & Hoek (1978) with additional data from the KTB site (Kontinentale Tiefbohrprogramm der Bundesrepublik Deutschland; Brudy et al., 1997) and the SAFOD pilot hole (San Andreas Fault Observatory at Depth; Hickman & Zoback, 2004). Different model-of-reference stress states are plotted. Figure after Gritto et al. (2014) and Hergert & Heidbach (2011, their Figure 2).

reservoir impoundment, (2) earthquakes triggered by other earthquakes and (3) in situ stress measurements in deep boreholes (see Townend & Zoback, 2000, for references). The stress ratio k for minimum and maximum stresses can then be calculated using Coulomb frictional-failure theory, defining a pore-fluid pressure regime (e.g. hydrostatic) and assuming that the vertical stress is a principal stress. This approach yields stress magnitudes consistent with observations from deep boreholes (Townend & Zoback, 2000). In addition to the stress magnitudes, the orientation of maximum and minimum horizontal stress can be determined as a tectonic component from the local plate motions and ridge push direction in the case of Scandinavia (Lund et al., 2009). It is important to calibrate this background stress field with data from nearby deep boreholes to correctly estimate the overall stress field. This approach nevertheless yields a stress ratio k that is constant with depth.

When the stress field is to be calculated in more detail, for example, when the contribution of tectonic forces is of interest or the focus lies on the upper 2 or 3 km of the crust, the contributing stress field components can also be calculated separately.

23.3.2 The Overburden Stress

The stress field in the absence of tectonic processes, topographic influence or man-made stresses may be referred to as 'reference stress field' (e.g. Hergert & Heidbach, 2011), 'initial stress state' (e.g. Buchmann & Conolly, 2007) or 'overburden stress'. We here use the third term to indicate that it is purely caused by the weight of the overburden. A simple, synthetic overburden stress is often calculated by taking the overburden load as the vertical stress and calculating the horizontal stress by a pre-defined k-ratio, often $k = 1$, to describe a lithostatic stress state (e.g. Wu & Hasegawa, 1996a,b; Steffen et al., 2019). Yet, data from tectonically quiet areas (e.g. the drill site of the Kontinentale Tiefbohrprogramm der Bundesrepublik Deutschland (KTB); Brudy et al., 1997, Figure 23.2) commonly show relatively high horizontal stresses and a depth-dependency of the stress ratio k, seen especially in the upper 3 km. This is not described by the simple overburden load nor by the Coulomb frictional-failure theory described above. It has been a topic of several studies to explain these relatively high near-surface horizontal stresses.

McCutchen (1982) showed that including the sphericity in the model of a self-gravitating, elastic Earth leads to horizontal stresses exceeding the vertical ones (cf. Figure 23.2). But their analysis could not explain the high amplitude of the near-surface horizontal stresses, for example up to 10 MPa in Scandinavia (Stephansson, 1993). The linear-elastic McCutchen model was extended by Sheorey (1994) to include the thermal properties of the lithosphere (cf. Figure 23.2). The resulting formula $k = 0.25 + 7E(0.001 + 1/z)$, with E being the Young's modulus in GPa and z being the depth in metres, fits many observed data convincingly well at medium-to-larger depth.

To find a more realistic overburden stress, some authors embedded their smaller-scale model into a larger box with tilted sides to represent the Earth's spherical shell (Hergert & Heidbach, 2011; Gritto et al., 2014; Gradmann et al., 2018). Other authors have made use of the formula by Sheorey (1994) or McCutchen (1982) and calculated a respective 'effective Poisson's ratio' $\nu_k = \frac{k}{1+k}$ and directly imposed a respective overburden stress field (Buchmann & Connolly, 2007; Gradmann et al., 2018). Imposing the overburden stress field on the model domain (often called 'pre-stressing', Buchmann & Conolly, 2007) is a first step to establishing the initial stress state before adding additional stress field components. The latter are commonly added in the form of boundary conditions.

23.3.3 Tectonic Stress

Tectonic stresses generally refer to all stresses related to plate motions (Figure 23.3). Nevertheless, we like to differentiate here between the primarily horizontal tectonic stresses induced by plate motions and the stresses related to vertical and bending motions of the lithosphere, here termed 'geodynamic stresses'. The latter are often significantly smaller than the horizontal plate motion stresses.

In a simplified scenario, tectonic stresses can be considered as uniform horizontal stresses (or forces) induced by plate motion and are often modelled as such (Jarosiński et al., 2006). More realistically, these stresses are depth dependent. Various attempts have been made by applying velocities or forces along model boundaries to create a tectonic stress field in conjunction with the overburden stresses (e.g. Zoback & Townend, 2001; Hergert & Heidbach, 2011).

Ridge push is an important tectonic component acting on the Atlantic margin of Scandinavia. Ridge push force (RPF) stems from the difference in density distribution between the elevated oceanic crust and mantle at the Mid-Atlantic Ridge and the cooler,

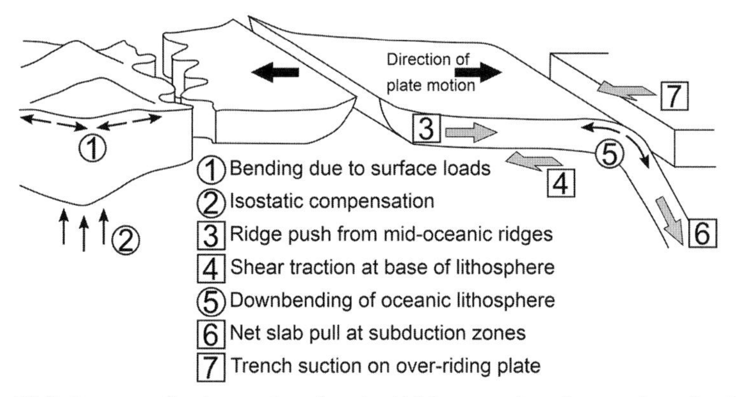

Figure 23.3 Sources of primary, broad-scale (thick arrows) and secondary, local (thin arrows) tectonic stresses in the lithosphere (adapted from Zoback et al., 1989).

lower-lying crust and mantle of the margin (e.g. Turcotte & Schubert, 2014). This force is thus strictly speaking not a tectonic force (which should stem from dynamic movement of the plates), yet it is suitable for local and regional studies to model ridge push as an external horizontal force. The RPF has been well studied and analytical calculations exist (e.g. Turcotte & Schubert, 2014). It is commonly calculated as a depth-independent force although a depth-dependence may play a large role (e.g. Naliboff et al., 2012). The magnitude of the RPF offshore Norway has been estimated to be up to 30 MPa (Fejerskov & Lindholm, 2000). The stress of the RPF (if applied as a stress boundary condition at a uniform angle) leads to an overall more compressive stress state. In addition, it induces a rotation of the maximum horizontal stress to the direction perpendicular to the assumed ridge orientation. Even if the RPF is considered not to decrease with depth, it will barely affect the stress field at depth greater than 5 km (Gradmann et al., 2018). Here, absolute stresses are comparatively large and the relative contribution of RPF is small. Closer to the surface, where absolute stresses are comparatively small, RPF can have a large effect and major variations in the stress regime are possible. It can be argued that the RPF is mainly supported by the crust and thus its effects on the stress field are largest where the crust is thin (the continental margin), but it decays over the thickening crust in the Scandinavian hinterland.

23.3.4 Geodynamic Stress

Geodynamic motions may include mantle upwelling, plate bending at subduction zones or be induced by plate drift across the ellipsoidal Earth (e.g. Mondy et al., 2018; Buffett & Becker, 2012). These motions can induce long-wavelength bending stresses in the lithosphere. Bending stresses within the lithosphere at subduction zones increase linearly from the surface to a depth between 30 and 40 km and can reach values of more than 600 MPa (Buffett & Becker, 2012). The amount of stress induced by bending increases with the age of the lithosphere as it becomes thicker and stronger with time. In addition, plate velocities and the curvature of the plate are also important constraints. However, these stresses are not relevant for Northern Europe due to the absence of an active subduction zone, but play, for example, an important role when analyzing the effect of glacially induced stresses in Alaska (Sauber et al., see Chapter 20).

23.3.5 Thermal Stress

Temperature or pressure variations can lead to stress changes if the rock column is not able to expand or contract freely. Due to the variability of rock properties, materials react differently to temperature variations, which can induce a large differential stress. In addition, temperature variations can also induce stresses due to changes in the expansion coefficient. An example is the formation of a dyke, where the material in the dyke cools down while the material surrounding the dyke gets warmer. Differential stresses due to thermal origin can be larger than 300 MPa (Lister, 1986). In Northern Europe, plate formation and intrusions of magma into the crust finished more than 300 Ma ago with minor exceptions, and thus thermal stresses are not expected to play a role in this area and are neglected when modelling the respective stress field.

23.3.6 Topographic and Other Gravitational Stresses

Gravitational stresses result from the material properties of the lithosphere itself. The density contrast along the surface (topography) is primarily studied. The resulting stresses are commonly considered of importance over a horizontal extent and depth similar to the topographic difference. This may be on the order of 1,000 m for the Norwegian coast, but the respective signal may be lost on a scale of all of Scandinavia. However, other studies point out that the wavelength – and not simply amplitude – of the topographic anomaly, determines the depth extent of influence (e.g. Martel, 2016). Other interfaces such as the base of sedimentary basins or the Moho and even the lithosphere-asthenosphere boundary may also play a role (Pascal, 2006; Maury et al., 2014; Gradmann et al., 2018) and respective stresses can be as large as 60 MPa. Stresses due to surface topography can vary from extensional stresses for elevated areas to compressional stresses within basins. Density variations within the crust also affect the magnitude of the entire stress field, with stress changes of up to 15 MPa between crusts of high and low density (Pascal, 2006). Gravitational stresses also depend on the lithospheric thickness and its thermal state, which can have an effect of up to 14 MPa (Pascal, 2006). These stresses are straightforward to implement by employing gravity as a body force. This force depends solely on the material's density, however, elastic parameters (Poisson's ratio and Young's modulus) also play a role on how the stresses are distributed.

23.3.7 Fault Structures

The inclusion of existing faults and related lithological changes can significantly alter the stress field in both orientation and magnitude. The stress field around faults can be modified by pre-existing fault movements (Barton & Zoback, 1994) leading to a change in the stress field magnitude and also a local change of the stress regime. In addition, fault movement may offset geological layers and rocks with different densities. Thus, the overburden stress field is modified by the altered distribution of rock masses at the same depth. When the maximum and minimum horizontal stresses are similar in their magnitude, the fault zone has a larger effect on the local stress field (Yale, 2003). Besides, while smaller faults have a local effect only, larger fault blocks can alter the tectonic stress field even kilometres away (Yale, 2003). The perturbed stress field can then be more prone to glacially triggered fault reactivation and earthquakes than would be the case under normal conditions. Thus, the implementation of faults in numerical models is important when the local stress field is analyzed (e.g. Henk, 2020) with respect to the reactivation potential of glacially induced faults. We use the term 'glacially induced fault' for the reactivated fault and glacially triggered faulting (GTF) when referring to the mechanism in accordance with Steffen et al. (see Chapter 1).

An additional effect arises due to the sudden release of stress in the form of an earthquake and the change in the stress field at the same and subsequent faults, which are able to increase or decrease the stability of surrounding faults. These stress changes, calculated as Coulomb failure stress (Harris, 1998), are known to be large enough to induce earthquakes along nearby faults or, in other cases, stabilize surrounding faults (e.g. Amini et al., 2018).

23.3.8 Loading/Unloading Stresses

Loading and unloading of rock or ice masses can have different effects on the in situ stress system. If unloading (e.g. erosion) occurs on a sufficiently small scale, it will not cause flexural isostatic adjustment but merely a reduction of the overburden load, namely a reduction of the vertical stress. The remaining horizontal stress results in an overall compressive stress regime. The high horizontal stresses may diminish due to frictional rock failure (fracturing, spalling), decay over time due to plastic processes or remain locked in. Locked-in horizontal stresses may also be created by overconsolidation and erosion (Amadei & Stephansson, 1997, their Chapter 2, and references therein).

If a larger amount of mass is moved (ice or eroded rock), the change in mass distribution will be isostatically compensated, namely by uplift in case of mass removal. This will induce bending stresses at the top and base of the elastic layer (the lithosphere in the rheological sense). Flexural uplift will impose tensional stresses near the surface and compressional stresses near the base of this layer. The latter are small compared to the existing overburden stress at several tens of kilometres depth and will barely modify the existing stress field, let alone bring it to failure. Near the surface, however, the bending stresses induced by sediment removal can have significant effects and may dominate the local stress field (Stein et al., 1989; Gradmann et al., 2018).

To quantify the flexural response to loading, the elasticity of the bending layer (the lithosphere) is needed. If the temporal evolution is to be considered, the viscoelastic behaviour of the asthenosphere must be included as well. The most comprehensive approach is used for GIA models (see next section). For studies at smaller scale, simple flat Earth models are sufficient. In a static model, the bending can even be calculated separately and applied as a boundary displacement condition.

In northern Norway, where extensional earthquakes have been observed (Fjeldskaar et al., 2000; Byrkjeland et al., 2000; Janutyte & Lindholm, 2017; Michalek et al., 2018) in an overall compressive stress field (Zoback, 1992; Fjeldskaar et al., 2000; Byrkjeland et al., 2000; Heidbach et al., 2018), the scenario of flexural extension due to local, coastal erosion has been proposed (Gradmann et al., 2018). In a similar setting of a glaciated margin, Stein et al. (1989) calculated the 2D bending stresses induced by the glacial sedimentary load (onshore erosion was not considered). In both studies, the calculated effects on the stress field were much larger than what could be confirmed by stress measurements in offshore boreholes or reconciled with other stress field estimates. Stein et al. (1989) suggested that stress dissipation and viscoelastic effects will over time reduce the bending stresses significantly. Over long timescales, stresses are usually considered transient – yet many unexplained stress configurations exist that may have their origin in the earlier formation of structure and stresses. Stress dissipation will thus not only play a role in unloading/loading but also in correct estimation of the other stress field components, in particular the overburden stress. Other loading/unloading mechanisms include volcanism or glacial melting (see below).

23.3.9 Glacially Induced Stresses

Ice mass changes have been well studied in Northern Europe and North America, and their effect on the stress field over time has been subject to several recent studies (e.g. Wu & Hasegawa, 1996b; Johnston et al., 1998; Lund et al., 2009; Steffen et al., 2019; Sauber et al., see Chapter 20). For the GIA-induced stresses, a much more detailed approach has been developed to calculate the bending stresses, as is outlined below.

Glacially induced stresses are determined as part of the modelling of GIA (Figure 23.4; Wu et al., see Chapter 22, and references therein), meaning the (primarily vertical) displacement of the lithosphere. A GIA model consists of an Earth model that is loaded with an ice model. The Earth model can be separated in several elastic and viscoelastic layers, which cover depths from the Earth surface to the core-mantle boundary. Furthermore, the stress development in the lithosphere due to glacial loading/unloading also depends on the stress migration from the mantle to the lithosphere (Steffen et al., see Chapter 2). This stems from the relaxation of the mantle and is therefore dependent on its viscosity structure. In addition, GIA stresses depend on the lithospheric structure (thickness, density variation) and ice-loading history (e.g. Wu & Hasegawa 1996a,b; Lund et al., 2009; Steffen et al., 2019). The latter is tightly linked to sea level changes, which result in an additional change of masses. These changes in ocean load are expressed in markers that can be observed and can serve as important constraints when modelling glacially induced stresses in coastal areas.

The magnitude of GIA stresses can be as large as 30 MPa (Wu & Hasegawa, 1996b), which is usually insufficient to create new faults but can rather reactivate pre-existing faults. A variation with depth can be seen as well (Figure 23.4). The magnitude of the

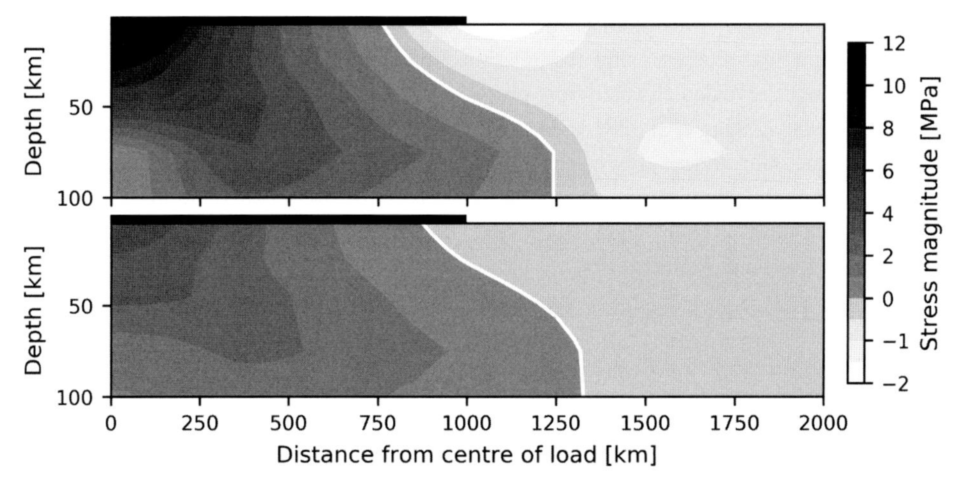

Figure 23.4 Glacially induced horizontal stresses at the end of the deglaciation (top panel) and 5,000 years after it (lower panel). The black bars mark the location of the ice sheet at glacial maximum. The same three-dimensional model setup as in Wu et al. (see Chapter 22) is used. A 90-km thick lithosphere (purely elastic) is loaded with a parabolic ice sheet of 2,000-m thickness at the centre of the load at glacial maximum. The mantle is divided into two different viscoelastic layers with viscosities of $5{\cdot}10^{20}$ Pa·s and $2{\cdot}10^{21}$ Pa·s.

compressive horizontal stress decreases from the surface to the base of the lithosphere and also with time once the loading is removed. The GIA stresses can then be used to analyze the reactivation potential of glacially induced faults. However, as GIA stresses are not the only stresses in the lithosphere in currently and formerly glaciated regions, stresses due to other sources, as described above, have to be included when evaluating fault stability (e.g. Wu et al., 1999; Lund et al., 2009; Steffen et al., 2014; Steffen et al., 2019).

Several GIA modelling techniques are available, however, only a few of them can also model the accompanying stress changes. In general, GIA model results are usually constrained by observations related to variations in the displacement, Earth rotation and gravitational field in order to find an Earth and ice model that fits the observations. GIA stress changes have received only little attention as the comparison to observations is problematic due to the difficulties in the timing of glacially triggered earthquakes. Nevertheless, the recent advances in mapping and dating of glacially induced faults as shown in the Chapters 3 to 21 (within this book) offered the possibility of also making use of the changes in the glacially induced stresses to compare them with the reactivation time of earthquakes and identify suitable GIA models (Steffen et al., 2019). Nevertheless, the large spread in timing, number and locations of earthquakes makes it still challenging to use these observations as an additional constraint.

23.3.10 Changes in Pore-fluid Pressure

A well-known factor in controlling the seismicity in specific regions and therefore their stress changes is the change in pore-fluid pressure. Several effects can change pore-pressure stresses, for example, precipitation, ground water movement, permafrost and ice-sheet melting. Numerical studies using the relationship between glacial unloading and pore-pressure changes revealed elevated pore pressures up to 150 km behind the ice margin that exist up to today (Zhang et al., 2018). However, these pore pressures are small compared to the existent tectonic stresses, but large enough with more than 0.27 MPa at depths of up to 5 km to change the stability of a region and induce earthquakes (Zhang et al., 2018). The effect of pore-fluid pressure changes on GTF has only poorly been explored so far and leaves room for many more studies.

23.4 Summary

Modelling of the Earth's in situ stresses has been a subject of research for several decades. It has seen many advances (3D, full Earth models, computational power) and changes in focus (GIA, seismicity, geotechnical) and the studies and approaches are numerous. Nevertheless, the concept of additive stress field components (regional and local) is always applied. Further research may better link the models to observations, increase resolution, employ more realistic rheologies and improve the 3D understanding of stress field evolution as well as stress dissipation.

Acknowledgements

We are thankful to Björn Lund and an anonymous reviewer for their constructive reviews.

References

Amadei, B. and Stephansson, O. (1997). *Rock Stress and Its Measurement*. Chapman & Hall, London.

Amini, S., Roberts, R., Raeesi, M. et al. (2018). Fault slip and identification of the second fault plane in the Varzeghan earthquake doublet. *Journal of Seismology*, **22**, 815–831, doi.org/10.1007/s10950-018-9734-0.

Barton, C. A. and Zoback, M. D. (1994). Stress perturbations associated with active faults penetrated by boreholes: possible evidence for near-complete stress drop and a new technique for stress magnitude measurement. *Journal of Geophysical Research*, **99**, 9373–9390, doi.org/10.1029/93JB03359.

Bell, J. S. and Wu, P. (1997). High horizontal stresses in Hudson Bay, Canada. *Canadian Journal of Earth Sciences*, **34**(7), 949–957, doi.org/10.1139/e17-079.

Brown, E.T. and Hoek, E. (1978). Trends in relationships between measured in-situ stresses and depth. *International Journal of Rock Mechanics and Mining Sciences & Geomechanics Abstracts*, **15**(4), 211–215, doi.org/10.1016/0148-9062(78)91227-5.

Brudy, M., Zoback, M. D., Fuchs, K., Rummel, F. and Baumgartner, J. (1997). Estimation of the complete stress tensor to 8 km depth in the KTB scientific drill holes: implications for the crustal strength. *Journal of Geophysical Research*, **102**, 18453–18475, doi.org/10.1029/96JB02942.

Buchmann, T. J. and Connolly, P. T. (2007). Contemporary kinematics of the Upper Rhine Graben: a 3D finite element approach. *Global and Planetary Change*, **58**(1), 287–309, doi.org/10.1016/j.gloplacha.2007.02.012.

Buffett, B. and Becker, T. W. (2012). Bending stress and dissipation in subducted lithosphere. *Journal of Geophysical Research – Solid Earth*, **117**, B05413, doi.org/10.1029/2012JB009205.

Byrkjeland, U., Bungum, H. and Eldholm, O. (2000). Seismotectonics of the Norwegian continental margin. *Journal of Geophysical Research: Solid Earth*, **105**(B3), 6221–6236, doi.org/10.1029/1999JB900275.

Fejerskov, M. and Lindholm, C. (2000). Crustal stress in and around Norway: an evaluation of stress-generating mechanisms. *Geological Society, London, Special Publications*, **167**(1), 451–467, doi.org/10.1144/GSL.SP.2000.167.01.19.

Fjeldskaar, W., Lindholm, C., Dehls, J.F. and Fjeldskaar, I. (2000). Postglacial uplift, neotectonics and seismicity in Fennoscandia. *Quaternary Science Reviews*, **19**(14), 1413–1422.

Gradmann, S., Olesen, O., Keiding, M. and Maystrenko, Y. (2018). The regional 3D stress field of Nordland, northern Norway – insights from numerical modelling. In O. Olesen et al., eds., *Neotectonics in Nordland – Implications for petroleum exploration (NEONOR2)*. NGU Report 2018.010, 215–240.

Gritto, R., Dreger, D. Heidbach, O. and Hutchings, L. (2014). *Towards the Understanding of Induced Seismicity in Enhanced Geothermal Systems*. Technical Report DE-EE0002756, Array Information Technology, Greenbelt (MD), United States, doi.org/10.2172/1154937.

Grünthal, G. and Stromeyer, D. (1992). The recent crustal stress field in central Europe: trajectories and finite element modeling. *Journal of Geophysical Research*, **97**(B8), 11805–11820, doi.org/10.1029/91JB01963.

Harris, R. A. (1998). Introduction to special section: stress triggers, stress shadows, and implications for seismic hazard. *Journal of Geophysical Research Solid Earth*, **103**(10), 24347–24358, doi.org/10.1029/98JB01576.

Hast, N. (1958). The measurements of rock stress in mines. *Swedish Geological Survey Publications* 52-3. Stockholm, Sweden.

Heidbach, O., Hergert, T., Reiter, K. and Giger S. (2014). *Local Stress Field Sensitivity Analysis – Case Study Nördlich Lägern*. Technical Report NAB 13-88, NAGRA – Nationale Genossenschaft für die Lagerung radioaktiver Abfälle.

Heidbach, O., Rajabi, M., Cui, X. et al. (2018). The World Stress Map database release 2016: crustal stress pattern across scales. *Tectonophysics*, **744**, 484–498, doi.org/10.1016/j.tecto.2018.07.007.

Henk, A. (2020). Numerical modelling of faults. In D. Tanner and C. Brandes, eds., *Understanding Faults – Detecting, Dating, and Modeling*. Elsevier, Amsterdam, pp. 147–165, doi.org/10.1016/B978-0-12-815985-9.00004-7.

Hergert, T., Heidbach, O., Reiter, K., Giger, S. B. and Marschall P. (2015). Stress field sensitivity analysis in a sedimentary sequence of the Alpine foreland, northern Switzerland. *Solid Earth*, **6**(2), 533–552, doi.org/10.5194/se-6-533-2015.

Hergert, T. and Heidbach, O. (2011). Geomechanical model of the Marmara Sea region – II. 3-D contemporary background stress field. *Geophysical Journal International*, **185**(3), 1090–1102, doi.org/10.1111/j.1365-246X.2011.04992.x.

Hickman, S. and Zoback, M. (2004). Stress orientations and magnitudes in the SAFOD pilot hole. *Geophysical Research Letters*, **31**, doi.org/10.1029/2004GL020043.

Janutyte, I. and Lindholm, C. (2017). Earthquake source mechanisms in onshore and offshore Nordland, northern Norway. *Norwegian Journal of Geology*, **97**(3), 227–239.

Jarosiński, M., Beekman, F., Bada, G. and Cloetingh, S. (2006). Redistribution of recent collision push and ridge push in Central Europe: insights from FEM modelling. *Geophysical Journal International*, **167**, 860–880, doi.org/10.1111/j.1365-246X.2006.02979.x.

Johnston, P., Wu, P. and Lambeck, K. (1998). Dependence of horizontal stress magnitude on load dimension in glacial rebound models. *Geophysical Journal International*, **132**, 41–60, doi.org/10.1046/j.1365-246x.1998.00387.x.

Klemann, V. and Wolf, D. (1998). Modelling of stresses in the Fennoscandian lithosphere induced by Pleistocene glaciations. *Tectonophysics*, **294**(3-4), 291–303, doi.org/10.1016/S0040-1951(98)00107-3.

Lister, C. R. B. (1986). Differential thermal stresses in the Earth. *Geophysical Journal International*, **86**(2), 319–330, doi.org/10.1111/j.1365-246X.1986.tb03831.x.

Lund, B. (2005). *Effects of Deglaciation on the Crustal Stress Field and Implications for Endglacial Faulting: A Parametric Study of Simple Earth and Ice Models*. SKB Technical Report TR-05-04, Swedish Nuclear Fuel and Waste Management Co., Stockholm, 68 pp.

Lund, B., Schmidt, P. and Hieronymus, C. (2009). *Stress Evolution and Fault Stability during the Weichselian Glacial Cycle*. SKB Technical Report TR-09-15, Swedish Nuclear Fuel and Waste Management Co., Stockholm, 106 pp.

Martel, S. (2016). Effects of small-amplitude periodic topography on combined stresses due to gravity and tectonics. *International Journal of Rock Mechanics and Mining Sciences*, **89**, 1–13, doi.org/10.1016/j.ijrmms.2016.07.026.

Maury, J., Cornet, F. H. and Cara, M. (2014). Influence of the lithosphere–asthenosphere boundary on the stress field northwest of the Alps. *Geophysical Journal International*, **199**(2), 1006–1017, doi.org/10.1093/gji/ggu289.

McCutchen, W. R. (1982). Some elements of a theory for in-situ stress. *International Journal of Rock Mechanics and Mining Sciences & Geomechanics Abstracts*, **19**(4), 201–203, doi.org/10.1016/0148-9062(82)90890-7.

Michalek, J., Tjåland, N., Drottning, A. et al. (2018). Report on seismic observations within the NEONOR2 project in the Nordland region, Norway (August 2013–May 2016). In O. Olesen et al., eds., *Neotectonics in Nordland – Implications for Petroleum Exploration (NEONOR2)*. NGU Report 2018.010, 63 pp.

Myrvang, A. M. (1993). Rock stress and rock stress problems in Norway. In J. A. Hudson, ed., *Comprehensive Rock Engineering, Vol. 3*. Pergamon Press, Oxford, pp. 461–471.

Mondy, L. S., Rey, P. F., Duclaux, G. and Moresi, L. (2018). The role of asthenospheric flow during rift propagation and breakup. *Geology*, **46**(2), 103–106, doi.org/10.1130/G39674.1.

Naliboff, J. B., Lithgow-Bertelloni, C., Ruff, L. J. and de Koker, N. (2012). The effects of lithospheric thickness and density structure on Earth's stress field. *Geophysical Journal International*, **188**(1), 1–17, doi.org/10.1111/j.1365-246X.2011.05248.x.

Pascal, C. (2006). On the role of heat flow, lithosphere thickness and lithosphere density on gravitational potential stresses. *Tectonophysics*, **425**, 83–99, doi.org/10.1016/j.tecto.2006.07.012.

Pascal, C., Roberts, D. and Gabrielsen, R. H. (2010). Tectonic significance of present-day stress relief phenomena in formerly glaciated regions. *Journal of the Geological Society*, **167**, 363–371, doi.org/10.1144/0016-76492009-136.

Reiter, K. and Heidbach O. (2014). 3-D geomechanical-numerical model of the contemporary crustal stress state in the Alberta Basin (Canada). *Solid Earth*, **5**(2), 1123–1149, doi.org/10.5194/se-5-1123-2014.

Sheorey, P. (1994). A theory for in situ stresses in isotropic and transverseley isotropic rock. *International Journal of Rock Mechanics and Mining Sciences and Geomechanics*, **31**(1), 23–34, doi.org/10.1016/0148-9062(94)92312-4.

Steffen, H. and Wu, P. (2011). Glacial isostatic adjustment in Fennoscandia – a review of data and modelling. *Journal of Geodynamics*, **52**, 169-204, doi.org/10.1016/j.jog.2011.03.002.

Steffen, R., Eaton, D. W. and Wu, P. (2012). Moment tensors, state of stress and their relation to post-glacial rebound in northeastern Canada. *Geophysical Journal International*, **189**, 1741-1752, doi.org/10.1111/j.1365-246X.2012.05452.x.

Steffen, R., Wu, P., Steffen, H. and Eaton, D. W. (2014). On the implementation of faults in finite-element glacial isostatic adjustment models. *Computers & Geosciences*, **62**, 150–159, doi.org/10.1016/j.cageo.2013.06.012.

Steffen, H., Steffen, R. and Tarasov, L. (2019). Modelling of glacially-induced stress changes in Latvia, Lithuania and the Kaliningrad District of Russia. *Baltica*, **32**(1), 78–90, doi.org/10.5200/baltica.2019.1.7.

Stein, S., Cloetingh, S., Sleep, N. and Wortel R. (1989). Passive margin earthquakes, stresses and rheology. In S. Gregersen and P. Basham, eds., *Earthquakes at North-Atlantic Passive Margins: Neotectonics and Postglacial Rebound*, NATO ASI Series C 266, Springer, Dordrecht, pp. 231–259.

Stephansson, O. (1993). Stress in the Fennoscandian Shield. In J. A. Hudson, Ed., *Rock Testing and Site Characterization*. Pergamon Press, Oxford, pp. 445–459, doi.org/10.1016/B978-0-08-042066-0.50024-0.

Townend, J. and Zoback, M. D. (2000). How faulting keeps the crust strong. *Geology*, **28**(5), 399–402, doi.org/10.1130/0091-7613(2000)28<399:HFKTCS>2.0.CO;2.

Turcotte, D. and Schubert, G. (2014). *Geodynamics*, 3rd ed. Cambridge University Press, Cambridge.

Wu, P. and Hasegawa, H. S. (1996a). Induced stresses and fault potential in eastern Canada due to a disc load: a preliminary analysis. *Geophysical Journal International*, **125**(2), 415–430, doi.org/10.1111/j.1365-246X.1996.tb00008.x.

Wu, P. and Hasegawa, H. S. (1996b). Induced stresses and fault potential in eastern Canada due to a realistic load: a preliminary analysis. *Geophysical Journal International*, **127** (1), 215–229, doi.org/10.1111/j.1365-246X.1996.tb01546.x.

Wu, P., Johnston, P. and Lambeck, K. (1999). Postglacial rebound and fault instability in Fennoscandia. *Geophysical Journal International*, **139**, 657–670, doi.org/10.1046/j.1365-246x.1999.00963.x.

Yale, D. P. (2003). Fault and stress magnitude controls on variations in the orientation of *in situ* stress. In M. S. Ameen, ed., *Fracture and In-Situ Stress Characterization of Hydrocarbon Reservoirs*. Geological Society, London, Special Publication, Vol. 209, pp. 55–64, doi.org/10.1144/GSL.SP.2003.209.01.06.

Zhang, Y., Person, M., Voller, V. et al. (2018). Hydromechanical impacts of Pleistocene glaciations on pore fluid pressure evolution, rock failure, and brine migration within sedimentary basins and the crystalline basement. *Water Resources Research*, **54**, doi.org/10.1029/2017WR022464.

Zoback, M. L., Zoback, M., Adams, J. et al. (1989). Global patterns of tectonic stress. *Nature*, **341**, 291–298, doi.org/10.1038/341291a0.

Zoback, M. L. (1992). First and second order patterns of stress in the lithosphere: the World Stress Map Project. *Journal of Geophysical Research*, **97**, 11703–11728, doi.org/10.1029/92jb00132.

Zoback, M. D. and Townend, J. (2001). Implications of hydrostatic pore pressures and high crustal strength for the deformation of intraplate lithosphere. *Tectonophysics*, **336**, 19–30, doi.org/10.1016/S0040-1951(01)00091-9.

Zoback, M. L. and Zoback, M. D. (2015). Lithosphere stress and deformation. In G. Schubert, ed., *Treatise on Geophysics. Crust and Lithosphere Dynamics, Vol. 6*, Elsevier, Amsterdam, pp. 253–273.

Part VII

Outlook

24

Future Research on Glacially Triggered Faulting and Intraplate Seismicity

ODLEIV OLESEN, HOLGER STEFFEN AND RAIMO SUTINEN

ABSTRACT

Glacially triggered faulting (GTF) is the release of glacially induced stresses along pre-existing faults or weakness zones. It has been generally thought that such glacially induced faults (GIFs), especially the so-called postglacial faults in northern Fennoscandia, were developed during a short period of time towards the end of and shortly after deglaciation. This concept is now challenged by new results from Fennoscandia documenting several episodes of fault rupture within the past 14,000 years. We speculate that some of these ruptures at known (or potential) GIFs may not be due to GTF but may contain a signature of tectonically driven intraplate seismicity. Glacially triggered faulting cannot be totally ignored for these episodes because the ongoing rebound of the lithosphere is continuously increasing glacially induced stresses that, eventually, can be released, under favourable conditions. As those conditions can only be described by a complex 4-dimensional model, simple identification of GIFs or categorization as GIFs is hampered. Precise dating of younger fault ruptures is especially important to produce the necessary spatiotemporal image. The planned DAFNE drilling and subsequent in situ observations of the Pärvie GIF combined with numerical modelling will contribute to an improved understanding of glacially triggered faulting. This will be important for assessing the seismic hazard of both northern Fennoscandia and other intraplate regions.

24.1 Introduction

Understanding of glacially triggered faulting (often constrained to be postglacial) has increased steadily over more than a century (Steffen et al., see Chapter 1). An especially intense period of geological and geophysical investigations in the 1970s and 1980s provided a picture of the development and distribution of faults that could be explained with the Mohr-Coulomb failure theory as outlined by Johnston (1987) and supported by numerical modelling (Wu & Hasegawa, 1996). Briefly summarized, those authors thought that the ice sheets of the last glaciation in Northern Europe and North America suppressed the stress release of accumulated strains in the underlying lithosphere until a brief period (of 1,000–2,000 years) shortly before and after deglaciation, when these stresses were released along remarkable ruptures and accompanying large-magnitude earthquakes. A set of mostly NE-SW-trending postglacial fault ruptures (PGFs, now often referred to as glacially induced faults, GIFs) were first discovered in Fennoscandia (Kujansuu, 1964; Lundqvist & Lagerbäck, 1976; Olesen, 1988). The term PGF emphasized the time of origin even though Lundqvist and Lagerbäck (1976) demonstrated that some segments of the Pärvie Fault ruptured prior to its deglaciation. In northern North America, similar prominent faults have not yet been found.

By contrast, historical and present-day seismicity in these formerly glaciated areas is generally low and consistent with their intraplate setting. In the area of the Fennoscandian GIFs no earthquake above magnitude 5 has been measured by seismometers (Lindholm, 2019; Gregersen et al., see Chapter 10). Nonetheless, earthquakes of minor magnitude occasionally occur near the mapped GIFs. In north-eastern Canada, although there are clusters of earthquake activity on Baffin Island and along the Boothia–Ungava zone (site of the M6.3 Ungava earthquake and its surface rupture, see Adams et al., 1991), there is very low seismicity in the areas of maximum land uplift, i.e. west and east of Hudson Bay.

In the following, we highlight the recent findings that challenge the generally accepted picture of glacially triggered faulting. We will discuss whether glacially induced faults should still be termed as such and how the faults and recorded seismicity in formerly glaciated regions may help to advance our knowledge of intraplate seismicity.

24.2 Extending the Timing and Spatial Distribution of Glacially Triggered Faulting

Radiocarbon dates of buried organic matter indicate episodic Fennoscandian GIF activity after the deglaciation of the Late Weichselian. The oldest rupture is the Isovaara–Riikonkumpu Fault Complex in western Finnish Lapland (Ojala et al., 2018) dated at 11.3 thousand calibrated or calendar years before 1950 (ka BP). There is also morphological evidence of subglacial seismically induced landforms (Sutinen et al., 2014, 2019) suggesting that the Younger Dryas (12.8–11.5 ka BP) and possibly also the preceding Bølling–Allerød interstadial (14.7–12.9 ka BP) were seismically active periods. This would be slightly earlier than predicted by the model of Wu et al. (1999; also see Chapter 22), which suggests that the Fennoscandian fault instability started around 13 ka BP but reached maximum instability in between 11 and 9 ka BP.

Recent radiocarbon dates have also revealed three palaeolandslide episodes attributed to earthquakes in northern Finland: between 9 and 11 ka BP, between 5 and 6 ka BP and between 1 and 3 ka BP (Ojala et al., 2018). Trenching of three different fault elements of the Stuoragurra Fault Complex and radiocarbon dating of buried and deformed organic material have further revealed a late Holocene age for the ruptures (between around 0.6 and 4 ka BP, Olesen et al., see Chapter 11). None of those dates indicate faulting associated with the deglaciation. If these dates are correct, faulting occurred at least 5,000 years after deglaciation and was not an immediate result of the rapid rebound following deglaciation. These results bring further information on the age constraints for these GIFs (e.g. Lagerbäck & Sundh, 2008), if they can still be named as such. There is a need for more detailed palaeoseismological studies in addition to seismological and geodetic studies of the GIFs in northern Fennoscandia to follow up these new findings. Detailed studies of fault scarps in Sweden and Finland also provide evidence of multiple rupture events even before the Late Weichselian (Smith et al., see Chapter 12; Sutinen et al., see Chapter 13). Precise dating of the fault ruptures, or the earthquakes they produced, is especially important to produce a spatio-temporal image of Late Pleistocene faulting and Holocene faulting.

With the help of new elevation data (LiDAR DEM), the length and number of GIFs have increased (Palmu et al., see Chapter 5; Olesen et al., see Chapter 11; Smith et al., see Chapter 12). Many GIFs have been reported in north-western Russia in recent years (Nikolaeva et al., see Chapter 14). New GIFs were discovered a few hundred kilometres south of the Lapland postglacial fault province (called Lapland Province in the following) in northern Fennoscandia (Smith et al., 2014; Palmu et al., 2015; Steffen et al., see Chapter 1; Smith et al., see Chapter 12; Sutinen et al., see Chapter 13). Moreover, GIFs were reported at the edge of (Brandes et al., 2018; Sorgenfrei–Tornquist Zone) and even outside of the former ice margin, e.g. in Germany (Brandes et al., 2012; Müller et al., see Chapter 16).

The displacement of the peripheral GIFs is much smaller than on the Lapland faults, likely due to the much smaller glacially induced stresses at the periphery compared to those in the uplift centre (Brandes et al., 2015; Steffen et al., 2019; Wu et al., see Chapter 22); there are no impressive metre-high scarps in the landscape. Nonetheless, these findings can help us to reconsider and re-evaluate intraplate seismicity – at least in Europe but also with an eye on global implications.

24.3 Glacially Induced Faults and the Role of Intraplate Seismicity

Two intriguing questions arise out of the compilation of glacially triggered faulting in the present book: (1) How do we distinguish glacially triggered earthquakes from earthquakes caused by other mechanisms? (2) Would the Fennoscandian and Canadian shields have been aseismic without the numerous Pleistocene glaciations? All Precambrian shields around the world have present-day seismicity, and Holocene fault scarps resulting from large-magnitude earthquakes occur regardless of whether the shield has been glaciated or not (e.g. Precambrian areas in non-glaciated regions such as Australia, India and China). In this context it is interesting that Calais et al. (2016) concluded that the seismicity in North America is higher to the south of the glaciated part than inside the area that was covered by the Laurentide ice sheet, perhaps with the exception of the Lower St. Lawrence Seismic Zone in Eastern Canada. We also notice that maximum present-day seismicity occurs outside the area of maximum change in Coulomb failure stress in Fennoscandia (Figure 22.3 in Wu et al., see Chapter 22). We may even ask ourselves if the title of the present book is still appropriate in view of these new findings.

Large mid-continental earthquakes around the world show complex patterns in time and space that do not fit existing seismotectonic models of steady stress accumulation and release (Clark et al., 2012; Calais et al., 2016; Liu & Stein, 2016). Individual faults tend to fall into thousands of years inactivity after a period or just one event (so-called one-off event) of faulting. Large-magnitude earthquakes seem to migrate from one fault to another. This model is also applicable to the GIFs within the Lapland Province and can explain the significantly younger ages of the northernmost GIFs in Norway and Finland (Olesen et al., see Chapter 11; Ojala et al., 2019) compared with the south-eastern faults in northern Sweden along the Gulf of Bothnia (Lagerbäck & Sundh, 2008).

Calais et al. (2016) argue that the lithosphere can build up elastic strain over long periods of time. Ridge push from the NE Atlantic spreading ridge and loading/unloading along the Fennoscandian margin during Pleistocene glacial erosion are candidates for such stress build-up (Gradmann et al., 2018). A recent aeromagnetic survey (Dumais et al., 2020) over the Knipovich Ridge offshore northern Fennoscandia shows that the spreading velocity has varied along the ridge over the last 20 Ma (including a ridge jump). The release of the intraplate stress may be triggered by local and transient stress changes caused by erosion, fluid migration or ice loading, resulting in the occurrence of seismicity. Such a paradigm shift has significant bearings on the seismic hazard estimates in previously glaciated areas such as Northern Europe and North America (Liu & Stein, 2016; Calais et al., 2016).

The known GIFs in northern Fennoscandia seem to a varying degree to be dependent on pre-existing faults; but how does young faulting reactivate pre-existing zones of weakness? Some of the GIFs seem also to break fresh bedrock without any pre-existing weakness zones (e.g. parts of the Pärvie Fault, Juhlin et al., 2010). Such information is of high societal importance for the hazard assessments and mitigating the effects of future large earthquakes on large mines, tailing dams, bridges, nuclear waste repositories, petroleum installations, carbon capture storage (CCS) and hydropower dams in formerly glaciated regions such as northern Fennoscandia. It also has consequences for assessments of geothermal potential and groundwater vulnerability.

The Nordland coast in northern Norway is the most seismically active area in northern Fennoscandia (Lindholm, 2019; Gregersen et al., see Chapter 10). Earthquake focal mechanisms exhibit patterns of extension normal to the coast (Janutyte & Lindholm, 2017) in contradiction to the first-order regional stress pattern reflecting compression from ridge push. While the regional stress field is considered to arise from the interaction of ridge push and glacial isostatic adjustment (GIA), the local stress field results mainly from gravitational stresses (e.g. from topography) and the flexural effects of erosion and sediment deposition. Gradmann et al. (2018) modelled the in situ rock stress in the Nordland area using finite element modelling and concluded that redistribution of sediments and basement rocks during Pleistocene glaciations is the main driver for the coastal extension. The distance from the Pärvie Fault to the Nordland seismicity area is around 150 km, and that also constitutes the distance between the individual fault complexes within the Lapland Province. Olsen and Høgaas (2020) have reported a total of approximately 20 liquefaction structures in Holocene sand within the same area. Undeformed littoral sand on top of each site at various elevations indicate formation of the mosaic sand at different Holocene ages from site to site. The largest earthquake (M = 5.8–5.9) on mainland Northern Europe occurred in 1819 in the same area (Bungum & Olesen, 2005; Mäntyniemi et al., 2020). The Nordland area is therefore a candidate for future discoveries of postglacial faults.

The underlying mechanism for the GIFs and present-day seismicity is most likely something other than GIA, with GIA serving only locally as a trigger for reactivation. Which among the faults or the episodes of fault reactivation of a certain fault are related to GIA can only be speculated. It is very likely that reactivations present at the end of the deglaciation are due to GIA. Modelling supports this conclusion, as GIA stresses can reach 6 MPa in the Lapland Province (Wu et al., see Chapter 22), which is sufficient even for non-optimally oriented faults to be reactivated.

The time frame of the Norwegian and Finnish faults should be subject to future combined GIA and stress modelling (e.g. Gradmann & Steffen, see Chapter 23). Ongoing rebound could provide a small increment of additional stress to reactivate a fault, as shown e.g. by Brandes et al. (2015) for the Osning Thrust: large GIA-induced stresses at the end of deglaciation were released and the fault was then stable for several thousands of years. Ongoing rebound of the lithosphere forced the stress build-up that was then presumably released during the historical earthquakes in the vicinity of the Osning Thrust. Stress release by the historical earthquakes led again to stable conditions at the fault (Brandes et al., 2015). As rebound continues, new stress accumulation might be possible. However, GIA modelling analysis suggests this to be unlikely for this particular location. The situation might be different in the Lapland Province.

In addition, Coulomb stress transfer may play a role at individual faults or even for the whole Lapland Province. GIA-induced stress released at one fault could transfer the stress to other faults, which are then primed to move at a later stage. Such a model was tested for major faults in Germany (Brandes et al., 2015), but the amount of stress release during a plausible earthquake and the distribution of faults did not result in significant stress transfer; the faults were too far away from each other, given the magnitude of the Coulomb stresses. The situation in the Lapland Province is different, partly because much larger GIA stresses have accumulated (and some have already been released) since the end of the deglaciation and partly because of the closer spacing of the faults.

One must consider that the stress drop due to stress release during an earthquake does not mean all stresses are released and hence stable conditions prevail thereafter. Firstly, GIA stresses can accumulate again during rebound of the lithosphere, which may lead to another reactivation of the same fault, but then produce a subsequent (possibly smaller magnitude) earthquake. This might explain the episodic postglacial reactivation of faults in the Lapland Province. Another consequence is that part of today's seismicity could be related to GIA, depending on the uplift behaviour of the lithosphere. Secondly, Coulomb stress transfer, which depends on fault distribution, fault parameters and stress release during an earthquake on a particular fault, can also lead to fault reactivation. This results in a very complex picture of fault activity, especially in postglacial times, because the trigger of an earthquake might be a prior earthquake and might or might not be a consequence of GIA. This is important because, thirdly, tectonic background stresses might also build up during a glaciation (Johnston, 1987). During some tens or hundreds of thousand years of glaciation this effect might saturate, so that despite the restraining weight of the ice load, earthquakes would begin to occur under the ice. For the long-lasting ice sheet of Antarctica, this saturation may have reverted the continent to a typical intraplate seismicity level

(Lough et al., 2018). It is thus difficult to say if a fault is a pure glacially induced fault or if it is due to tectonically driven intraplate seismicity intermittently overprinted by GIA.

There is further no guarantee that all GIFs will be preserved for many hundreds or even thousands of years. The 1989 Ungava earthquake in northern Quebec was, for instance, a magnitude 6.3 earthquake that ruptured the top 5–6 km of the crust and formed a 10-km-long fault rupture with a scarp up to 1.3 m high (Adams et al., 1991). Adams and Brooks (see Chapter 19) note that the long-term discernibility of the Ungava rupture will be low because it ruptured through hummocky terrain without clear markers, extended under lakes for more than half its length, seldom exposed offset bedrock, and much of its discernibility was due to torn peat and newly exposed lichen-free boulders, which will re-vegetate in a few tens to hundreds of years. Poor discernibility might explain the very low number of confirmed GIFs away from the Lapland Province, although many geological indications such as soft-sediment deformation structures have been found (Sandersen et al., see Chapter 15; Müller et al., see Chapter 16; Pisarska-Jamroży et al., see Chapter 17; Bitinas et al., see Chapter 18).

There are, however, indirect methods for identifying palaeoearthquakes that should have associated large-scale surface ruptures. Numerous synchronous landslides in flat terrain and liquefaction phenomena are such secondary effects. Many hundred landslides have for instance been mapped within a 50 × 50-km area in the northern part of the Permian Oslo Rift in south-eastern Norway (Løvø, 2019). Dating of the landslides reveals clusters of dates, inferred to represent earthquakes during the Holocene (Mangerud et al., 2018; Lars Olsen, personal communication, 2020). Palaeoliquefaction structures along the coast of Mid- and Northern Norway (Olsen & Høgaas, 2020) occur in the area of highest present-day seismicity. We suggest that such palaeoseismological mapping constitutes a key to defining areas prone to future large-magnitude earthquakes, as such earthquakes are too rare in intraplate environments for the historical record to properly sample them. More detailed studies of LiDAR data and multi-beam echo-sounding bathymetric data may perhaps lead to new discoveries of GIFs in areas with abundant palaeoseismological phenomena (e.g. Adams & Brooks, see Chapter 19). These complementary methods should be applied in other formerly glaciated areas such as North America and the coastal areas of Antarctica and Greenland, where indications for glacially triggered faulting also have been found (Steffen & Steffen, see Chapter 21).

24.4 A Look into the Near Future

Currently, systematic in situ information at depth is lacking for GIFs, which the new DAFNE project (Drilling Active Faults in Northern Europe; Ask et al., see Chapter 9) will address. The proposal has been accepted for ICDP funding, and the current plan is to drill the first site in 2022. The new borehole information and monitoring data will be important in advancing our understanding of GIFs and related intraplate seismicity.

The fundamental DAFNE objective is to make in situ observations of the Pärvie postglacial fault followed by a programme of borehole monitoring and numerical modelling. Borehole measurements will document rock mechanics and thermal and hydrogeological regimes that can be used to calibrate and test current hypotheses for GIA and intraplate seismicity. With such constraining information, the distal effects on the passive margin development along the western Fennoscandian Shield during the Pleistocene can be modelled with a higher degree of certainty. Such data can only be achieved by drilling and are expected to provide a better understanding of large glacially triggered earthquakes and subsequent aftershocks.

The DAFNE scientific goals are of great relevance to society because they will improve assessment of seismic hazard to mines, tailing dams and hydropower dams near GIFs in Fennoscandia and other previously glaciated regions. This hazard assessment will also improve prediction of the future behaviour of bedrock during forthcoming glacial cycles, in particular to ensure safe disposal of toxic and nuclear waste in crystalline bedrock.

24.5 Conclusions

Dating records indicate episodic Fennoscandian GIF activity (11.3–0.6 ka BP) during and after the Late Weichselian deglaciation. That the GIFs in northern Fennoscandia occurred up to 10,000 years after the ice vanished, together with possible GIFs further south, even beyond the periphery of the formerly glaciated area in Europe, has challenged the idea that the GIFs in Europe exist only in the far north and that they were reactivated by GIA-induced stresses only at the beginning of the local deglaciation.

More detailed palaeoseismological studies combined with additional seismological and geodetic studies of the GIFs in northern Fennoscandia are needed to provide a detailed spatio-temporal image of the young crustal deformation of Fennoscandia. Slow slip processes such as low-frequency earthquakes have been observed along plate boundaries in recent years (Peng & Gomberg, 2010; Araki et al., 2017) and should be looked for within the seismically active Lapland Province. The relationship between pre-existing fault zones and GIFs must also be known to make possible estimation of seismic hazard.

One of the goals of future research will be to find a unified and consistent model for all confirmed and currently suggested GIFs and to test if the model is compatible with the GIA-fault modelling by Steffen et al. (2014a,b) who predict faulting to have occurred multiple times in a pre-stressed region beneath the centre of the ice sheet. The stress contribution from ridge push, the Scandes mountains and Pleistocene loading/unloading must be incorporated into such modelling. Earthquakes may be triggered by mechanisms other than GIA, such as local and transient stress changes caused by erosion and fluid migration. Such a model must explain the uniform NE-SW orientation and approximate 100–150-km spacing of the faults within the Lapland Province in addition to the observed timing of the ruptures. Construction of this model will benefit from results of the planned drilling into the Pärvie Fault. The model's results will also provide insight into possible glacially triggered faulting and corresponding seismic and tsunami hazard in regions of current ice melting such as Svalbard, Greenland and parts of Antarctica (see e.g. Steffen et al., 2020).

Acknowledgements

We would like to thank John Adams, Christian Brandes and Conrad Lindholm for their constructive reviews as well as Rebekka Steffen for comments on an earlier version of this manuscript.

References

Adams, J., Wetmiller, R. J., Hasegawa, H. S. and Drysdale, J. (1991). The first surface faulting from a historical intraplate earthquake in North America. *Nature*, **352**, 617–619, doi.org/10.1038/352617a0.

Araki, E., Saffer, D.M., Kopf, A. et al. (2017). Recurring and triggered slow-slip events near the trench at the Nankai Trough subduction megathrust. *Science*, **16**, 1157–1160, doi.org/10.1126/science.aan3120.

Brandes, C., Winsemann, J., Roskosch, J. et al. (2012). Activity along the Osning Thrust in Central Europe during the Lateglacial: ice-sheet and lithosphere interactions. *Quaternary Science Reviews*, **38**, 49–62, doi.org/10.1016/j.quascirev.2012.01.021.

Brandes, C., Steffen, H., Steffen, R. and Wu, P. (2015). Intraplate seismicity in northern Central Europe is induced by the last glaciation. *Geology*, **43**, 611–614, doi.org/10.1130/G36710.1.

Brandes, C., Steffen, H., Sandersen, P. B. E., Wu, P. and Winsemann, J. (2018). Glacially induced faulting along the NW segment of the Sorgenfrei-Tornquist Zone, northern Denmark: implications for neotectonics and Lateglacial fault-bound basin formation. *Quaternary Science Reviews*, **189**, 149–168, doi.org/10.1016/j.quascirev.2018.03.036.

Bungum, H. and Olesen, O. (2005). The 31st of August 1819 Lurøy earthquake revisited. *Norwegian Journal of Geology,* **85**, 245–252.

Calais, E., Camelbeeck, T., Stein, S., Liu, M. and Craig, T. J. (2016). A new paradigm for large earthquakes in stable continental plate interiors. *Geophysical Research Letters*, **43**, 10,621–10,637, doi.org/10.1002/2016GL070815.

Clark, D., McPherson, A. and Van Dissen, R. (2012). Long-term behaviour of Australian stable continental region (SCR) faults. *Tectonophysics*, **566**, 1–30, doi.org/10.1016/j.tecto.2012.07.004.

Dumais, M. A., Olesen, O., Gernigon, L., Johansen, S. and Brönner, M. (2020). Delineating the geological settings of the southern Fram Strait with state-of-the-art aeromagnetic data. In H. A. Nakrem and A. M. Husås, eds., *34th Nordic Geological Winter Meeting January 8th–10th 2020, Oslo, Norway. Abstracts and Proceedings of the Geological Society of Norway,* **1**, 2020, p. 51.

Gradmann, S., Olesen, O., Keiding, M. and Maystrenko, Y. (2018). The regional 3D stress field of Nordland, northern Norway – insights from numerical modelling. In O. Olesen, I. Janutyte, J. Michálek et al., eds., *Neotectonics in Nordland – Implications for petroleum exploration (NEONOR2).* NGU Report 2018.010, pp. 215–240.

Janutyte, I. and Lindholm, C. (2017). Earthquake source mechanisms in onshore and offshore Nordland, northern Norway. *Norwegian Journal of Geology*, **97**(3), 227–239, doi.org/10.17850/njg97-3-0.

Johnston, A. C. (1987). Suppression of earthquakes by large continental ice sheets. *Nature*, **330**, 467–469, doi.org/10.1038/330467a0.

Juhlin, C., Dehghannejad, M., Lund, B., Malehmir, A. and Pratt, G. (2010). Reflection seismic imaging of the end-glacial Pärvie Fault system, northern Sweden. *Journal of Applied Geophysics*, **70**, 307–316, doi.org/10.1016/j.jappgeo.2009.06.004.

Kujansuu, R. (1964). Nuorista siirroksista Lapissa [Recent faults in Lapland]. *Geologi*, **16**, 30–36 (in Finnish).

Lagerbäck, R. and Sundh, M. (2008). Early Holocene faulting and paleoseismicity in northern Sweden. *Sveriges Geologiska Undersökning*, **C836**, 80 pp.

Lindholm, C. (2019). Jordskjelv i Norge [Earthquakes in Norway]. In S. Bjøru, H. Wiig, V. Woldsengen and S. Engen, eds., *Fjellsprengningsdagen* [Rock Blasting Day]. Oslo, 21 November 2019. Norsk forening for fjellsprengningsteknikk, Norsk Bergmekanikkgruppe & Norsk Geoteknisk Forening, pp. 8.1–8.13, nff.no/wp-content/uploads/sites/2/2020/05/198431-Fjellsprengningsbok-2019-minnepenn-web.pdf

Liu, M. and Stein, S. (2016). Mid-continental earthquakes: spatiotemporal occurrences, causes, and hazards. *Earth-Science Reviews*, **162**, 364–386, doi.org/10.1016/j .earscirev.2016.09.016.

Lough, A. C., Wiens, D. A. and Nyblade, A. (2018). Reactivation of ancient Antarctic rift zones by intraplate seismicity. *Nature Geoscience*, **11**(7), 515–519, doi.org/10.1038/ s41561-018-0140-6.

Lundqvist, J. and Lagerbäck, R. (1976). The Pärve Fault: a late-glacial fault in the Precambrian of Swedish Lapland. *Geologiska Föreningens i Stockholm Förhandlingar*, **98**, 45–51, doi.org/10.1080/11035897609454337.

Løvø, G. (2019). Østlandet ble rystet av kraftige jordskjelv: – Må ha blitt oppfattet som ragnarok av forfedrene våre [Eastern Norway was shaken by powerful earthquakes: must have been perceived as Ragnarök by our ancestors]. forskning.no/geofag-norges-geologiske-undersokelse-partner/ostlandet-ble-rystet-av-kraftige-jordskjelv-ma-ha-blitt-oppfattet-som-ragnarok-av-forfedrene-vare/1354661.

Mangerud, J., Birks, H. H., Halvorsen, L. S. et al. (2018). The timing of deglaciation and sequence of pioneer vegetation at Ringsaker, eastern Norway – and an earthquake-triggered landslide. *Norwegian Journal of Geology*, **98**, 315–332, doi.org/10.17850/ njg98-3-03.

Mäntyniemi, P. B., Sørensen, M. B., Tatevossian, T. N., Tatevossian, R. E. and Lund, B. (2020). A reappraisal of the Lurøy, Norway, earthquake of 31 August 1819. *Seismological Research Letters*, **91**, 2462–2472, doi.org/10.1785/0220190363.

Ojala, A. E. K., Markovaara-Koivisto, M., Middleton, M. et al. (2018). Dating of paleo-landslides in western Finnish Lapland. *Earth Surface Processes and Landforms*, **43**(11), 2449–2462, doi.org/10.1002/esp.4408.

Ojala, A. E. K., Mattila, J., Hämäläinen, J. and Sutinen, R. (2019). Lake sediment evidence of paleoseismicity: timing and spatial occurrence of late- and postglacial earthquakes in Finland. *Tectonophysics*, **771**, 228227, doi.org/10.1016/j.tecto.2019.228227.

Olesen, O. (1988). The Stuoragurra fault, evidence of neotectonics in the Precambrian of Finnmark, northern Norway. *Norsk Geologisk Tidsskrift*, **68**, 107–118.

Olsen, L. and Høgaas, F. (2020) *"Shaken, Not Stirred": Mosaic Sand – A Semi-liquefaction Phenomenon Originating from Strong Earthquakes*. NGU Report 2020.020, 32 pp.

Palmu, J.-P., Ojala, A. E. K., Ruskeeniemi, T., Sutinen, R. and Mattila, J. (2015). LiDAR DEM detection and classification of postglacial faults and seismically-induced landforms in Finland: a paleoseismic database. *GFF*, **137**(4), 344–352, doi.org/10.1080/ 11035897.2015.1068370.

Peng, Z. and Gomberg, J. (2010). An integrated perspective of the continuum between earth-quakes and slow-slip phenomena. *Nature Geoscience*, **3**(9), 599–607, doi.org10 .1038/ ngeo940.

Smith, C. A., Sundh, M. and Mikko, H. (2014). Surficial geology indicates early Holocene faulting and seismicity, central Sweden. *International Journal of Earth Sciences*, **103**(6), 1711–1724, doi.org/10.1007/s00531-014-1025-6.

Steffen, R., Steffen, H., Wu, P. and Eaton, D. W. (2014a). Stress and fault parameters affecting fault slip magnitude and activation time during a glacial cycle. *Tectonics*, **33**, 1461–1476, doi.org/10.1002/2013TC003450.

Steffen, R., Wu, P., Steffen, H. and Eaton, D. W. (2014b). On the implementation of faults in finite-element glacial isostatic adjustment models. *Computers & Geosciences*, **62**, 150–159, doi.org/10.1016/j.cageo.2013.06.012.

Steffen, H., Steffen, R. and Tarasov, L. (2019). Modelling of glacially-induced stress changes in Latvia, Lithuania and the Kaliningrad District of Russia. *Baltica*, **32**(1), 78–90, doi.org/10.5200/baltica.2019.1.7.

Steffen, R., Steffen, H., Weiss, R. et al. (2020). Early Holocene Greenland-ice mass loss likely triggered earthquakes and tsunami. *Earth and Planetary Science Letters*, **546**, 116443, doi.org/10.1016/j.epsl.2020.116443.

Sutinen, R., Aro, I., Närhi, P., Piekkari, M. and Middleton, M. (2014). Maskevarri Ráhhpát in Finnmark, northern Norway – is it an earthquake-induced landform complex? *Solid Earth*, **5**, 683–691, doi.org/10.5194/se-5-683-2014.

Sutinen, R., Andreani, L. and Middleton, M., (2019). Post-Younger Dryas fault instability and de-formations on ice lineations in Finnish Lapland. *Geomorphology*, **326**, 202–212, doi.org/10.1016/j.geomorph.2018.08.034.

Wu, P. and Hasegawa, H. S. (1996). Induced stresses and fault potential in eastern Canada due to a disc load: a preliminary analysis. *Geophysical Journal International*, **125**, 415–430, doi.org/10.1111/j.1365-246X.1996.tb00008.x.

Wu, P., Johnston, P. and Lambeck, K. (1999). Postglacial rebound and fault instability in Fennoscandia. *Geophysical Journal International*, **139**, 657–670, doi.org/10.1046/j.1365-246x.1999.00963.x.

Index